Migration of Freshwater Fishes

Migration of Freshwater Fishes

Martyn C. Lucas
Department of Biological Sciences, University of Durham, UK
m.c.lucas@durham.ac.uk

Etienne Baras
*Department of Ethology and Animal Psychology,
University of Liège, Belgium*
e.baras@ulg.ac.be

with contributions from

Timothy J. Thom
Department of Biological Sciences, University of Durham, UK

Annie Duncan
*Royal Holloway Institute of Environmental Research,
University of London, UK*

and

Ondrej Slavík
T.G. Masaryk Water Research Institute, Prague, Czech Republic

b
Blackwell
Science

© 2001 by
Blackwell Science Ltd
Editorial Offices:
Osney Mead, Oxford OX2 0EL
25 John Street, London WC1N 2BS
23 Ainslie Place, Edinburgh EH3 6AJ
350 Main Street, Malden
   MA 02148 5018, USA
54 University Street, Carlton
   Victoria 3053, Australia
10, rue Casimir Delavigne
   75006 Paris, France

Other Editorial Offices:

Blackwell Wissenschafts-Verlag GmbH
Kurfürstendamm 57
10707 Berlin, Germany

Blackwell Science KK
MG Kodenmacho Building
7–10 Kodenmacho Nihombashi
Chuo-ku, Tokyo 104, Japan

Iowa State University Press
A Blackwell Science Company
2121 S. State Avenue
Ames, Iowa 50014–8300, USA

The right of the Author to be identified as the Author of this Work has been asserted in accordance with the Copyright, Designs and Patents Act 1988.

All rights reserved. No part of this publication may be reproduced, stored in a retrieval system, or transmitted, in any form or by any means, electronic, mechanical, photocopying, recording or otherwise, except as permitted by the UK Copyright, Designs and Patents Act 1988, without the prior permission of the publisher.

First published 2001

Set in 10/13 pt Times
by Sparks Computer Solutions Ltd, Oxford
http://www.sparks.co.uk
Printed and bound in Great Britain by
MPG Books Ltd, Bodmin, Cornwall

The Blackwell Science logo is a
trade mark of Blackwell Science Ltd,
registered at the United Kingdom
Trade Marks Registry

DISTRIBUTORS

Marston Book Services Ltd
PO Box 269
Abingdon
Oxon OX14 4YN
(*Orders:* Tel: 01235 465500
         Fax: 01865 465555)

USA and Canada
   Iowa State University Press
   A Blackwell Science Company
   2121 S. State Avenue
   Ames, Iowa 50014-8300
   (*Orders:* Tel: 800-862-6657
            Fax: 515-292-3348
            Web www.isupress.com
            email: orders@isupress.com

Australia
   Blackwell Science Pty Ltd
   54 University Street
   Carlton, Victoria 3053
   (*Orders:* Tel: 03 9347 0300
            Fax: 03 9347 5001)

A catalogue record for this title
is available from the British Library

ISBN 0-632-05754-8

Library of Congress
Cataloging-in-Publication Data
is available

For further information on
Blackwell Science, visit our website:
www.blackwell-science.com

This book is dedicated to Annie Duncan, whose wide-ranging knowledge of freshwater ecology in both temperate and tropical climates added much to the development of this book. Annie was closely involved in our original review work of fish migration, but early in the writing of this book she suffered a severe stroke and subsequently died.

Contents

Preface		xiii
Acknowledgements		xv

1 Migration and Spatial Behaviour — 1
- 1.1 Introduction — 1
- 1.2 Fish migration – from phenomenology to functional biology — 4
  - 1.2.1 Environment-related criteria and definitions — 5
  - 1.2.2 Diadromy and types of diadromous migrations — 6
  - 1.2.3 Superimposed migratory patterns: added complexity or just simple adaptive responses? — 8
  - 1.2.4 From adaptation to the migration continuum concept: do scale and salinity matter when defining migration? — 9
  - 1.2.5 From the restricted movement paradigm towards a general definition of migration — 12

2 The Stimulus and Capacity for Migration — 14
- 2.1 Introduction — 14
- 2.2 Stimuli for migration — 15
  - 2.2.1 Internal factors — 16
  - 2.2.2 External factors — 25
- 2.3 The capacity for migration — 38
  - 2.3.1 Overview of muscle structure and function — 39
  - 2.3.2 Swimming performance: how fast can a fish swim? — 40
  - 2.3.3 Relationships between swimming speed and endurance — 41
  - 2.3.4 A metabolic approach to swimming costs — 43
  - 2.3.5 How far can a fish migrate? — 45
  - 2.3.6 Constraints on early and late migrants — 47
  - 2.3.7 Implications of migration costs on size at first sexual maturity: when time matters — 49
  - 2.3.8 A tentative synthesis and conclusion — 50
- 2.4 Piloting, orientation and navigation — 52
  - 2.4.1 Landmarks and surface topography — 55
  - 2.4.2 Celestial cues — 55

	2.4.3	Currents	57
	2.4.4	Electric and magnetic fields	58
	2.4.5	Olfaction and gustation	58
	2.4.6	Other cues	59
	2.4.7	Homing, memory and imprinting	61

3 Types of Migration — 66
- 3.1 Introduction — 66
- 3.2 Migrations at the seasonal and ontogenetic scale — 67
  - 3.2.1 Feeding migrations — 67
  - 3.2.2 Refuge-seeking migrations — 75
  - 3.2.3 Spawning migrations — 81
  - 3.2.4 Post-displacement movements, recolonisation and exploratory migration — 84
- 3.3 Diel horizontal and vertical migrations — 87

4 Effects of Climate on Patterns of Migratory Behaviour — 93
- 4.1 Introduction — 93
- 4.2 Arctic and subarctic regions — 94
- 4.3 Temperate regions — 99
- 4.4 Tropical regions — 107
  - 4.4.1 Introduction — 107
  - 4.4.2 Setting the scene: what makes tropical freshwater systems similar or different to those of temperate areas? — 108
  - 4.4.3 Influences of predation pressure on fish migrations — 112
  - 4.4.4 Influence of dissolved oxygen on fish migrations — 115
  - 4.4.5 Other environmental factors shaping habitat use and seasonal migrations — 120
  - 4.4.6 Life history, breeding systems and migration patterns of tropical fish — 121
  - 4.4.7 Conclusion — 135

5 Taxonomic Analysis of Migration in Freshwater Fishes — 137
- 5.1 Introduction — 137
- 5.2 Lampreys (Petromyzontidae) — 139
- 5.3 Sharks and rays (Elasmobranchii) — 142
- 5.4 Sturgeons (Acipenseridae) — 142
- 5.5 Paddlefishes (Polyodontidae) — 148
- 5.6 Gars (Lepisosteidae) and bowfins (Amiidae) — 149
- 5.7 Bonytongues, mooneyes, featherfin knifefishes, elephant fishes (Osteoglossiformes) — 149
- 5.8 Tenpounders and tarpons (Elopiformes) — 151
- 5.9 Freshwater eels (Anguillidae) — 151
- 5.10 Anchovies, shads, herrings and menhaden (Clupeiformes) — 155
- 5.11 Milkfish (Chanidae) — 159

5.12	Carps and minnows (Cyprinidae)	159
5.13	Suckers (Catostomidae)	169
5.14	Loaches (Cobitidae) and river loaches (Balitoridae)	171
5.15	Characins (Characiformes)	172
5.16	Catfishes (Siluriformes)	180
5.17	Knifefishes (Gymnotiformes)	183
5.18	Pikes and mudminnows (Esociformes)	184
5.19	Smelts (Osmeridae)	186
5.20	Noodlefishes (Salangidae and Sundsalangidae)	187
5.21	Southern smelts and southern graylings (Retropinnidae)	187
5.22	Galaxiids, southern whitebaits and peladillos (Galaxiidae)	189
5.23	Salmons, trouts, chars, graylings and whitefishes (Salmonidae)	190
5.24	Trout-perches (Percopsidae) and pirate perch (Aphredoderidae)	198
5.25	Cods (Gadidae)	198
5.26	Mullets (Mugilidae)	200
5.27	Silversides and their relatives (Atheriniformes)	202
5.28	Needlefishes, half-beaks and medakas (Beloniformes)	203
5.29	Killifishes, livebearers, pupfishes and their relatives (Cyprinodontiformes)	204
5.30	Sticklebacks and their relatives (Gasterosteiformes)	204
5.31	Scorpionfishes (Scorpaenidae and Tetrarogidae)	206
5.32	Sculpins (Cottidae)	206
5.33	Snooks (Centropomidae)	208
5.34	Temperate basses (Moronidae)	209
5.35	Temperate perches (Percichthyidae)	213
5.36	Sunfishes (Centrarchidae)	214
5.37	Perches (Percidae)	216
5.38	Snappers (Lutjanidae)	221
5.39	Drums (Sciaenidae)	221
5.40	Tigerperches (Terapontidae)	222
5.41	Aholeholes (Kuhliidae)	222
5.42	Cichlids (Cichlidae)	222
5.43	Southern rock cods (Bovichthyidae) and sandperches (Pinguipedidae)	224
5.44	Fresh and brackish water dwelling gobioid fishes (Eleotridae, Rhyacichthyidae, Odontobutidae and Gobiidae)	224
5.45	Fresh and brackish water dwelling flatfishes (Pleuronectidae, Soleidae, Achiridae, Paralichthyidae and Cynoglossidae)	227

6 Methods for Studying the Spatial Behaviour of Fish in Fresh and Brackish Water — 230

6.1 Introduction — 230
6.2 Capture-independent methods — 234
    6.2.1 Visual observation — 234
    6.2.2 Resistivity fish counters and detection of bioelectric outputs — 235
    6.2.3 Hydroacoustics — 236

	6.3	Capture dependent methods	244
		6.3.1 Variations in density and catch per unit effort	244
		6.3.2 Marks and tags	247
		6.3.3 Types of marks and tags	247
		6.3.4 Electronic tags – telemetry	256
		6.3.5 Passive electronic tags	256
		6.3.6 Signal propagation and detection of battery-powered transmitters	260
		6.3.7 Transmitter positioning	261
		6.3.8 Telemetry of intrinsic and extrinsic parameters	261
		6.3.9 Attachment methods	263
		6.3.10 Limitations of telemetry systems	266
		6.3.11 Archival tags	267
	6.4	Choice of methods in fish migration studies	268
		6.4.1 Functional delimitation of fisheries districts	269
		6.4.2 Lateral and longitudinal migrations of large catfishes in a South American assemblage	269
		6.4.3 Fidelity of fish to spawning grounds	270
7	**Applied Aspects of Freshwater Fish Migration**		**271**
	7.1	A broad view of the impact of man's activities on freshwater fish migration	271
	7.2	Impact of man's activities on the diversity of fish assemblages in different geographic regions, focusing on damming	273
		7.2.1 Eurasia	273
		7.2.2 North America	275
		7.2.3 Australasia and Oceania	277
		7.2.4 Tropical South America, Africa and Asia	278
	7.3	Other impacts on fish migration resulting from man's activities	282
		7.3.1 Entrainment and impingement	282
		7.3.2 Hydropeaking, changes in temperature and oxygen	285
		7.3.3 Culverts and road crossings	288
		7.3.4 Changes in genetic diversity and life history	289
	7.4	Mitigation of hazards and obstacles to fish migration	291
		7.4.1 Fish passes	293
		7.4.2 Fish ladders	294
		7.4.3 Fish locks and elevators	296
		7.4.4 Shipping locks and elevators	297
		7.4.5 Nature-like fish passes	298
		7.4.6 Typical downstream passes: surface bypass systems	299
		7.4.7 Attracting and deterring fishes	300
	7.5	Installation, monitoring and efficiency of fish passes	307
	7.6	Conclusions	309
8	**Conclusion**		**311**

References	314
Geographical index	393
Taxonomic index	398
Subject index	412

Preface

The sight of adult salmon leaping at a waterfall, in order to make their way upstream to spawn, has generated a sense of awe in many a human observer, as well as a desire to understand the nature of the migration. The salmon's large size and impressive swimming performance, its high economic value and the conspicuousness of the migration have stimulated much research on the migratory behaviour of salmonids. This has resulted in a wealth of knowledge about the diadromous migrations of salmon, trout and char species and the migrations of freshwater-resident salmonid species. As well as providing perspectives on the functional biology of migration, these studies have been integral to the rehabilitation of many northern hemisphere rivers from human damage, and the provision of appropriate fish passage facilities.

Far less attention has been paid to elucidating the patterns of movement of other fishes inhabiting fresh water. In some cases this has been due to a perception that freshwater-resident (holobiotic) species, such as many of the cyprinids, do not move between habitats and the term 'non-migratory freshwater fishes' has been widely applied to describe these fishes. Other taxa such as anguillid eels and many lampreys are fully accepted to be migratory, but some aspects of their behaviour are less well understood because of their cryptic habits. Tropical river ecosystems are still poorly understood and we are just beginning to discover the nature and significance of the migratory habits of many species.

In recent years, several unifying themes have developed with regard to freshwater systems. The river continuum concept seeks to explain the physical, chemical and biological basis for observed patterns of productivity and nutrient cycling along the course of a river. The ideas of longitudinal, lateral and vertical connectivity in river systems, and their interactions over time (the extended serial discontinuity concept), provide a basis for considering the dynamic nature of aquatic habitats within these systems. Together with these developments, recent studies have increasingly demonstrated the widespread existence of spatio-temporal variations in the abundance and distribution of freshwater fishes, often on a seasonal or ontogenetic basis, for spawning, feeding and refuge. The use of methods such as radio-tracking and shallow-water hydroacoustics has enabled the quantification of changes in the use of space over time and in response to changes in environmental variables. In some cases these movements may be hundreds or thousands of kilometres, in others just a few metres, but their existence is indicative of a presumed evolutionary fitness benefit in doing so – in many cases these movements are fundamental for the completion of lifecycles. A reassessment of river catchment management processes is currently under way, and this is partially in response to an increasing awareness of the importance of migratory behaviour to fish species which were previously regarded as non-migratory. Fish communities play a key role in the functioning of aquatic ecosystems and movement between habitats is an important component of this, notably as regards energy transfer between ecosystems. However, there is an increasing realisation that management practices such as provision of fish passage facilities suitable for salmonids are often inappropriate for the majority of other fish species.

With the current re-evaluation of the significance of fish migration within freshwater environments, this book seeks to redress the balance of information available concerning fish migration in fresh and brackish water habitats. We review the occurrence and nature of migratory behaviour of fishes in fresh and brackish water, and the implications of such behaviour for ecologically sensitive catchment management. We examine these issues for all lifecycle stages and consider longitudinal, lateral and vertical movement patterns. In this context we include reference to salmonids, but given the vast literature and excellent reviews of this taxonomic group, we provide a limited review of them. To a lesser extent the same is true of our treatment of other diadromous fishes. We also provide a chapter on methods suitable for studying spatial behaviour of all fresh and brackish water fishes in the natural environment. Our aim is to provide a single source for a range of widely dispersed information, to which the river manager, scientist or student can refer to obtain information, advice and current opinion on the migratory behaviour of most taxonomic groups of fishes occurring in fresh water. We are aware of the enormity of this task and that we will, inevitably, have failed in places; yet we hope that we have provided a more balanced taxonomic and geographic perspective than has been available for many years. Where we refer to taxonomic names we have relied upon the November–December 2000 issue of the 'FishBase' database. Any errors and omissions within this book are our responsibility. We shall be grateful to anyone who takes the trouble to point these out to us. We have made all reasonable efforts to obtain permission to reuse previously published material.

During early research for this book it was discovered that one of the first 'classic' books on fish migration, *Migrations of Fishes*, published in 1916, was written by Alexander Meek, Professor of Zoology at the University of Durham, and Director of the Dove Marine Laboratory. Furthermore, G. Denil built the first example of his new fishway design (the 'Denil fish pass') on the River Meuse in Liège in 1908. Therefore, it seems fitting that the study of fish migration should, in some small part, once more be associated with the Universities of Durham and Liège.

Acknowledgements

Martyn Lucas and Etienne Baras extend their gratitude to their respective partners, Jo and Dominique, for all their patience and encouragement during the gestation of this book. We thank Dominique for checking all reference citations and compiling the reference section. Without their support and help, this book would never have been completed.

Etienne Baras is a Research Associate of the Belgian FNRS and acknowledges their support. We are grateful to all those colleagues who convinced us that the preparation of such a book would be useful to a wide audience of people with an interest in fish migration, conservation and management of freshwater ecosystems. The initial stimulus for this book began with a review of the migration of British non-salmonid freshwater fishes, funded by the Environment Agency (England and Wales) and carried out by the University of Durham and Royal Holloway Institute of Environmental Research, in close collaboration with the T.G. Masaryk Water Research Institute, Prague and the University of Liège. During our discussions with many colleagues at conferences and meetings it became clear that it would be useful for such a source of information to be published more widely, particularly if it could be expanded to reflect a broader range of taxa over a larger geographic area! The support of the Environment Agency for preliminary information collection on European non-salmonid species is gratefully acknowledged by M.L. Many people provided useful discussion and information, or gave permission to include their published material, including Kim Aarestrup, Miran Aprahamian, Bill Beaumont, Jose Bechara, Gordon Copp, Fred DeCicco, Paul Frear, Scott Hinch, Anders Koed, Jan Kubecka, Rosemary Lowe-McConnell, James Lyons, Steve McCormick, Bob McDowall, Scott McKinley, J. Mallen-Cooper, Mary Moser, Tom Northcote, Didier Paugy, Jean-Claude Philippart, Carol Richardson-Heft and John Simmonds. Many others were kind enough to send us copies of reprints and other information; thank you all for your help. We are grateful to Nigel Balmforth of Blackwell Science Ltd, Tom Fryer and Heather Addison for their assistance and endless patience.

Publishers and authors are warmly thanked for giving their permission for the use of material from their publications as follows:

Fig. 1.2 Northcote, T. G. (1978) Migratory strategies and production in freshwater fishes. In: *Ecology of Freshwater Production* (ed. S. D. Gerking), pp. 326–359. Blackwell, Oxford.

Fig. 1.3 McDowall, R. M. (1997a) The evolution of diadromy in fishes (revisited) and its place in phylogenetic analysis. *Reviews in Fish Biology and Fisheries,* **7**, 443–462.

Fig. 2.3 Richardson-Heft, C. A., Heft, A. A., Fewlass, A. & Brandt, S. B. (2000) Movement of largemouth bass in northern Chesapeake Bay: relevance to sportfishing tournaments. *North American Journal of Fisheries Management*, **20**, 493–501.

Fig. 2.4 Lowe, R. H. (1952) The influence of light and other factors on the seaward migration of the silver eel (*Anguilla anguilla* L.). *Journal of Animal Ecology*, **21**, 275–309.

Fig. 2.6 Videler, J. J. (1993) *Fish Swimming*. Chapman & Hall, London.

Fig. 2.7 Brett, J. R. (1964) The respiratory metabolism and swimming performance of young sockeye salmon. *Journal of the Fisheries Research Board of Canada*, **21**, 1183–1226.

Fig. 2.8 Saldaña, J. & Venables, B. (1983) Energy compartimentalization in a migratory fish, *Prochilodus mariae* (Prochilodontidae), of the Orinoco River. *Copeia*, **1983**, 617–625.

Fig. 3.2 Koed, A., Mejlhede, P., Balleby, K. & Aarestrup, K. (2000) Annual movement and migration of adult pikeperch in a lowland river. *Journal of Fish Biology*, **57**, 1266–1279.

Fig. 3.3 Knights, B. C., Johnson, B. L. & Sandheinrich, M. B. (1995) Responses of bluegills and black crappies to dissolved oxygen, temperature, and current in backwater lakes of the Upper Mississippi River during winter. *North American Journal of Fisheries Management*, **15**, 390–399.

Fig. 3.4 Weinstein, M. P. (1979) Shallow marsh habitats as primary nurseries for fishes and shellfish, Cape Fear River, North Carolina. *Fishery Bulletin*, **77**, 339–357.

Table 4.1 Ruffino, M. L. & Isaac, V. J. (1995) Life cycle and biological parameters of several Brazilian Amazon fish species. *Naga*, **18** (4), 41–45.

Fig. 4.1 Beddow, T. A., Deary, C. A. & McKinley, R. S. (1998) Migratory and reproductive activity of radio-tagged Arctic char (*Salvelinus alpinus* L.) in northern Labrador. *Hydrobiologia*, **371/372**, 249–262.

Fig. 4.3 Oxford Scientific Films.

Fig. 4.4 Paugy, D. & Lévêque, C. (1999) La reproduction. In: *Les Poissons des Eaux Continentales Africaines – Diversité, Ecologie, Utilisation par l'Homme* (eds C. Lévêque & D. Paugy), pp. 129–151. IRD Editions, Paris.

Fig. 4.5 Pavlov, D. S., Nezdoliy, V. K., Urteaga, A. K. & Sanches, O. R. (1995) Downstream migration of juvenile fishes in the rivers of Amazonian Peru. *Journal of Ichthyology*, **35**, 227–248.

Fig. 4.6 Toledo, S. A., Godoy, M. P. de & Dos Santos, E. P. (1986) Curve of migration of curimbata, *Prochilodus scrofa* (Pisces, Prochilodontidae) in the upper basin of the Paraná River, Brazil. *Revista Brasiliera Biologica*, **46**, 447–452.

Fig. 5.2 Auer, N. A. (1996) Importance of habitat and migration to sturgeons with emphasis on lake sturgeon. *Canadian Journal of Fisheries and Aquatic Sciences*, **53** (Suppl. 1), 152–160.

Fig. 5.3 Tsukamoto, K., Nakai, I. & Tesch, W. V. (1998) Do all freshwater eels migrate? *Nature*, **396**, 635–636.

Fig. 5.4 Irving, D. B. & Modde, T. (2000) Home-range fidelity and use of historic habitat by adult Colorado pikeminnow (*Ptychocheilus lucius*) in the White River, Colorado and Utah. *Western North American Naturalist*, **60**, 16–25.

Fig. 5.7 McDowall, R. M. (1988) *Diadromy in Fishes: Migrations Between Freshwater and Marine Environments*. Croom Helm, London.

Fig. 5.8 DeCicco, A. L. (1989) Movements and spawning of adult Dolly Varden charr (*S. malma*) in Chukchi Sea drainages of northwestern Alaska: evidence for summer and fall spawning populations. *Physiology and Ecology Japan*, Special Volume **1**, 229–238.

Fig. 5.10 Carmichael, J. T., Haeseker, S. L. & Hightower, J. E. (1998) Spawning migration of telemetered striped bass in the Roanoke River, North Carolina. *Transactions of the American Fisheries Society*, **127**, 286–297.

Fig. 5.11 Reynolds, L. F. (1983) Migration patterns of five fish species in the Murray-Darling River system. *Australian Journal of Marine and Freshwater Research*, **34**, 857–871.

Fig. 5.12 Bell, K. N. I. (1999) An overview of goby-fry fisheries. *Naga*, **22** (4), 30–36.

Fig. 6.2 MacLennan, D. N. & Simmonds, E. J. (1992) *Fisheries Acoustics*. Chapman & Hall, London.

Fig. 7.1 Halls, A. S., Hoggarth, D. D. & Debnath, K. (1998) Impact of flood control schemes on river fish migration in Bangladesh. *Journal of Fish Biology*, **53** (Suppl. A), 358–380.

Fig. 7.2 Dadswell, M. J. & Rulifson, R. A. (1994) Macrotidal estuaries: a region of collision between migratory marine animals and tidal power development. *Biological Journal of the Linnean Society*, **51**, 93–113.

Fig. 7.3 Bechara, J. A., Domitrovic, H. A., Quintana, C. A., Roux, J. P., Jacobo, W. R. & Gavilán, G. (1996) The effects of gas supersaturation on fish health below Yaciretá Dam (Paraná River, Argentina). In: *Ecohydraulics 2000* (eds M. Leclerc *et al.*), Volume A, pp. 3–12. INRS-Eau, Québec.

Fig. 7.4 Mallen-Cooper, M. (1992) Swimming ability of juvenile Australian bass, *Macquaria novemaculeata* (Steindachner), and juvenile barramundi, *Lates calcarifer* (Bloch), in an experimental vertical-slot fishway. *Australian Journal of Marine and Freshwater Research*, **43**, 823–834.

Fig. 7.5 Kusmic, C. & Gualtieri, P. (2000) Morphology and spectral sensitivities of retinal and extraretinal photoreceptors in freshwater teleosts. *Micron*, **31**, 183–200.

Chapter 1
Migration and Spatial Behaviour

1.1 Introduction

The migrations of fish in freshwater environments have played an important role in the settlement of human populations, and even to the casual observer, the movement of large aggregations of fish in shallow water or at obstructions is an astonishing sight. Indeed it is often only when such movements are obstructed and fish aggregate that they are noticed. Large-scale movements and migrations of fishes have been recognised through history and have encouraged the publication of a range of detailed books at least as early as Meek (1916). Migratory movements of fish need not involve their aggregation in high concentration, but such movements tend to follow particular pathways at a regular periodicity and so there is, inevitably, a concentration in space and time, to a variable extent, depending on environmental harshness. Before the advent of efficient fishing gear and fish location devices (especially sonar), the detection and capture of fish at sea and in large lakes was relatively inefficient. By contrast, the concerted movement of large numbers of fish through a restricted space at a particular time of year provided the advantage needed to enable the capture of significant numbers of fish by humans using simple nets, spears and traps (Fig. 1.1). From a more modern fisheries viewpoint, Brett (1986) makes clear the significance of salmon (Salmonidae) migrations as an easily utilised resource:

> 'As a fishery, salmon are ideal. They comb the ocean for its abundant food, convert this to delectable flesh, and return regularly in hordes to funnel through a limited number of river mouths exposing themselves to the simplest method of capture – a gill net or seine net.'

Humans have exploited migratory fishes in fresh water as food for several thousand years, utilising sites where fish congregated at natural obstructions or in shallow water. There are numerous prehistoric sites where bone and scale remains of fishes such as salmon and sturgeon (Acipenseridae) have been found in association with human communities. Prehistoric art involving fish is widespread and leaping salmon are among the most common depictions of fish in temperate parts of the northern hemisphere – perhaps reflecting their importance or the conspicuous nature of their migrations. There are frequent reports of the exploitation and management of migratory fish species in fresh water, including reference to management of English Atlantic salmon *Salmo salar* fisheries in the thirteenth century (Regan 1911).

Fig. 1.1 Trapping of migratory beluga sturgeons *Huso huso* on the River Danube at the Iron Gate, Romania, in the early nineteenth century (original illustration by Ludwig Ermini, reproduced from Waidbacher & Haidvogl, 1998).

Exploitation of lampreys (Petromyzontidae) occurred in Scandanavia in the fifteenth century (Sjöberg 1980) and it is said that King Henry I of England died from a surfeit of lampreys. All of this gives the impression that, of migratory fish in fresh water, it is those which move between the sea and fresh water (diadromous fishes) which are of greatest significance as a resource. This may be true at high latitudes (McDowall 1988), but in inland tropical regions many freshwater-resident fishes are utilised as food (Welcomme 1985; Lucas & Marmulla 2000). The operation of floodplain fisheries which utilise these is strongly dependent on the migratory tendencies of many of these species (Lowe-McConnell 1975; Welcomme 1979, 1985). As waters subside during the dry season of many tropical floodplain rivers, fish which have moved out onto the floodplain to exploit feeding opportunities often become trapped in shrinking pockets of water, providing easy fishing for local human populations, as well as for hosts of fish-eating birds such as egrets. More advanced fishing with gill nets or baited lines within river channels and backwaters still relies on an intimate knowledge of the seasonal and daily movements of the fish species that utilise them.

While man has exploited fish populations, he has had other direct interactions with them. Burning of forest to produce agricultural land over the last few thousand years has altered catchment characteristics, while irrigation and dam building in some of the earliest organised societies in Africa and Asia has altered catchment hydrology (Baxter 1977; Dudgeon 1992). Nevertheless, these influences on freshwater fish communities are small in comparison to the local and global influences that have occurred in the last 300 years.

Typically, fish form the most mobile components of freshwater communities. Locomotor muscle normally comprises 50–60% of body mass in fishes and, in combination with fins, provides the necessary power and stability for moving and maintaining position in a wide variety of freshwater habitats. However, conditions in fresh water are often highly variable in space and time. Fish behaviour is particularly influenced by factors such as flow, temperature and water quality, and habitat use may alter with changes in environmental conditions (Northcote 1978; Garner 1997). Movement is one of the main options available to fish when responding to changes in their environment. That predictable and synchronised movements of fish should occur, is therefore not surprising. Yet, despite this, the movements of many freshwater fish species have not been studied in detail (Northcote 1998). Until relatively recently, many freshwater-resident fish in rivers were regarded as non-migratory and considered to be relatively static populations, with their longitudinal location in the river defined by habitat preferences, leading to zonation (Huet 1949).

In our opinion, much of the development of knowledge concerning fish migration in fresh water has been strongly influenced by three factors – a sense of awe, the conspicuousness of migratory behaviour and the commercial importance of fishes. The spectacular sight of salmon fighting their way through rapids and leaping at waterfalls in order to spawn hundreds of kilometres upstream without feeding en route has undoubtedly inspired study of the physiology of swimming performance, particularly in 'athletic' fishes (Beamish 1978; Brett 1986). The sight of thousands of adult salmon returning from the sea and congregating in a river has stimulated research on homing and orientation mechanisms (Hasler 1971, 1983). The economic importance of salmon has necessitated study of their migratory habits and how to maintain or enhance the numbers of fish available for exploitation (e.g. Smith 1994). However, for many species of fish utilising freshwater habitats, local movements of a few kilometres or a few hundred metres may be of very substantial importance for survival and reproduction (Northcote 1978, 1998; Jordan & Wortley 1985). This situation is often the antithesis of that described for salmon and similar fishes: the swimming performance of these fishes is often poor or unexceptional, they and their migrations are often small and inconspicuous, and they are often of little commercial importance. The factors which determine the selective 'fitness' of spatial behaviour to fishes may be very different to those by which we have traditionally categorised that behaviour – economics is certainly not one of them!

Before continuing, we must consider the term 'freshwater fish'. Myers (1938) distinguished between 'primary' freshwater fishes, whose members are strictly confined to fresh water, and 'secondary' freshwater fishes whose members are generally restricted to fresh water but may enter salt water (usually considered to be 25–35‰ salinity). Most families of primary freshwater fishes display a long evolutionary history of physiological intolerance to, and an inability to survive in, full-strength seawater. A development of this approach considers fishes in fresh water in relation to their zoogeography and dispersal modes, of which two main types can be considered, 'euryhaline marine fishes' such as the bull shark *Carcharinus leucas*, which are primarily marine but are capable of entering and living in fresh water for long periods, and 'obligatory freshwater fishes' that normally must spend at least part of their life in fresh water (Moyle & Cech 2000). Of the latter category, two main groups can be considered: freshwater dispersants and saltwater dispersants. Freshwater dispersants are those fish species or families whose members are generally incapable of dispersing long distances through salt water including, for example, most cypriniforms, characiforms and gymnotiforms. The zoogeographic distributions of these fishes are best described by freshwater routes of dispersal and through geological influences including plate tectonics. Freshwater dispersants include Myers' 'primary' and 'secondary' freshwater fishes, but the former group contains many fishes such as certain cichlids that are tolerant to a wide range of salinities, while the distribution of the latter is best explained through freshwater dispersal, despite their ability to enter seawater. Saltwater dispersants are those fishes whose distribution patterns can be explained largely by their movements through salt water. Diadromous fishes are those which move between fresh water and seawater, spending different parts of their lifecycle in these environments, and include families such as the Petromyzontidae and Salmonidae. In many cases parent stocks of these diadromous groups have given rise to freshwater-resident species. The same is true of many predominantly marine groups of fishes which have freshwater-resident representatives, such as the Clupeidae and Cottidae.

In the context of this book, and in the interests of inclusivity (notwithstanding the comments made above), we regard freshwater fishes as all those which live in or regularly enter fresh or low-salinity brackish water (less than about 10‰). This is partly because brackish water and estuarine areas are an integral component of freshwater drainage basins and a substantial number of predominantly freshwater species migrate into brackish inland seas such as the Baltic, Black and Caspian Seas. It also reflects the fact that, from an applied viewpoint, many of the problems relating to obstruction of migration in truly freshwater habitats occur also in estuarine conditions.

1.2 Fish migration – from phenomenology to functional biology

Definitions of migration in animal populations abound and have been examined on numerous occasions (Heape 1931; Baker 1978; Dingle 1980, 1996), and they have evolved in the light of biological concepts. Many animals remain within a localised area of one or more habitats but exhibit daily and seasonal patterns of movement associated with variable resource utilisation. Where fish or other animals restrict their activities to a well-defined region of space, this is commonly termed the 'home range' or 'home area'. Longer range exploratory movements and dispersal processes may enable areas with better resource characteristics to be discovered and utilised, and may result in the adoption of new home ranges. Generally it is agreed that migrations are synchronised movements by species that are large relative to the average home range for that species and which occur at specific stages of the lifecycle. Heape (1931) considered that migration is a cyclical process which 'impels migrants to return to the region from which they migrated' and identified gametic migration (for reproduction), alimental migration (to locate food or water), and climatic migration (to find more appropriate climatic conditions). Implicit in this definition was the notion of homing which, however, is not a compulsory trait of migratory behaviour.

Climatic migrations have been referred to in articles dealing principally with temperate climates as 'winter' migrations (Nikolsky 1963; Harden Jones 1968), although Northcote (1978) pointed out that this was an inappropriate term for tropical and subtropical climates where wet and dry seasons are of greater significance to lifecycle processes. Also, synchronised movements may occur to areas that are less accessible to predators. In many marine fishes the lifetime track of movement often resembles a triangle (Harden Jones 1968). Eggs and larvae, having poor locomotor capabilities, are carried passively on ocean currents to nursery areas. The juveniles move to the adult feeding grounds and are recruited to the adult stock. At maturity they move to the spawning grounds, usually against the oceanic currents

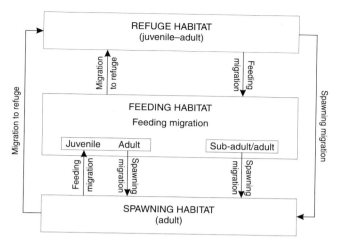

Fig. 1.2 Schematic representation of migration, based on movements between three functional habitats: refuge, feeding and spawning. Redrawn from Northcote (1978).

that the young and spent fish utilise to arrive in areas of good habitat. The areas utilised by the different life stages are characterised by habitat differences.

Northcote (1978, 1984) argued that three functional types of habitat can be recognised: one for reproduction, one for feeding and one for refuge in periods of unfavourable conditions. These habitats need not be the same at different stages of the lifecycle. Individual fish can minimise reduction of their genetic fitness if they move between these habitats at the right times during their lifecycle (Fig. 1.2). Based on this there are three principal functional categories of migration:

(1) Reproductive (spawning) migration
(2) Feeding migration
(3) Refuge migration

Northcote (1978, 1984) provides a convenient definition of migration equating to the following: those movements that result in an alternation between two or more separate habitats, occur with a regular periodicity within an individual's lifetime, involve a large proportion of the population and involve directed movement at some stage of the lifecycle. Seasonal movements are included in this definition so long as they are between distinct habitats and fulfil the other criteria. Although Landsborough Thompson (1942), who mainly studied bird spatial ecology, specifically discounted seasonal movements from his definition of migration, these are often intrinsically associated with the above-mentioned lifecycle processes and are of high fitness value. Dingle (1980) defined migration as 'specialised behavior especially evolved for the displacement of the individual in space' and makes clear the existence of selective forces for the development and maintenance of such behaviour.

1.2.1 Environment-related criteria and definitions

With regard to fish migration there has justifiably been a great deal of emphasis placed upon the significance of movement between salt water and fresh water from an evolutionary and physiological viewpoint (Myers 1938, 1949; McDowall 1988, 1997a). In the broadest sense, two major types of lifecycle may be recognised depending on whether a fish species spends its entire life in seawater or fresh water (holobiotic lifestyle), with similar and consistent constraints on osmoregulatory processes, or whether it travels between environments with different salinities (amphibiotic lifestyle), sometimes greater in osmotic concentration than its tissues, sometimes lower, thus involving seasonal or ontogenetic changes in its osmoregulatory function. From a more practical point of view, three types of patterns of fish migration may be recognised in terms of the biomes used:

(1) Oceanodromy – migrations occurring entirely at sea, e.g. plaice *Pleuronectes platessa*.
(2) Potamodromy – migrations occurring entirely in fresh water, e.g. Colorado pikeminnow *Ptychocheilus lucius*.
(3) Diadromy – migrations occurring between freshwater and marine environments, e.g. Atlantic salmon *Salmo salar*.

The occurrence and nature of oceanodromous and diadromous migration is of direct significance to the existence and history of exploitation by many of the world's most valuable fisheries (Harden Jones 1968; McDowall 1988). Thus there is good reason to understand why attention should have been focused on the spectacular transoceanic migrations of tunas (Scombridae), salmons and anguillid eels (Anguillidae) rather than freshwater-resident fishes moving relatively short distances. The terms 'pseudo-migratory' and 'semi-migratory' have been used to describe the extensive movements of several species of cyprinids, hundreds of kilometres between inland seas such as the Caspian and the middle reaches of rivers draining these seas, because the fishes tended not to move into the more saline, offshore areas of the seas (Vladimirov 1957; Nikolsky 1963).

1.2.2 Diadromy and types of diadromous migrations

Myers (1949) considered the groups of freshwater fishes in terms of their functional responses to varying levels of salinity, and included diadromous fish as one grouping, essentially similar in definition to that given above. The term 'euryhaline', which is used to describe an organism that is tolerant when exposed to frequent variations in salinity, is sometimes wrongly equated with 'diadromous', which as McDowall (1988) and several other authors have pointed out is incorrect. Some, though by no means all, diadromous fishes are euryhaline. Many diadromous fishes are amphihaline, able only to cope with changes in salinity during specific time windows of physiological responsiveness (Fontaine 1975). The smolting process in diadromous salmonids is a good example of this. Conversely, many euryhaline fishes such as the viviparous blenny *Zoarces viviparous* are often estuarine residents and do not exhibit well-defined migrations.

Essentially for purposes of clarity, McDowall (1988) regarded diadromy as typically occurring between fresh water and full-strength seawater. The occurrence of large areas of brackish water of varying salinity makes the division of a diadromous lifecycle between fresh and sea water appear too precise, although in a zoogeographical context the ability to move across large stretches of full-strength seawater is significant (Myers 1949; McDowall

1988, 1997a). Another possible confusion in McDowall's working definition of diadromy originates from the belief that movements between salt water and fresh water are obligatory and predictable. However, there is a wide range of typically freshwater species which make facultative use of brackish or marine environments and vice versa. That such movements are not obligatory and do not therefore strictly meet the definition of diadromy does not make them of lesser adaptive significance. Instead it reiterates the existence of a range of strategies for euryhaline and amphihaline fishes which may enhance their lifetime reproductive fitness in a range of conditions.

In several cases there are clear examples where migratory behaviour changes with latitude. For example, populations of striped bass *Morone saxatilis* typically adopt a largely marine existence, roaming extensively, but return to rivers to spawn, whereas at the southern and northern limits of their range they mostly complete their lifecycles in rivers (Coutant 1985). Neither strategy is superior to the other in terms of inclusive fitness. Each has been selected for under differing environmental conditions. Several species of fish, including Atlantic salmon *Salmo salar*, northern pike *Esox lucius* and Eurasian perch *Perca fluviatilis*, which exhibit a feeding migration into the upper Baltic Sea where salinities are rarely higher than 10‰, may not be truly diadromous yet their migratory behaviour allows much greater growth potential than would be possible if all fish were resident in the small tributaries in which they were spawned (Johnson & Müller 1978a; Müller & Berg 1982). Microchemical analysis of otoliths of European eels *Anguilla anguilla* from the North Sea and Japanese eels *Anguilla japonica* from the East China Sea strongly suggest that a substantial proportion of large, subadult yellow eels have never entered fresh water (Tsukamoto *et al.* 1998). Many male Atlantic salmon *Salmo salar* maturing at a juvenile size ('precocious parr') in otherwise migratory stocks never move to sea and the level of diadromy in stocks of brown trout *S. trutta* is phenomenally plastic (Guyomard *et al.* 1984; McDowall 1988; Northcote 1992).

While acknowledging the issues of graded rather than abrupt changes in salinity in aquatic environments, and the behavioural plasticity of fish species, the concept of diadromy has been widely accepted for many years. Depending on the life stage showing responsiveness to changes in salinity, and direction of migration, diadromy usually is divided into three classes (Fig. 1.3), originally introduced in their current context by Myers (1949), and defined by McDowall (1997a) as:

'(i) Anadromy: Diadromous fishes in which most feeding and growth are at sea prior to migration of fully grown, adult fish into fresh water to reproduce; either there is no subsequent feeding in fresh water, or any feeding is accompanied by little somatic growth; the principal feeding and growing biome (the sea) differs from the reproductive biome (fresh water).

'(ii) Catadromy: Diadromous fishes in which most feeding and growth are in fresh water prior to migration of fully grown, adult fish to sea to reproduce; there is either no subsequent feeding at sea, or any feeding is accompanied by little somatic growth; the principal feeding and growing biome (fresh water) differs from the reproductive biome (the sea).

'(iii) Amphidromy: Diadromous fishes in which there is a migration of larval fish to sea soon after hatching, followed by early feeding and growth at sea, and then a migration of small postlarval to juvenile fish from the sea back into fresh water; there

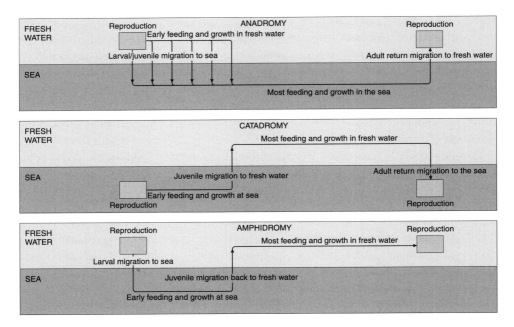

Fig. 1.3 Schematic representation of anadromy, catadromy and amphidromy. Redrawn from McDowall (1997a).

is further, prolonged feeding in fresh water during which most somatic growth from juvenile to adult stages occurs, as well as sexual maturation and reproduction; the principal feeding biome is the same as the reproductive biome (fresh water).'

Anadromy and catadromy are widely accepted terms although there has been refinement of the definitions to those given above. The nature and occurrence of amphidromy has been more widely debated (Balon & Bruton 1994; McDowall 1997a). Originally considered to encompass freshwater and marine forms, with the first migration after hatching being to freshwater and marine environments respectively (McDowall 1988), freshwater amphidromy is now regarded as being very rare, possibly occurring in just a few species of fish such as the Pacific staghorn sculpin *Leptocottus armatus* (McDowall 1997b), but probably absent. Although there have been suggestions that marine amphidromy is merely a special case of anadromy, in which the feeding interval at sea is limited to the larval phase (Balon & Bruton 1994), it is true that, unlike anadromous species, the bulk of somatic growth in amphidromous fishes occurs in fresh water.

1.2.3 Superimposed migratory patterns: added complexity or just simple adaptive responses?

Definitions of anadromy, catadromy and amphidromy relate to predictable reciprocal movements between biomes occurring at specific stages of the lifecycle. There are, however, further possible migration patterns which may be superimposed upon these characteristic patterns. Diadromous fishes may also exhibit repeated seasonal patterns of migration between habitats for feeding or refuge which reflect the Northcote (1978) definition of migration as applied to fishes but do not necessarily meet the lifecycle-based definitions applied to fishes (Myers 1949; McDowall 1997a) and other animals such as birds (Landsborough Thompson 1942). For example, mugilids such as thin-lipped mullet *Liza ramada* are regarded as facultatively catadromous, with much of the somatic growth often occurring in fresh water but with adults spawning at sea (McDowall 1988). In the Loire River, France, and throughout much of southern Europe their migratory behaviour may be much more complicated (Sauriau *et al.* 1994). Spawning occurs over the continental shelf and juveniles migrate shorewards, often into moderately brackish water, but appear not to enter low-salinity or fresh water until they are several years old (Torricelli *et al.* 1982; Sauriau *et al.* 1994). Subadults and adults make annual migrations into brackish and often fresh water in spring to feed principally on algae (Hickling 1970; Sauriau *et al.* 1994; Almeida 1996). They remain and grow over summer and return to the sea in autumn, when those that are sexually mature spawn. Juvenile Atlantic sturgeon *Acipenser oxyrinchus* commonly migrate into brackish or fresh water in summer, especially in the southern regions of their range on the eastern coast of North America and the Gulf of Mexico, many years before they are sexually mature (Kieffer & Kynard 1993; Foster & Clugston 1997). Therefore synchronised, seasonal migration patterns of presumed fitness benefit may be superimposed upon diadromous lifecycles.

Similar patterns of behaviour can be applied to wholly freshwater fish populations. The Colorado pikeminnow *Ptychocheilus lucius* of the Colorado River catchment, western United States and Mexico typically migrates to fast-flowing areas with a cobble substratum to spawn in early summer (Tyus 1985, 1990; Irving & Modde 2000). The larvae are displaced downstream and settle wherever there is available near-bank lentic habitat, especially in floodplain areas (Tyus 1991). The postspawners return to adult feeding habitat in deep pools which may be separated from spawning areas by about 150–300 km (Tyus 1985, 1990; Irving & Modde 2000). The young develop in nursery areas and as their swimming performance increases they leave this habitat and occupy faster, shallow-water areas. There are therefore distinct larval migrations (feeding and refuge combined) and adult migrations (separate spawning and feeding movements). In many European rivers, fishes exhibit a seasonal pattern of spring-upstream, autumn-downstream movement. These movements often involve habitat shifts from slower, deepwater habitat in autumn to faster, shallow-water habitat in spring and they occur commonly in juvenile and subadult fishes as well as in adults (Lucas *et al.* 1998, 1999). While upstream movements by adults might be associated with spawning, similar movements by juvenile fish are clearly not. For juveniles, such movements may reflect a migration to summer feeding areas, an information-gathering movement with juvenile fish following experienced adults to spawning areas (Dodson 1988), or an upstream redistribution movement, offsetting downstream displacement in high flows (Lucas *et al.* 1998). Although the last hypothesis seems most plausible, particularly since many of these rivers are largely channelised and offer little refuge on the floodplain, this does not exclude the other possibilities.

1.2.4 *From adaptation to the migration continuum concept: do scale and salinity matter when defining migration?*

It is perfectly true that, by virtue of the scale of marine environments, the oceanodromous and diadromous migrations displayed by many fish species are spectacular in scale (Harden Jones

1968; McDowall 1988; Lutcavage *et al.* 1999). Additionally, the possible long distances of movement, particularly within marine environments, may require reliable orientation or navigation mechanisms to enable survival and reproduction. Long diadromous migrations are often conspicuous, and many of the species involved have historically provided important sources of food through interception fisheries sited in the lower reaches of rivers. Studies of oceanodromy and especially diadromy have provided rich fields of research in fish ecology, evolution and physiology, and will continue to do so. However, concentration on these fields has tended to detract from the significance of migratory processes for wholly freshwater fishes (Northcote 1998).

However, distance travelled does not make oceanodromous or diadromous migrations more important in functional terms than less obvious potamodromous migrations. Synchronised seasonal refuge migrations of a few hundred metres within lakes or rivers may be just as important to lifetime fitness as long-distance migrations to or from the sea. They can just be viewed as movements between habitats, which are useful to the completion of the lifecycle, (almost) irrespective of the distance travelled and salinities crossed. For other species of fish one habitat may fulfil all purposes throughout the lifecycle, or may fulfil several but not all requirements, a fact established by Nikolsky (1963). For some species of fish, including many salmonids, the spawning and principal feeding areas are clearly separable geographically, but the refuge and feeding areas (either in freshwater or marine habitats) are usually similar. Many European rheophilic cyprinids such as barbel *Barbus barbus*, dace *Leuciscus leuciscus* and chub *Leuciscus cephalus* often use the same areas of river for spring spawning and summer feeding but move to a different refuge habitat in unfavourable conditions, most commonly in winter. In other rivers or streams exhibiting a different mosaic of habitats, the winter and summer habitats of these species can be similar whereas spawning places are located further upstream. Bullhead *Cottus gobio*, stone loach *Barbatula barbatula* and gudgeon *Gobio gobio* broadly use the same habitats for all three purposes and consequently do not need to migrate significant distances. Different species occupying different parts of a river catchment will undertake these migrations at different times of the year with varying levels of duration and extent. Figure 1.4 shows a 'typical' temperate, meandering lowland river with different types of migration superimposed onto the diagram.

The nature and extent of migration may be influenced by biotic environmental factors such as predation risk as well as abiotic environmental factors. The definition of types of fish migration by reference to the salinity through which they swim is a somewhat arbitrary method of categorisation, even though marine and freshwater environments represent very different biomes. Such categories might also be in terms of water velocity, depth, physical structure or the combination of these and other physical (e.g. temperature, dissolved oxygen) and biotic factors which combine to define habitat, and may prove more critical than variations in salinity for the completion of a lifecycle.

An illustration of this broad statement concerns the behaviour of the central mudminnow *Umbra limi* in temperate North American ponds which freeze in winter (Magnuson *et al.* 1985). When freezing occurs, oxidation of sediments removes oxygen from the deeper water first and there is negligible replacement from the air, due to ice cover, or from photosynthesis. Mudminnows move up in the water column and lie just beneath the ice, often utilising oxygen from small air pockets where oxygen becomes depleted last. Similar behaviour may be exhibited by other species such as yellow perch *Perca flavescens*, while centrarchids such as

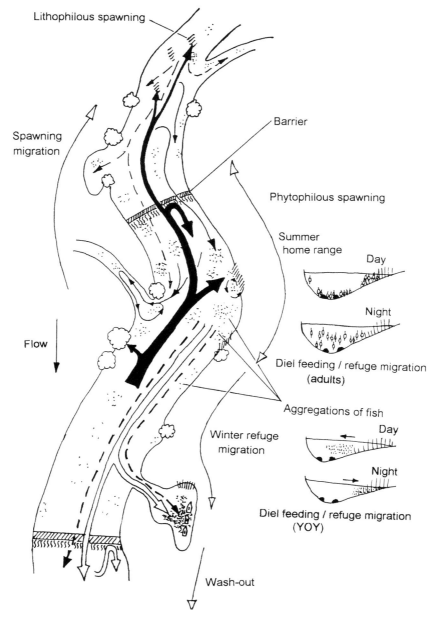

Fig. 1.4 Schematic diagram illustrating typical patterns of freshwater fish movement in a regulated lowland European river. Continuous lines indicate movement of adult fish; broken lines indicate movement of young-of-the-year fish. The bold solid line represents the main spring spawning (adults)/redistribution (juveniles) migration, while the wide, open line represents the main winter refuge migration (adults, juveniles). The small figures on the right-hand side represent typical diel distributions of fish for a typical cross-section of the river channel. For simplicity the figure shows limited on- and off-floodplain movement, which is usually widespread in unregulated rivers.

bluegill *Lepomis macrochirus* often remain over a greater depth range. Coldwater fish kills may result from the lack of oxygen, but mudminnows rarely die, as a result of their seasonal movement. Similar examples can be found in the literature dedicated to lacustrine or riverine young-of-the-year (YOY) fish, that make seasonal and daily migrations between warm, superoxygenated waters in shallow inshore habitats, and offshore areas (Baras & Nindaba 1999a, b). Do mudminnows and young fish move between two places that can just be classed as microhabitats or mesohabitats, or does this movement represent migration? In both examples provided, the distance travelled is short, whereas the ecological significance of the movement is high. Hence, from an ecological and functional viewpoint, such movements are clearly of as great an adaptive advantage to survival in this environment as is transoceanic migration to pelagic marine fishes such as tuna.

1.2.5 From the restricted movement paradigm towards a general definition of migration

Confusion in the perception of fish migrations is currently reflected in the management policies of river environments, since the term 'migratory' as addressed in freshwater fisheries legislation often pertains only to economically important and usually diadromous species such as salmonids, sturgeons and angullid eels. The actions of regulatory bodies in fisheries have tended to reflect this view, partly through a lack of information on the behaviour of a wide range of freshwater fish species. This is despite the fact that dramatic declines in some species of potamodromous freshwater fishes in river systems have been at least partly due to blockage of migration routes and damage to habitats associated with particular lifecycle stages (Baçalbasa-Dobrovici 1985; Backiel 1985; Dudgeon, 1992; Cowx & Welcomme 1998; see also section 7.2).

Why should opinion concerning many freshwater fish species be slow to change? In part it must relate to pioneering studies throughout the 1950s, based principally on stream-dwelling fishes including centrarchids, that provided substantial evidence for very limited movement by these species, and of the existence of small, stable home ranges (Gerking 1950, 1953). These observations, together with a variety of studies on other freshwater fish species, were incorporated in Shelby Gerking's (1959) important review, *The restricted movement of fish populations*. In the light of more recent works including those of Northcote (1978, 1984, 1997, 1998) and Welcomme (1979, 1985) potamodromy has been given a greater significance. The relative inadequacy of techniques used to investigate the migration of freshwater fishes also contributed to strengthen the idea that freshwater fishes were just 'resident', which is nowadays increasingly viewed as a misplaced paradigm (Gowan *et al*. 1994). However, the transition from discovery to education and management takes time, and old ideas tend to prevail, especially when confusion is involuntarily encouraged by the maintenance of potentially misleading phraseology (e.g. 'non-migratory' fishes when referring to holobiotic freshwater fishes) in many recent scientific publications, including some very high-profile articles. Further obstacles to this transition relate to the fact that humans perceive the aquatic environment differently from fish species. For humans examining migration in aquatic environments, the primary cues perceived are distance and salinity (probably because we happen to drink fresh water exclusively). Perception of the environment by fishes includes many other variables, among which temperature, water velocity, and chemical cues are important. From a consideration of spatial and salinity axes, the movements of central mudminnows *Umbra limi* and YOY fish described above would not appear to be indicative of migration.

Looking along depth, velocity, temperature and oxygen axes, these can be regarded as migration, as defined below.

Animal migration or, in this case fish migration, may be regarded as

'a strategy of adaptive value, involving movement of part or all of a population in time, between discrete sites existing in an n-dimensional hypervolume of biotic and abiotic factors, usually but not necessarily involving predictability or synchronicity in time, since interindividual variation is a fundamental component of populations'.

There is clearly great plasticity and adaptability of migratory behaviour between freshwater and saline environments in many fish species, and the movement of 'primary' freshwater species into brackish and coastal waters deserves greater consideration than it has received to date. It is now increasingly apparent that some riverine populations of European freshwater cyprinids and percids spend substantial periods of their lifecycle in estuarine and coastal brackish water environments, and that the migrations involve rhythmic periodicity (e.g. Nikolsky 1961; Lehtonen *et al.* 1996; Kafemann *et al.* 2000; Koed *et al.* 2000; K. Aarestrup [pers. comm.]). It is not the intention of this book to provide a definitive review of the biology of all diadromous fishes, and we make relatively limited reference to the migratory behaviour of salmonids (diadromous or freshwater-resident), given the vast amount of information available and well reviewed. For these subjects the reader is encouraged to consult texts which are, or include, detailed examinations of salmonid migration and applied issues (Northcote 1978, 1992, 1995, 1997; Dadswell *et al.* 1987; McDowall 1988; Groot & Margolis 1991; Høgåsen 1998) or of other diadromous fishes (Tesch 1977; Dadswell *et al.* 1987; McDowall 1988, 1997a, b). Instead we hope to provide a broader picture of migratory behaviour across the wide range of fish species found in fresh and brackish water.

Chapter 2
The Stimulus and Capacity for Migration

2.1 Introduction

In order to pass on their genetic characteristics to offspring, fishes, like all other organisms must survive, grow and reproduce. They must, therefore, obtain food of sufficient quality and quantity for body maintenance and growth, they must be able to tolerate or avoid unfavourable periods and must find conspecifics with which to spawn. When conditions are stable and near-optimal for survival, growth and reproduction, fish should remain in that vicinity. Establishment and use of a restricted home range is a key component of such a strategy of resource utilisation under favourable conditions (Gerking 1959). However, environmental conditions in most freshwater systems are rarely constant, and optimal habitat may vary as conditions change. Also, for a given species, habitat preferences or tolerance range may change with developmental stage. From an ecological viewpoint, such variations contribute to reduce intraspecific and interspecific competition through spatial, behavioural and temporal partitioning of resources (Ross 1986). Tracking of optimal or tolerable conditions may also involve a trade-off between factors such as temperature and dissolved oxygen levels, as has been implicated for striped bass *Morone saxatilis* (Coutant 1985). As a corollary, during the course of its life history a fish normally requires several functional units in terms of habitat or microhabitat conditions (Fig. 2.1) and so well-defined, somewhat predictable, movement between these is common.

Implicit in this overview is that fish can identify cues that indicate when the environment becomes less suitable and that they have the physical capacity of migrating to less uncomfortable places. When environmental changes are of a rhythmic nature, their repeated occurrence reinforces the response to a stimulus or set of stimuli, and promotes the emergence of 'predictive' behaviours, as fish orientate almost immediately to the places where less uncomfortable conditions can be found (e.g. predictive behavioural thermoregulation; Neill 1979; Baras 1992). This further implies the notions of memory and navigation, which may enable fish to minimise the energetic costs resulting from swimming and migration. Migration can thus be viewed in terms of a bias in movement towards a goal (Dodson 1988) that minimises discomfort in a variable environment. Migratory movements that minimise the immediate discomfort of fishes bring a tactical, short-term advantage. However, as they may affect survival, growth, energy storage and, ultimately the probability of relaying genetic information to the next generation (genetic fitness), they also have a strategic connotation. Within an evolutionary perspective, traits that ultimately enhance lifetime reproductive success will

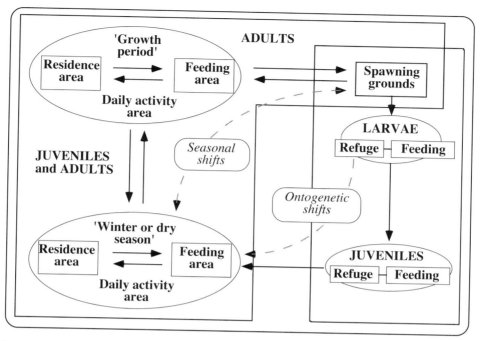

Fig. 2.1 Schematic illustration of the functional life unit concept in fish.

normally be selected for. Therefore, migratory behaviour, stimuli that trigger this behaviour, return to places previously identified as suitable, and capacities to navigate and relocate these places will be selected for, provided that they give an advantage over the non-migratory lifestyle, and provided that they can be relayed to the next generation. One question that is pertinent to these issues is 'how can a fish relay some form of behavioural homeostasis between generations?' We will return to this issue at the end of the chapter.

The present chapter examines the two key elements involved in migration by freshwater fishes: factors which stimulate migration and the capacity of fishes to migrate. The former is of great interest to those who manage freshwater habitats and fisheries since it is important to understand how change to environmental conditions may influence the behaviour of freshwater fishes and the resultant effects on populations. The latter is of fundamental importance in understanding mechanisms of migration, as well as applied issues such as enhancing fish passage at obstructions. Of course, there is a duality between stimulation of migration and the migratory response itself, which demands an understanding of the processes together rather than in isolation.

2.2 Stimuli for migration

Behaviour is the outcome of internal and external cues that interact to stimulate a response. Individual fish may respond differently to the same stimulus on different occasions because of motivational (non-structural) or structural changes which directly affect its capacity to

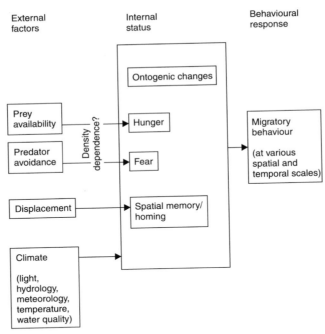

Fig. 2.2 Flow diagram of the nature and influence of internal and external factors which may stimulate migratory behaviour.

act (Colgan 1993). In this context, hormonal changes are highly important (Hasler & Scholz 1983; Høgåsen 1998). Figure 2.2 summarises those internal and external cues which may stimulate a fish to migrate. This section outlines some of the environmental factors which are thought to act as stimuli for the onset and maintenance of migratory behaviour, together with some suggestions of their function. Issues associated with mechanisms that utilise such stimuli for orientation and navigation are dealt with later.

2.2.1 Internal factors

2.2.1.1 Genetic and ontogenetic factors

The occurrence of migratory behaviour has a partial genetic basis in many freshwater-resident fish populations, although it is clear from many studies that the genetic 'signal' for migratory behaviour may be strongly influenced by environmental and developmental factors. Jennings *et al.* (1996) demonstrated that young walleye *Stizostedion vitreum*, produced from broodstock with a history of migrating into a river to spawn, and translocated to a new environment, still displayed greater migratory tendencies at maturity than a 'non-migratory' strain also released into the same lake at the same time. In diadromous species of salmonids, life history and migration characteristics often differ between river populations (Brannon 1984) and are associated with genetic factors (Jonsson 1982; Näslund 1993). However, anadromous and freshwater-resident salmonids also originate from the same gene pool in many populations (Nordeng 1983; Hindar *et al.* 1991; Jones *et al.* 1997). Environmental and biotic factors can have an overriding influence on lifecycle characteristics including the existence and extent of migratory behaviour of salmonids (Thorpe 1987; Morita *et al.* 2000), and may also do so on a wide variety of other freshwater fish species (Gross 1987; Tsukamoto *et al.* 1987, 1998; Pender & Griffin 1996). Evidence of genetically based variations in life history and migratory behaviour of the three-spined stickleback *Gasterosteus aculeatus* has also been obtained (Snyder & Dingle 1989; Snyder 1991), although lifecycle characteristics in this species are also strongly influenced by environmental conditions (Wootton 1976, 1984).

Ontogenetic changes in motivational and structural responses to stimuli result from both maturation, which involves intrinsic processes, and environmental experience (Colgan 1993). For example, seasonal changes in the motivation to feed in juvenile Atlantic salmon *Salmo salar* are associated with different patterns of growth and maturation (Metcalfe *et al.* 1986). The most obvious ontogenetic change in behaviour is that related to spawning activity, which has a marked impact on the migratory behaviour of many fish species (see also section 2.3.7). Additionally, the movement of young-of-the-year (YOY) fish from the spawning grounds is also an important ontogenetic change. In many circumstances this is regarded as a passive movement and for those fish which have pelagic eggs, such as many characins (Characiformes), this is certainly true (Pavlov *et al.* 1995; Araujo-Lima & Oliveira 1998). In rivers, buoyant eggs and larvae are carried downstream until they are deposited into lentic habitats, as seen in many taxa of tropical freshwater fishes (section 4.4.6). However, the suggestion that passive drift is dictated by flow is oversimplistic. Developmental responses play a strong part in determining the nature and extent of drift, especially at high water temperatures that facilitate rapid ontogeny of sensory and swimming capacities. Development of sensory capabilities, even when associated with little locomotor apparatus, may allow larvae to regulate patterns of drift in a controlled fashion. For example, in the walleye *Stizostedion vitreum*, larval migration typically occurs at an age of 2–6 days after hatching and is almost exclusively a nocturnal event, with larvae settling on the bottom at dawn and moving up into the water column at dusk (Corbett & Powles 1986; Mitro & Parrish 1997).

Larvae of tropical characiforms and pimelodid catfishes have also been found to drift predominantly at night in clear rivers of the Amazon River basin (Pavlov *et al.* 1995), whereas they drift throughout the day and night in turbid rivers, suggesting that such nocturnal migratory behaviour serves to reduce predation risk from visual predators (Robertson *et al.* 1988; Pavlov *et al.* 1995). Changes in the phototactic reaction during ontogeny may account for the timing of drift, and as to why some species drift further than others. Legkiy and Popova (1984) examined the ontogenetic variations of photoreaction in larvae of European minnow *Phoxinus phoxinus* and roach *Rutilus rutilus*. They found that roach were strongly phototactic from the start of exogenous feeding whereas larvae of minnow initially showed a negative phototactic response, then became positively phototactic at the time of filling their swimbladders, and gradually turned to a negative phototactic response at the end of the larval stage. These findings parallel the field studies of Pavlov *et al.* (1981) who observed that young roach drifted in greater numbers and at an earlier life stage than minnows.

Linfield (1985) showed that, in eastern English lowland rivers, larger fish were found in the upper reaches of rivers and smaller younger fishes in the lower reaches and argued that this was largely a result of predominantly downstream movement of YOY fish, followed by progressive net upstream movement of older fish. However, Lucas *et al.* (1998) found a more complicated pattern in the River Nidd, northern England, with the broadest range of size classes being found 7–24 km from the confluence with the Yorkshire Ouse, but more restricted size ranges upstream and downstream of this section. They argued that these differences might have been partly related to the existence of weirs, which restricted upstream movement of fish. Similar patterns to those reported by Linfield have been found for Arctic grayling *Thymallus arcticus* during summer in Alaskan streams (Hughes & Reynolds 1994), but in this case there appears to be a different cause. Arctic grayling exhibit seasonal migration to overwintering, spawning and summer feeding areas, but during summer, stream populations are characterised by a clear size gradient, with the largest fish furthest upstream. Through a fish removal experiment, Hughes and Reynolds concluded that there was competition for optimal feeding habitat in headwaters and that large, dominant fish secured these, while small, immature Arctic grayling were restricted to downstream areas and moved upstream in a progressive migration as they became larger, better competitors.

2.2.1.2 Hunger and metabolic balance

Many fish species migrate in search of food, sometimes over considerable distances and at increased risk of predation. The stimulus to migrate in search of food involves both a gastric factor based on gut fullness and a systemic factor reflecting metabolic balance. There are, however, few studies of the impact of hunger on the migratory behaviour of freshwater fishes. Thomas (1977) showed that, under laboratory conditions, the acceptance and rejection of food items during a meal have marked and opposite influences on behaviour in the three-spined stickleback *Gasterosteus aculeatus*. After accepting a food item, a stickleback searched more intensively in the immediate vicinity, but after rejecting a food item, was more likely to leave the area. Thomas (1977) argued that, in addition to the effects of satiation extending over an entire meal, acceptances and rejections result in respective short-term positive and negative changes in feeding motivation. These changes are adaptive if prey are patchily distributed, as is often the case in freshwater systems that present a mosaic of microhabitats. Similar arguments have been used by Charnov (1976), who proposed that animals leave a foraging place when the return rate falls below a certain value (marginal value theorem). These concepts and behaviours may account for the periodic long-distance movements of some species between locations where movements are normally short (Langford 1981; Chapman & Mackay 1984; Schulz & Berg 1987; Hockin *et al*. 1989; Baras 1992).

In lake environments, many pelagic fishes exhibit diel migratory behaviour during the summer growth season, moving between the warm surface water layer and cool, deep water. There may be several reasons for this, including following the movement of prey animals, especially zooplankton, or predator avoidance. However, when foraging, fish adopt a cyclical behaviour pattern of vertical migration for feeding and digestion, for which there may be additional advantages. Feeding in warm, surface layers allows high activity through the attendant high metabolic rate, leading to rapid stomach filling and digestion of food. Subsequent rest in cooler, deeper layers allows minimisation of maintenance energy costs, until the next feeding episode. This idea was first expounded by Brett (1971) for juvenile sockeye salmon *Oncorhynchus nerka* in lake environments such as Babine Lake, Canada, where juvenile sockeye salmon make twice daily vertical migrations from cold hypolimnial water (5–9°C) to warm epilimnial water (12–18°C) (Narver 1970). However, such behaviour may also be relevant to a wide variety of fish species occurring in thermally heterogeneous environments. Biette and Green (1980) and Diana (1984) showed that, for non-satiation rations, fish on cyclical temperature regimes grew faster than those on constant warm regimes. Where adequate food is available at all temperatures, Diana concluded that fish should feed and stay at the optimum temperature for growth. Where food availability is limited to temperatures above optimum, fish should feed there but rest at cool temperatures. Where food is limited to cool temperatures below the optimum for growth there may be no advantage in cyclic temperature shuttling compared with remaining at constant cool conditions. However, these generalisations ignore the effects of increased risk from predation under certain conditions, and this typically applies to the inshore–offshore migrations of YOY fish in lakes and rivers (Tonn & Paszkowski 1987; Persson 1991; Gliwicz & Jachner 1992; Copp & Jurajda 1993; Baras & Nindaba 1999b).

Notwithstanding the importance of feeding migrations, swimming and feeding are somewhat opposed activities, notably because aerobic metabolism is constrained within finite limits, and energy dedicated to foraging or digestion cannot be invested into swimming or vice versa (see also section 2.3). Evidence shows that there is a threshold level of energy reserves above which maturing salmonids become anorexic (Kadri *et al.* 1995; Tveiten *et al.* 1996). In species that do not eat during their spawning migration, the distance travelled is closely dependent on stored energy, and there are obviously good reasons for not starting a migration before energy reserves are replenished (see section 2.3.5), and the degree of replenishment may serve as an internal threshold to migration. Saldaña and Venables (1983) used this functional interpretation to account for why some coporo *Prochilodus mariae* with high energy content started an upstream migration in the Orinoco River, Venezuela, whereas others with lower energy content remained in a floodplain lake during the same season. Metabolic rates, in combination with food availability and temperature, have been invoked as a stimulus for migration of brown trout *Salmo trutta*. Ovidio *et al.* (1998) proposed that mature brown trout were undertaking their upstream spawning migration at a time when temperature became suboptimal and no longer enabled fast growth. Forseth *et al.* (1999), who investigated the energy allocations in age 2+ and 3+ brown trout of different sizes with a ^{133}Cs tracing method for measuring food consumption, found that fast growers migrated earlier. They suggested that fast growers, with a higher metabolic rate, were more energetically constrained by limited food resources than slow growers, and had to leave their juvenile habitat earlier for energetic reasons.

2.2.1.3 Homing

Homing of adult fish to spawning areas in which they originated brings individual fish back to an environment which is known to be suitable for reproduction at a time when other sexually mature fish will also be present (Wootton 1990). It is evident, therefore, that the ability to home can be an important strategy in maintaining an individual's genetic fitness. The ability to home to a particular natal spawning location after migrations of thousands of kilometres has been well documented for salmonid species (Hasler 1983; Stabell 1984; Dittman & Quinn 1996). By extension, the return to a previously occupied spawning site has also been named reproductive homing (*sensu lato*), and documented in many non-salmonid iteroparous fish species (e.g. white bass *Morone chrysops*, Hasler *et al.* 1969; white sucker *Catostomus commersoni*, Werner 1979; European minnow *Phoxinus phoxinus*, Kennedy 1977; common carp *Cyprinus carpio*, Otis & Weber 1982; largemouth bass *Micropterus salmoides*; Mesing & Wicker 1986; barbel *Barbus barbus*; Baras [unpubl.]). Levels of reproductive homing to the natal stream often exceed 95% for a number of populations of several species of anadromous salmonids (Hasler & Scholz 1983). However, straying may have significant fitness value also as an alternative life-history strategy, particularly in circumstances where the natal spawning habitat is unsuitable at the time of migration, or is inherently unstable or unpredictable, resulting in heavy mortality of progeny in some years (Whitman *et al.* 1982; Quinn 1984).

Measurement of the level of homing to natal spawning areas is difficult to obtain in many freshwater fish species which move away from spawning areas early in life (and at a small size). Alternative measures of homing precision may examine the level of repeat migration into home areas. Although data are limited, it would appear that homing precision of salmonids to spawning areas is generally higher than for non-salmonid fishes in temperate regions. Homing precision for several salmonids has been reported as: rainbow trout *Oncorhynchus mykiss*, 94% (Lindsey *et al.* 1959), brook trout *Salvelinus fontinalis*, 99.5% (O'Connor & Power 1973) and brown trout *Salmo trutta*, 100% (Stuart 1957). Comparable figures for non-salmonids are: white sucker *Catostomus commersoni*, 85.2% (Werner 1979) and roach *Rutilus rutilus*, 83.5–92.0% (L'Abée-Lund & Vøllestad 1985). In the latter case, roach tagged in two tributaries of Lake Årungen, Norway, exhibited considerable repeat homing to spawn in these streams, after which the larvae drifted downstream to the lake and grew to maturity, although it is not known whether these returned to spawn in their natal streams. In the muskellunge *Esox masquinongy*, the return rate to spawning places previously occupied was lower (about 20%), but some fish were reported to home successfully after up to 6 years (Crossman 1990). Possibly the best documented example of reproductive homing in non-salmonid fishes is the Colorado pikeminnow *Ptychocheilus lucius*, which homes consistently to preserved spawning grounds in the Yampa and Desolation Canyons, US, approximately 150 km distant from adult feeding grounds (Tyus 1990), or over 300 km in the case of fish from the White River (Irving & Modde 2000).

Whelan (1983) showed that aggregations of common bream *Abramis brama* at a single spawning site on the River Suck, Ireland broke down into four separate shoals after spawning which returned to their respective feeding grounds. Similar post-reproductive homing behaviour has been documented in many species, sometimes to the nearest metre (e.g. barbel *Barbus barbus*, Baras 1992; European grayling *Thymallus thymallus*, Parkinson *et al.* 1999). Homing to places other than the spawning site is frequent in species that exhibit diel foraging migrations (e.g. barbel, Pelz & Kästle 1989; Baras 1992; burbot *Lota lota*, Carl 1995). In Lake Årungen, Norway, after making a spawning migration into tributaries, many adult roach carry out a second migration, for feeding, into the streams 1–2 months after spawning, again displaying a strong tendency for use of their 'home' stream (L'Abée-Lund & Vøllestad 1987). The homing behaviour of the dace *Leuciscus leuciscus*, a small cyprinid, is so strong over short distances that they can return to the same small refuge area and occupy the same position in the shoal relative to other recognisable fish (Clough & Ladle 1997). The ability to return to a home area or territory has the same fundamental selective advantage for fishes as described above, in that it has provided the animal with appropriate habitat in the past, even if the distances and complexity of orientation required for return may be less than for the long distances of many diadromous migrations.

Translocation of fishes outside of their established or presumed home range has provided one of the key approaches for testing hypotheses of homing and migration behaviour in fishes (Gerking 1950, 1953, 1959; Funk 1957; Hasler *et al.* 1958). Homing following translocation has been demonstrated in many freshwater fish species including gudgeon *Gobio gobio* (Stott *et al.* 1963), common bream *Abramis brama* (Malinin 1971), flathead catfish *Pylodictis olivaris* (Hart & Summerfelt 1973), European minnow *Phoxinus phoxinus* (Kennedy & Pitcher 1975), smallmouth bass *Micropterus dolomieui* (Ridgway & Shuter 1996), largemouth bass *Micropterus salmoides* (Peterson 1975; Richardson-Heft *et al.* 2000), roach *Rutilus rutilus* (Goldspink 1977), the cichlid *Pseudotropheus aurora* (Hert 1992), American eel *Anguilla rostrata* (Parker 1995) and barbel *Barbus barbus* (Baras 1997). Similarly, return to home range has been noted after displacement by high floods (e.g. common bream and northern pike *Esox lucius*, Langford 1981; barbel, Lucas 2000). Richardson-Heft *et al.* (2000) radio tagged and streamer tagged largemouth bass in northern Chesapeake Bay, Maryland, US and examined the effects of translocation. They showed that 33% and 43% of bass, translocated 15–21 km from sites of capture on the western and eastern shores respectively, returned to their capture areas, by comparison to cross-bay movements of 4% and 6% for control groups released at the site of capture (Fig. 2.3).

Homing precision, an indicator of site fidelity, may be linked to both the ability to orientate and to a specialist rather than a generalist lifestyle. Hodgson *et al.* (1998) evaluated the homing tendency of largemouth bass *Micropterus salmoides*, smallmouth bass *Micropterus dolomieui* and yellow perch *Perca flavescens* utilising the same littoral habitat in a small (8 ha), unexploited oligotrophic lake in Wisconsin, US. Fish were captured by electric fishing in the margins at night, individually marked, and homing tendency was estimated as the probability of a fish being recaptured at the original site after translocation to a single release site. All three species exhibited a significant homing tendency, with yellow perch showing the highest site affinity, followed by largemouth bass and then smallmouth bass, reflecting exploitation of spatial resources in different ways by the three species. Nevertheless, in Lake Opeongo, Ontario, 15 out of 18 acoustically tagged smallmouth bass translocated an average of 6.7 km from capture sites within established home ranges took an average of 11 days to return to home ranges (Ridgway & Shuter 1996). Homing also varies with territorial habits and travel time, as exemplified by the study of the rock-dwelling cichlid *Pseudotropheus aurora* around Thumbi West Island, Lake Malawi (Hert 1992). Hert observed that, despite their small size (less than 10 cm in length), males could home from distances as far as 2.5 km, but that homing success was inversely proportional to translocation distance. Male *P. aurora* have strong territorial habits, and they may spend up to 1.5 years in the same territory. As a territorial species, it is rather unlikely to find a suitable unoccupied territory, and establishing a new territory in an unfamiliar environment may be harder than to regain its initial home. Increasing homing failure with increasing translocation distance in *P. aurora* at the site in question is probably not a matter of distance, or capacity to orientate, as this species rarely ventures in deep water, and would eventually find its way home by swimming around the island. However, the increased distance requires longer travel time, resulting in a greater probability that another fish will settle in the empty territory, so reducing the chance of the original fish to regain its territory (Hert 1992). As a corollary, homing success may be underestimated

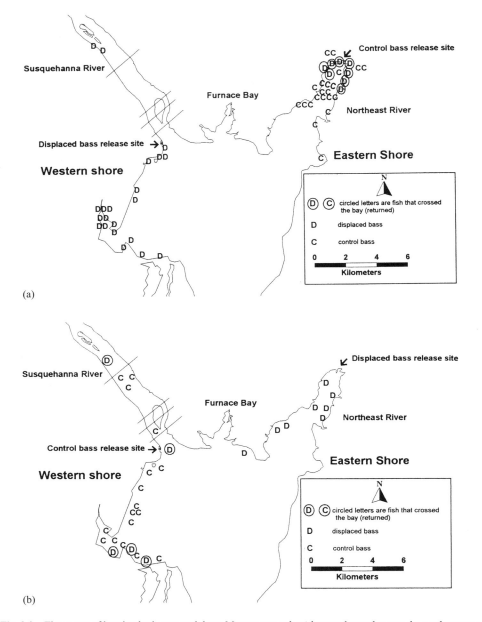

Fig. 2.3 The extent of homing by largemouth bass *Micropterus salmoides* translocated across the northernmost part of Chesapeake Bay, Maryland, by comparison to control fish, released where they were captured, for (a) treatment fish captured on the eastern shore and released on the western shore and (b) treatment fish captured on the western shore and released on the eastern shore. Reproduced from Richardson-Heft *et al*. (2000).

in experimentally translocated territorial fishes, as some fish manage to home but fail to regain their territory.

For species exhibiting aggregative behaviour, such as many cyprinids, there is no need to regain a territory. The stimuli behind homing behaviour can thus be of a different nature, including search for a shoal or specific partners. Kennedy and Pitcher (1975) demonstrated homing of European minnows *Phoxinus phoxinus* in a two-compartment tank after reciprocal transfer of two shoals. They found that individuals would still home even if half of each shoal was transferred. They showed that the strength of homing depended on the length of time the fish spent in the tank, arguing that fish needed to gain information about their environment before they could home. Baras (1997) conducted a field translocation experiment, where a shoal of barbel *Barbus barbus* was split into three groups that were released at different distances from their home, and radio-tracked. He found that the main criterion for residence area selection by a translocated fish was the presence of a shoal of at least 10 conspecifics within a suitable habitat, irrespective of whether these fish were or were not members of the shoal to which it initially belonged. Straying barbel selected residence areas in an unfamiliar stretch of the stream, but with a higher availability of potential feeding areas. In gregarious species, resident conspecifics may enhance the exploitation of an unfamiliar environment by a naïve fish, and this may trade off against the benefit provided by home site fidelity, depending on the availability of food resources.

2.2.1.4 Individual differences in behaviour: a matter of internal or external factors?

As illustrated by the example of homing and non-homing barbel *Barbus barbus*, individuals within a population may vary in their ability or motivation to migrate. Stott (1967) showed that populations of both gudgeon *Gobio gobio* and roach *Rutilus rutilus* consisted of a static component and a more mobile component. It was argued that the mobile component failed to secure or accept a home range and could be considered to be the exploratory element of the population. Such a population structure has been suggested in a wide variety of freshwater-resident species from marking studies, including northern pike *Esox lucius* (Mann 1980), chub *Leuciscus cephalus* (Libosvárskí 1961; Nicolas *et al.* 1994), barbel (Hunt & Jones 1974), European minnow *Phoxinus phoxinus* (Kennedy & Pitcher 1975) and pikeperch *Stizostedion lucioperca* (Fickling & Lee 1985). However, there is increasing evidence that this delimitation partly originates from the limited suitability of some methods for monitoring fish migration, and that mobility within a population can be viewed as a continuum. For example, radio-tracking studies on barbel and chub have demonstrated a continuum of annual ranges of movement between individuals ranging from less than 1 km to more than 30 km (Baras 1992; Lucas & Batley 1996; Lucas *et al.* 1998).

Interindividual differences in behaviour and mobility may relate to internal factors, notably as regards reproductive homing, as defined earlier, that implies the fish returns to the place where imprinting occurred (see section 2.4.7). In this context, the decision to migrate is intimately dependent on how far the spawner has moved from its birthplace, and for some species which drift during the early life stages, this may be affected by the distance drifted. While investigating the phototactic responses of roach *Rutilus rutilus* larvae, Legkiy & Popova (1984) found that, at all ontogenetic stages examined (D and E), drifting larvae had a stronger phototactic response than those captured in inshore habitats. They suggested this difference in phototactic response might partly account for the subdivision of roach populations into 'sedentary' and mobile groups. Variations in migratory behaviour may also depend on strictly external factors, such as the homogeneity and structure of the environment. Bruylants *et al.* (1986) studied the behaviour of Eurasian perch *Perca fluviatilis* in two sections of the Kleine Nete, a eutrophic canalised lowland river in northern Belgium. One section was homogenous in habitat characteristics with respect to depth, substrate and current and the other was a pool and riffle system, exhibiting substantial habitat heterogeneity. They found evidence of mobile and static components to the population, and that perch in the homogenous section were more mobile than in the heterogeneous section, suggesting that lack of suitable habitat may be responsible for the failure of some fish within the population to adopt a home range. Similarly, Baras (1992) compared the daily and seasonal migrations of barbel *Barbus barbus* in different streams and pool–riffle sequences, and found that distances travelled were chiefly a matter of habitat structure and distribution. The general pattern of greater mobility of fish in homogeneous aquatic environments seems widespread and is almost certainly associated with greater dispersal of key microhabitats in homogeneous river or lake environments by comparison to more heterogeneous environments.

2.2.1.5 Predator avoidance

There is clear evidence that many fish species and life stages use movement to refuge areas as a method of avoiding predators. Although predators clearly are external factors, predator avoidance behaviour may be innate or may be influenced by experience and learning. There are, however, few studies which have examined the mechanisms by which fish assess the threat posed by, and respond to, predators under natural conditions. It may be anthropomorphic to suggest that fishes can display 'fear', but many teleosts, especially those of the superorder Ostariophysi, do exhibit characteristic behavioural responses to release of alarm substance ('*Schreckstoff*') into the water (Smith 1992). Such alarm substances are released when the epidermal tisues are damaged, as may occur during an attack by a predator. There is debate about the influence of alarm substances in natural environments (Magurran *et al.* 1996), but many fishes exhibit rapid associative learning to avoid or retreat from apparent predators. Laboratory studies have demonstrated that shoaling cyprinids alter their foraging behaviour following experience of predators such as northern pike *Esox lucius* (Pitcher *et al.* 1986). Although in laboratory environments of limited size, prey fish do not necessarily fully retreat from predators and may engage in close-range predator inspection, these processes may substantially influence their movements in natural systems.

Several studies have examined the behaviour and habitat selection of prey fish in response to introduction of piscivorous predators at the mesocosm level (e.g. Savino & Stein 1982) and natural environment level (He & Kitchell 1990; He & Wright 1992). The latter authors demonstrated that following introduction of northern pike into Bolger Bog, a small 1 ha lake in Wisconsin, US, with an outflow channel connected to a nearby stream, they recorded reduced catch per unit effort (CPUE) of the prey fish community. Part of the reason for this was direct consumption of prey fish by northern pike, but a number of small-bodied, soft-rayed species, such as northern redbelly dace *Phoxinus eos*, golden shiner *Notemigonus crysoleucas* and fathead minnow *Pimepheles promelas*, susceptible to predation by northern pike, exhibited clear emigration from the lake to the stream in each of three years after northern pike were introduced, and by comparison to the control year. He and Wright (1992) produced evidence to show that over 4 years the community changed from one dominated by small, soft-rayed fishes towards one in which larger-bodied, spiny-rayed fishes were more important. They argued that under persistent predation by pike, the fish assemblage of Bolger Bog would eventually become an *Esox*–centrarchid assemblage. However, many aquatic systems remain in a state of dynamic flux through periodic physical or biotic perturbations, which may encourage migratory behaviour of prey fishes, where the advantages of movement for reproduction, feeding or non-predator refuge outweigh the disadvantages of elevated predation risk. Evidence to support the role of predator avoidance in habitat selection, mediated through movement, is also provided through studies of habitat segregation between young northern pike and their adult conspecific predators (Grimm 1981). Predation pressure may also provide a significant trigger to migratory behaviour in tropical floodplain environments (section 4.4).

2.2.2 External factors

2.2.2.1 Light

Day–night alternation has a direct synchronising influence on fish physiology and behaviour, but it also strongly influences the variations of temperature, oxygen and other physicochemical variables. The activity of planktonic prey in lakes, or macro-invertebrates in rivers and streams, also is dependent on day–night alternation. Finally, light intensity determines the risk of being preyed upon by visual predators. Therefore, it may be difficult to determine whether light intensity has a direct or mediated influence on fish migration in natural environments. Whatever the exact cause, the migration of many freshwater fish species has been found to have a diel component (Table 2.1).

Regularity both in time and space has been demonstrated for several cyprinid fish species that make daily migrations in between discrete habitats. Clough and Ladle (1997) showed that in summer adult dace *Leuciscus leuciscus* did not forage actively during the day, moved shortly before dusk to riffle habitats, then homed at dawn to the same daytime site, where they occupied the same position in the shoal relative to other recognisable fish. Similar cases of repeated homing behaviour day after day have been demonstrated for barbel *Barbus barbus*, during summer (Pelz & Kästle 1989; Baras 1992), with some individuals showing remarkably constant departure times with respect to sunset time, and light intensity (Baras 1992, 1995b). Schulz and Berg (1987) demonstrated that common bream *Abramis brama* show rhythmic diel migrations between the littoral and pelagic zones of Lake Constance, Germany. Sanders (1992) attributed the higher numbers of species, individuals and greater biomass of fish in night-time electric-fishing catches on Ohio rivers to diel movements from offshore to nearshore waters during the evening twilight period. These kinds of diel migrations between littoral and pelagic zones may occur in larger rivers as shown by Kubecka and Duncan (1998a) in the River Thames, England. Here, the greatest activity of the larger fish (mainly roach *Rutilus rutilus*, dace *Leuciscus leuciscus*, gudgeon *Gobio gobio* and Eurasian perch *Perca fluviatilis*) followed immediately after the onset of dusk and continued at the surface of the open river and in the littoral zone until dawn as light intensities increased. During daylight hours fish activity was not detectable acoustically as the larger fish were near the bottom. The diel periodicity of most fish movements is usually interpreted as an anti-predator response, and this notably applies to the diel migrations of YOY fish between inshore and offshore habitats in lakes and rivers (Cerri 1983; Hanych *et al.* 1983; Fraser & Emmons 1984; Copp & Jurajda 1993; Baras & Nindaba 1999b).

Table 2.1 The effect of light–dark cycles on movements and activity of several European freshwater fish species.

Species	Effect	References
Sea lamprey *Petromyzon marinus* River lamprey *Lampetra fluviatilis*	Avoid light in early part of freshwater spawning migration. Diel pattern of movement varies with season.	Hardisty (1979) Claridge *et al.* (1973)
European eel *Anguilla anguilla*	Yellow eels predominantly nocturnal swim faster during day. Silver eels most active at night.	Tesch (1977); McGovern & McCarthy (1992)
Northern pike *Esox lucius*	Migration of spawning adults into tributaries greatest at night. Emigration of juveniles mostly on sunny days.	Clark (1950); Franklin & Smith (1963) Franklin & Smith (1963)
European grayling *Thymallus thymallus*	Peak movements of grayling fry out of nursery stream occur at start and end of night.	Bardonnet *et al.* (1991)
Dace *Leuciscus leuciscus*	Adults show little activity during day and night in summer. Rapid movement at dawn and dusk between discrete day and night habitats. Juveniles move into and out of bays in response to predation risk at different light intensities.	Clough & Ladle (1997) Baras & Nindaba (1999a)
Chub *Leuciscus cephalus*	Juveniles move into and out of bays and from littoral to pelagic zones in response to predation risk at different light intensities. Adults principally move through fish passes at night during spawning migration. Positive correlation between photoperiod and migration intensity.	Schulz & Berg (1987); Baras & Nindaba (1999a) Lucas *et al.* (1999); Lucas (2000)
Gudgeon *Gobio gobio*	Vertical migration in large rivers – more abundant near bottom during day and in surface at night.	Copp & Cellot (1988)
Barbel *Barbus barbus*	Diel movements between refuge and forage areas. Attempt to cross weirs at night only during spawning migration. Seasonal variation in peaks of activity in early morning and late evening in summer. Dormant in winter.	Baras (1995b) Lucas & Frear (1997) Lucas & Batley (1996)

Whether diel movements are the expression of circadian rhythms synchronised by marked variations of light intensity at sunrise and sunset, or the adaptive response to particular light levels with associated foraging benefits or predation risks, is still debated, although there is evidence that fish adapt their diel patterns to light level. For example, YOY cyprinids on cloudy summer days are far less clumped in inshore areas than on days with bright illumination (Baras & Nindaba 1999a). Anguillid eels are typically nocturnal, and do not leave their daytime shelter before sunset, except on cloudy days (Parker 1995; Baras *et al.* 1998). McGovern and McCarthy (1992) showed that the swimming speeds of the European eel *Anguilla anguilla* tended to be higher in yellow eels that moved during the day, except during overcast weather, when speeds were similar to those during night-time. Their activity pattern is also deeply modified under moonlit conditions, including in reaches upstream of any tidal influence, suggesting that light intensity is the prevailing influence for diel migration. Because of the lunar cycle, moonlight can have a strong influence on the periodicity of upstream and downstream migrations. For example, the peak in the seaward migration of coho salmon *Oncorhynchus kisutch* smolts was found to coincide with the new moon (Mason 1975). Catches of silver European eels *A. anguilla*, which normally migrate during nighttime (Tesch 1977), are lower than average on full moon days (Todd 1981), as their migratory activity ceases when the moon appears above the horizon and shines on the river (Lowe 1952; Vøllestad *et al.* 1986) (Fig. 2.4). Similarly, anadromous lampreys (Petromyzontidae) are known to avoid light, hiding under rocks or river banks in the daytime, only resuming their upstream movement during the night (Hardisty 1979), but more rarely under moonlit conditions (Malmqvist 1980; Sjöberg 1980). Commercial fisheries in eastern Europe made use of this behaviour by illuminating rivers with lamps, leaving only a narrow, dark corridor through which the lampreys swam and were trapped (Abakumov 1956). Moon phase also strongly influences diel vertical and horizontal migratory behaviour of fishes in some lake systems (Luecke & Wurtsbaugh 1993; Gaudreau & Boisclair 2000). For amphidromous or anadromous species returning to rivers lunar phasing is frequent (e.g. gobioids, section 5.44; Manacop 1953; Bell 1999), but it is generally interpreted within the context of variations in tide height rather than as consequence of variable intensity of moonlight.

Greater turbidity can act in a similar way to overcast weather. Adult Atlantic salmon *Salmo salar*, which tend to migrate in rivers at night, often enter estuaries and migrate upstream during daytime in turbid water, frequently associated with high flows (Hellawell *et al.* 1974; Potter 1988). Similarly, diurnal downstream movements in turbid waters were observed for young pink salmon *Oncorhynchus gorbuscha* and chum salmon *O. keta*, whereas these species normally move at night (Neave 1955). Pavlov *et al.* (1995), who investigated the diel periodicity of drift in larvae of characids and pimelodids in the Amazon River basin, found that migration was chiefly nocturnal in clear rivers, whereas there was no marked diel periodicity in turbid rivers. De Graaf *et al.* (1999) also found that larvae of monsoon-spawning strategists drifted throughout the day and night in the turbid waters of Bangladesh rivers during the monsoon period. In some other cases, the migration of fishes under the threat of sight-feeding predators has been found to be diurnal, as has been reported for smolts of salmonids at high latitudes (Thorpe & Morgan 1978). At high latitudes, nights are extremely short during the periods of seaward movement by smolts, and northern pike *Esox lucius* are reported to have difficulty capturing smolts during bright, daytime conditions due to ripple effects in shallow water (Bakshtansky *et al.* 1977). The number of migrants has also been invoked as a factor behind the diurnal migration of species or life stages reputed as nocturnal. For example, Jellyman and Ryan (1983) observed that *Anguilla* spp. elvers migrated during daytime when they were in high concentrations (see also section 2.2.2.5, density-dependent factors). Smolts of Atlantic salmon were found to move downstream during daytime while schooling, whereas isolated individuals moved during night-time until they joined a school (Hansen & Jonsson 1985), possibly because schools offered sufficient protection against sight-feeding predators (Pitcher 1986).

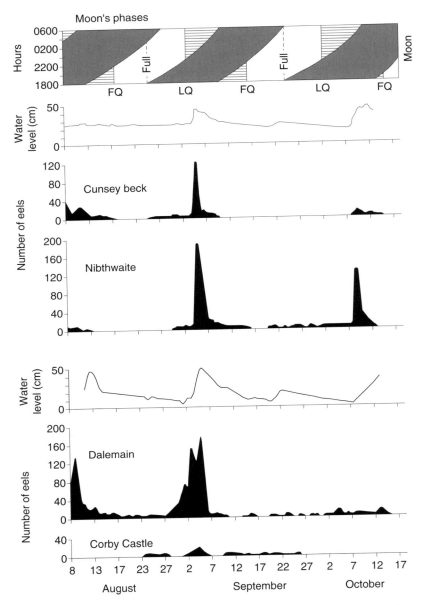

Fig. 2.4 Relationships between the seaward runs of silver European eels *Anguilla anguilla*, moon phases and water level in several northwest English streams. Modified from Sinha and Jones (1975) after Lowe (1952).

2.2.2.2 Temperature

As ectotherms, fish (excluding the few homeothermic marine fish species) are generally more active at higher temperatures within their normal range of tolerance. Conversely, the action of temperature on the pattern of movement by fish may be masked at comfortable temperatures, and become predominant outside this range. These arguments have been put forward to account for the abrupt shift of brown trout *Salmo trutta* and other fishes to nocturnal activity during winter (Heggenes *et al.* 1993) and for the progressive changes in the diel activity pattern of barbel *Barbus barbus* throughout the year (Baras 1995b). Adult barbel, which typically are dusk and dawn feeders at temperatures greater than 10–12°C, progressively shift to an increasingly diurnal pattern as temperature decreases, and they move to their foraging places at the time of the day when water temperature is the closest to their thermal optimum. Where the environment is thermally heterogeneous, variations of temperature beyond the fish's tolerance or preference range may trigger short migrations which are usually interpreted in the context of behavioural thermoregulation. This applies to the diel vertical migrations of juvenile Bear Lake sculpin *Cottus extensus* (Neverman & Wurtsbaugh 1994), to the winter aggregation of YOY cyprinids in habitats warmed by heated effluents (Brown 1979), and to diurnal offshore movements of YOY cyprinids in rivers when temperature exceeds 31–32°C in marginal habitats (Baras [unpubl.]). Tanaka *et al.* (2000) used archival tags to investigate the homing migrations of chum salmon *Oncorhynchus keta* in coastal waters, and found returning adults made dives as deep as 100 m when the surface water temperature was high (about 20°C). They interpreted this vertical migration as a balance between acquiring directional cues in surface water and behavioural thermoregulation in deep water.

The periodicity of long upriver spawning migrations is also influenced by temperature, for direct and indirect reasons, among which is the thermal dependence of swimming speed and endurance (see sections 2.3.3 and 2.3.4). In many species, the periodicity of reproduction is intimately dependent on temperature (Scott 1979; Lam 1983; Rodriguez-Ruiz & Granado-Lorencio 1992; Trebitz 1992). In some species, small changes in temperature may strongly affect the behaviour of spawners (e.g. *Lampetra* spp., Sjöberg 1977), especially when responses occur at precise temperature thresholds (e.g. barbel *Barbus barbus*, Baras & Philippart 1999). In temperate climates, for seasonal total spawners rather than fractional spawners, the need to reach the spawning grounds by a given time of year may be a sufficient pressure for their migration to be influenced by temperature too. This is certainly the case for species making relatively short spawning migrations (Table 2.2). At high latitudes, increasing temperature occurs during spring in association with advancing photoperiod and increasing productivity, and these may also act as stimuli, together with or in place of temperature. However, there is wide evidence that temperature acts as a stimulus for migration (Northcote 1984; Jonsson 1991). Malmqvist (1980) showed that, in one year of their study, upstream migration in brook lamprey *Lampetra planeri* was primarily triggered by a threshold temperature of 7.5°C. Additionally, increased temperature was indirectly responsible for decreased dissolved oxygen concentrations in summer, which stimulated larvae to drift or actively swim from streams into lakes. For those species spawning during early spring in temperate or cold areas, there are further constraints for starting a spawning migration at a very low temperature. Clark (1950) found that spawning northern pike *Esox lucius* began their movements into the feeder streams of Lake Erie, Ohio when water temperatures were 0°C and ice covered the pools. No spawning activity took place, however, until temperatures rose to 8°C. Franklin and Smith (1963) found that slightly higher temperatures of 2–3°C were required for the onset of the spawning migration of northern pike in the feeder streams of Lake George, Minnesota.

Various studies have shown that at the glass eel stage, ascent of European eels *Anguilla anguilla* into fresh water may be initiated by temperatures of around 6–8°C (Deelder 1952; Creutzberg 1961; Tesch 1971). At the elver stage, migratory activity depends on temperature

Table 2.2 The effect of temperature on spawning migrations and activity of several temperate, European freshwater fishes.

Species	Effect	References
Brook lamprey *Lampetra planeri* River lamprey *L. fluviatilis*	Upstream migration in adult brook lamprey triggered by threshold of 7.5°C. Spawning in both species usually commences at 11°C.	Malmqvist (1980); Sjöberg (1980)
Sea lamprey *Petromyzon marinus*	Upstream migration at 8–15°C. Spawning usually commences at 15°C.	Sjöberg (1980); M. Lucas (unpubl. data)
European eel *Anguilla anguilla*	Ascent of glass eel into freshwater initiated by temperatures of 6–8°C. Yellow eel migration only occurs above 10°C. Effect of temperature decreases upstream due to increasing number of older eels. Silver eel migration delayed by summers that extend into autumn and also inhibited by extremely low temperatures.	Deelder (1952); Creutzberg (1961); Tesch (1971) Tesch (1977); Moriarty (1986) White & Knights (1997) Frost (1950); Tesch (1977)
Northern pike *Esox lucius*	Adults begin movements into feeder streams in USA at 0–3°C.	Clark (1950); Franklin & Smith (1963)
Barbel *Barbus barbus*	Adult maximum movements at temperatures of 10–22°C while cold (pre-spawning) and hot (summer) conditions characterised by stability. Mean daily local activity of barbel linearly correlated with monthly water temperatures. Onset of spawning typically 14–18°C, but varies. Migration through fish ladders in R. Meuse occurs at 13–15°C and in the Dordogne and Garonne rivers at greater than 11°C.	Baras & Cherry (1990); Baras (1992) Lucas & Batley (1996) Hancock *et al.* (1976); Baras (1994); Prignon *et al.* (1998); Travade *et al.* (1998)
Dace *Leuciscus leuciscus*	Migration of dace through fish pass in R. Meuse occurs at 10–15°C. Activity of dace attempting to pass weir on R. Nidd increased with temperature and was maximal at 12–15°C.	Prignon *et al.* (1998); Lucas *et al.* (2000)
Roach *Rutilus rutilus*	Upstream migration in tributary of Lake Årungen started at 6–10°C. Migration through fish pass in R. Meuse occurs at 10-15°C and in the Dordogne and Garonne rivers at greater than 11°C.	Vøllestad & L'Abée-Lund (1987); Prignon *et al.* (1998); Travade *et al.* (1998)
White bream *Blicca bjoerkna*	Migration through fish pass in R. Meuse occurs at 10–15°C.	Prignon *et al.* (1998)
Common bream *Abramis brama*	Migration through fish pass in R. Meuse occurs at 10–15°C.	Prignon *et al.* (1998)
Chub *Leuciscus cephalus*	Migration through fish pass in R. Meuse occurs at 10–15°C.	Prignon *et al.* (1998)

(Mann 1963; Larsen 1972; White & Knights 1997; Baras *et al.* 1998). Tesch (1977) showed that migratory activity of European eels in the River Elbe, Germany, declined at temperatures below 10°C. Moriarty (1986) observed that the onset of migration of small yellow eels *A. anguilla*, in the River Shannon, Ireland, was correlated with water temperatures of 13–14°C. For the same species, White and Knights (1997) found a similar relationship between temperature and migration of elvers and yellow eels at the tidal and lower non-tidal limits of the River Severn, England. At the Ampsin navigation weir on the River Meuse, Belgium, Baras *et al.* (1996b) found that, except for the early migration in April, the effect of temperature on eel migration was highly variable, probably due to the unusual temperatures resulting from the warm effluent from the Tihange power plant. White and Knights (1997) argued that, because of this relationship between temperature and the migration of elvers and juvenile eels, global warming may be partially responsible for the current downward trend in eel recruitment (Moriarty 1990). Temperature also plays a significant role in the onset of the seaward migration of adult silver eels. In the Elbe estuary Tesch (1977) showed that in years with extended summers, migration was delayed. He argued that minimum threshold temperatures had to be exceeded to initiate downstream migration and Vøllestad *et al.* (1986) showed a similar pattern in the Imsa River, Norway, where migration occurred between 9 and 12°C although no threshold temperature was observed. It is also possible that extremely low temperatures cause a cessation in migratory behaviour in silver eels. In the River Bann, Northern Ireland, Frost (1950) showed that eel migration ceased with the onset of frost and Tesch (1972) showed that eels released into brackish water at temperatures of 6°C did not actively migrate.

2.2.2.3 Hydrology and meteorology

Discharge-related events are important stimuli for initiation of migration in many fish. There is widespread evidence to show that many adult temperate fishes, especially salmonids, are stimulated to move upstream on spawning migrations during or following periods of high flow (Banks 1969; Northcote 1984; Hawkins & Smith 1986; Jonsson 1991). In many rivers originating in upland regions or occurring at high latitudes, the spring elevation in temperature brings thawing, resulting in increased river discharge. These conditions often encourage longitudinal migrations of spring/summer spawning species to appropriate spawning areas (Northcote 1984) as occurs for the quillback *Carpiodes cyprinus* which migrate from Dauphin Lake, Manitoba, into the Ochre River to spawn (Parker & Franzin 1991). Razorback sucker *Xyrauchen texanus* migrate to spawning areas as spring river flows are increasing in the Colorado River system, US (Tyus & Karp 1990; Modde & Irving 1998), while in the same system Colorado pikeminnow *Ptychocheilus lucius* normally migrate to spawning areas after the spring increase in discharge has peaked (Tyus 1990). Many spring-running stocks of adult anadromous salmonids also enter and partially ascend rivers during periods of elevated discharge associated with snow melt, but remain in the river before spawning in autumn (e.g. Hawkins & Smith 1986; Laughton & Smith 1992). Spring elevations in river discharge resulting from rainfall at this time of the year or from snow melt often cause inundation of floodplain habitat in lowland areas. This acts as a stimulus for many species with lentic larval and juvenile stages to make lateral migrations onto the floodplain through ditches and backwaters for spawning (Kwak 1988). Such species include northern pike *Esox lucius* (Clark 1950; Masse *et al.* 1991), spotted gar *Lepisosteus oculatus* (Snedden *et al.* 1999) and some centrarchids (Raibley *et al.* 1997b). Clark (1950) observed that the main factor controlling the movement of northern pike into feeder streams of Lake Erie was the level of ice cover on the stream riffles. When no ice was present spawning fish were seen in early February. Franklin and Smith (1963) also showed that northern pike did not enter feeder streams until there was sufficient clearance between the inshore ice and the bottom to allow access to the stream. The presence of ice, especially frazil ice which can build up in large aggregations below dams due to the supercooling of that water, can also influence movements of fish. Frazil ice tends to accumulate in slow pools, building up from the surface downwards so that it can occupy much of the depth. Such 'hanging dams' can cause dramatic increases in near-bottom water velocity, and may result in the emigration of overwintering fishes from these pools (Brown *et al.* 2000).

Elevations in river discharge in tropical regions occur with relatively little seasonal change in water temperature and are responsive to the timing of the rainy season(s). Increases in river discharge result in widespread inundation of the floodplain, as observed to a lesser extent in many intact floodplain ecosystems at temperate latitudes. Flooding of low-lying areas results in dissolving of mineralised nutrients from the ground, triggering a burst of primary productivity, fuelling aquatic food chains. The cyclical advance and retreat of water on floodplains, with its associated enhancement of aquatic productivity and habitat diversity, is known as the 'flood-pulse concept' (Junk *et al.* 1989; Bayley 1995) and it is this hydrological phenomenon that drives the behaviour of many tropical river–floodplain ecosystems, including the movements of fishes onto and from the floodplain (Lowe-McConnell 1975, 1987; Welcomme 1979, 1985, section 4.4). In many cases, rainfalls or thunderstorms set the migration pattern. Spawning of the cyprinids *Catla catla*, *Labeo rohita* and *Cirrhinus mrigala* in the freshwater tidal zone of the Halda River, Bangladesh, are timed precisely to the slowest flow during the peak of high tide during a driving rainstorm. When rain is delayed, these cyprinids may fail to spawn (Jhingran 1975; Tsai *et al.* 1981). In these habitats many fish species rely on the first seasonal rainstorm to start their migration as they quickly attempt to make use of new resources in the floodplain, or escape isolated marshes. However, moving on the first rainfall and crossing recently flooded land can prove a very risky gamble, especially when the first rainfall is a short one. Rainboth (1991) reported the migration of small *Puntius* and *Esomus* spp. (Cyprinidae) in shallow water across a recently inundated soccer field during a 30-min cloudburst, and their failure (and death) as water percolated into the soil! Many African catfishes exhibit the same behaviour, but with lesser risks, as they can succeed in making terrestrial migrations while air-breathing with their suprabranchial organs and walking with their pectoral spines, provided the slope is low enough. Rainfalls induce changes in ambient noise, habitat availability, water level, current, transparency, conductivity, oxygen content and temperature, making it extremely difficult to determine which of these factors, if any, is prevalent in shaping migration patterns. Kirshbaum (1984) experimentally demonstrated that the sexual maturation of mormyrids could be induced by increased water levels and reduced conductivity. Similar observations were made by Bénech and Ouattara (1990) for the characin *Brycinus leuciscus* in the Niger River basin.

In some cases, very high discharges inhibit the initiation or continuation of migration of salmonids and other fishes (Northcote 1984; Jonsson 1991; Lucas *et al.* 1999). Malmqvist (1980) showed that the upstream migration of brook lampreys *Lampetra planeri* was inhibited by high flows during periods of heavy rain, probably because of the energetic cost of swimming against strong flows. The upstream migration of elvers of the European eel *Anguilla anguilla* is inhibited by high flows (Sörensen 1951). During their non-migratory stage, yellow eels of *A. anguilla* still make sporadic movements during periods of unstable weather conditions, and there is strong evidence that eel may be flushed downstream during freshets (Baras *et al.* 1998). Tesch (1977) argued that during flood conditions the area of riverbed available to eels for foraging would be increased with eels moving to take advantage of this. As water levels recede eels must leave these new areas or risk being stranded in unsuitable conditions. LaBar *et al.* (1987) provided some evidence for this increased use of space during flood conditions. They radio-tracked European eels in a small lake in southwestern Spain and showed that eels used a larger area in rainy weather than did those tracked during drier, more stable conditions. In the Elbe greater numbers of European eels were caught during periods of high flow (Lühmann & Mann 1961). This pattern is shown in Figure 2.4, since elevated water level generally reflects increased discharge, although the relative importance of increased water level, discharge and mean water velocity in stimulating eel migration are difficult to distinguish (Tesch 1977). Vøllestad *et al.* (1986) supported this view, finding that the migration of silver eels in the River Imsa, Norway, started earlier in autumns with high water discharge. Deelder (1954) found that the direction of migration of silver eels was also influenced by the direction of water flow. White and Knights (1997), however, found no relationship between eel migration and current velocity or tidal cycles.

Montgomery *et al.* (1983) showed that six fish species, including salmonids, cyprinids and the sea lamprey *Petromyzon marinus*, simultaneously emigrated from the Rivière à la Truite, Quebec, as water levels and discharge declined indicating the importance of migration as a strategy for avoiding drought conditions or warming. Although seasonal fluctuations in stream discharge have an important influence on spatial distribution of many fishes, and can act as important cues for migratory behaviour, in some environments there is little movement in response to major seasonal changes in discharge. Meador and Matthews (1992) sampled fish communities monthly at seven sites for a year in Sister Grove Creek, an intermittent prairie stream in north central Texas. Despite drastic seasonal fluctuations in discharge, abundance of individual species varied more between sites than temporally at individual sites, suggesting that hydrological changes did not have dramatic effects on fish movement in this community. Except for a few downstream movements caused by displacement due to high flow conditions, Baras and Cherry (1990) found no relationship between discharge conditions and movement of barbel *Barbus barbus* in the River Ourthe, Belgium, essentially as movements were promoted by an increase in water temperature. Investigations in a channelised and dammed environment also revealed that increasing temperature was the key factor stimulating barbel migration, but that high flow was important in attracting barbel to the bypass (Baras *et al.* 1994a). Slavík (1996a), on the other hand, observed the passage of many barbel through a fish ladder in the River Elbe, Czech Republic, after rain and this was associated with considerably increased conductivity and decreased water transparency (see also section 7.7).

2.2.2.4 Water quality

Changes in water quality may be due to natural factors such as deoxygenation of water in inundated forest areas, as leaf litter is decomposed, or acid flushes on naturally acidic upland areas with a geology characterised by little buffering capacity. Other variations in water quality may be due to anthropogenic influences such as pollution from discharge of organic material from wastewater discharges. Frequently organic and nutrient enrichment may exacerbate natural cycles in water quality deterioration such as oxygen depletion in thermally stratified or ice-covered lakes. In many cases, deterioration of environmental conditions on a seasonal or diel cycle triggers emigration of fish from what was previously adequate habitat (Box 2.1). Those fish which are able to tolerate hypoxic water do not need to emigrate from such conditions and a wide variety of tropical floodplain fishes which are regularly exposed to hypoxic conditions fall into this category (section 4.4). In other cases, there is clear evidence of a trade-off in terms of avoiding hypoxic conditions set against increased food availability (Suthers & Gee 1986; Rahel & Nutzman 1994) or refuge from predators (Suthers & Gee 1986).

Some studies using angler catch data have demonstrated very low catch rates immediately below sewage outfalls and have interpreted this as a movement response away from areas of

---

Box 2.1 When the going gets tough – the not-so-tough move out

Many temperate fishes prefer lentic water conditions, especially in winter when low temperatures limit metabolism and swimming performance. However, organically enriched backwaters may suffer from oxygen depletion, especially when there is little mixing of water or when extended ice cover occurs and prevents oxygen transfer at the air–water interface. When this occurs, those fishes that cannot tolerate hypoxia often move to better oxygenated areas, even if other environmental conditions are apparently suboptimal. Knights *et al.* (1995) showed how two species of centrarchids, black crappie *Pomoxis nigromaculatus* and bluegills *Lepomis macrochirus*, overwintering in backwater lakes on the upper Mississippi River, preferred water with almost imperceptible flow, temperature greater than 1°C and dissolved oxygen greater than 2 mg O_2 per litre. However, oxygen became depleted in backwater areas due to high biological oxygen demand and no replacement of oxygen from river flow. When oxygen levels dropped below 2 mg O_2 per litre, radio-tagged bluegills and black crappie moved out of these areas to sites with increased river throughflow, characterised by lower temperatures and slightly higher water velocities.

Similar behaviour was shown by radio-tagged largemouth bass *Micropterus salmoides* in a regulated, backwater complex of the upper Mississippi River, which exited the lake complex in winter as oxygen levels declined below 6 mg O_2 per litre. The fishes moved into a connection with the main river channel which remained oxygenated, but when a water control structure to the backwater complex was opened and oxygen levels rose, the bass returned, travelling up to 6 km in a matter of a few days (Gent *et al.* 1995). However, these and similar studies have shown that centrarchids rarely move into the main river channel in winter, even when backwater levels fall or conditions become hypoxic, although other species such as northern pike *Esox lucius* will do so. Under circumstances of winter drawdown of backwater levels, or deoxygenation, where no appropriate lentic refuge habitat is available, large fish kills of centrarchids may ensue.

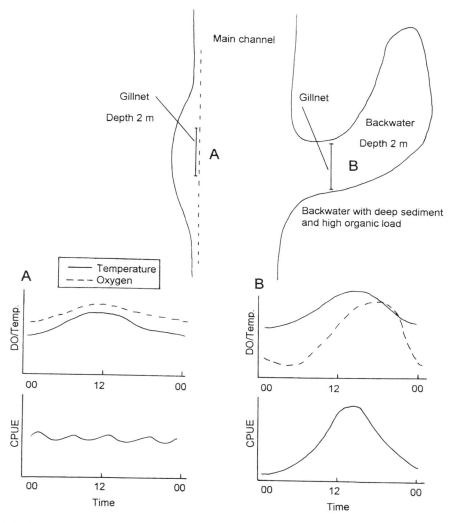

Fig. 2.5 Schematic illustration of the diel pattern of fish movements between the main channel and an organically enriched backwater of the Vltava River, Prague, Czech Republic, in relation to water quality during summer. Fish tend to aggregate in the warm backwater, oxygenated due to high photosynthetic activity, but leave at night when oxygen levels decline due to the high biochemical oxygen demand. Based on Slavíc & Bartos 2001.

poor water quality (Cowx 1991). However, Duncan and Kubecka (1993) reported aggregations of large fish attracted to the actively discharging sewage outfalls on the River Thames in England. Organic pollution has been demonstrated to be responsible for movement of European grayling *Thymallus thymallus* out of large stretches of the River Rhone, France (Roux 1984). Hendry *et al.* (1994) demonstrated that roach *Rutilus rutilus* colonising the Salford Docks, adjacent to the Manchester ship canal, England, were only able to do so during a period in winter when oxygen concentrations were adequate in the ship canal, due to high flows resulting in dilution of pollutants, improved mixing and cool temperatures. Libosvárski *et al.* (1967) found that low dissolved oxygen in two Czech brooks polluted with sewage effluent resulted in low abundance of fish in stretches some considerable distance downstream of the pollution source. They showed that the occurrence of fish in affected areas adjacent to a repopulation source changed according to variations in toxicity. This would suggest that fish move into the polluted areas when conditions are favourable and out again when conditions were poor.

Distribution and abundance of fish in the main river and in backwater sites of the Vltava below Prague, Czech Republic is strongly influenced by water quality (Slavík & Bartos 1997, 2001). In summer, diel fluctuations in water quality were important in stimulating diel migrations of fishes, mostly cyprinids, between the main river and backwaters. Fish tended to aggregate in the backwater during the day when photosynthetic action oxygenated the water but moved to the main channel at night when oxygen levels declined (Fig. 2.5). Similar arguments have been evoked for the emigrations of YOY cyprinids from littoral habitats in the River Ourthe, that became strongly hypoxic at night (Baras 1995a).

Carline et al. (1992) showed that, in response to spring acid flushes in streams on the Northern Appalachian Plateau, Pennsylvania, US, radio-tagged brook trout *Salvelinus fontinalis* moved downstream when dissolved aluminium levels reached toxic concentrations. Most fish moved downstream a few hundred metres, which was insufficient to escape the acid conditions, and substantial mortality occurred, although some fish utilised groundwater seep refuges with higher pH and lower dissolved aluminium concentrations than in the main channel (Carline et al. 1992; Gagen et al. 1994). Unhindered migration is necessary if fish are to recolonise areas affected by pollution incidents. Lelek and Köhler (1989) showed that the reduced abundance of European eels *Anguilla anguilla* in the southern part of the upper Rhine, after a fish-kill caused by a pollution incident from a large chemical factory (Sandoz AG, Basel, Switzerland), was quickly compensated by immigration from the tributaries and side-streams.

2.2.2.5 Food availability

The food resources in most natural waters vary continuously and the many generalist feeders respond by shifting from one feeding mode to another (e.g. from pelagic to benthic feeding, from particulate to filter feeding) or by migrating to other habitats. Additionally, it was argued earlier that diel migration in many species is the result of a compromise between the need to avoid predation by occupying refuge habitats in the daytime and the need to find food. It is likely, therefore, that prey availability will have a significant impact on the movements of fish at a variety of spatial and temporal scales.

In many cases such movements may not constitute directed migrations. Several studies have shown that adult northern pike *Esox lucius* generally make short movements within one habitat for a period of days followed by rapid long-distance movements between habitats (e.g. Chapman & Mackay 1984; Lucas et al. 1991). These movements are likely to be in response to fluctuations in prey availability, short movements being undertaken when prey is abundant followed by long-distance movements to find a new patch of prey, and conform to models of foraging in environments with patchy food distribution (Charnov 1976). In other cases, the spatiotemporal signal of food availability may be stronger. Pervozvanskiy et al. (1989) showed that northern pike fed on migratory salmon smolts in the Keret River, Russia but were restricted to particular reaches by their swimming ability. In Denmark, there is evidence that pikeperch *Stizostedion lucioperca*, but not northern pike, follow sea trout *Salmo trutta* smolts migrating through reservoirs (Jepsen *et al.* 2000). Similarly, large pimelodid catfishes in South American ecosystems undertake upstream feeding migrations as they track groups of migratory curimatids on their way to spawning areas (Reid 1983).

Common bream *Abramis brama* demonstrate a wide range of migratory behaviour in response to changes in food availability. Schulz and Berg (1987) argued that diel migrations of common bream in lakes enabled the favourable use of different resources: dominant benthic organisms in the littoral zone and increased zooplankton abundance in the pelagic zone. However, at other times tagged fish would join aggregations of hundreds of common bream responding to high abundance of plankton or emerging insects. Using echosounder surveys, Duncan and Kubecka (1996) detected a large aggregation of fish in a reach of the River Thames as they rose to feed on a mass emergence of mayflies (Ephemeroptera) in July. Over a distance of approximately 2 km, mean fish densities were as high as 0.2 fish m^{-3}, compared to 0.01–0.03 fish m^{-3} earlier in the night. It appeared that fish had moved from elsewhere in the river, attracted by the emergence of the mayflies. Hockin *et al.* (1989) also demonstrated that grass carp *Ctenopharygodon idella* movements consisted of short distance movements (<10 m) within restricted feeding habitats together with long-distance movements (>20 m) between such areas. From these studies it is clear that short diel foraging movements, together with longer distance movements between prey patches and migrations between spawning and feeding habitats, all play an important role in enhancing growth and survival. The lower the abundance of prey, the greater the stimulus to migrate, and the longer the feeding migration. Applying this general principle to life-history strategies may account for the latitudinal differences in the relative occurrence of anadromy and catadromy across taxa. In temperate areas, rivers exhibit poor productivity and the sea is highly productive due to good oceanic mixing and high nutrient availability, whereas in the tropics the reverse is true. These traits may have represented sufficient evolutionary pressures to favour the greater occurrence of anadromy at high latitudes, and catadromy in the tropics (Gross 1987; Gross *et al.* 1988).

Density-dependent factors, particularly associated with food availability, influence fish movements and many studies on juvenile salmonids have demonstrated clear relationships between food availability, territory size and the measured or presumed level of emigration (e.g. Slaney & Northcote 1974; Dill *et al.* 1981; Egglishaw & Shackley 1985; Elliott 1986). Elliott (1986) showed that, of brown trout *Salmo trutta* fry emerging from nests in Black Brows Beck, northern England, about 80% rarely fed, rapidly lost condition once the remaining yolk was exhausted and drifted downstream in a moribund state. Of the remaining fry the proportion which migrated away varied from less than 1% to 12% and was dependent on the numbers of fry per nest. Knights (1987) suggested that for European eels *Anguilla anguilla*, increasing density and competition might increase migration, with low densities suppressing the need to migrate. A lack of juvenile eel recruitment (i.e. immigration) results in low population densities in the upper reaches of rivers and a greater proportion of older female eels (Aprahamian 1988; Naismith & Knights 1993). These females may then form an important component of the breeding stock when they eventually return to spawn (Knights *et al.* 1996). Baras *et al.* (1996b) argued that European eels in the River Meuse, Belgium, migrated in waves that were seemingly independent of environmental conditions, except for the first seasonal wave, which clearly was related to an increase in water temperature. It is possible that these waves may have been the result of density-dependent factors which cause yellow eels to migrate after aggregating in large groups similar to the aggregations of elvers which congregate before starting their movement into inland waters (Deelder 1958). In a mark recapture study, Downhower *et al.* (1990) showed that movements of bullhead *Cottus gobio* in a small French stream were density dependent and that increased dispersal occurred at high densities.

2.2.2.6 Interactions between stimuli

Determining which of several environmental cues triggers fish migration has been debated for decades, notably with respect to freshwater and anadromous salmonids (Banks 1969; Jonsson 1991). Several authors highlighted the role of high river-discharge events (e.g. Baglinière *et al.* 1987; Brown & MacKay 1995) whereas others suggested that temperature variations were prevalent (e.g. Clapp *et al.* 1990; Meyers *et al.* 1992). Jonsson (1991) suggested that these different mechanisms corresponded to adaptations to local environmental factors, ensuring that the migration starts at the right time. More recently, Ovidio *et al.* (1998) proposed an alternative interpretation, based on an integration of stimuli, and somewhat similar to the law of summation of heterogeneous stimuli that was originally proposed by Tinbergen (1951). Migration would start when the forces promoting residency are outweighed by those stimulating the fish to move, almost irrespective of the internal or external nature of these stimuli. Ovidio *et al.* (1998) found that the spawning migration of brown trout *Salmo trutta* in the Belgian Aisne stream took place when both flow and temperature varied considerably between consecutive days, within a restricted thermal range (10–12°C). They interpreted these as cues of environmental unpredictability, stimulating trout to trade off reproduction against feeding and growth, which was no longer optimal at temperatures below 12°C. They further argued that reliance on a combination of stimuli was a more efficient strategy than responding to a single cue, which could occur on several occasions outside the breeding season, and could cause the fish to decrease its fitness. This interpretation is consistent with the notion of associative memory, imprinting and learning that will be developed in section 2.4.7.

2.3 The capacity for migration

The capacity for migration relies on the integration of locomotor activity and associated energy provision combined with the ability to orientate in the direction of the overall migration goal. These subjects form major study areas within fish biology and the review of these subjects that is provided here is relatively brief. For further information, the reader is recommended to consult other detailed accounts of fish locomotion and swimming physiology (Beamish 1978; Brett & Groves 1979; Videler 1993; Webb 1994) and orientation, navigation and homing (Harden Jones 1968; Hasler 1966, 1971; Leggett 1977; McKeown 1984; Dodson 1988; Quinn 1991; Dittman & Quinn 1996). For diadromous fishes, changes in osmoregulatory physiology are an integral component of the capacity for migration between marine and freshwater biomes and, although we recognise the importance of this issue, it is outside the objectives of this book. Excellent recent reviews on this subject are presented by Wood and Shuttleworth (1995) and by Høgåsen (1998), the former providing a comprehensive text on ionic regulation in fishes and the latter specifically reviewing physiological changes associated with migration of diadromous salmonids.

Freshwater fishes have a wide range of body forms, energy metabolism strategies and oxygen uptake/transport strategies. This results in diversity in swimming modes and performance, from the sluggish, serpentine locomotion of eels and lampreys, to the phenomenal acceleration during prey capture but poor sustained swimming performance of esocids, and the high, sustained swimming performance of rheophilic species such as salmonids (Webb 1984, 1994). Although downstream migration may be achieved with little expenditure of energy, by passive drift on currents, the capacity to migrate in an upstream direction requires that the fish swims faster than the water velocity, necessitating substantial energy expenditure from locomotor activity. For those fishes that spawn buoyant eggs in river systems, such as many of the characiforms in tropical regions, currents provide an effective means of dispersal. However, for this to be a successful life-history strategy, these fishes must normally return upstream before or at reproductive maturation, to ensure that the larvae arrive in appropriate rearing conditions (Northcote 1984; see also section 4.4). Upstream spawning migration is not compulsory though, and some species show no upstream movement at spawning, or even some downstream movement, depending on the distribution of feeding and spawning areas in the river system (see section 5.15).

2.3.1 Overview of muscle structure and function

Swimming muscle typically represents about 50–60% of a fish's muscle mass and is packed into a series of myotomes, separated from one another by sheets of connective tissue. Each myotome is shaped like a W on its side, and successive myotomes fit into one another in a system which promotes efficient power transmission. The precise myotome morphology and muscle fibre packing arrangements within myotomes vary between fishes (Videler 1993), but this description will suffice here. Swimming muscle comprises two main types of muscle fibre, oxidative, slow-contracting 'red' fibres and glycolytic, fast-contracting 'white' fibres. Red muscle's colour reflects its high concentrations of oxygen-binding blood pigments: myoglobin within the muscle and haemoglobin within the rich capillary blood supply. Most freshwater fishes' red muscle constitutes a relatively small proportion of muscle mass, being lowest (3–5% of muscle mass) in sedentary species such as esocids and cottids and highest (15–20% of muscle mass) in fishes such as salmonids and clupeids which are capable of sustained, rapid swimming. Many fish species have 'pink' or other muscle fibres which have characteristics that are intermediate to classical white and red muscle fibres (Johnston *et al*. 1977).

From an energy perspective, swimming can be viewed as the way the fish transforms its energy into distance moved. For a given muscle fibre type, the power generated for swimming is proportional to the muscle mass utilised, but the mass of muscle required increases as an approximately cubic function of the swimming speed. The result of this is that at low swimming velocities only a small amount of muscle needs to be used, but it must be able to contract repeatedly for long periods, which may amount to much of the fish's lifetime in pelagic species such as the cyprinid bleak *Alburnus alburnus* and tropical freshwater clupeids such as the Lake Tanganyika sardine *Limnothrissa miodon*. This necessitates aerobic metabolism, using oxygen transported to the muscle via the circulatory system to oxidise (mostly) lipid respiratory substrate within the abundant mitochondria of red muscle fibres. For relatively high swimming speeds, the power demands exceed those that can be supplied by red muscle, and white muscle is recruited to provide additional power output. White muscle is poorly vascularised and contracts using energy from anaerobic conversion of glycogen to lactate. Glycogen stores are rapidly depleted during fast swimming and at maximum speeds are largely exhausted in a few tens of seconds (Wardle 1975; Beamish 1978), so high-speed sprints can only be attained for short periods.

2.3.2 *Swimming performance: how fast can a fish swim?*

For fishes employing the trunk musculature for swimming, the distance moved forward is the product of the length moved per stride (S) and tail beat frequency (TBF). The maximum stride length (i.e. when all swimming muscles are used) is typically species-specific, as it depends on the proportion of swimming muscles, but also on other intrinsic variables such as body shape, skin, drag, tail size, fin structure and amplitude of lateral movements. It can be as high as 1.2 times body length (BL) for specialised swimmers with reduced drag, such as the marine Xiphiidae or Istiophoridae (swordfishes and billfishes), but it usually ranges between 0.5 and 1.0 BL in most fish species (average of 0.71 BL, as determined by Videler 1993, who compiled information on 45 species and a variety of size ranges). Because under steady swimming conditions stride length is relatively constant for a given species (except for larvae), tail beat frequencies are used to modify speeds. It should be noted, however, that during slow or non-steady swimming, there may be deviation from this relationship and tail beat amplitude may vary dramatically (Webb 1986; Videler 1993). For a fish swimming by using the trunk musculature, each stride requires the contraction of muscles on the left and right sides of the body, and the minimum stride period is thus twice the muscle contraction time (Wardle 1975). Muscle contraction time decreases with increasing temperature, and increases with increasing body size (Bainbridge 1958; Wardle 1975, 1977), these two values being defined by the indicators $Q_{10°C}$ and Q_{10cm}, which give the ratio of the muscle twitch frequencies for each difference of 10°C and 10 cm, respectively. Values of about 2 for $Q_{10°C}$ are common as they reflect the rate of speeds of enzyme-catalysed chemical reactions. Values of Q_{10cm} determined over a wide range of fish species vary very little (from 0.85 to 0.90, averaging 0.89; Videler 1993). The maximum tail beat frequency (TBF_m) at any temperature and fish size can thus be calculated from a single benchmark (i.e. TBF of fish of size BL_i at temperature T_j) while using the equation (Videler & Wardle 1991):

$$TBF_m = TBF_{BLi,Tj} [0.89^{(BL-i)/10} (2.00^{(T-j)/10})].$$

This general equation accounts for why the relationship between the maximum swimming speed (m s^{-1}) is a logarithmic function of size, and how it varies with temperature. These considerations are of particular relevance to an understanding of the seasonality of migration in thermally variable environments, and to the management of fish pass structures. In view of the relative dearth of knowledge on the actual swimming capacities of most freshwater fish species, further consideration should be given to kinematics, which provide information on stride frequency and length, as a key set of techniques for analysing the functional mechanisms behind fish migration (review in Videler 1993). However, although measurements of swimming speed made in still water confirm Wardle's predictions, these do not necessarily apply in turbulent conditions or in complex habitats.

2.3.3 Relationships between swimming speed and endurance

Knowledge of maximum swimming speed is important as it delimits the places to which, and times of the year when, a fish can travel and negotiate natural or man-made obstacles with fast currents. However, a fish cannot travel at maximum speed for periods of time greater than a few tens of seconds, simply because the white muscle's energy reserves are rapidly depleted (Wardle 1975; Beamish 1978). Reconversion of lactate to glycogen, together with removal of associated physiological disturbances such as tissue acid–base imbalance, requires aerobic energy expenditure, described as repayment of the 'oxygen debt' incurred during anaerobically-fuelled swimming, and can last as long as 24 h after a vigorous burst of such activity. Use of white muscle by fish should thus be restricted in time for physiological reasons. From combined telemetry of heart rate and electromyogram interference signals in lake-dwelling northern pike *Esox lucius*, Lucas *et al.* (1991) were able to demonstrate that even in these ambush predators, anaerobic swimming bursts accounted for less than 1% of daily energy costs. Although anaerobic contraction of white muscle is energetically inefficient at the biochemical level, at the whole animal level it is economic because fast sprinting is rarely required and because white muscle has low maintenance metabolism costs (Itazawa & Oikawa 1986).

Red muscles, by virtue of their aerobic metabolism, are almost inexhaustible, but they are slower (about 1 : 2.5 contraction frequency ratio), and develop lower power outputs (about 1 : 5 ratio) than white muscles (Altringham & Johnston 1986). This is partly the reason why fish have been viewed as two-gear animals, cruising somewhat leisurely most of the time, but being capable of sprinting occasionally. However, many fishes have a number of intermediate gaits, afforded by variations in swimming style and muscle fibre use, which give greater flexibility in swimming performance (Webb 1994). In spite of their cytological and histological differences, white and red muscles share a common trend, as their power output is maximal when the contraction frequency (V) is 20–40% of the maximum achievable frequency at a given temperature (V_{max}). While investigating the swimming performance of common carp *Cyprinus carpio*, Rome *et al.* (1990) found that carp start recruiting white muscle when the V/V_{max} of the red muscle exceeds its maximum efficiency. As a corollary, the faster the swimming speed, the greater the proportion of white (and pink) muscle fibres recruited, and the shorter the time over which this swimming speed can be employed. The greater the proportion that red muscle constitutes as a total of muscle mass, the faster the swimming speed that can be sustained wholly aerobically. Although there is a continuum between swimming speed and endurance, several key swimming speeds have been recognised in fish, based on the relationship between metabolism and the cost of swimming (Fig. 2.6):

- the optimum swimming speed (U_{opt}), i.e. the speed at which the energetic cost of transport is lowest (Tucker 1970; Brett & Groves 1979);
- the maximum sustained speed (U_{ms}), i.e. the maximum speed achievable while using aerobic muscle only. At values above U_{ms}, anaerobic white muscle is recruited, oxygen debt is incurred and the fish becomes exhausted;

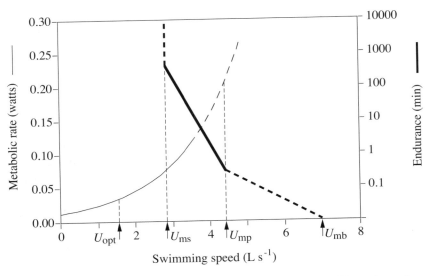

Fig. 2.6 Variation of metabolic rate with increasing swimming speed in a 0.188-m, 50-g sockeye salmon *Oncorhynchus nerka*. At speeds greater than the maximum sustained speed (U_{ms}), anaerobic white muscles become increasingly recruited, and endurance decreases steeply. U_{opt} is the optimum swimming speed, U_{mp} is the maximum prolonged speed, and U_{mb} is the maximum burst speed. Redrawn after Videler (1993), based on data of Brett (1964).

- the maximum prolonged speed (U_{mp}), where endurance is reduced to a fraction of a minute;
- the absolute or maximum burst (sprint) speed (U_{mb}, as depicted above).

In addition to these definitions, mention must be made of another derivation of swimming performance, the critical swimming speed U_{crit}, the velocity at which a fish becomes fatigued after incremental increases in swimming velocity, usually achieved by increasing water velocity (Brett 1964; Beamish 1978). Although arguably less informative in terms of swimming physiology than the other speed definitions outlined above, U_{crit} is a useful measure of prolonged swimming performance and is relatively easily and widely measured. Most information concerning swimming performance, especially U_{crit} and U_{mp}, has been obtained from empirical measurements made in swimming flumes or in rotating annular tanks (Beamish 1978). This approach has been used to derive swimming performance data for a wide variety of freshwater fish species (Beamish 1978; Videler 1993), but has been of limited value for fishes which are not motivated to swim continuously at significant speeds. This problem occurs at the individual and species level but is most common for sedentary species such as northern pike *Esox lucius* and bluegill *Lepomis macrochirus* which swim in an inconsistent manner and seek shelter in the slightest current eddy of an experimental apparatus. Benthic fishes such as cottids use their body morphology to remain in contact with the bottom, even at water velocities higher than those which could be sustained (see also section 4.4.3 for other adaptations to fast currents). As for burst speed, maximum sustained and prolonged speeds also are influenced by fish size. Values of Q_{10cm} for U_{ms} or U_{mp} are usually lower than for U_{mb} (Videler 1993), implying that the decrease in sustained or prolonged swimming speed with increasing size is steeper than for maximum speed. Water temperature also influences U_{ms}

and U_{mp} but no consistent trend emerges from the few experimental data available (Videler 1993), essentially as different fish species have different thermal metabolic optima.

2.3.4 *A metabolic approach to swimming costs*

For fishes, the rate of aerobic metabolism is most conveniently assessed through measuring the oxygen consumption rate in a respirometer (Brett & Groves 1979). Aerobic metabolic costs may be divided into several components. For a given temperature, basal metabolic rate (*BMR*) refers to the energy cost of maintaining the fish in a state of readiness for higher rates of energy expenditure. Activity and feeding both exert additional energy costs, and so when *BMR* is measured, the fish should be in a resting, postdigestive state. Basal metabolic rates are easy to measure for homeotherms, during inactivity within the thermoneutral zone, and temperature-specific *BMR* can be measured at rest for sedentary fishes such as northern pike *Esox lucius*. However, many fishes rarely display extended periods of rest. The use of a flume respirometer enables measurement of oxygen consumption rate at defined swimming speeds for a fish and from this an extrapolation can be made to estimate *BMR*. This specialised estimate of *BMR* is termed standard metabolic rate (*SMR*) and it should be noted that *SMR* is not necessarily equivalent to *BMR* (Brett & Groves 1979; Lucas *et al.* 1993b), although the terms are frequently used interchangeably and in some cases may be similar in magnitude (Schurmann & Steffenson 1997).

Despite recent debate, for most, probably all fishes, the sustainable upper limit of aerobic metabolism occurs during maximal aerobically fuelled exercise (Brett 1964; Brett & Groves 1979). Brett termed this sustained upper limit of metabolism, active metabolic rate (*AMR*). For clarity, following Videler (1993), we consider the metabolic rate during swimming to be the active metabolic rate (*AMR*), while the maximum sustained value of active metabolism is AMR_{max}. Active metabolic rate is a function of swimming speed (*U*), and is given by $AMR = a + b U^x$, where 'a' is an estimate of *SMR*. From this power function, the value of U_{opt} can be deduced after differentiation with respect to *U*, and annulation, and gives an estimate of the minimal gross cost for locomotion. Weihs (1973, in Videler 1993) predicted from theoretical grounds that the energy required by propulsion while swimming at U_{opt} was equal to *SMR* (thus implying that $AMR_{opt} = 2\ SMR$, and b = 1). In many fish species, the maximum sustained aerobic metabolic rate at the species' optimum temperature (i.e. for which *AMR* minus *SMR* is maximised) is about 4–8 times higher than *SMR*, and sometimes up to 16 times higher (sockeye salmon *Oncorhynchus nerka*, Brett 1964). In comparison, the level of metabolism during anaerobic bursts of speeds can attain 100 times *SMR* (Brett 1972). Assuming Weihs' prediction is correct, it might be tempting to suggest that for fish showing no other activity than swimming, the active metabolism at speeds above U_{opt} doubles each time the swimming speed increases by a margin of 1 U_{opt}, thus producing rough cost estimates of 4, 8 and 16 *SMR* while swimming at 2, 3 and 4 U_{opt}, respectively. The equation above would thus become $AMR = 2^x\ SMR$ (or more generally, $AMR = m^x\ SMR$), where *x* is the ratio of the swimming speed to U_{opt}. To test for this hypothesis, we used Brett's (1964) classical data on the oxygen consumption of an 18.8-cm (55 g) sockeye salmon swimming at speeds of 1 to 4 BL s^{-1} at 5, 10 and 15°C (Fig. 2.7). We assumed U_{opt} was the swimming speed corresponding to twice the *SMR* for each of the three tested temperatures, and plotted the increases in oxygen consumption against the corresponding increases in swimming speed (*x* values). The

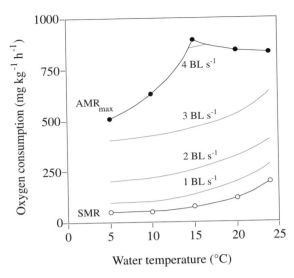

Fig. 2.7 Oxygen consumption rate of a 0.188-m, 50-g juvenile sockeye salmon *Oncorhynchus nerka* at different water temperatures, in relation to swimming speed (body lengths per second: BL s^{-1}). *SMR* and AMR_{max} are the standard metabolic rate and maximum active metabolic rate respectively. Redrawn from Brett (1964).

m-values producing the best fit were obtained after linearisation of the equation, and were 1.990, 2.016 and 2.025 at 5, 10 and 15°C, respectively. This supports the idea that active metabolism almost doubles for each increment of 1 U_{opt}. As the ratio between AMR_{max} and *SMR* ranges from 4 to 8 in most fishes, U_{ms} should normally range from 2 to 3 U_{opt}. Calculations of U_{opt} and associated energy costs (deemed to approximate to *SMR*, see above) provide an approach, using limited data, to determining the capacities of fish species to migrate.

Videler (1993), who compiled existing information on U_{opt} in different fish species of masses from less than 10 mg to over 1 kg, proposed a general model for U_{opt} against body mass (*M*), which stands as:

$$U_{opt} = 1.10\, M^{0.14}\ (r^2 = 0.80)$$

with units of BL s^{-1}. This implies that the relative optimum swimming speed increases with decreasing body mass but that (expectedly) large fish at U_{opt} have higher absolute swimming speeds than smaller fish. A dimensionless estimator of the cost of transport (*COT*) at U_{opt} is frequently used to compare locomotion costs between species (synoptic tables in Videler 1993). However, fish of increasing length have higher absolute swimming speeds and greater bulk, both of which increase the drag, and the power required for overcoming this drag. For these reasons, it may be best to use a length- and weight-dependent estimate, $COT \times BL \times BW$, which represents the total energy (in joules) needed by the fish to swim its body length (*BL*) at U_{opt}. Using the same data base as that for modelling U_{opt}, Videler (1993) found that $COT \times BL \times BW$ in fish was best described by the following model:

$$COT \times BL \times BW = 0.5\, M^{0.93}\ (r^2 = 0.99),\ \text{where } M \text{ is the fish mass (kg)}$$

Dividing $COT \times BL \times BW$ by the fish's body length gives the amount of work invested by the fish to swim over 1 m (work per metre, WPM, J m^{-1}). Work per metre is a more practical estimator of the capacity of fish to migrate over certain distances, as it combines the metric system used by biologists and river managers, to the energy used by the fish. For example, U_{opt} for a 100-g (about 18 cm) and a 1-kg (about 38 cm) South American *Prochilodus* can be estimated at 0.28 and 0.42 m s^{-1}, respectively (1.5 and 1.1 BL s^{-1}). The corresponding energetic cost (WPM) needed for migrating at U_{opt} in still water for these two fish would be as low as 0.32 and 1.32 J m^{-1}. If these two fish travel at the same ground speed against a water current of 0.5 m s^{-1}, the corresponding energetic costs become 5.25 and 10.34 J m^{-1}. Also, the 18-cm fish could not negotiate areas with currents faster than about 1 m s^{-1} without incurring an oxygen debt (assuming U_{ms} is about 3 U_{opt} in these species).

Implicit in aerobic metabolism is the requirement to extract sufficient oxygen from the water, and swimming performance can be restricted by low ambient oxygen concentration. Oxygen solubility in water decreases with increasing temperature and salinity, and hypoxic conditions may prevent full use of metabolic scope. Such conditions are common in bottom water layers in aquatic habitats where the water column is poorly mixed, especially in tropical environments, although most tropical fish species have adapted to low oxygen values through a series of physiological and morphological adaptations (see section 4.4.4). Other factors influencing the energy cost during migration include variation in osmoregulatory costs for diadromous species crossing haloclines and the ability of a substantial number of fish species to offset gill ventilation costs by employing ram gill ventilation instead of branchial ventilation during swimming. For striped bass *Morone saxatilis*, ram ventilation provides an increase in energetic efficiency of 8.1% during swimming at 1.6 BL s^{-1} (Freadman 1981).

2.3.5 How far can a fish migrate?

The previous sections have described how fast fish can swim, their endurance while swimming at different speeds, and how energetically expensive it is to swim. Given a clearer picture of the 'engine' of the fish, it is appropriate to examine its 'fuel tank', whether it can be refilled during the journey and whether travelling at an economic pace can enable it to reach its destination within a given time. Fish can mobilise energy from protein, carbohydrate and lipids, the latter representing the most efficient storage form, as the energy density is about twice that of carbohydrate and protein (37 versus about 20 kJ g^{-1}). Some species such as the Atlantic and Pacific salmons (*Salmo* and *Oncorhynchus* spp.) do not eat at all during their upstream migration in rivers, implying that they can travel until they run out of fuel, and that energy management is the key to their migration, especially over long distances (Bernatchez & Dodson 1987). Species that eat during upriver migration can partly or totally replenish their reserves, but feeding imposes further constraints, notably with respect to the time dedicated to foraging, especially in unknown environments. Also the costs of foraging and digestion reduce the available metabolic scope for directed swimming. This could substantially delay migration, and cause the fish to arrive late at its destination, putting it at a significant disadvantage, especially when migrating to spawning grounds. For these reasons, most species making extensive migrations, and especially those relying on low-energy diets, replenish their energy reserves very little during their upstream migration. Therefore, starting a migration with the maximum fuel reserve can be viewed as an optimum strategy, provided sufficient energy can be stored within a period of time short enough not to delay the start of the migration.

Fish species moving over considerable distances have generally been found to accumulate significant amounts of energy in lipid droplets within the muscles and lipid deposits around the digestive tract (Idler & Bitners 1958; Saldaña & Venables 1983; Leonard & McCormick 1999). Prior to their riverine spawning runs, sockeye salmon *Oncorhynchus nerka* have a tissue energy content of 7.8 kJ g^{-1} (Brett 1983), and values greater than 8.0 kJ g^{-1} have been documented for the Venezuelan coporo *Prochilodus mariae* (Curimatidae) in the Orinoco River (Saldaña & Venables 1983) (Fig. 2.8). Both values are above the range reported by Kitchell *et al.* (1977) for 81 fish species. These provide estimates of the maximum energy reserve available for upriver migration and spawning. The proportion of energy effectively used during migration varies substantially depending on distance travelled and water velocity, and it also differs between lifestyles.

To our knowledge, the almost complete depletion of energy reserves which happens in some marine pleuronectiforms after spawning (muscle containing 95% water, 3% protein and 0.05% lipid; also dubbed 'jellying'; Jobling 1995) has never been observed in freshwater fish species making long potamodromous migrations, possibly because such an extreme investment would compromise the chances of successful spawning. Nevertheless, the cost in energy of spawning runs (migration plus spawning) ranges from 75 to 82% in semelparous Pacific salmons *Oncorhynchus* spp., and is about 60% in Atlantic salmon *Salmo salar* (Glebe & Leggett 1981b). American shad *Alosa sapidissima* use as much as 70–80% of their stored energy for the spawning runs in the warm St Johns River, Florida, US, where there are no repeat spawners (Leggett & Carscadden 1978). In northern populations of the Connecticut

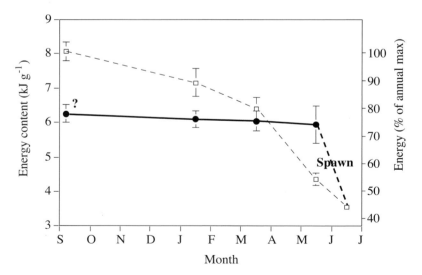

Fig. 2.8 Total energy content of, and seasonal utilisation of energy by, female coporo *Prochilodus mariae* in the Orinoco River system, Venezuela. The thick solid curve corresponds to fish remaining in floodplain lagoons, and the thin dotted curve to fish undertaking potamodromous migrations. Lagoon residents invest much more energy into reproductive products than migratory fish, but overall energetic expenses are similar. Values are means ± 2 standard errors for a typical female coporo of 1 kg wet weight. Redrawn after Saldaña and Venables (1983).

River, New England, US, where American shad show repeated spawning, only 35–60% of the energy is used (Glebe & Leggett 1981a; Leonard & McCormick 1999). The boundary between semelparous and iteroparous lifestyles seemingly lies within a 60–70% depletion range (Glebe & Leggett 1981a; Wootton 1990). Regarding the division of energy between migration and spawning, sockeye salmon invest 55% into migration in the Skeena River, Canada and another 23% into spawning (Brett 1983). *Prochilodus mariae* ascending the Orinoco River, Venezuela, invests 46–47% into migration and only 9% into gonads and spawning, whereas populations trapped in lagoons (and making no upstream migration) invest up to 36% into gonads and spawning (Saldaña & Venables 1983). Recently, Unwin *et al.* (1999) documented post-maturation survival, and repeat maturation in captive male chinook salmon *Oncorhynchus tshawytscha*, with greater energetic investment into reproductive products, and lower energy storage in the muscles, suggesting that shift to iteroparity was possible in this typically semelparous species in conditions where energy expenditure was low.

Swimming in still water is relatively inexpensive, and even fish as small as 50 g could travel over 500 km in lakes within less than 7 weeks, while swimming at U_{opt} for 12 h per day. The same applies to those fish species or life stages taking advantage of tidal transport to move upstream in estuaries (see section 2.4.3). Returning to the *Prochilodus* example given in section 2.3.4, we can examine the influence of energy reserves on migration limits. Assuming that both the 100-g and the 1000-g fishes can store equivalent energy reserves relative to their body weight, and that they expend 47% of their energy while migrating (in accordance with Saldaña & Venables 1983, for *P. mariae*), the maximum distances they can travel at U_{opt} in still water are 1188 and 2880 km, respectively. Corresponding estimates for ground speed = U_{opt} against a constant water current of 0.5 m s^{-1} are 426 and 1312 km. The times needed to migrate over these distances (U_{opt} being 0.28 and 0.42 m s^{-1}) are 423 and 868 h respectively (about 36 and 72 days for fish migrating 12 h per day). Constraints on migration in flowing waters are much greater, and they vary with spawning season and latitude.

2.3.6 *Constraints on early and late migrants*

Spring spawners at temperate latitudes (e.g. most cyprinids, shads) spawn under conditions of decreasing water flow and rising temperature. As a result of low temperature, early migrants have lower swimming capacities, and swimming speeds are likely to be slow. Ground speed is even slower as the current is fast, and on some occasions too fast to enable the fish to continue its migration while using aerobic swimming only. Conversely, late spring migrants have greater swimming capacities, and they normally face slower currents, implying they can travel much faster, but with greater energy expenditure as a result of the higher temperature. Nevertheless, assuming there is an optimum time window for seasonal spawners, late migrants take the risk of arriving late on the spawning grounds if water currents at this time of the year are faster than usual, and thus of incurring a loss of genetic fitness as their genes will be less represented in the next generation. Swimming faster (e.g. shifting from U_{opt} to nearer U_{ms}) also is an alternative, but it implies much greater pre-migration energy investment, and a greater depletion of energy reserves, which could compromise the possibility of repeated spawning, thus also causing a loss of fitness (e.g. American shad *Alosa sapidissima* in the Connecticut River; Glebe & Leggett 1981a).

At temperate latitudes, autumn spawners (e.g. many salmons) face the opposite challenge. Summer migrants travel under conditions of relatively low water velocity and high water temperature, giving them some extra capacity to travel further while using aerobic metabolism only, at a greater energetic cost per stride, but at a potentially lower cost per unit distance travelled (as regards ground speed). Provided that there are no obstacles (either natural or man-made) on their migration route, early migrants could arrive in the vicinity of their spawning grounds much earlier than the time for spawning, imposing further maintenance metabolism costs while waiting for the spawning period. This appears somewhat paradoxical for species that do not feed during their spawning migrations. However, this may not be a great disadvantage if fish can find places where temperature and water currents are low enough to minimise these maintenance costs through behavioural thermoregulation (e.g. Berman & Quinn 1991). Late migrants reduce residence costs, but they need to travel quickly, against fast currents, and at lower temperatures, thus implying greater energy expenditure per unit distance travelled (as regards ground speed). Species such as Atlantic salmon *Salmo salar* may enter rivers in every month of the year and remain in the river for up to a year before spawning (Shearer 1992). Spring-running Atlantic salmon are usually larger and so have greater energy reserves to enable river residence for many months without feeding. Early migrants seemingly take a gamble on temperature conditions, and late migrants on water velocity, although there may be other reasons for early entry such as the possibility of reduced predation risk by comparison to coastal habitats. The longer the upstream migration and the steeper the slope, the greater the advantage brought about by an early start. When the journey is short, a later start is advantageous in energy terms, provided maximum water velocities do not exceed the maximum prolonged swimming or burst swimming capacities of the fish. Implicit in this assumption is that the range of suitable strategies for diadromous species is broader for small than for large river basins. Implicit too is that a migration strategy adapted to a particular river basin may prove partly or totally inefficient in another basin with a different length, water velocity profile and climate, and that this may have represented a sufficient environmental pressure behind the selection of homing behaviour and orientation capacities in diadromous species.

Seasonal migrations in tropical rivers resemble, to some extent, those of autumn spawners in temperate systems. Most seasonal strategists spawn at the onset of the flood season, or under maximum floods (section 4.4), suggesting that late migrants also take a gamble on water velocity. Because water temperature shows little seasonal fluctuation in the tropics, late migrants would not appear to suffer a decrease in their aerobic swimming performance throughout the season (provided the oxygen level is sufficient), and late starters might not be disadvantaged. This may be an incorrect viewpoint, for several reasons. Variations of water levels in tropical rivers are generally of greater amplitude than in their temperate counterparts (chapter 4), so the gamble on water velocity may be more serious than in temperate areas. By virtue of size-dependent swimming speeds, it also is proportionally more serious for small than for large fishes. The dilemma between starting upstream migrations earlier or later can also be viewed from a different viewpoint, i.e. what is the advantage brought about by staying longer in the 'summer' residence?

If food is abundant, temperature and oxygen conditions are suitable to permit fast growth, and energy reserves are not fully replenished, there is a definite advantage for postponing departure time. Conversely, if the summer environment becomes limiting, with little chance of the fish increasing its energy reserves, this advantage no longer exists. After spawning, many adults of tropical fish species move into recently flooded lowland or floodplain lakes, where productivity is high and gives them the opportunity of replenishing the energy reserves that have been depleted during the previous spawning migration. This also applies to immature fishes, which drift, as eggs or larvae, onto the recently inundated floodplain. Decreasing water levels and phytoplankton biomass on the one hand, and increasing fish biomass, as a result of growth, on the other hand, make the environment increasingly unfavourable for growing or storing energy reserves. This trade-off might be viewed as a continuum, but this may be erroneous for fishes in floodplain lakes. During summer, receding waters make these areas increasingly isolated from the main channel, until the connection is no longer functional, the precise timing of this being dependent on floodplain topography, climate and associated flow regime. Adults could not then migrate until the connection became functional again during relatively high flows. This would leave a relatively short period of time for those migrants to reach upstream spawning grounds, presenting a severe energy challenge in view of the distances to be travelled and velocities to be traversed. The study of Saldaña & Venables (1983) on *Prochilodus mariae* in the Orinoco River basin, suggested that fish which had stored enough energy prior to isolation would leave floodplain lakes early, and undertake a long upriver migration, whereas those which had stored little energy would remain in the isolated lake and undertake no migration at all, eventually spawning in the lake itself. These arguments tend to refute the relevance of the late starting strategy in tropical freshwater environments, and further account for why migrations in the tropics are so seasonally constrained (section 4.4.6).

2.3.7 Implications of migration costs on size at first sexual maturity: when time matters

In systems where suitable spawning places are abundant and scattered throughout tributaries, migrating at a slow swimming speed is acceptable for small fish, as distances travelled may be short. In systems where spawning places are scarcer, or in species for which the upstream migration compensates the drift of pelagic eggs, such as for many characiforms, a minimum upstream migration distance is required, and this imposes additional constraints on the size at first sexual maturity.

The previous paragraphs have highlighted how expensive migration is in flowing waters, and how important energy reserves and their use are in shaping life-history strategies. Small fish expend less energy per body length swum, but also have lower absolute swimming capacities than larger conspecifics. The increase of swimming costs against increasing water currents is thus steeper in small than in large fish, and while negotiating fast currents, small fish may spend more energy than larger fish travelling at the same ground speed. The corresponding decrease of the energy reserve is even steeper as the energy reserve is proportional to body mass, which increases as an approximately cubic function of body length. Swimming faster or against fast currents can thus be achieved, but only at the expense of maximum distance travelled, and long distances (the maximum distance being the energy reserve divided by travel cost at U_{opt}) can only be travelled at the expense of swimming speed, and thus at the expense of travel time. Therefore, small fish could be forced by bioenergetic constraints to migrate at such a slow pace that they could not arrive in time for the spawning period.

This can be illustrated using the *Prochilodus* spp. referred to earlier (calculations in sections 2.3.4 and 2.3.5). Most *Prochilodus* spp. in South American rivers make upstream migrations that generally range from 250 to 700 km (see section 4.4.6) and start during the late dry season. Water velocity is relatively low during this period of the year, and it ranges from 0.10 to 0.75 m s^{-1}, depending on stream order (estimated from Colombian rivers and streams where another *Prochilodus* sp., *P. magdalenae*, is found). Later, water velocity is two to three times greater as the rainy season progresses and flood increases. We can examine the possibilities and constraints for fish of different body weights (160, 180 and 200 g; A, B and C, respectively) travelling an upstream distance of 250 km. For this calculation, we assume that the frequency distribution of water velocities along the path during the late dry and early rainy season is given by equivalent proportions of stretches at 12.5, 25, 37.5, 50 and 75 cm s^{-1}, which is deemed to reflect variations in the river slope along the river course and pool–riffle sequences. The body lengths of fish A, B and C are 21, 22 and 23 cm, and their U_{opt} are 30.4, 31.0 and 31.6 cm s^{-1}, respectively, representing only a 4% advantage for fish C over fish A as regards swimming performance. In view of the maximum water velocities on the proposed migration path, all three fishes could negotiate it by using aerobic swimming only (assuming $U_{ms} = 3 U_{opt}$), but none could do it at U_{opt} throughout. The smaller the difference between swimming speed and U_{opt}, the lower the cost of swimming and the longer the distance that can be travelled, but the longer it takes (see sections 2.3.4 and 2.3.5 for approaches to calculating swimming speeds, together with energy depletion constraints). The fastest average ground speed at which fish C could travel to cover 250 km in the conditions above is 0.28 km h^{-1}, *versus* 0.16 and 0.03 km h^{-1}, for fishes B and A respectively. Assuming all three fishes migrate for 12 h per day, the corresponding travel times are 74, 130 and 651 days, respectively. The estimate for fish C is consistent with travel times reported for *Prochilodus lineatus* in the Paraná River basin (about 2.5–3.0 months; Toledo *et al.* 1986). Fish A would be unable to reach the spawning ground within time. Fish B could theoretically do so but probably not on a 12 h per day basis. The long duration of the journey also implies it would have to migrate under higher floods, and thus incur greater swimming costs that would slow down its progression, and shorten the distance travelled. This demonstrates how a 2-cm size difference can have a large influence on the viability of migration. A similar situation applies to small differences in water temperature, as swimming speed and swimming cost both increase with increasing temperature.

Henceforth, we suggest that bioenergetic constraints, probably more so than size-related differences in swimming speed, have represented a selective pressure behind the size at first maturity in fish species making long upriver migrations. In some poeciliid species, the approximate age at maturity is determined by a single sex-linked locus, and selection may thus operate more readily.

2.3.8 A tentative synthesis and conclusion

The muscle mass, proportion of red muscle, and energy reserves govern a fish's capacity for migration, which is further influenced by ambient temperature and adverse currents, and thus by the time of the year when migration is made. From a functional viewpoint, the selection of the appropriate time for migration implies that stimuli be identified and correctly interpreted, the nature and level of these stimuli being theoretically dependent on their predictive value regarding the set of environmental changes that affect the capacity for migration. If we accept that stimuli and capacity for migration have been selected during the co-evolution of the species and its environment, then their present and future adequacy also depends on how the environment changes, as a result of natural or human-induced causes.

Global warming in general, or local warming resulting from channelisation, industrial and urban effluents, may strongly affect the swimming capacity of fish, their use of energy reserves for migration, and the relevance of thermal cues. Similarly, flood pulses in regulated rivers may be confusing for migratory species relying on hydrological cues. Finally, man-made obstacles have contributed to reduce the capacity for migration, in a dramatic way when differences in height or adverse currents in fish passage facilities exceed the species' burst swimming speed or distance over which fish can swim at burst speed. These have reshaped the geographical distribution of some migratory species in the same way as natural obstacles did in the past (e.g. Atlantic salmon *Salmo salar* and allis shad *Alosa alosa* populations in many western European rivers). Similar situations are occurring or expected in many tropical systems undergoing hydroelectric development. Undoubtedly fish passage facilities help to alleviate this problem (Clay 1995), but the steep slopes of many designs impose additional swimming costs, and greater use of energy reserves that may jeopardise the success of the migration, especially when the entrance of the fish pass requires a long searching time, or when the obstacle cannot be cleared at the first attempt. Probably the best illustration of such challenging obstructions and fish passes is the reach of the Fraser River on the Canadian west coast, known as Hell's Gate. This region of rapids and extremely fast water velocity has probably greatly inhibited upstream migration by adult salmonids of several species, including sockeye salmon *Oncorhynchus nerka*, in previous centuries, but landslips resulting from construction activities of the Canadian Northern Railway in the early twentieth century seem to have exacerbated the problem (Smith 1994). Despite provision of a fishway, movement of sockeye salmon through this section remains a problem. Studies using electromyogram telemetry (see chapter 6) have provided a wealth of information on the cost of migratory behaviour at the individual fish level (Hinch *et al.* 1996; Hinch & Bratty 2000), and provided some evidence that passage through this section of river is strongly dependent on the swimming strategy employed. They demonstrated that salmon swimming close to U_{crit} for periods as short as 10 min in downstream turbulent waters without entering the fishway eventually failed, disappeared downstream and died (Fig. 2.9).

To date, there is a dearth of knowledge on the swimming capacity of the vast majority of freshwater fish species, and especially those in the tropics, whereas this knowledge is increasingly needed for a better understanding of their migratory capacities, and how man-made changes are likely to affect them. Kinematics and respirometer-swim chambers are of great interest for measuring their burst speed and sustained swimming capacities. Telemetry techniques offer great potential for determining the actual costs of migration (Box 2.2) and functionality of fish passes. Nevertheless, more straightforward, simple and rapid approaches (determination of fish muscle mass, proportion of red muscle, and energy reserves) are extremely relevant as they may give a broad picture of likely speeds and endurance, by comparison with existing models. These measurements would also be relevant to determining the adaptive potential of fish cultured for restocking purposes, and to give a more objective statement on the relevance of culture environments for restocking purposes.

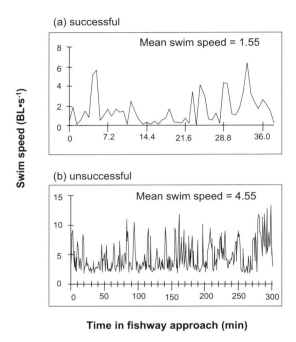

Fig. 2.9 Examples of electromyogram records of adult sockeye salmon *Oncorhynchus nerka* which (a) successfully and (b) unsuccessfully migrated through the Hell's Gate region of the Fraser River, Canada. Figure provided by S. Hinch and based on data provided in Hinch and Bratty (2000).

2.4 Piloting, orientation and navigation

Movement by a fish within a small stream is somewhat constrained by the limited scale of the environment, particularly in terms of depth and width in what is essentially a linear environment. By contrast, for fishes carrying out oceanodromous and diadromous migrations, the marine environment represents a much greater challenge for successful movement to a defined habitat or a specific locality. Therefore, although the capacity to direct movement towards particular spatial goals is of great significance to holobiotic freshwater fishes as well as for marine and diadromous fishes, the greatest research interest has been applied to the last two categories, because of the large scale of many oceanodromous migrations, the issue of how fishes may home to areas hundreds or thousands of kilometres from where they originated, the perceived relative lack of sensory clues for migration in marine environments and the high economic value of many of these fishes. Indeed, methods by which fishes exhibit movement towards particular spatial goals are highly variable and may be strongly influenced by spatial scale.

Directional responses of fish to various external stimuli may be broadly grouped into three categories (Baker 1978; McKeown 1984). The first category, known as 'piloting', involves reaching a spatial goal by reference to familiar landmarks, which may be identified using one or more sensory mechanisms. The concept of cognitive mapping involves integration of the goal location with a knowledge of spatial relationships between particular landmarks. The second category is 'compass orientation', in which the goal is reached by orientating on a

Box 2.2 Measuring the energetics of migration

Perspectives of energy costs of migration may be made by employing a number of approaches. For those species which undertake migrations in fresh water without feeding (e.g. some lampreys and salmonids), samples of fish may be taken at the beginning of the migration and analysis of body tissue energy content carried out, with this process repeated at successive stages along the migration route (Idler & Bitners 1958). These proximate analysis techniques are generally only useful where migration behaviour and rate is similar between individuals of a species, since with current technology, body composition analysis can only be carried out once on each fish! Bioenergetic modelling, using a combination of input data derived from laboratory measurements and field records, is another useful approach. However, due to the difficulty in providing accurate (and sometimes appropriate) data for input into such models, they are best used with a range of values to make predictions and generate hypotheses about migration energetics (see our predictions on migration of *Prochilodus* earlier in this section). Doubly labelled water techniques, used widely for energy expenditure estimates of free-living terrestrial animals, are inappropriate for fishes due to their high rates of water turnover (Lucas *et al.* 1993a).

Recently physiological telemetry techniques have proved valuable for estimating energy expenditure of fishes in the natural environment (Lucas *et al.* 1993a). The advantage of such techniques is that, provided the behaviour of the fish is not unduly affected by capture and transmitter attachment, the behaviour of individual fishes, and resultant influence on energy expenditure, can be monitored. Physiological telemetry requires calibration of the telemetered variable against energy expenditure (usually as oxygen consumption rate) under a range of conditions and for fishes of the ontogenetic stage and size to be studied. The most widely used physiological telemetry approaches for estimating energy expenditure during migration is electromyogram telemetry from the trunk musculature (McKinley & Power 1992; Demers *et al.* 1996; Hinch *et al.* 1996; Hinch & Bratty 2000; Fig. 2.9). This has the advantage that the relationship between tail beat frequency (as identified from EMGs) and oxygen consumption rate is unlikely to be strongly influenced by environmental factors. Physiological variables such as heart rate may be influenced by factors such as water oxygen content, and while cardiac output is usually closely correlated with oxygen supply to the tissues, heart rate is only a good indicator of metabolic rate if its contribution to increased oxygen delivery at varying metabolic rates is a constant proportion or predictable (Lucas *et al.* 1993a; Lucas 1994; Armstrong 1998). Heart rate telemetry currently remains the only physiological telemetry technique which can be used to provide direct estimates of total aerobic metabolic costs including basal, feeding and activity metabolism. For those fish species that feed during migrations, so long as appropriate calibrations can be achieved and environmental conditions monitored during tracking, the full respiratory budget can be deduced from heart rate telemetry.

particular compass bearing without reference to local landmarks. The goal is reached when the compass direction leads to the goal or a familiar area containing the goal. In the simplest form the return compass bearing may be the opposite of that on an outward journey or migration, involving route reversal. The third category includes 'true navigation' in which the spatial goal can be achieved by orientating in the appropriate direction in unfamiliar territory, when moving in a new and unfamiliar direction (Baker 1978). In most cases navigation refers movements to a 'home' site and when in unfamiliar territory, it involves measurement of local conditions and comparison with home conditions through extrapolation of the cognitive map. There is good evidence for piloting and orientation in fishes, but less evidence for true navigation (McKeown 1984; Dodson 1988). In many freshwater systems of limited size, piloting and orientation alone are likely to provide adequate spatial clues for the attainment of a spatial goal.

For at least one stage of a migratory cycle, movements may involve little spatial goal-seeking behaviour. Eggs and larvae may drift passively in rivers according to the vagaries of currents, some accumulating in appropriate nursery habitat such as lentic backwaters, while others may fail to arrive in suitable habitat before energy reserves are exhausted. Even here though, a minor rheotropic response (orientation with respect to current), involving slight active movement at a vector off the current direction, may greatly increase the probability of arriving in river edge habitat with lower water velocity and perhaps appropriate microhabitat. Postlarval, juvenile and adult fishes generally have adequate sensory and muscular development to provide sufficient orientation responses to at least enable search for appropriate microhabitat conditions. By contrast the orientation and homing capabilities of various diadromous fishes, particularly various salmonids, are highly impressive. The most studied of these fishes are the Pacific salmons *Oncorhynchus* spp. and Atlantic salmon *Salmo salar* which may migrate several thousand kilometres in the ocean, to return to the stream in which they were spawned (Hasler 1966; Harden Jones 1968; Hasler & Scholz 1983; Stabell 1984; Dittman & Quinn 1996). Increasing evidence of the precise homing of such fish through the late nineteenth century (Buckland 1880) and the early part of the twentieth century led to the 'parent stream theory' (Davidson 1937) and increased interest in the orientation mechanisms of fishes.

Fish biologists have sometimes been guilty of searching for detailed mechanisms of explaining highly directional spatial behaviour and migration while ignoring other simpler explanations for some conspicuous migrations (Leggett 1977; Dodson 1988). It is important to note that the efficiency of many migrations is not always as great as might be thought from catches or return rates. Telemetry of fish locating river mouths (Quinn *et al.* 1989), moving through estuaries (Moser *et al.* 1991; Moser & Ross 1994) or when displaced in lakes or rivers (Baras 1997) has shown that in many cases the migration paths are convoluted and/or energy cost of travel per unit distance is high, thus reducing the maximum distance that can be travelled.

Within river environments, essentially linear systems where unidirectional currents provide a valuable cue, highly advanced orientating mechanisms may be unnecessary for enabling holobiotic freshwater fishes to direct their movement towards particular habitats or conditions. Although movement through inundated floodplains may require better orientation capabilities, river channels allow large-scale movements in one spatial dimension only. Even in marine, open-water environments where an advanced orientation capability would seem necessary it has been demonstrated that random search, with a few biologically sound assumptions of tolerance to environmental variables (depth, temperature, etc.), combined with the sensory capability of halting or limiting movement when preferred habitat is located, is adequate to explain observed spatial and temporal phenomena of migrations of several marine species as well as the marine homing phase of salmon migration (Saila 1961; Harden Jones 1968; Leggett 1977).

A wide variety of cues exist for piloting, orientation and navigation, including physical landmarks, currents, celestial characteristics, polarised light, geomagnetic and geoelectric fields, odours and other chemical factors, temperature, salinity and social interaction. These features, the reactions of fish to them and their value in fish migrations have been widely reviewed (Harden Jones 1968; Leggett 1977; Hasler 1971; Northcote 1984; Quinn 1991; Dittman & Quinn 1996). Some cues such as the characteristic physicochemical boundaries of ocean fronts, although important for the ocean migrations of diadromous fish species, are outside the scope of this book.

2.4.1 Landmarks and surface topography

In fresh water, visual cues are undoubtedly important for piloting in familiar areas, using local landscape features. Little work has been done on the importance of visual cues during migration outside of day-to-day home ranges. The general view is that visual cues are unlikely to be important during long-distance or open-water orientation, but that if a fish returns to a familiar area, for example during a homing migration, then underwater landmarks may be important (Cunjak & Power 1987). Long-lived fishes in freshwater systems, especially rivers, may have ample opportunity to build and repeatedly use spatial maps of areas traversed during seasonal migrations between discrete habitats. The significance of such mechanisms has been greatly downplayed in the assessment of migration processes involved in long-distance lifecycle-related migrations of fishes such as those of salmon and anguillid eels. Many freshwater species such as the green sunfish *Lepomis cyanella* and the European minnow *Phoxinus phoxinus* appear to recognise their home range using visual information, and these cues are known to be important for intertidal fishes also (Hasler 1966). There is some experimental evidence that blinded translocated adult salmonids and shads (Clupeidae) find it more difficult to locate home rivers than intact fishes (McKeown 1984), although in such experiments it has to be recognised that for many fishes vision is fundamental to many behaviours such as predator avoidance and feeding as well as directed movement, and so the context of such experiments must be considered.

2.4.2 Celestial cues

There is a considerable body of evidence to suggest that fishes can utilise the sun for orientation during migration (Hasler 1971; Box 2.3), but it is also clear from many experiments that most fishes can orientate towards a home site or other goal in the absence of sunlight (Hasler 1971; Leggett 1977; McKeown 1984). Orientation using the sun may rely on deriving information from changes in the sun's azimuth (the sun's angle relative to the horizontal plane) or altitude (the angle of the sun in the vertical plane). However, the sun itself is often obscured by cloud cover and therefore does not necessarily provide a reliable orientation cue.

Box 2.3 Homing on a sunny afternoon – Arthur Hasler's experiments on sun-orientation

Among the first work on solar orientation in open-water conditions was that carried out by Arthur Hasler's team in homing experiments using white bass *Morone chrysops*, translocated from spawning grounds in Lake Mendota (Hasler *et al.* 1958). By attaching a small float with a tether greater than the water depth to a fish, they could follow the fish's movements from a boat. They demonstrated rapid and relatively precise open-water orientation towards the 'home' spawning grounds after displacement, but fish appeared to have difficulty in orientating during overcast conditions or when eye caps were applied.

Observing a towed float is not an ideal way of studying fish behaviour and, in later experiments on white bass, Hasler's team used acoustic telemetry (Hasler *et al.* 1969). In spring and summer of 1965, 26 white bass were tagged with ultrasonic transmitters and their movements tracked following displacements of 1.6 km to the middle of Lake Mendota, during the spawning season, and others 3.1 km, 2 months after the spawning season to the same release point. Fish were tracked for a mean period of 7.3 h and exhibited steady swimming towards the 'home' site for both spawning and nonspawning groups. Some open-water orientation was clearly demonstrated and supported the use of a sun-compass mechanism. However, the work by Hasler's team also suggested that other mechanisms such as olfaction were used in addition to sun-orientation by white bass (Hasler *et al.* 1958, 1969; Hasler 1971) and given the limited distances of displacement it is quite conceivable that piloting using local landmarks was also important in homing (see also Dat *et al.* 1995).

Polarised light is common in nature and is maximal at dawn and dusk when as much as 60% of light at the water's surface may be polarised. Several studies have demonstrated that fishes can detect and discriminate different planes of polarised light, and orientation to polarised light has also been demonstrated, although it is unclear how important such mechanisms might be in the natural environment (Leggett 1977). The night sky may also provide celestial cues to fish for determining compass direction although there is only limited evidence of its significance (McKeown 1984).

Levin and his co-workers (Levin & Belmonte 1988; Levin & Gonzáles 1994; Levin *et al.* 1998) investigated the swimming orientations of South American microcharacids (genera *Aphyocharax*, *Cheirodon*, *Hyphessobrycon*, *Moenkhausia* and *Roeboides*) in experimental enclosures. They provided evidence that the sun compass mechanism includes both diel and seasonal components, consisting of internal relays that control the vertical and horizontal changes in sun position that take place at these two temporal scales. This causes changes in the swimming direction at different sun angles resulting from different water column heights, and could account for the opposing directions of migrations during the dry and rainy season. They further suggested the mechanism had an inertial basis, i.e. fish resetting (or zeroing) their mechanism during periods of swimming at a constant angle to the sun (or light), this capacity enabling them to maintain a constant direction in the absence of the solar cue, and thus to swim in turbid waters or in deep water layers.

2.4.3 Currents

In river environments one of the most obvious cues that may be used for biasing movement towards a spatial goal is the water current, which may provide visual (optomotor response), tactile and inertial cues. Currents provide strong, broadly unidirectional cues in rivers, although in lakes, estuaries and coastal environments such cues may be ambiguous or of small amplitude and fishes may require additional stimuli for concerted movement in a single direction. Rheotropic (orientation to current) responses are probably of substantial importance in almost all freshwater fishes and innate rheotactic (movement) responses have been demonstrated for a wide variety of freshwater fish species (Northcote 1984), although work has concentrated on salmonids. Experiments with rainbow trout *Oncorhynchus mykiss* from lake outlet and inlet stocks demonstrated the existence of differential responses mediated through genetic and environmental factors (Northcote 1962; Kelso *et al*. 1981), with similar results found for a range of salmonids (Rayleigh 1971). Dodson and Young (1977) demonstrated that positive rheotropic behaviour of adult common shiners *Notropis cornutus* could be modified by temperature and photoperiod, and suggested that the interactive influence of environmental conditions other than currents was important for stimulating upstream migration to spawning gravels in this species. Influence of environmental factors such as temperature and light conditions on rheotactic behaviour appears widespread in freshwater fishes (Northcote 1984).

In brackish or tidal environments movements are more variable. Holobiotic freshwater fish species that migrate into tidal stretches of the lower reaches of rivers (often during winter) must remain within a limited area to prevent movement into fully saline water, requiring a degree of current stemming or utilisation of low flow pockets during ebb tides, with the mean position varying little on a tide-to-tide basis, although longitudinal distance moved on each tidal cycle might be substantial. There have been few studies of the mechanisms involved for 'primary' freshwater species, but such behaviour also occurs in thin-lipped mullet *Liza ramada* during summer feeding episodes in tidal waters (Almeida 1996) and with migratory salmonids during holding periods in estuaries (Priede *et al*. 1988). Acoustic telemetry suggests that this behaviour involves orientation into the flow and travel up and down the estuary during each tidal cycle. Downstream movement in estuaries may involve passive drift (e.g. Fried *et al*. 1978) or might utilise selective tidal transport (e.g. McCleave & Arnold 1999), involving stemming or seeking shelter from flood tides and swimming or drifting with the ebb tide. However, tidal stemming during both flood and ebb tides, resulting in extended, energetically expensive estuarine holding, has been reported for coho salmon *Oncorhynchus kisutch* smolts (Moser *et al*. 1991). Upstream movement can also be achieved by selective tidal stream transport, by stemming ebb tides (or taking refuge) and swimming or drifting on flood tides. Alternatively upstream progress may be made by continuous swimming in an upstream direction during both ebb and flood phases. The latter is energetically costly, especially for smaller species, and river entry utilising selective tidal stream transport has been reported for European eel *Anguilla anguilla* elvers (Creutzberg 1961), adult American shad *Alosa sapidissima* (Dodson & Leggett 1973) and larvae of flounder *Platichthys flesus*

(Bos 1999). In some, perhaps frequent, cases, fish entering rivers may swim vigorously against the ebb and flood tides in a highly rheotactic, but energy inefficient, manner as found by acoustic tracking of adult American shad, striped bass *Morone saxatilis* and shortnose sturgeon *Acipenser brevirostrum* (Moser & Ross 1994), resulting in a long holding period before moving upriver on one or more flood tides. They suggest that a high energetic cost of estuarine migration may be offset against a need for osmoregulatory acclimation although Priede (1992) provides evidence from salinity telemetry of adult Atlantic salmon *Salmo salar* that movement from full-strength seawater to fresh water can be extremely rapid.

2.4.4 Electric and magnetic fields

When water moves across the earth's magnetic field, electric currents may be induced, which although tiny, may be of sufficient magnitude to enable detection by migrating fishes (Rommel & McCleave 1972). Such cues are probably of greatest significance for large-scale movements in oceanic environments, where high precision may be unimportant and where other cues may be less available. In most cases fish do not appear to be influenced by electric fields generated by their own swimming, but their orientation can be influenced by electromagnetic stimuli of the magnitude found in the natural environment (McCleave & Power 1978; Quinn & Brannon 1982; McKeown 1984) and magnetic compass orientation (or navigation) occurs. Magnetic material, capable of transducing inducted currents, has been isolated from discrete tissues in the nose region of several fish species (e.g. Kirschvink *et al*. 1985). In rainbow trout *Oncorhynchus mykiss*, magnetoreceptor cells, containing chains of single-domain magnetite crystals (Fe_3O_4 of biogenic origin), occur within a discrete sublayer of the olfactory lamellae. These cells are closely associated with a branch of the trigeminal nerve that responds to changes in magnetic field intensity (but apparently not direction) and are in accordance with behavioural observations to fluctuations in magnetic field. These studies (Walker *et al*. 1997; Diebel *et al*. 2000) provide a framework at the anatomical, physiological and behavioural level for explaining the sensory basis of the mechanism used by fishes to aid navigation in relation to the earth's magnetic field.

Electroreceptors of electric fishes such as the mormyrids and gymnotoids might also be used to detect the earth's magnetic field. However, these fishes principally use electrodetection in combination with distinct electric organ discharges (EODs) from specialised organs for localised electrolocation and communication (Heiligenberg 1991, see also section 4.4.4). However, there is good evidence to show that a variety of electric fish species, most of which are nocturnally active, use electrolocation to return precisely to home refuges before dawn (Friedman & Hopkins 1996). In most cases such observations reflect piloting by electrolocation rather than by more familiar visual or olfactory discrimination of landmarks.

2.4.5 Olfaction and gustation

Olfaction is an extremely important sense for fishes and in the late nineteenth century differences in stream odour were suggested as a mechanism for Atlantic salmon *Salmo salar* returning to home streams (Buckland 1880). Some of the earliest research which showed the potential significance of olfaction in migration was the demonstration that bluntnose minnows *Hyborhynchus notatus* could be trained to discriminate between water rinses from different aquatic plants (Walker & Hasler 1949) and that this species could differentiate between water from two streams (Hasler & Wisby 1951). A large body of evidence, particularly on Pacific salmon *Oncorhynchus* species, has grown to support the importance of olfaction in fish migration and homing (Hasler 1966; McKeown 1984; Smith 1985), although more recent studies have laid the emphasis on gustation, instead of, or in addition to olfaction (e.g. Keefe & Winn 1991).

The extreme chemosensitivity of many fishes is a key feature in use of 'odours' for homing, although some early studies which recorded detection thresholds as low as 3.5×10^{-18} for β-phenylethyl alcohol for trained European eels *Anguilla anguilla*, have not been supported by more advanced electrophysiological recording techniques (Hara 1993). Nevertheless, a variety of naturally occurring substances, including amino acids, bile acids and salts, and many hormones, are potent olfactory stimuli, and in some cases are detectable at concentrations of 10^{-16} M, implying detection of just a few molecules to stimulate an olfactory response (Hara 1993). In a series of experiments Hasler's group provided evidence that coho salmon *Oncorhynchus kisutch* become imprinted on the odour of their home stream during a narrow time window of sensitivity as juveniles (see also section 2.4.7), usually immediately prior to emigration as smolts. They showed that this could be achieved, using synthetic compounds such as morpholine, for hatchery-reared coho salmon stocked in the North American Great Lakes (Hasler & Scholz 1983). When adults returned to spawn, imprinted fish moved into streams in which morpholine had been released, with a homing efficiency of about 95%.

Two principal methods of olfactory imprinting are possible. The first relies on detecting a specific combination of compounds originating from the local home stream environment, the second on species-specific compounds. Organic compounds associated with catchment characteristics have been suggested as odour cues for migrating American eel *Anguilla rostrata* elvers (Sorensen 1986; Tosi & Sola 1993) and other fishes (Hasler 1966). Species-specific or population-specific pheromones produced by conspecifics living in the 'home' habitat have also been suggested as being important (Nordeng 1971, 1977; Box 2.4) and Døving *et al.* (1974) showed that the olfactory bulbar neurones of Arctic char *Salvelinus alpinus* responded differentially to mucus from different populations of this species. These strategies have been described as alternative hypotheses, but as Hara (1993) points out, they are in no way mutually exclusive, and these odours may be interactive, reinforcing one or more components.

2.4.6 Other cues

Diadromous fish species pass between fresh water and seawater and it might be expected that an ability to detect differences in salinity would be an important component in orientation mechanisms of these fishes. Some Pacific salmon *Oncorhynchus* spp. are able to detect NaCl concentrations as low as 5×10^{-3} M, and they exhibit salinity preferences dependent on species, developmental stage and other environmental conditions which may be used as a migration cue (Baggerman 1960; McInerney 1964). Similar observations have been made for anadromous three-spined stickleback *Gasterosteus aculeatus* (Baggerman 1957). Fish may also respond to other physicochemical factors such as oxygen (see section 4.4.4), carbon dioxide or pH (Jones *et al.* 1985), turbidity levels and variations of pressure. For example, swimming activity and orientation of black crappie *Pomoxis nigromaculatus* (Guy *et al.*

Box 2.4 Show me the way to go home – a role for species and population-specific pheromones?

There is now compelling evidence that adult sea lampreys *Petromyzon marinus* from the North American Great Lakes rely on innate attraction to pheromones from larvae in streams, for finding spawning streams. Teeter (1980) showed that washings of larval sea lampreys were highly attractive to adult sea lampreys. Recent studies have shown that bile products, particularly petromyzonol sulphate and allocholic acid, of juvenile sea lampreys are potent olfactory stimuli in conspecific adults at concentrations as low as 10^{-12} M (Li *et al.* 1995). Naturally occurring mixtures of these chemicals may well provide a species-specific odour cue. There is good evidence from experiments of emigrating juvenile sea lampreys tagged with coded wire tags and subsequent recapture of adults, that the returning adult sea lampreys do not home to their natal streams (Bergstedt & Seelye 1995). As a result, the great bulk of evidence is in place to demonstrate how a species-specific odour cue enables orientation of adult sea lampreys to streams containing conspecifics, but does not necessarily result in natal homing.

Although pheromone hypotheses have been put forward for natal homing of salmonid species, and species-specific pheromones have been identified for salmonids (Stabell 1984, 1992; Hara 1993), these cannot explain the high precision of homing to natal streams unless population-specific differences in pheromone production and migratory response occur, other home stream odour cues are used, or other cues such as landscape features assist in home stream selection. Døving *et al.* (1974) showed that individual cells of the olfactory lobe of Arctic char *Salvelinus alpinus* responded differently to chemicals in the mucus from several different populations. Careful studies have recently shown that juvenile coho salmon *Oncorhynchus kisutch* can discriminate between the chemical emanations of similarly aged salmon from their own and other populations (Courtenay *et al.* 1997). Odour production appeared to vary in populations and some components of these odours appeared to occur in faeces. Such population-specific odours could act as pheromones to enable homing to natal streams if present in effective concentrations for olfactory (or gustatory) stimulation, and if recognised by returning adults. However, odour recognition by juveniles might play other roles such as territoriality or sibling recognition.

1992) and sole *Solea solea* (Macquart-Moulin *et al.* 1988) has been shown to be influenced by barometric pressure variations. Fish are ectothermic organisms, and they can detect differences in water temperature as small as 0.03°C, suggesting temperature gradients may act as a directional cue (e.g. diel migration of YOY cyprinids to inshore habitats; Baras 1995a).

Travelling in groups or shoals may assist social transmission of migration routes and directions, particularly where migrations include members of several age groups (Stasko 1971). There is now increasing evidence to demonstrate the significance of information transfer for spatial mapping and learning of migration routes (e.g. Helfman & Schultz 1984), and of informed leaders influencing foraging movements (e.g. Reebs 2000).

As illustrated by this brief overview, the range of cues used by fishes during their migration is wide. Not all species can use the same cues as efficiently, as sensitivity clearly is species-specific (Harden Jones 1968; Leggett 1977; McKeown 1984). Much research has been focused on the response of fish to particular cues, but there is growing evidence that cues may influence each other, and that experience of one cue may modify a fish's response to another cue (see also section 2.2.2.6). Also, the sensitivity to orientation cues is 'status-dependent', as internal factors, such as thyroid hormones may alter the responsiveness of fish, especially as regards prolonged navigation and homing.

2.4.7 Homing, memory and imprinting

The return of fish to previously occupied feeding, resting or spawning places, and associated orientation/navigation mechanisms, intimately rely on the possibility of memorising characteristic features of the home area. For short-term homing, as during the diel movements of cyprinids in between resting and foraging sites, or intertidal fishes between pools, there is no need to invoke complex mechanisms other than a topographic short-term memory involving the visual or olfactory identification of landmarks and surface topography. As early as 1951, Aronson (in Gibson 1986) provided evidence that a single tide was enough for an efficient topographic memory in the small tropical goby *Bathygobius soporator*, and that this knowledge was retained over at least 40 days. This type of topographic memory should really be renamed 'mid-term' memory, with truly short-term memory applying better to the memory retention of food patches explored, which can be as short as 0.5–5.0 min in intertidal fishes, and corresponds to the period of time during which food distribution might remain stable in an intertidal environment (e.g. fifteen-spined stickleback *Spinachia spinachia*; Hughes & Blight 1999). Similar schemes where the exploitation of the environment was aided by the fish's experience and mid-term memory have been demonstrated in freshwater fish species (e.g. three-spined stickleback *Gasterosteus aculeatus*, Beukema 1968; barbel *Barbus barbus*, Baras 1992).

Reproductive homing, which involves the return of spawners to their birthplace, or their consistent fidelity to a dedicated spawning ground over several reproductive seasons, is deemed to go far beyond the capacities of mid-term memory. It requires reliable information storage in a long-term memory, access to stored information, and the capacity to compare environmental cues with stored information. Cooper and Hasler (1974) showed from recordings of electrical activity in the olfactory bulbs of coho salmon *Oncorhynchus kisutch*, imprinted with morpholine, released in the wild, and recovered as upstream migrants, that olfactory information could be retained for 18 months, thus demonstrating the existence of a long-term olfactory memory in salmonids. There is a good deal of evidence that fish cannot access and 'write' information in their long-term memory at any time. This led to the transfer of the 'imprinting' concept, originally developed for the visual sense in birds, to olfaction in salmonids (Hasler & Wisby 1951), with the smoltification period being regarded as the imprinting period (e.g. Hasler & Scholz 1983). Hatchery-reared salmonids released at a pre-smolt stage in a stream indeed show homing behaviour towards their release site (review in Stabell 1984). Probably because fish science has been driven by studies of salmonids for a while, it has been erroneously thought that imprinting was taking place during the smoltification period exclusively, and that only salmonids had this ability.

The so-called 'pink problem' is a good illustration of how limiting this view can be. The pink salmon *Oncorhynchus gorbuscha* migrates to the sea immediately after hatching, and shows homing behaviour, even though it is known to stray more frequently than other salmons (review in Heard 1991). More or less complex hypotheses, including associative homing, institutional learning and genetic homing have been put forward to account for this deviation from the standard homing pattern, but none seems truly satisfactory (see Stabell 1984). Also, there has been increasing evidence that young salmonids translocated at different life stages could home to their birthplace, suggesting there is a genetically-based component to homing behaviour. Notwithstanding the relevance of the genetic control of homing ability (Bams 1976), its exclusive role became more doubtful when translocated salmonids were recovered as adults in the ponds of the hatchery where they were born (e.g. Icelandic farms; Gudjonsson 1970; Isaksson *et al.* 1978). Also, there have been numerous reports that hatchery-reared salmonids demonstrate a lower return rate than wild fish. It had been thought for a while that this originated from a lower survival rate of hatchery fish in the sea, but further experiments where this variable was investigated (Jessop 1976) did not support this hypothesis. Hence, the lower return rate of hatchery-reared salmon can be regarded as a disturbance of their navigational ability (Stabell 1984), and thus of the imprinting process. However, this again is paradoxical if based on the assumption that there is a single imprinting period taking place during the smoltification process.

More recently, it has been proposed that fish can be imprinted at a very early developmental stage (basically when the sensory nervous system becomes operational; Baras 1992), enabling them to memorise the environmental conditions of their birthplace, generalised later by Cury (1994) as the '*éternel retour*'. Starting from this viewpoint, all fish species could demonstrate some long-term memory, and homing behaviour, removing the so-called 'pink problem'. Salmonids undergoing smoltification would be given the opportunity of a second imprinting period, enabling a reinforcement of the initial 'memory' if environmental conditions were similar, and generating a different mnesic trace if they smoltified in a different set of conditions. This functional hypothesis may account for why the pink salmon *O. gorbuscha* shows a lower return rate than other salmonids, and why naturally propagated salmonids stray less frequently than hatchery fish, and also why homing of diadromous salmonids is so spectacular and accurate. The notion of multiple imprinting periods, and underlying mechanisms (Baras 1996), based on neural networks and calculatory energy (Hopfield 1982; Hopfield & Tang 1986) is consistent with the irreversible connotation of imprinting (Lorenz 1970), as it explains strayers as fish with multiple memories, not as fish with altered imprinting.

The multiplicity of imprinting periods may sound problematic at first sight, but it is not if we examine the factors that condition the access ('write or read') to a long-term memory. Hasler & Scholz (1983) demonstrated that olfactory imprinting in salmonids was greatly facilitated by an injection of thyroid-stimulating hormone (TSH), which mediated the action of triiodothyronine on the brain (Scholz *et al.* 1985). Increased plasma concentrations of thyroid hormones are also known to facilitate the response to visual and olfactory stimuli (Hara *et al.* 1966; Oshima & Gorbman 1966a, b). Elevated concentrations of thyroid hormones can thus be viewed as enhanced conditions for a 'read' (high responsiveness) or 'write' (imprinting) access to the long-term memory, and there is convincing evidence from the literature that there are several rather predictable periods of this kind during the fish's life (see review in Iwata 1995 for salmonids). Smoltification of salmonids perfectly fits this definition, so does the metamorphosis of milkfish *Chanos chanos* (Grace-de-Jesus 1994) and the downstream migration of the fourspine sculpin *Cottus kazika* (Mukai & Oota 1995). Early life-history stages are also characterised by high levels of thyroid hormones, either of maternal origin (Tagawa *et al.* 1990) or produced by the embryo itself (e.g. *Oncorhynchus* spp., Greenblatt *et al.* 1989).

Thyroid hormones are also known to increase during the return migration of adult salmonids (e.g. Atlantic salmon *Salmo salar*, Youngson & Webb 1993), and during the reproductive season of most fish species (Sage 1973; Dickhoff & Darling 1983), including lampreys (e.g. *Geotria australis*, Leatherland *et al.* 1990). The prespawning season is also associated with high levels of sex hormones (oestradiol-17β and testosterone), which are known to be correlated with a high responsiveness to stimuli, including home stream water (Hasler & Scholz 1983). The role of cortisol has been invoked too, notably by analogy to the stress response during predator encounters, and associated learning of who is friend or foe. However, regarding wild fish, there are obvious difficulties for discriminating between naturally high levels of cortisol, or levels elevated by the capture of the fish. Also, it seems that cortisol alone cannot guarantee a reliable access to long-term memory, as salmonid presmolts injected with adrenocorticotrophic hormone (ACTH) could not be imprinted on morpholine or phenethyl alcohol, whereas those given ACTH + TSH demonstrated this ability (Hasler & Scholz 1983).

Homing by fishes has most frequently been comprehended in terms of spatial locations, essentially because this is the way human observers perceive aquatic environments. The transfer of the imprinting concept from the visual sense in birds to the olfactory sense in salmonids also raised some problems of a similar nature. Imprinting in birds has often been considered in terms of a moving object on a stationary background, implying a notion of relativity, which was somewhat ignored in the transposition to fish. Further questions have been asked as to how fish could select one or a few odorant cues among the many odours or potential cues at hand. Imprinting in fish, birds and probably other vertebrates can all be viewed in a homogenous way by invoking the notions of associative memory, signal integration and 'gestalt', assuming that it is the process that matters, and not the senses underlying this process.

Relying on the concept of dynamic and associative memory, mnesic traces can be viewed as a network of connected pieces of information, which can be recalled during key periods (see above) when signals integrated from neural afferents bring back a 'prototype picture' resembling more or less closely the mnesic trace. Using the gestalt concept (e.g. if we are shown a drawing of a one-eared cat, we can identify it as cat, despite the fact that a 'real' cat normally possesses two ears), there is no need for each and every piece of information to be present or strictly identical to the mnesic trace to recall this memory. As a corollary, the multiplicity of pieces of information at the time of imprinting can promote homing ability, even though there have been some changes between the environmental conditions that governed imprinting, and the present environment. Implicit too is the idea there is no particular need to 'select' particular pieces of information at the time of imprinting (imprinting basically is involuntary, even though it has an adaptive value). Within an evolutionary context, the 'selection' of cues by fishes may be regarded as due to specific receptors or discrimination capacities being selected during the evolution of the species or populations exposed to different environmental pressures and stimuli. The so-called 'innate' component of homing might simply reflect fish having evolved slightly different sensory capacities, and being capable of sensing different concentrations or combinations of environmental cues. Such genetically based traits facilitate homing to the native stream, and conversely homing promotes their selection, provided the environment shows a sufficient degree of stability.

Until this stage, the imprinting mechanism underlying homing behaviour has been debated within the context of returning to previously occupied places, i.e. spatial homing. In the introduction of this book, we defined migration on a broader scale, as a movement in an n-dimensional hypervolume of biotic and abiotic factors, which has the merit of taking into account factors that are obscure to the observer but obvious or important for fish. The definition of homing could be revised in a similar way, and thus become 'any return to a particular point or area within this hypervolume' (Baras 1992), and this may apply to environmental factors such as temperature or day-length. In this way, the environmental stimuli triggering migration or spawning may also have been elicited through an imprinting mechanism similar to that invoked in spatial homing. Regarding thermal stimuli, there is some experimental evidence that fish, such as the cyprinid, barbel *Barbus barbus*, having been incubated as eggs and reared during the larval stage may spawn as adults at temperatures higher than normal (Poncin 1988), which contrast with the remarkable constancy of the thermal stimulus triggering spawning in wild populations (Baras & Philippart 1999). As for spatial homing, the inscription of different mnesic traces during successive imprinting periods could generate some form of ambiguity. With respect to thermal memories, ambiguity would only take place in environments where temperatures corresponding to different mnesic traces would exist simultaneously, whereas in more homogeneous environments, ambiguity would only exist along time. For temperate spring spawners, it would cause the lowest thermal memory to be selected (Baras 1996), provided the associated environmental conditions do not jeopardise the survival of offspring, as seems to be the case for barbel (Baras & Philippart 1999). Further debates on the relevance and limitations of this 'phenotypically driven thermal homing' within the context of postglacial recolonisation, and adaptation to global warming, can be found in the last-mentioned article.

This functional hypothesis, and especially the capacity of early imprinting and the multiplicity of imprinting periods, can be viewed as a conservative and adaptive mechanism. It is conservative as it enables fish species to relay some form of stability (or 'homeostasis') of behavioural traits between generations in spite of environmental variations. It is adaptive, as homeostasis is relayed on an environmental basis, not on a genetic basis, even though homing behaviour may favour genetic differentiation into different populations, and eventually speciation. Such behaviour could also reinforce selection of physiological attributes associated with local populations and characteristic environmental conditions. As such, this mechanism may apply to many fish species, and especially to those with an iteroparous lifestyle, for which each return of a spawner to a previously occupied place could reinforce a mnesic trace, and thus the probability that the fish homes during the next spawning season.

However, this interpretation fails to account for the presumed homing behaviour of fish with pelagic eggs that start drifting within the first minutes or hours after being spawned (e.g. many tropical characiforms and probably pimelodid catfishes, see sections 4.4.6 and 5.16), long before the sensory system becomes operational and permits any inscription of memories of environmental characteristics. For these riverine species, rheotaxis and information transfer should probably be invoked for homing mechanisms. Some observations of large juveniles accompanying mature adults during the spawning runs might tend to support this statement (see section 5.15 on characiforms). Implicit in this hypothesis is that severe disturbance to population structure (e.g. the elimination of 'experienced' spawners by chemical pollution during the spawning period) might have an impact much greater than for species with non-pelagic eggs, but this remains to be investigated.

Chapter 3
Types of Migration

3.1 Introduction

Migration of fishes has traditionally been associated with three purposes: spawning, feeding and refuge (Heape 1931; Northcote 1978). Although classification of migration in this way provides a helpful framework, migration can simultaneously provide improvement in conditions for more than one of these factors. The definition of migration provided in chapter 1 does not delimit the temporal or spatial scale of migration to particular sets of conditions. Therefore, in discussing types of migration the reader should be aware that classification of migration types as follows allows ease of discussion, but does not always reflect the interactive nature of multiple stimuli for migration. The factors affecting, and timescale of, migration change with ontogeny. For larval and juvenile fishes with tiny energy reserves and high susceptibility to predation the diel scale of movement is, in many aquatic systems, far more important to survival and growth than it is for larger fishes.

Dispersal is often regarded as distinct from migration since in the terrestrial environments in which most development of such ideas initially occurred, dispersal movements (resulting in reduced density of individuals) tend to occur as an equidirectional outward movement from a central concentration. Within aquatic environments with distinct currents, passive dispersal results in clear directional movement of much of the population, though to different degrees, resulting in apparent migration as well as effective dispersal. Equally, for fishes with adequate swimming capacity, migration against currents may allow utilisation of areas with richer resources or reduced competition. Such behaviour is often regarded as 'exploratory' and is often seen at its most significant in unpredictable climates where rainfall may be highly variable between years, resulting in streams becoming discontinuous series of pools (e.g. Mediterranean or sub-Sahelian flow regimes). When floods reconnect these pools, large concentrations of juvenile as well as adult fishes often move upstream (Cambray 1990). The ability to disperse or display exploratory migratory behaviour is often critical to fishes for access to new spawning habitat, for maintenance of populations in areas unsuitable for reproduction, for access to prey, or avoidance of predators. Barriers to dispersal may delay or preclude recovery of fish assemblages following disturbance and increase extinction risk by fragmentation (Detenbeck *et al.* 1992).

3.2 Migrations at the seasonal and ontogenetic scale

3.2.1 Feeding migrations

The characteristics of feeding migrations are highly variable between freshwater habitats and different fish taxa. They may involve a variety of phases at different parts of the lifecycle and may vary in timing and intensity between ontogenetic stages. As Northcote (1978) points out, the variety of migration types across the range of geographic zones and taxonomic groups is so wide as to make identification of clear trends difficult. Migrations between habitats that appear associated with growth and development, mediated through feeding, often appear as a characteristic of ontogenetic changes rather than as clear responses to improved feeding opportunities. In many cases this may actually be true, since movements of fishes to growth or nursery areas may be more closely related to optimising growth and survival potential than purely to maximising food intake. Growth may be enhanced by reducing competition for available food sources, seeking optimal thermal environments for efficient growth, and limiting energy expenditure in activity as well as a range of other options. Appropriate energetic and migratory strategies are strongly influenced by the prevailing biotic and abiotic environmental conditions. While one response of juveniles in food-limited nursery habitat may be to migrate away to more productive environments, an alternative is to adopt an energetically conservative strategy, limiting activity costs by remaining in the same place, and attaining sexual maturity at a small size, as occurs in a wide variety of fish species (Gross 1987, 1991; Tsukamoto *et al.* 1987).

Although a variety of life-history strategies may result in equal fitness benefit for a fish species, a comparison of the geographic distribution of anadromy and catadromy provides circumstantial evidence that migratory traits have evolved in many species as a response to productivity gradients between freshwater and marine environments (Gross 1987; Gross *et al.* 1988). There are far more anadromous than catadromous fish species at high latitudes where the marine environment, characterised by well-mixed nutrient-rich water, has higher productivity. By contrast, the relative incidence of catadromy is greater at low latitudes where primary production in fresh waters tends to be greater than in the ocean, where in most cases surface waters are stratified and nutrient-poor. This provides an appealing and convincing explanation for the relative incidence of anadromy and catadromy. On this basis, one might predict that anadromy would be relatively common in rivers adjacent to subtropical or tropical upwelling (nutrient-rich) oceanic areas. The available evidence suggests that it is not (Gross 1987; Gross *et al.* 1988), although we are unaware of any detailed analysis of this issue. Also, amphidromy is common in the temperate Southern Hemisphere, yet this lifecycle involves the main growth phase and maturation in fresh water.

Nevertheless, other factors also indicate that spatiotemporal patterns of productivity tend to be important in influencing feeding migrations. Potamodromous migrations in fresh water also often track spatiotemporal variations in productivity. Lateral movements of fishes into inundated tropical floodplains are associated with a pulse of production in these habitats, resulting from nutrient release (see chapter 4). The areas of highest productivity and food availability are not constant, but are associated with the land–water ecotone, which can move tens of kilometres as the water level rises. The highest densities of fish are usually in the shallow waters at the edge of the land–water ecotone and track its movement (Bayley 1995). At temperate latitudes, spring/summer peaks in abundance of food in warm, well oxygenated, shallow bays of lakes are exploited by fishes such as perch *Perca* spp. moving inshore from deep, unproductive overwintering habitat (Allen 1935; Thorpe 1974; Craig 1977, 1987).

3.2.1.1 Drift and movements of early life-history stages

A wide variety of guilds occur for modes of reproduction and deposition of eggs in particular conditions (Balon 1975). For example, eggs may be retained within a particular habitat until hatching in the case of adhesive eggs that are attached to plants or bottom material, as is common in the cyprinids. In other cases the eggs may be deposited in spawning pits or 'redds' where the eggs and larvae may be retained until they actively escape, as in many salmonids. Production of buoyant or neutrally buoyant eggs is a common strategy in many fish species which reproduce in marine environments (Harden Jones 1968), but is less common in freshwater environments. This may be because, within river systems, buoyant eggs may be rapidly swept downstream and out of appropriate nursery areas. Nevertheless, release of buoyant eggs may provide an extremely efficient form of dispersal of the progeny, particularly in tropical environments where the high temperatures enable rapid egg development and hatching or where spawning occurs in backwater or delta areas where water currents are slower. Buoyant or semi-buoyant eggs are produced by several taxa including the goldeye *Hiodon alosoides* (Hiodontidae) found as far north as the subarctic region of North America (Donald & Kooyman 1977a, b), the freshwater drum *Aplodinotus grunniens* (Sciaenidae) from temperate North America (Hergenrader *et al.* 1982), and characins such as *Prochilodus* (Curimatidae) from tropical South America (Bayley 1973; Lowe-McConnell 1975, 1987) and *Alestes baremoze* (Alestiidae) in the Chari River (Chad) (Carmouze *et al.* 1983).

In some anadromous and virtually all amphidromous fish species, eggs or larvae are swept into estuarine or marine waters, where they grow and develop (McDowall 1988, 1997a). Fertilised eggs of the anadromous hilsa shad *Tenualosa ilisha* drift several hundred kilometres down rivers to tidal reaches where the hatchlings grow (Ganapati 1973). Newly hatched larvae of the ayu *Plecoglossus altivelis* drift downstream to estuaries or to lake rearing environments (Tsukamoto *et al.* 1987). However, for those fish species with buoyant, drifting eggs and larvae that are intolerant of high salt levels, the distance which must be travelled upstream by adults to assist survival of the young to a stage where they can orientate and maintain position in slower-water microhabitats, may be large (Bayley 1973; Reynolds 1983; Northcote 1984). An egg travelling downstream at 0.5 m s^{-1} and which takes 10 days to develop into a swimming postlarva, will travel 432 km. Eggs of many tropical species hatch faster, but larvae keep on drifting over several days, and estimated distances of 500–1300 km have been suggested (Araujo-Lima & Oliveira 1998; see also section 4.4). For marine amphidromous fish species feeding does not begin until the larva enters brackish or marine environments and so a long duration of drift increases the risk of predation. This is believed to be a key reason for explaining why the majority of amphidromous fishes reproduce in short river systems or migrate relatively short distances up rivers (Iguchi & Mizuno 1999).

Many freshwater fish species laying negatively buoyant, buried or adhesive eggs still have young which may drift downstream on currents. In the case of some taxa such as most of the salmonids, by the time the late alevins or young fry emerge from the gravel in which they were spawned, they have relatively well-developed locomotor capabilities and are capable of directed movement (e.g. brown trout *Salmo trutta*, Roussel & Bardonnet 1999), albeit of limited magnitude (Northcote 1984). The behaviour of the emerging young of salmonids is variable between and also within species (Northcote 1978, 1984). In some species, such as pink salmon *Oncorhynchus gorbuscha* and chum salmon *O. keta*, the fry migrate downstream towards the sea almost immediately after emergence. In Atlantic salmon *Salmo salar* and anadromous forms of brown trout *S. trutta* there may be an initial movement from the spawning site to feeding areas, followed one or more years later by physiological transformation to a smolt and migration to sea. Similar patterns are illustrated by a variety of *Oncorhynchus* species. Although upstream migration of young salmonids soon after emergence is relatively rare, it can and does occur, particularly in lake outlet, stream-spawning populations (Godin 1982; Roussel & Bardonnet 1999).

There is great plasticity in the migratory behaviour of stocks of a wide variety of salmonids, influenced both by genetic and environmental factors, with many populations existing as freshwater residents and others having anadromous and holobiotic freshwater components. Arctic grayling *Thymallus arcticus* and European grayling *T. thymallus* are restricted to fresh water and display a range of migratory behaviour in association with ontogeny and growth phase (Northcote 1995). Bardonnet *et al.* (1991) found that in June and July the young of European grayling in the River Suran, France, moved away from microhabitats and low velocities associated with banks into the channel and areas with higher velocities. This was then followed by a downstream migration out of this spawning and nursery area. This downstream migration ended in the complete desertion of the Suran by young-of-the-year fish. Scott (1985) demonstrated a similar pattern of movement of young European grayling in the River Frome, southern England. Valentin *et al.* (1994) showed that 2-month-old European grayling actively sought refuge sites during periods of high discharge in artificial stream channels. This strongly suggests that young European grayling probably make an active decision to allow themselves to drift or move downstream.

By contrast to salmonids, on hatching, many fish species, including most cyprinids, have larvae with very poorly developed locomotor capabilities which may drift for substantial distances on currents. Where floodplains are intact, many of these taxa are either spawned or accumulate in lentic water in backwaters, edge habitat and seasonally inundated ponds, although where the river channel and floodplain have become disconnected by anthropogenic disturbance, lentic habitat is often limited to river channel margins. Although many species of freshwater fishes have larvae which drift, these may not necessarily move long distances if directed movements, even small ones, can be made towards river edge habitat or the boundary layer over the river bottom. Such movements may be considered as analogous to 'selective tidal stream transport' with the movement in one direction only, but controlled by small active movements into higher velocity areas resulting in large-scale passive movements. Controlled drift appears to play a key role in dispersing young fishes into appropriate nursery areas (Brown & Armstrong 1985; Mitro & Parrish 1997) and disruption to natural flow regimes can have a major influence on these patterns (Box 3.1).

A variety of studies have used ichthyoplankton nets to study drift behaviour of larval and juvenile fishes (e.g. Penáz *et al.* 1992; Box 3.1), while some have assessed the significance of downstream migration of young fishes from studies of impingement at water intakes (Hergenrader *et al.* 1982; Solomon 1992). Penáz *et al.* (1992) used a 0.5 mm mesh size ichthyoplankton net to determine the downstream drift of larval and juvenile fish at two sites 5 km

Box 3.1 To drift or not to drift? That is the question

Robinson *et al.* (1998) hypothesised that passive drift of larvae could lead to clumping of longitudinal distributions of larvae, these settling out in depositional habitat, whereas active movement or behaviour combined with drift would result in more even distributions or the presence of larvae in some non-depositional habitats, due to additional cues such as food availability, predators, substrate type and the presence of cover (Scheidegger & Bain 1995). Robinson *et al.* studied larval drift using ichthyoplankton nets for several species native to the Colorado River system, where cold water releases and water level fluctuations from regulating dams appear to have caused population declines in many native species. Humpback chub *Gila cypha*, flannelmouth sucker *Catostomus latipinnis*, bluehead sucker *Catostomus discobolus* and speckled dace *Rhinichthys osculus* in the lower 14.2 km of the Little Colorado River spawned primarily during March–June, typical of most Colorado River basin cyprinids and catostomids, resulting in eggs hatching at the end of the spring high flow period. The Little Colorado River is an unregulated tributary in the upper part of the Colorado River system.

Generally egg drift densities did not vary across river channel, but larvae of all species except speckled dace protolarvae were denser in the margins, showing active movement to these microhabitats. This has been attributed to habitat selection by larvae and has been observed in larvae of other species such as channel catfish *Ictalurus punctatus* (Armstrong & Brown 1983; Brown & Armstrong 1985). In Robinson *et al.*'s study, only speckled dace exhibited diel variations in drift, with more drifting at night; others showed no significant variation. Before closure of Glen Canyon Dam, annual peak Colorado River flows (April–July) typically coincided with decreasing flows in the tributaries and timing of reproduction. Such flows ponded the tributary mouths which probably helped to retain larvae, preventing them from being forced out into the fast canyon waters until more fully developed. Before impoundment, near-shore habitats would have been stable but are now subjected to daily oscillations in flow associated with patterns of dam water release, washing fish downstream. The Colorado River water is also often colder due to hypolimnetic water releases. These factors probably substantially increase mortality of young-of-the-year fish (see also Chapter 7).

apart on the French upper River Rhone in the old bypassed river bed between a dam and its powerhouse. Sampling was conducted in August and the main drift occurred at twilight and during the night hours. Only 84 fish per 24 h period were caught at the upper site where few backwaters existed compared with 271 fish per 24 h period at the lower site adjacent to a natural floodplain, showing the importance of the latter for providing spawning sites and nursery areas as well as a source of recruitment for riverine fish. Underyearling roach *Rutilus rutilus* formed 67% of the 'drift' at the upper site together with chub *Leuciscus cephalus* (13%) and nase *Chondrostoma nasus* (6%) whereas the composition at the lower site contained more rheophilic cyprinids: chub (40%), roach (36%) and barbel *Barbus barbus* (10%).

Baras and Nindaba (1999a, b) used prepositioned electric fishing frames to examine seasonal variations in the abundance of larvae and juveniles of chub, dace *Leuciscus leuciscus* and nase in inshore bays of the River Ourthe, Belgium. Larvae essentially remained inside the bay, whereas juveniles progressively moved into the stream, but some returned to the bay at some periods of the day. From late summer onwards, the number of juvenile dace and nase decreased progressively, as they dispersed further into the stream. Chub, which were spawned later (May–June versus March–April for dace and nase), and reached smaller body size at the end of summer, were captured in equivalent amounts until the first winter floods, which caused their dispersal downstream, until they found more structured shelters where they could overwinter in mixed shoals together with other cyprinids, including roach, nase and dace (Baras *et al.* 1995). Barbel also disperse in the stream during late summer and early autumn, but they obviously select offshore overwintering places (Baras & Nindaba 1999b).

Lightfoot and Jones (1996) observed the longitudinal dispersion of young roach in the River Hull, northeast England, during June and July 1973 while they grew from 7.5 to 29 mm in length in a nursery area close to the spawning sites. The smallest fish were confined to the shallow margins and among *Sparganium* sp. weed beds where the current velocities were lowest. As the fish grew larger, they extended their range into deeper water with fewer plants and greater flows where they could maintain station. At about 29 mm in length, the fry became scarce locally, left the nursery area and dispersed downstream. In the Great Ouse, eastern England, shallow water with a coarse substratum, zero velocity and floating or submerged plant cover was the preferred habitat of 0+ roach during August and September (Garner 1995). The scarcity of such conditions might be the cause of observed downstream dispersal by older fry. Garner *et al.* (1995) showed that weed beds provided young fish with both high food densities of large cladoceran species and refuge during periods of elevated flow. Cutting vegetation (largely *Nuphar lutea*) significantly reduced the availability of the preferred and more nutritious large cladoceran species which supported optimal growth and the fry turned to less nutritious '*aufwuchs*' with a subsequent reduction in growth (Garner *et al.* 1995).

Upstream migrations of juvenile fishes to freshwater growing areas are most well characterised in several catadromous groups, particularly the anguillid eels, and in marine amphidromous fish species in which after a brief period of growth at sea, the young migrate back upstream into fresh water where most growth takes place (McDowall 1988). On completion of their oceanic migration leptocephali of the European eel *Anguilla anguilla* metamorphose into transparent glass eels which migrate into estuaries. They then undergo a transition phase as they adjust to fresh water, metamorphose into the pigmented elver stage and commence feeding. Some of these may stay in the estuary or join coastal stocks, others migrate upstream during their first year in fresh water or as juveniles in subsequent years (White & Knights 1997). As the eels move upriver they become more pigmented (Tesch 1977) and generally are fully pigmented at 7–8 cm in length. Only small numbers of eels migrating upriver are greater than 20–30 cm in length (Tesch 1966, 1977; Larsen 1972), although upstream migrants of greater size can be found in longer river systems (e.g. mean size >30 cm in the River Meuse; Baras *et al.* 1996b).

While the larvae of many tropical freshwater fishes also drift with currents, the adults of many species migrate laterally into flooded areas to spawn (Welcomme 1979, 1985). The eggs are therefore deposited in appropriate backwater habitat. On hatching, a combination of slow currents and active locomotion, assisted by the fast rate of development at high temperatures, assist dispersal of young throughout inundated areas. In some mouth-brooding fishes such as several of the cichlid taxa, including *Pseudocrenolabrus multicolor*, the eggs or young are carried laterally into the flooded areas by mouth-brooding females from stream spawning areas, where they grow and later return to river habitats (Welcomme 1969). Similar parent-assisted migrations of young take place in lakes, with young of some species being moved from spawning areas to nursery and feeding habitat, as occurs for *Oreochromis variabilis*. Finally, there are several cichlid species in the African Great Lakes, which make no migration, as young develop close to the adults, these species being found sometimes only on one side of a small island.

3.2.1.2 Feeding migrations of subadults and adults

Feeding migrations in subadult and adult freshwater fishes occur on a broad range of spatial and temporal scales. Many of the long-lived iteroparous fishes that occur in the temperate region exhibit some form of spring migration from wintering areas to spawning areas and remain within or close to the latter for feeding and growth during much of the summer period, during which restricted movements are often exhibited, associated with occupation of a home range (Fig. 3.1). This behaviour, which may be interspersed with sporadic longer distance movements to new locations, is shown by a broad range of species including salmonids (e.g. Craig & Poulin 1975), cyprinids (Lelek 1987; Liu & Yu 1992; Baras 1992; Lucas & Batley 1996; Philippart & Baras 1996), centrarchids (Gerking 1959; Scott & Crossman 1973; Mesing & Wicker 1986; Langhurst & Schoenike 1990) and percids (Allen 1935; Craig 1987). Where autumn spawners such as salmonids exhibit iteroparous lifecycles they may migrate away from the spawning areas to discrete feeding areas or remain close to the spawning area for feeding (Ovidio 1999). Typically territorial species, like brown trout *Salmo trutta*, can show almost no feeding migration at all as they seemingly adapt, at least partly, to variations

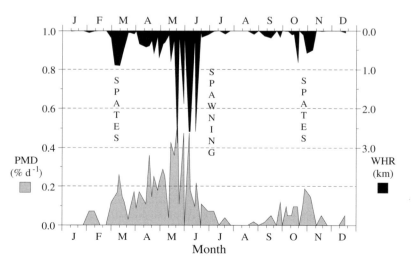

Fig. 3.1 Annual mobility cycle of radio-tracked barbel *Barbus barbus* in the River Ourthe, southern Belgium. WHR is the weekly home range (distance between the most upstream and downstream locations), and PMD is the probability of moving from a daytime refuge to another within 24 h. Modified from Baras (1992).

in prey abundance through time budgeting, as demonstrated by recent investigations combining surveys of local drift and telemetry of trout behaviour (Giroux *et al.* 2000). In other cases, as for species such as goldeye *Hiodon alosoides* (Hiodontidae) in the Peace-Athabasca system, Canada, the adults migrate from wintering sites to spawning areas and remain to feed in shoals, and do not adopt localised home ranges (Donald & Kooyman 1977 a, b).

Tracking and mark recapture studies on adult European cyprinids such as common bream *Abramis brama* (Whelan 1983; Caffrey *et al.* 1996), barbel *Barbus barbus* (Baras 1992; Lucas & Batley 1996; Philippart & Baras 1996), chub *Leuciscus cephalus* (Allouche *et al.* 1999), roach *Rutilus rutilus* (Baade & Fredrich 1998) and dace *Leuciscus leuciscus* (Clough & Ladle 1997; Clough & Beaumont 1998) have generally shown limited movements within summer home ranges of less than 3 km (sometimes much less) and occasional longer distance movements. Shoaling, limnophilic species such as common bream tend to be more nomadic than rheophilic species such as barbel (Langford *et al.* 1979; Whelan 1983), particularly in relatively homogeneous river environments. L'Abée-Lund and Vøllestad (1987) found that roach migrating into small tributaries of Lake Årungen, southern Norway, fed during a migration into the streams for spawning in May, but many roach also exhibited a second migration 1–2 months later for feeding. The authors suggest that this behaviour may have been to make use of the higher quality of food temporarily available in the streams while water levels were still high enough to make access possible, since the assimilation efficiency for the plant/detritus-dominated lake diet is much lower than for the animal-dominated stream diet.

Studies of northern pike *Esox lucius* behaviour, mostly carried out in lake and reservoir environments, have shown pike to be relatively sedentary outside the spawning season except for sporadic long-distance movements (Malinin 1972; Vostradovsky 1975 (in Raat 1988); Diana 1980; Chapman & Mackay 1984; Cook & Bergersen 1988). Vostradovsky (1975, in Raat 1988) found that northern pike which exhibited these longer movements showed higher daily gains in weight than 'resident' pike, which they argued was due to a greater chance of encountering prey. Bregazzi and Kennedy (1980) also attributed observed movements of northern pike to the movements of prey species in Slapton Ley, southern England. However, these movements are not highly directed and could be considered as wide-ranging foraging movements, rather than migrations. Pervozvanskiy *et al.* (1989) argued that because of high flow conditions on riffles in the Keret River, northern pike foraging on migratory salmon were unable to migrate over long distances.

Jepsen *et al.* (2000) radio-tracked adult northern pike, pikeperch *Stizostedion lucioperca* and seaward migrating sea trout *Salmo trutta* smolts in a shallow Danish reservoir in order to obtain information on predator–prey interactions. Female pikeperch spent more time near to the sluice outlet (where emigration of smolts was delayed) during the smolt run (May) than at other times, and appeared to actively hunt the smolts, while male pikeperch remained stationary and were presumed to be guarding their nests. Most tagged adult northern pike were engaged in spawning during the smolt run and only a few moved to the sluice outlet area. Stomach sampling of northern pike and pikeperch in this and previous studies in the same reservoir and inflowing river demonstrated high mortality of smolts (85–95%) due to predation by pike and pikeperch, both for radio-tagged and non-radio-tagged smolts. In the unobstructed lower section of the River Gudenå, Denmark, downstream of the power station at Lake Tange, radio-tracked pikeperch exhibit a clear seasonal migration, moving downstream into the lower river and estuary in autumn, and upstream in spring (Koed *et al.*

2000). These movements appear to follow the seasonal movements of the major cyprinid prey fish such as bleak *Alburnus alburnus* and roach *Rutilus rutilus* in this system (Fig. 3.2), although incontrovertible evidence is difficult to obtain. Upstream movements in spring are also probably associated with spawning.

In harsh arctic and subarctic freshwater systems, fishes often exhibit migrations between separate feeding and refuge areas (also see chapter 4), involving upstream migration in species such as Arctic grayling *Thymallus arcticus* (Craig & Poulin 1975; Hughes & Reynolds 1994; Northcote 1995) and migration to sea in species such as Arctic char *Salvelinus alpinus* (Moore 1975a, b). Adult and juvenile Arctic grayling leave overwintering areas after ice break-up. Juveniles and subadults exhibit extensive upstream feeding migrations to rivers, tributaries and lakes. Adults migrate to spawning areas in tributaries and subsequently to feeding areas which may or may not be in the same stream. In many Alaskan streams, the feeding migration of Arctic grayling is associated with competition for preferred feeding areas, resulting in a decrease in modal length with increasing distance from stream source (Hughes & Reynolds 1994). European grayling *Thymallus thymallus* in similar climates also exhibit upriver migrations from downstream overwintering areas to feeding areas opened up by icemelt and increasing spring river flow (Zakharchenko 1973). Although potamodromous feeding migrations occur for grayling species and a variety of other arctic and subarctic freshwater species, at least 30 of over 40 freshwater fish species (mostly salmonids) in northwestern Canada and Alaska make anadromous feeding migrations, mostly remaining in coastal areas (McPhail & Lindsey 1970). In many cases the period of feeding at sea is only 1–2 months, but this provides a much richer food supply, resulting in greater growth and more rapid sexual maturity than in holobiotic freshwater forms of the same species.

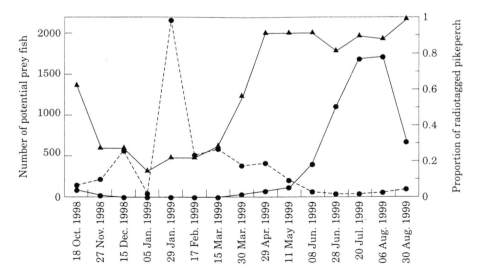

Fig. 3.2 Seasonal variation in the proportion of radio-tagged pikeperch *Stizostedion lucioperca* in the upper region (23.1–36.5 km above the tidal limit, bounded at the upper limit by a dam) of the River Gudenå, Denmark (—▲—), in relation to CPUE of potential prey fishes in the upper section (—●—) and a lower section (---●---, 0–9 km above the tidal limit). Reproduced from Koed *et al.* (2000).

In tropical freshwater environments, feeding migrations of adult and subadult fishes tend to be closely related to the pattern of rainfall and seasonal inundation of floodplain habitat (Lowe-McConnell 1975, 1987; Welcomme 1979, 1985). Many fish species, especially characins (Characiformes), move into flooded forest regions in South American tropical floodplain systems, to utilise the allochthonous food resources available there (Goulding 1980). Some characins, particularly those of the genus *Prochilodus*, and some pimelodid catfishes exhibit large upstream longitudinal migrations for spawning and then migrate back downstream over hundreds of kilometres to feeding areas, often near river confluences or in floodplain lakes (Bonetto *et al*. 1969; Bayley 1973; Lowe-McConnell 1975, 1987). Many other tropical species show the same pattern (see chapter 4). Although most predators show home-range behaviour near confluences, some large South American pimelodids undertake upstream feeding migrations as they hunt the shoals of characins migrating upriver for spawning. Within lake environments conspicuous feeding migrations may also be apparent. In Lake Sibaya, South Africa, the cichlid *Oreochromis mossambicus* migrates inshore to feed in shallow water, often in newly flooded areas during and following the rainy season and later, during the dry season, moves offshore into deeper water (Bruton & Boltt 1975).

3.2.2 Refuge-seeking migrations

Maintenance of freshwater-resident or diadromous fish populations in freshwater systems where conditions become unfavourable for survival necessitates tolerance to those conditions, or movement to better conditions. As well as enhancing survival probability, migrations to refuges may also result in conservation of somatic energy reserves, enhancing growth and reproductive output when more favourable conditions return. Refuge-seeking migration is therefore a significant strategy in minimising loss of fitness during stressful conditions occurring at a timescale shorter than the generation period, including seasonal fluctuations and irregular events such as unusually severe floods and droughts. Migratory behaviour has also been important over the geological timescale for maintaining freshwater fish populations during periods of changing climate, for example during glacial periods. During glaciations, freshwater fishes in north–south running drainage basins such as the Mississippi would have been able to shift annual longitudinal or lateral migrations southwards along the catchment (Hocutt & Wiley 1986). This is believed to be a key reason why north-south running river catchments in temperate regions tend to have higher diversity of freshwater fish species than east-west aligned catchments (Hocutt & Wiley 1986). Edwards (1977) observed that Texas populations of the Mexican tetra *Astyanax mexicanus*, the only characid native to the US, survived otherwise lethal winter conditions more than 600 km north of their natural range, by migrating to spring heads consistently having temperatures within the tetra's thermal range.

Seasonal changes in the habitats of otherwise sedentary yellow European eels *Anguilla anguilla* are probably to avoid unfavourable conditions in winter. Surface ice, cold water and ground ice formation are all conditions which eels avoid (Tesch 1977). In rivers, brackish areas and tidal waters, eels move to quiet backwaters and channels where the water is deep enough to buffer the effects of winter. In the River Hunte, Germany, during winter, Lübben and Tesch (1966) found European eels at depths of 2–2.5 m, 5 km from where they were first captured the previous summer and where they were again captured the following summer. Aker and Koops (1973) found that in the River Eider, a North Sea coastal river, the autumnal migration of European eels was directed downstream in the middle reaches of the river and upstream in the coastal regions. They argued that both populations were migrating to a common area in which to spend the winter. Similar movements to overwintering habitats were observed by McGovern and McCarthy (1992). Such movements may, however, be dependent on the suitability of habitats. If suitable refuges are available within an individual's normal home range then migration is unnecessary. For example, Baras *et al.* (1998) radio-tracked European eels which utilised gaps in stone walls along riverbanks which provided refuges of up to 1 m into the riverbank. Yellow eels occupying these habitats did not, therefore, need to move to seek refuge.

In many temperate rivers, during autumn many fishes move downstream from shallow areas, that are warm and productive in summer but which are often associated with high current velocity in winter, to deeper, slower pools further downstream. Such migrations may not always be in a downstream direction, depending on the spatial distribution of appropriate refuge habitat, but downstream movements appear to be most common. Neither are seasonal refuge-seeking movements as conspicuous or concerted in time and space as spawning migrations appear to be, and they strongly depend on habitat structure. In highly structured riverine habitats, many barbel *Barbus barbus* were found to occupy refuges in winter that were identical to their summertime resting place (e.g. River Ourthe, Belgium, Baras 1992; Philippart & Baras 1996). By contrast, Lucas & Batley (1996) showed that barbel in the Nidd, northern England, tended to move downstream in autumn and winter, although there was substantial variation in the extent of this behaviour by individuals. They argued that barbel may either be displaced or seek refuge downstream during high flow conditions which occurred frequently in the Nidd in the autumn and winter. The movements of radio-tracked barbel in response to high flow events in autumn were different to those in summer (Lucas & Batley 1996; Lucas 2000). In summer, high flow events often resulted in downstream movement of several kilometres, but when flows subsided was usually followed by a rapid return to the home area. In autumn, a downstream movement occurred in response to a higher proportion of high flow events and return movement upstream was rare, resulting in a step-like pattern of downstream movement. Downstream wintering migrations in rivers are common in a wide variety of temperate freshwater fishes including grass carp *Ctenopharyngodon idella* in eastern Russia (Nikolsky 1963), common bream *Abramis brama* in the upper River Don, Russia (Federov *et al.* 1966), mountain whitefish *Prosopium williamsoni* in the Clearwater River, Idaho, US (Pettit & Wallace 1975) and pikeperch *(Stizostedion lucioperca)* in the River Gudenå, Denmark (Koed *et al.* 2000). In the latter case, some radio-tagged pikeperch moved into the estuary, although it is thought that this downstream migration was also associated with following cyprinid prey fish downstream into this area. In other cases there may be a less clear downstream movement of fish, but concerted movements within a stretch of river, resulting in distinct aggregations in favoured areas such as deep 'holes' and river confluences (Lucas *et al.* 1998). Similar aggregating behaviour in favoured areas is also characteristic in lakes (Hasler 1945; Johnsen & Hasler 1977).

In temperate and arctic climates, the metabolic scope, swimming capacity and digestive ability of many fishes is severely reduced during the low temperatures of winter. Under these circumstances feeding activity may be very low or non-existent, even when plentiful food is available. The low temperature compromises the ability of many fishes, especially juveniles and small species, to maintain position in the river channel at elevated flows. Many fishes from temperate climates move out of river channel habitats and into backwaters over winter or during periods of high flow and/or low temperature, particularly those fishes with a preference for relatively high temperature, such as many centrarchids and cyprinids. Deep, slow pools within the river channel, as described above, can also serve this purpose. Overwintering in backwaters has been described for a wide variety of species including centrarchids such as black crappie *Pomoxis nigromaculatus* and bluegill *Lepomis macrochirus* (Knights *et al.* 1995), and cyprinids such as roach *Rutilus rutilus* and common bream *Abramis brama* (Jordan & Wortley 1985; Copp 1997). Knights *et al.* (1995) showed how several centrarchids chose backwaters with extremely slow flow and only moved from these when oxygen levels fell too low (Fig. 3.3; see also Box 2.1). Jordan and Wortley (1985) suggest that seasonal movements of a high proportion of adult cyprinids, especially roach, in the river systems of the Norfolk Broads, England, must explain the variable results from an extensive series of fish surveys carried out between 1978 and 1984, using quantitative techniques described in Coles *et al.* (1985). During winter months, the mean fish biomass from the open waters of the rivers and broads was less than 1 g m^{-2} compared with 9.4 g m^{-2} during the summer. At certain sites adjacent to rivers connected to broads, very large winter aggregations of fish were found, with densities up to 36.7 fish m^{-2} and biomass up to 1787 g m^{-2}. These aggregations were found in off-river and off-broad dykes, often associated with boat moorings. The fish were largely adult roach, small common bream and some roach-bream hybrids. At two sites, the roach were 3+ or older, which were scarcely caught in summer surveys. Jordan and Wortley (1985) suggest that these exceptionally high winter densities explain the relative lack of fish in open waters in winter and that they must result from adult migration to the winter refuges offered by off-river areas such as particular boatyards.

The importance of such off-channel waters and marinas has also been demonstrated for YOY European, lowland river fishes, particularly cyprinids (Copp 1997). YOY fishes, espe-

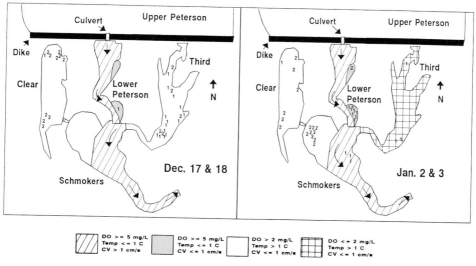

Fig. 3.3 Movements of radio-tracked bluegill *Lepomis macrochirus* (denoted by '1') and black crappie *Pomoxis nigromaculatus* (denoted by '2') in Mississippi River backwaters in relation to oxygen, water temperature and current velocity (CV). Reproduced from Knights *et al.* (1995).

cially young larvae smaller than 10 mm long, have very poor capabilities of locomotion and are swept away by water velocities of just a few centimetres per second (Houde 1969; Harvey 1987; Lightfoot & Jones 1996). Floods, particularly soon after hatching, may wash YOY fishes long distances downstream and away from appropriate rearing habitat if no appropriate refuges are available (Harvey 1987). In rivers or streams with little or no backwater available, most YOY fish also take advantage of rare riparian shelters with low water velocity, such as rootwads, where they aggregate in large numbers (as high as 1500 fish m^{-2}; Baras *et al.* 1995). In man-modified rivers, banks consolidated with boulder assemblages or gabions may serve the same purpose (Baras 1992).

Even where fishes normally remain within the river channel during winter, during flood events fishes often temporarily move out of the river channel, onto the inundated floodplain, into backwaters or tributaries, returning as water levels decline. This behaviour has been observed for a wide variety of fish species. Radio-tracked barbel *Barbus barbus* and chub *Leuciscus cephalus* which were found to enter small tributaries or moved on to recently inundated islands during temporary floods in early spring, subsequently homed to their original residence areas (Baras 1992). In tributaries of the Magdalena River, Colombia, temporary floods caused by tropical rainfalls in the Andean foothills can raise the water level by several metres within a few hours, and water velocity exceeds the swimming capacity of almost all fish species. During these spates, most fish species (mainly characins and catfishes) take refuge in inundation ditches ('*caños*') with near zero water velocity (Baras *et al.* 1997). Such behaviour may involve little longitudinal or lateral movement in response to elevated discharge, depending on habitat availability. Matheney and Rabeni (1995) noted this behaviour in adult northern hog suckers *Hypentelium nigricans* (Catostomidae), radio-tracked in the Current River, Missouri. This species exhibited very limited movements between pools and riffles, but moved into flooded riparian areas during floods. Adult dace *Leuciscus leuciscus*, radio-tracked by Clough and Beaumont (1998), did not exhibit large-scale movements in response to high discharge, but would on occasion take up positions out of the river channel, in the flooded riparian zone (Beaumont [pers. comm.]).

Although many temperate freshwater fishes retreat downstream to deeper, slower flowing areas of rivers, some species move upstream from feeding areas in brackish seas and overwinter in deep, slow stretches of rivers or in lakes. This behaviour has been noted for roach *Rutilus rutilus*, common bream *Abramis brama* and pikeperch *Stizostedion lucioperca* in rivers draining to the Caspian, Black and (formerly) Aral seas (Nikolsky 1963), and ide *Leuciscus idus* in the River Kävlingeån draining into the Baltic Sea (Cala 1970).

In arctic and subarctic environments a variety of migration patterns are observed that appear to be related to refuge-seeking behaviour, since they occur outside of the reproductive period and when food availability is extremely scarce. During autumn many populations of anadromous subadult and adult Arctic char *Salvelinus alpinus* leave their main growth and feeding habitat in the sea and migrate back to fresh water where they 'overwinter' in lakes and rivers, in fact spending as much as 11 months in fresh water (Mathisen & Berg 1968; Moore 1975 a, b). Reproduction also occurs during this period, but requires a small fraction of this time during which feeding opportunities are very limited. It is unclear what the benefits of such long-term residence in fresh water are, although in many cases, as winter approaches, the entrances to such rivers, particularly in delta areas, become icebound with dramatically reduced flows, making river entry difficult or impossible. Such migration may, therefore, have as much to do with being in the right place at the right time for spawning, as acting as a refuge to enhance survival and fitness. Spring-fed tributaries may provide refuge overwintering habitat that is critical for survival of anadromous and freshwater-resident Arctic char in some Alaskan rivers (Craig & Poulin 1975). In many cases Arctic grayling *Thymallus arcticus* exhibit autumn downstream migrations of up to 80 km from feeding areas in shallow tributaries and lakes to deeper areas of rivers and lakes, and become concentrated in small areas, often associated with springs (Craig & Poulin 1975; Northcote 1995). In some cases, as for the Okpilak River, Alaska, which drains into the Beaufort Sea, Arctic grayling, radio-tracked from summer feeding grounds, moved downstream and traversed estuarine reaches before swimming up the adjacent Hulahula River to overwintering sites, up to 101 km from the tagging site (West *et al*. 1992).

Juvenile salmonids that remain in fresh water also move into pools in winter and in many cases migrate into gravel and cobble substrates (Fraser *et al*. 1993), in some instances reaching a depth of 0.5 m. This provides a buffer to environmental changes in the surrounding stream environment including temperatures close to 0°C, or in more temperate environments to the effects of high water velocity during periods of high stream discharge combined with low temperatures. Under these circumstances some juvenile salmonids still exhibit daily activity, emerging from within the substratum to forage, but become nocturnal instead of exhibiting their 'normal' diurnal summer activity rhythm. It has been suggested that such shifts may be an adaptive response to continued predator risk, in some circumstances where the fish's escape response is compromised by low temperatures (Fraser *et al*. 1993; Heggenes *et al*. 1993; Valdimarsson & Metcalfe 1998; see also section 3.3).

A complex pattern involving both downstream and upstream refuge migrations has been observed in adult three-spined stickleback *Gasterosteus aculeatus* in Alaska (Harvey *et al*. 1997). In the winter, available habitat in Black Lake declines by up to 85% due to ice cover resulting in low dissolved oxygen levels, and stickleback move into the deeper Chignik Lake where they can find a more stable environment in which to overwinter. In May, they move up the Black River to avoid high discharges and low water temperatures caused by the June snowmelt.

In hot, dry regions, refuge migrations may occur as fishes move out of tributaries or floodplain areas into remaining sections of rivers during the dry season (Cambray 1990; Magalhães 1993). Such movements may also occur in more temperate conditions during infrequent drought periods (Gagen *et al*. 1998). Where surface water in stream beds dries up, some species may migrate into the hyporheic zone, in interstices between gravel or cobbles, with species such as bullhead *Cottus gobio* found up to 1 m down (Bless 1990). In arid areas some fishes, including lungfishes such as *Protopterus* spp., may migrate into the sediment and secrete a mucus cocoon enabling them to survive complete drying of the mud, while in a state of torpor.

Although at the seasonal or life-history scale many refuge migrations in freshwater and brackish environments appear to be associated with avoiding unfavourable physical conditions, there appear to be some cases where migratory behaviour may serve to provide appropriate conditions for recovery from activities such as spawning and/or reduce mortality from predation. The most common examples of this behaviour do indeed relate to postspawning behaviour in which downstream movements are reasonably frequent. In some instances, particularly for anadromous iteroparous fishes, such as some populations of shads *Alosa* spp.,

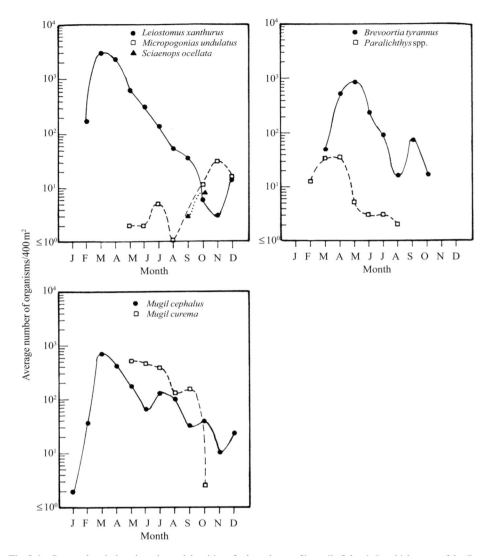

Fig. 3.4 Seasonal variations in estimated densities of selected taxa of juvenile fishes in brackish water of the Cape Fear Estuary, North Carolina, 1977. Redrawn from Weinstein (1979).

and some anadromous salmonids (where they do not exhibit complete semelparity) these may be seen as part of the cycle of downstream movements towards marine feeding areas. However, in several holobiotic, freshwater species there are examples of well-defined postspawning movements by species such as razorback sucker *Xyrauchen texanus* (Catostomidae; Tyus & Karp 1990; Modde & Irving 1998) and dace *Leuciscus leuciscus* (Cyprinidae; Clough *et al.* 1998). In both of these cases, the authors considered that movements were probably associated with recovery from spawning. Razorback sucker moved downstream an average of 20 km from spawning areas, mostly to slower water at the mouths of tributaries (Modde & Irving 1998). The study by Clough *et al.* (1998) on the River Frome, southern England, is notable because, although relatively few dace were radio-tracked immediately after spawning, most moved distances of a few kilometres from the main river into narrow, slow flowing side channels which were rarely used at other times of the year. Yet, observations showed that many untagged dace were found with the tagged fishes. Collections of these fish showed most to be spent and not actively feeding, with the guts of 66% of dace found to be completely empty, and Clough *et al.* (1998) considered their behaviour was probably a result of seeking habitat with low water velocity in order to minimise energy expenditure during recovery.

The movement of the young of a substantial number of marine-spawning fish species into estuarine and brackish marsh environments, which act as nursery habitats (e.g. Weinstein 1979; Marotz *et al.* 1990), may be a life-history response to the high predation pressure on young life stages that occurs in marine environments (Miller *et al.* 1985). In some cases, the same may also be true for the young of typically freshwater species such as largemouth bass *Micropterus salmoides*, which may be found in brackish environments (Meador & Kelso 1989). The number of fish species which enter brackish environments from freshwater and marine environments is relatively low and so, although brackish environments may be harsh habitats due to fluctuations in salinity, temperature and oxygen, they can provide a refuge from predation. Several of the creeks draining into the Cape Fear Estuary, North Carolina, have substantial freshwater input and are utilised extensively by postlarvae and juveniles of several species including Atlantic menhaden *Brevoortia tyrannus*, spot croaker *Leiostomus xanthurus*, striped mullet *Mugil cephalus*, white mullet *M. curema* and flounders *Paralichthys* spp., all of which are spawned at sea and migrate into the estuary (Weinstein 1979; Fig. 3.4). Standing stocks (all species, up to 400 fish m^{-2}) of most of these species were highest in the lowest salinity creeks, also frequented by some freshwater and euryhaline species. Although these species are often regarded as marine or euryhaline, such marsh systems appear important for population maintenance, and hence for some stocks at least, migration into brackish water may be a significant lifecycle component.

3.2.3 *Spawning migrations*

Because in many fish populations spawning occurs over a relatively limited portion of the year and requires aggregation of adults of both sexes (although this can vary from pairing to aggregations of tens or hundreds of thousands of fishes), migrations to spawning habitats are often more concerted and visibly apparent than feeding or refuge migrations. Many freshwater fish species exhibit upstream spawning migrations. Such migrations may help to offset downstream drift or translocation of young life stages to some degree, or enhance their dispersal over a range of appropriate habitat (Northcote 1978, 1984; Linfield 1985; see also section 3.2.1). In lithophilous spawning fishes that make nests in gravel or spawn on gravel substrates, such as most lampreys, sturgeons, salmonids, clupeids, osmerids, catostomids and some cyprinids, upstream movement through positive rheotaxis is an important migratory behaviour. To spawn, adult anadromous fishes must, of course, migrate some distance upstream, whether it is only to brackish water at the head of tide in the case of the European smelt *Osmerus eperlanus* (Wheeler 1969; McAllister 1984), or far upstream to headwater tributaries as in many sockeye salmon *Oncorhynchus nerka* stocks (Foerster 1968).

Many lithophilous salmonids, cyprinids and catostomids from temperate regions overwinter in slow, deep pools, or lakes, which are often several kilometres or tens of kilometres downstream from appropriate spawning habitat, requiring upstream migration in spring to utilise these habitats. However, this behaviour is not fixed and there are several examples where salmonids (Hartman *et al.* 1962), catostomids (Tyus & Karp 1990; Kennen *et al.* 1994; Modde & Irving 1998) and cyprinids (Tyus 1985, 1990) exhibit downstream migrations out of overwintering or feeding habitat to reach appropriate spawning areas. Such behaviour obviously relates to the spatial distribution of appropriate spawning habitat, relative to feeding/refuge areas of adults. In some cases where adults are distributed above and below spawning areas they may congregate through downstream migration of some and upstream migration of others (Whelan 1983; Tyus & Karp 1990; Modde & Irving 1998). The large spawning migrations of over 150 km displayed by Colorado pikemninnow *Ptychocheilus lucius*, in the Green and White rivers, of the Colorado River system in the midwest US (Tyus 1985, 1990; Irving & Modde 2000), appear to relate to a lack of appropriate spawning areas close to or upstream of feeding areas in this river since, by contrast, Colorado pikeminnow in the upper Colorado River appeared to spawn in the area over which they ranged in the remainder of the year and did not exhibit a well-defined spawning migration (McAda & Keading 1991).

A typical and relatively early historical example of the marked periodicity of spawning migration of fish other than salmonids has been provided by Lelek and Libosvárskí (1960), who used electric fishing in a fish pass to determine the migration of fish in the Dyje River, Breclav, in former Czechoslovakia. The whole pass was fished with the pass blocked off with a steel screen to prevent downstream movement. The pass was reopened at successive intervals to determine the number of fish per 6-hour period. Roach *Rutilus rutilus* and white bream *Blicca bjoerkna* were the main species in the pass. Of the 31 species in the river, 20 entered the ladder – roach, white bream, bleak *Alburnus alburnus*, nase *Chondrostoma nasus*, chub *Leuciscus cephalus*, ide *Leuciscus idus*, common bream *Abramis brama*, schneider *Alburnoides bipunctatus*, rudd *Scardinius erythrophthalmus*, Eurasian perch *Perca fluviatilis*, barbel *Barbus barbus*, vimba *Vimba vimba*, Danube bream *Abramis sapa*, whitefin gudgeon *Gobio albipinnatus*, blue bream *Abramis ballerus*, dace *Leuciscus leuciscus*, asp *Aspius aspius*, European eel *Anguilla anguilla*, tench *Tinca tinca* and wels *Siluris glanis*. Fish appeared in the ladder after 20 April when temperatures rose above 8°C. The maximum occurrence was from the end of April to the end of May. Water temperatures varied from 12 to 20°C. There was a mass occurrence of fish between 2 and 13 May 1958, when the average daily temperature during this period increased by 10°C in 10 days. The occurrence of fish in the pass after this was negligible with only nine individuals between July and October. It was argued that the presence in the pass of white bream and roach during this period of peak abundance was due to a spawning migration. Other species were not so numerous and so were not necessarily considered to be migrating, even though several species are now regarded as characteristically potamodromous.

Those freshwater fish species which produce buoyant or semi-buoyant eggs often exhibit long-distance upstream spawning migrations, which appear to be of high fitness value in providing adequate distance for eggs to develop, hatch and young to be carried to, or find appropriate nursery habitat as they are drifting downstream. These long-distance upstream migrations are characteristic of many species of *Alestes* (Alestiidae; Carmouze *et al.* 1983), *Prochilodus* (Curimatidae; Bayley 1973; Lowe-McConnell 1975, 1987), and pimelodid catfishes (Lowe-McConnell 1975, 1987; Barthem & Goulding 1997) in tropical African and

South American rivers. These migrations are also exhibited by some Australian percichthyid fishes such as golden perch *Macquaria novemaculeata* in response to high water levels, and for which upstream migrations of several hundred kilometres in the Murray-Darling River system are common, and upstream movements of 2300 km have been recorded (Reynolds 1983). This behaviour is also shown by many populations of anadromous species with semi-buoyant eggs such as the striped bass *Morone saxatilis*, although for this species the extent of such spawning migrations can vary dramatically (from head of tide to 300 km upstream) since the eggs are somewhat salt-tolerant and often reach brackish water before hatching (Setzler *et al.* 1980; Rulifson & Dadswell 1995).

Other fish species with buoyant eggs may not exhibit long-distance upstream spawning migration but instead migrate to backwaters, deltas or inundated floodplain areas to spawn, so that eggs are not normally washed long distances downstream. This is the case for species such as the goldeye *Hiodon alosoides*, a North American hiodontid, and the freshwater drum *Aplodinotus grunniens*, the only temperate North American freshwater sciaenid. In some groups of freshwater fishes the eggs that are laid are negatively buoyant or sticky, but the larvae are washed downstream and again, where the eggs are deposited in the main river channel, substantial upstream spawning migrations of adults may occur as in some lithophilous European cyprinids such as nase *Chondrostoma nasus* and vimba *Vimba vimba* (Lelek 1987; Penáz 1996; Zbinden & Maier 1996). This behaviour is also typical for many adult anadromous fishes, including many shads such as most temperate *Alosa* species (Loesch 1987), and the tropical hilsa *Tenualosa ilisha* (Jones 1957; Islam & Talbot 1968).

The most well-known catadromous fishes are the anguillid eels, and they are worthy of further consideration here. The large-scale migrations of European eels *Anguilla anguilla* and American eels *A. rostrata*, between their spawning grounds in the vicinity of the Sargasso Sea and freshwater feeding habitats are well documented (Harden Jones 1968; Tesch 1977). This review is primarily interested in migratory behaviour in the freshwater environment so attention will be focused on this stage of the eel lifecycle. The migration of silver European eels to their spawning grounds takes place in the late summer or autumn. The exact month, however, may vary as a result of a temporal shift from inland waters to coastal waters with the earliest migrations occurring furthest from the sea (Tesch 1977). There is also some migratory activity in the spring and it is argued that this is due to eels which are prevented from migrating in the autumn becoming inactive in the winter to resume their migration in the spring (Frost 1950). The migration of males and females do not coincide, which may be due to larger females coming from inland waters whereas the smaller males occur in coastal areas (Tesch 1977).

Silver eels drift downstream in the middle depths of rivers, often in groups (Tesch 1977). The distances covered by migratory silver eels vary depending on the individual's swimming capacity, swimming speed and current. Svedang and Wickstrom (1997) argued that the high proportion of lean silver European eels at a number of sites in Sweden refuted the hypothesis that eels must accumulate fat to a critical level before events associated with spawning are possible. They suggested that either many eels will not be able to spawn successfully or that the energy needs of migrating eels have been exaggerated. Svedang and Wickstrom (1997) argued that it was more likely that eel maturation is more flexible than previously thought. The transition from the growth phase to the migratory phase may be a stepwise process which can be arrested at various stages, as occurs for Atlantic salmon *Salmo salar* (Mills 1989), and probably during the upstream migration of yellow eels in long river systems, in view of the considerable size disparity of upstream migrants. Svedang and Wickstrom (1997) showed that landlocked eels could revert from the silver phase to yellow and resume feeding. However, Tsukamoto *et al*. (1998) have suggested that, on the basis of evidence from microchemistry of maturing silver eels, eels from fresh water may not contribute significantly to reproduction.

3.2.4 Post-displacement movements, recolonisation and exploratory migration

Periodic events such as floods, droughts or water quality disturbances (either natural events such as from volcanic activity or anthropogenic influences such as pollution) may depopulate areas of freshwater habitat (Detenbeck *et al*. 1992; Dolloff *et al*. 1994; Ensign *et al*. 1997). Recolonisation of these areas relies on the maintenance of pockets of fish, the presence of resistant lifecycle stages or immigration. Where fish remain, the rate of recovery may be strongly influenced by parental investment in offspring (Ensign *et al*. 1997). However, immigration may be an important factor and can result from regular lifecycle-scale migration events where fishes which have grown and matured elsewhere may return to spawn in natal habitat, as for many anadromous salmonids. Alternatively fish which may have emigrated from the area at the onset of poor conditions may return. However, in many cases, particularly where recovery takes several years, natural seasonal migratory or dispersal movements (Schlosser 1995) from outside the affected area may result in fishes finding appropriate habitat within the recovered section. This may be regarded as exploratory migration and appears to be important for recolonisation of damaged habitat (Detenbeck *et al*. 1992), of headwater streams under improved environmental conditions and greater food supply (Schlosser & Ebel 1989) or following reconnection after drought conditions (Cambray 1990).

Yellow European eels translocated from their home waters are capable of finding their way back (Mann 1965; Tesch 1966, 1977; Deelder & Tesch 1970). Most eels were capable of finding their way home at distances of 100 km. Beyond this distance, the percentage of successful returns was much smaller. However, some individuals were capable of homing from distances of up to 200 km (Tesch 1977). Similar post-displacement homing or directed behaviour, although less spectacular than in eels, has been noted for many freshwater fish species, either after displacement due to floods or by experimental translocation (Table 3.1). Such movements are thought to be important for fishes regaining the use of appropriate areas when conditions have improved once more, having an ecological role somewhat similar to that of postspawning homing (see section 2.2.1.3).

Langford (1979, 1981) showed from acoustic tracking data that in the River Witham, England, several common bream *Abramis brama* were flushed downstream when flows increased suddenly as hydraulic weirs were lifted after heavy rainfall. Some bream subsequently returned upstream over several kilometres. He also showed that northern pike *Esox lucius* in the River Thames, England, moved or were displaced up to 1.5 km downstream during major spates. Following these floods almost all fish returned upstream to their original location demonstrating a strong homing tendency after displacement. Slavík [unpubl.] found that 10% of fin-clipped minnows, *Phoxinus phoxinus*, washed downstream during twice-daily high flow events caused by discharges from a small hydroelectric plant on a tributary of the Vltava, Czech Republic, were displaced greater than 200 m downstream but returned to their original

Table 3.1 Homing movements by selected freshwater fishes following displacement by unfavourable conditions (e.g. high flow, low oxygen) or translocation. ? indicates no information available.

Species	Stage	Distance (km)	References
European eel *Anguilla anguilla*	Yellow eel	up to 200	Mann (1965); Tesch (1966, 1977); Deelder & Tesch (1970)
Northern pike *Esox lucius*	Adult	1.5	Langford (1981)
Common bream *Abramis brama*	Adult	1.5–60	Goldspink (1978); Langford (1979, 1981)
Barbel *Barbus barbus*	Adult	2	Baras & Cherry (1990); Baras *et al.* (1994a); Baras (1997); Lucas *et al.* (1998)
Gudgeon *Gobio gobio*	Adult	?	Stott *et al.* (1963)
Chub *Leuciscus cephalus*	Adult	1–13	Fredrich (1996); Lucas [unpubl.]
European minnow *Phoxinus phoxinus*	Adult	0.2	Kennedy & Pitcher (1975); Kennedy (1977); Slavik [unpubl.]
Roach *Rutilus rutilus*	Adult	?	Champion & Swain (1974)
Smallmouth bass *Micropterus dolomieui*	Adult	0.8–14.0	Ridgway & Shuter (1996)
Largemouth bass *Micropterus salmoides*	Adult	15–21 up to 6	Richardson-Heft *et al.* (2000) Gent *et al.* (1995)
Cichlid *Pseudotropheus aurora*	Adult (8 cm)	2–3	Hert (1992)

position once flows had subsided (Fig. 3.5(a)). Champion & Swain (1974) showed that the numbers of roach moving upstream through a fish trap on the River Axe (England) increased after floods in November 1965 and February 1969, which they argued was the result of their downstream displacement by the flood. However, no fish moved upstream after floods in April 1961 or December 1965. Baras *et al.* (1994a) argued that the presence of barbel *Barbus barbus* in the fish pass of the Ampsin-Neuville weir on the River Meuse, Belgium, in mid-April was not related to spawning since most individuals were immature. Since these captures followed high flow conditions and took place much before spawning, they were regarded as compensatory upstream movements of individuals flushed downstream during flow increases as found in the River Ourthe, Belgium (Baras & Cherry 1990) and River Nidd, England (Lucas *et al.* 1998). In the Nidd in summer these were usually brief and followed by a subsequent upstream homing migration to the location occupied prior to the high flow (Fig. 3.5(b)). In autumn and winter, however, successive downstream movements associated with high flow resulted in a stepwise pattern of downstream migration (Lucas *et al.* 1998; Lucas 2000).

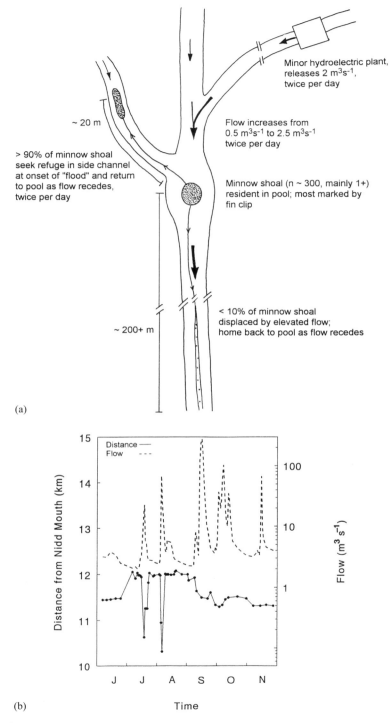

Fig. 3.5 Displacement and subsequent homing of (a) fin-clipped European minnow *Phoxinus phoxinus* in a Czech stream (schematic figure) (Slavík [unpubl.]) and (b) a radio-tagged barbel *Barbus barbus* in the River Nidd, northern England (redrawn from Lucas *et al.* 1998), following high discharge events.

Much knowledge of the capacity of fish for making directed or homing movement has been obtained through translocation experiments. Goldspink (1978) showed that marked common bream *Abramis brama* captured in the Zwartemeer and released in the Tjeukemeer, the Netherlands, left the lake, moved into the surrounding canals and then showed some homing behaviour once into the Ijsselmeer. The maximum distance travelled was 60 km. Baras (1997) examined the homing behaviour of barbel *Barbus barbus* outside the spawning season. He found that, after experimental translocation close to the site of capture, individual barbel homed almost immediately to their previous residence area. When translocated further, however, fish downstream of their capture site homed more accurately than those translocated at equivalent distances but upstream of the capture site. Malinin (1970, 1971) also found that common bream translocated downstream homed more frequently and more rapidly than those translocated upstream. It was argued that these differences might have been due to a lack of orientation cues for fish upstream of the capture site. However, other species like chub *Leuciscus cephalus* can show homing behaviour when translocated either upstream or downstream (over 2 km; Fredrich 1996). In some cases, homing does not take place when the translocation distance exceeds the fish's home range (e.g. largemouth bass *Micropterus salmoides*, Mesing & Wicker 1986). Regarding territorial species, the factor governing the capacity of homing after translocation can be time rather than distance, as empty territories are prone to be colonised in the meantime (e.g. the cichlid *Pseudotropheus aurora*, Hert 1992).

3.3 Diel horizontal and vertical migrations

Rhythmic diel movements are characteristic of many freshwater fish species, particularly juveniles or species that are small at adulthood. Except for during the spawning season, when they may also include a reproductive function, diel migrations are generally interpreted as a trade-off between foraging and escaping predators (Hall *et al.* 1979). On some other occasions, movements to a particular habitat can be further dictated by other purposes, and daily migrants leave this place for homeostatic reasons, essentially, as the changes of physicochemical conditions over the daily cycle no longer enable them to occupy these areas. These traits show considerable variations between ecosystems, fish assemblages, species and life stages, so that the separate analysis of influencing factors (feeding, predation and homeostasis) is problematic, making it difficult to obtain a comprehensive picture of the adaptive nature of these migrations.

In most rivers and streams, diel migrations have a longitudinal and/or lateral component, and sometimes an obvious vertical component, as shown by Kubecka and Duncan (1998a) through the acoustic monitoring of fish behaviour over a 24 h period in the littoral and open water (3 m deep) zones of the River Thames, England. At night larger fish moved to the surface and towards the littoral zone, returning to deeper layers during the day. The vertical movements of fish were more marked in the open water of the river where fish were oriented to the current. In the littoral zone movements were more random. For fishes inhabiting lakes, diel vertical migrations are more frequent and they generally occur at the same time as the diel migration of planktonic prey (e.g. sockeye salmon *Oncorhynchus nerka* and *Mysis relicta*; Levy 1991). Brett (1971) suggested three functions for this vertical migration. The first was that fish were following the vertical migration of zooplanktonic prey. The second was that during daylight fish move into darker water to avoid predation. The third was that fish are maintaining a homeostatic control over their rate of energy expenditure by moving after feeding into cooler deep waters where their rate of expenditure is reduced. Views and concepts have slightly changed since then, notably as the diel vertical migration of many zooplanktonic organisms is nowadays regarded as an anti-predator response which is mediated by fish kairomones, chemicals released by the fish and detected by zooplankton (e.g. Neill 1992; Loose *et al.* 1993; Lieschke & Closs 1999; Ringelberg 1999). Nevertheless, Brett's view remains valid; constraints upon the timing of foraging are just greater.

Although vertical migration has been suspected as the main type of diel migration in lacustrine environments, there is growing evidence that lateral movements between shallow edges and pelagic areas occur more frequently than thought initially (Naud & Magnan 1988; Gaudreau & Boisclair 1998). Kubecka (1993) provides some evidence that, during the non-spawning summer period, a wide variety of Palaearctic fish species of deep or large temperate lakes spend the day offshore and migrate inshore at night. There were five-fold differences between day and night catches in inshore seining in Loch Ness, Scotland and Lake Baikal, Siberia. Densities of fish were seventeen times greater in night catches than in day catches in the littoral zone of the Rimov Reservoir, Czech Republic, while, in the Thames valley reservoirs, England, fish were only caught at night (Kubecka 1993). Daily horizontal fish migrations between inshore and offshore zones were demonstrated in two Canadian lakes using acoustic techniques by Gaudreau and Boisclair (1998) and Comeau and Boisclair (1998). They showed, however, that movement occurred in the reverse direction, from the littoral to the offshore pelagic zone at night. Piscivorous fish were present in these lakes and the authors postulated that the reverse migration was associated with their presence. In two other lakes without piscivorous fish, the highest relative densities of fish in the pelagic zone occurred during the day. Predation has also been invoked as the key factor shaping the diel migrations of the YOY of many lacustrine fishes, which generally move inshore at night (e.g. Tonn & Paszkowski 1987; Persson 1991). Gliwicz and Jachner (1992) suggested further that such nocturnal inshore movements might also be elicited even in the absence of predators, and correspond to an adaptive behaviour driven by a 'ghost of predation'.

Lateral migrations are frequent in riverine fishes. Sanders (1992) showed that night electro-fishing catches in marginal habitat of Ohio rivers contained significantly higher numbers of species, individuals, weight and biological index scores than day catches. Catch differences were attributed to diel movements from offshore to nearshore waters during the evening twilight period. These movements were attributed to movements from deepwater refuge areas to marginal foraging areas. Roach *Rutilus rutilus* and chub *Leuciscus cephalus* 0+ fry in the Great Ouse, England, fed continuously during the day and night but fewer prey were caught at night because a proportion of the fry migrated offshore beyond the weed beds where food was less abundant but predators were also fewer (Garner 1996). Conversely, Copp (1990) interpreted a shift of juvenile roach in the upper River Rhone floodplain, from deeper water with macrophytes into shallower open waters, as a need for a refuge from fish predation. Similar inshore movements of fish at night were demonstrated for YOY fish by Copp and Jurajda (1993) who sampled two adjacent stretches of bank (one comprising shallow sand, one comprising steep boulders) at different times of the day. They showed that as light levels decreased numbers of whitefin gudgeon *Gobio albipinnatus* and roach decreased along the boulder bank as numbers increased along the sand bank suggesting a dusk migration to the sand bank – probably to avoid predation. This finding was corroborated by a significantly higher number of potentially piscivorous fish (Eurasian perch *Perca fluviatilis* and chub, about 80 mm) along the boulder bank at night. In contrast to these, 0+ juveniles of rheophilic cyprinids (barbel *Barbus barbus*, chub, dace *Leuciscus leuciscus* and nase *Chondrostoma nasus*) during summertime were found to move inshore in the morning, and into neighbouring shallow riffles at dusk (Baras & Nindaba 1999a, b), where they probably search for food, as these are the periods when 0+ dace are known to feed in the summer (Weatherley 1987). Here too, however, predation was invoked as the main factor behind this migration pattern, notably with respect to nase, which feeds on algae with no diel periodicity in abundance, and showed the same diel migration pattern.

Cerri (1983) argued that the potential success of predatory fish decreased with increasing light intensity. During periods of increased predator activity YOY fish may move to the shallow littoral zone (Schlosser 1988; Slavík & Bartos 2000) and occupy highly structured habitats which they use as refuges from predators (Hanych *et al.* 1983; Fraser & Emmons 1984). Highly structured habitats provide ideal refuges against predation, but their conspicuous nature can contribute to attract predators, and make it easier for them to encounter prey fish. Shallow bays and shorelines also constitute a size-limiting refuge against predators, with lesser possibilities of predators controlling the timing and nature of encounters with prey fish. Additionally, by virtue of their shallow depth and near zero water current, bays warm up considerably during the growing period and photosynthetic activity causes the water to become locally superoxygenated, resulting in high primary production, and high concentration of small planktonic prey (Reckendorfer *et al.* 1996). Warm temperature and superoxygenated water may also act as additional, physiological barriers against predators, which do not tolerate these conditions as well as larvae or small juveniles (e.g. Kaufmann & Wieser 1992).

Diel lateral movement patterns of fishes can change considerably with ontogeny, as exemplified by the contrasting diel patterns of abundance of yellow perch *Perca flavescens* in Lake St George, Ontario (Post & McQueen 1988). Early perch larvae remain offshore day and night, in contrast to larger larvae, which move offshore at night and move back inshore at dawn, and juveniles about 30 mm that are found in nearshore habitats throughout the day and night. In lakes, fish of any size can theoretically access any habitat, in contrast to rivers and streams, where water velocity can impose further restrictions. Recently, Schindler (1999) suggested that the age and size at which fish start making daily migrations, and leave safe inshore habitats for risky, but rich offshore places, was a matter of temporal constraint. By virtue of size-dependent overwinter mortality, fish must reach a sufficient size by the end of the growing season, and thus initiate diel littoral–pelagic migrations at proportionally smaller sizes as winter severity increases or as hatch date within a season increases. Early migrants (risk-takers) would thus incur a higher immediate risk, but survivors would receive a benefit that would be denied to most surviving care-takers (late migrants). As suggested by Schindler (1999), time constraints on diel migration behaviour are expected to be greater for the populations near the high latitudinal limits of a species' geographical distribution.

Riverine fishes also show ontogenetic variations in their diel migration patterns, but here, water currents may offset time constraints, notably because of size-related swimming capacities and abilities of capturing food items. Flore and Keckeis (1998) provided evidence that young nase *Chondrostoma nasus* were unable to capture passing prey at water velocities much greater than their critical swimming speed. For this reason probably, larvae of barbel *Barbus barbus*, chub *Leuciscus cephalus*, dace *L. leuciscus* and nase remain inshore day and night, whereas juveniles of increasing size become increasingly independent of inshore habitats (Fig. 3.6; Baras & Nindaba 1999a, b): the larger the YOY, the later they move inshore in the morning, and the earlier they move offshore in the afternoon, resulting in size-structured daily migrations and habitat use (see also Copp & Jurajda 1999). The shift from residency in bays to diel inshore–offshore movements in these species can thus be viewed as a matter of size, but rather independently of predation, whereas the size-structure of the diel migration presumably reflects a graded response to size-dependent threats posed by predators (Helfman

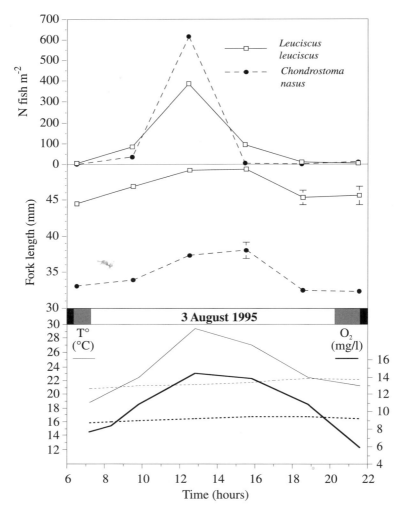

Fig. 3.6 Diel size-structured movements of 0+ juvenile cyprinids, the omnivorous dace *Leuciscus leuciscus*, and the herbivorous nase, *Chondrostoma nasus*, in the River Ourthe during summertime. The upper graph shows the diel variations of fish density in shallow bays. Values on the central graph are mean fork length, and error bars stand for 1 standard error. Plain and dotted lines on the lower graphs are temperature (or oxygen) in the bay and in the main stream, respectively. Drawn based on data from Baras and Nindaba (1999a, b).

1989). During early autumn, juvenile dace and chub move to calmer deeper habitats (Baras et al. 1995) where they may encounter nocturnal predators. The use of inshore bays at night during this period (Baras & Nindaba 1999a) may be a mechanism to avoid these predators.

Species with direct ontogeny, however, can make diel migrations from the moment they emerge from the spawning gravel. Roussel and Bardonnet (1999) experimentally demonstrated that within 1–2 days after emergence, young brown trout *Salmo trutta* rested in calm and shallow (2 cm deep, 0–2 cm s^{-1}) shoreline habitats at night, and moved during daylight to deeper and faster habitats (12 cm, 2–4 cm s^{-1}), where they held stationary swimming positions. Diel migrations were elicited in the absence of food and predators, suggesting there was an innate component behind this pattern. In the presence of bullhead *Cottus gobio* predators, young trout showed cryptic behaviour, and remained inshore through the day and night.

Feeding and escaping predation are requirements for maximising energy gain, growing and reaching a sufficient size for successful overwintering in temperate or arctic environments. Minimising energetic costs, and maintaining homeostasis, represent further conditions to maximising energy gain, but these have been found to be frequently traded off against the two aforementioned factors. For example, Slavík [unpubl.] found that twice-daily pulses of water from small hydroelectric power plants in a small Czech stream caused 90% of fin-clipped European minnows *Phoxinus phoxinus* to move into side-streams during these peaks in flow (see Fig. 3.5(a)). On warm spring or summer days, larvae and juvenile cyprinids resting in shallow shorelines can be exposed to extremely high temperatures (up to 32–33°C measured in May in the River Ourthe, Belgium, 50°N). Also, moving to or remaining in shallow environments increases the risk of death after the bays become isolated and dry up, or through predation by terrestrial predators. Furthermore, at night, shallow backwater environments often become hypoxic, particularly in eutrophic water bodies. Diel migration into and out of backwater habitats suffering large oxygen fluxes has been described in chapter 2 (Fig. 2.5; Slavík & Bartos 2001). In some vertical migrations in lakes, the risks of hypoxia are also traded off against the benefits of rich food availability. Rahel and Nutzman (1994) found that substantial numbers of central mudminnows *Umbra limi* moved into hypoxic hypolimnetic environments in order to feed on abundant *Chaoborus* which remain in the hypolimnion by day. Fishes might also utilise other hypoxia tolerant plankton and benthos such as chironomids and oligochaete worms containing blood pigments with a high oxygen affinity. The rather frequent occurrence of oxyphilic fish in anoxic or strongly hypoxic waters (*Perca flavescens*, Hasler 1945; Hubert & Sandheinrich 1983; *Oncorhynchus mykiss*, Luecke & Teuscher 1994, in Rahel & Nutzman 1994) suggests that this behaviour may be more frequent than thought initially. In other circumstances, it is predation risk that is traded off against homeostasis. This applies to many species inhabiting inundated tropical forest, where oxygen concentrations can fall below 0.5 mg O$_2$ per litre at night, and forces most fish to move to open habitats where they can utilise aquatic surface respiration at the air–water interface (e.g. Saint-Paul & Soares 1987; see section 4.4.4 for more information on tropical fishes).

In riverine environments, diel migrations by large juvenile or adult fishes may also include a longitudinal component. Clough and Ladle (1997) described highly regular movements of radio-tracked dace *Leuciscus leuciscus* during summer, between distinct daytime and nighttime sites in the River Frome, England. Daytime resting areas were usually characterised by open habitat without areas of macrophyte growth which northern pike *Esox lucius* might use as ambush sites. The distance travelled varies substantially depending on habitat structure and length of pool–riffle sequences. In long pool–riffle sequences of the rivers Nidda (Germany) and Ourthe (Belgium), barbel *Barbus barbus* occupied daily home ranges extending over several hundreds of metres (Pelz & Kästle 1989; Baras 1992), whereas their daily movements extended over no more than a dozen metres in shorter pool–riffle sequences. Some ambush predators, such as northern pike *Esox lucius*, can make no diel migration at all, and sit-and-wait predators feeding on drifting organisms, such as various trout species (e.g. *Salmo* and *Oncorhynchus*) may show time budgeting instead of space budgeting as an adaptive behaviour to variations in food availability (Young *et al.* 1997; Giroux *et al.* 2000).

Chapter 4
Effects of Climate on Patterns of Migratory Behaviour

4.1 Introduction

Patterns of migratory movement by fishes, wholly within fresh water or as part of a diadromous lifecycle, are strongly influenced by environmental conditions. These migrations are the result of key resources being separated in space and time. Arctic and subarctic climates are subject to extreme seasonal changes in photoperiod and very low winter temperatures. These result in severe winter conditions with negligible productivity and freezing of aquatic environments, especially in lacustrine habitats. Stream flow may decline to very low levels due to freezing of surface water, while oxygen may become severely depleted in ponds and lakes where oxygen diffusion from the air is prevented by ice, and oxidation of material occurs within the lake. There is a short window of opportunity for feeding, growth and reproduction in fresh water and a relative paucity of winter refuge areas. As a result migratory patterns of fishes in arctic and subarctic areas may be highly distinct in time and space.

Conditions at temperate latitudes are less severe, although there is distinct seasonality of photoperiod and temperature, which provide seasonal peaks in productivity. Winter freeze-up may occur in colder, continental temperate climates, but winter conditions tend to be associated with increased precipitation and lower evapotranspiration resulting in elevated river discharges and inundation of floodplain habitats. Water levels recede and temperature increases in spring, the precise timing depending on local climatic conditions, and results in a progressive reduction of floodplain habitat, reaching a minimum in late summer, when ephemeral ponds and streams may dry out. Seasonal inundation of floodplain may be very limited or unpredictable in heavily modified catchments where drainage has been increased and flood barriers have been developed. Migratory movements in temperate freshwater systems are closely associated with the seasonal optimality of conditions for growth, survival and reproduction of different lifecycle stages. Seasonal temperature variations in temperate systems play a key part in the metabolic capacity for growth, activity and migration.

Tropical and subtropical freshwater ecosystems are characterised by little variation in photoperiod and temperature. There is therefore relatively little seasonality in temperature-related variation in metabolic capacity for feeding and activity in tropical fishes. However, there may be intense seasonal variations in precipitation resulting in seasonal inundation of floodplain habitat during the wet season(s) and subsequent drying out during dry periods. Productivity in these systems is closely related to the pattern of flooding, and migration of fishes within tropical systems occurs as a mechanism of making efficient use of seasonally available resources. Additionally, movements of tropical fishes commonly occur to escape from areas that are drying out and where predation is high, or where oxygen levels decline below tolerable limits.

This chapter provides an overview of the patterns of migratory behaviour, from a functional standpoint, of fishes in arctic, temperate and tropical freshwater systems. As a wealth of information is available on fishes in temperate freshwater systems, with many examples given in other chapters, this chapter contains a greater level of detail on migratory behaviour in arctic/subarctic and, in particular, tropical systems. This also reflects the need for a review of fish migration, in tropical habitats in particular, notwithstanding the excellent investigations of Welcomme (1985) and Lowe-McConnell (1987), and that much study in the tropics has, to date, largely been undertaken from a fisheries or community viewpoint. Detailed information concerning individual species and specific taxonomic groups is considered in chapter 5, especially for temperate species.

4.2 Arctic and subarctic regions

Arctic and subarctic climates are characterised by extreme seasonal variations in the magnitude of solar radiation, with complete or near-complete darkness during winter and near-continuous daylight in summer. Temperatures are low, especially in winter when widespread freezing occurs, which may dramatically reduce stream and river flow. Ice-melt, usually in May and June, results in expansion of the number and size of streams and rivers and is usually associated with scouring of the riverbed and high levels of suspended solids. Since all known freshwater fishes are ectotherms, the low temperatures that characterise arctic and subarctic latitudes severely limit metabolic capacity for energy-intensive processes such as feeding, growth and activity. Feeding and growth in fresh water is limited to the short summer period. The river systems are usually subject to freezing, so that low temperatures and physical obstruction often limit migratory behaviour to relatively narrow time windows. Low temperatures reduce swimming performance, while near darkness may preclude foraging. Autumn movements in many species are therefore associated with location of refuge areas for over-wintering. These are often slow, deep sites, which may minimise energy expenditure. In these harsh environments with extensive and prolonged ice cover, freshwater fish optimise their use of clearly defined spawning, feeding and wintering areas within their range, even though these areas may be widely separated. Migration may occur through small river channels within delta systems, making affected species susceptible to overexploitation or damage to migration routes.

Most of the examples in the following discussion of behaviour of fishes in arctic and subarctic freshwaters originate from the more accessible North American literature, although extensive areas of arctic and subarctic freshwater habitat occur in Eurasia, especially in northern Russia. Nevertheless the main environmental features and patterns of behaviour appear to be similar in both geographical areas and a variety of anadromous species occur within both regions. The main groups of fishes which migrate within freshwater arctic and subarctic systems are salmonids, including the genera *Oncorhynchus, Salmo, Salvelinus, Coregonus, Stenodus* and *Thymallus*. Although *Thymallus* are exclusively freshwater the majority of these groups include anadromous forms, and have been discussed in detail elsewhere (e.g.

McDowall 1988; Northcote 1995). Many of the salmonid species occurring in arctic and subarctic areas are clearly anadromous and exhibit a seaward feeding migration of juveniles towards food-rich oceanic environments, although the duration of freshwater residence is often longer than at more temperate latitudes (Northcote 1978). At first this would seem surprising since the sea provides greater feeding opportunities and a less harsh environment, but this may be outweighed by increased predation risk for small fishes. Many other species move into estuarine or coastal environments. At high latitudes iteroparity is prevalent among the salmonid species to a much greater degree than in populations from lower latitudes, and a similar pattern is found for many other fish species.

Approximately 75% of the 40 or so freshwater fish species described by McPhail and Lindsey (1970) from the Arctic in northwestern Canada and Alaska exhibit a significant degree of diadromous behaviour, compared with about 21% of 181 species of freshwater fishes from the whole of Canada (Scott & Crossman 1973; section 5.1). Among the former are the well-known Pacific salmons including sockeye salmon *Oncorhynchus nerka*, coho salmon *O. kisutch*, chinook salmon *O. tshawytscha*, chum salmon *O. keta* and pink salmon *O. gorbuscha*, although only the last two species occur at latitudes above 70°N. Within the Mackenzie River basin there are long reaches without any major barrier and even small fish species such as the Arctic cisco *Coregonus autumnalis* and Arctic grayling *Thymallus arcticus* can move extensively (Bodaly *et al.* 1989). Most anadromous species, principally the salmonids Arctic char *Salvelinus alpinus*, Arctic cisco, broad whitefish *C. nasus*, lake whitefish, *C. clupeaformis*, least cisco *C. sardinella* and inconnu *Stenodus leucichthys*, move into fresh water during the short summer time window and spawn in early autumn (McPhail & Lindsey 1970; Reist & Bond 1988; Bodaly *et al.* 1989; Howland *et al.* 2000). However, many of these iteroparous salmonids remain in fresh water throughout winter and return to sea in about May, after an 8-month period of freshwater residence, to feed and restore body condition (Moore 1975a, b; Northcote 1978). The Arctic lamprey *Lampetra japonica* occurs throughout arctic drainages in Alaska, northwestern Canada and Russia and comprises anadromous and holobiotic freshwater forms (McPhail & Lindsey 1970). Anadromous Arctic lampreys exhibit conspicuous spawning migrations into the lower reaches of rivers such as the Yukon and Mackenzie in spring, whereas non-parasitic forms are most common from stream environments, often where access to marine environments is limited. Other than the Pacific salmons, *Oncorhynchus* spp., the Arctic lamprey is one of the fish species from this region that exhibits a semelparous rather than iteroparous life-history strategy.

River-dwelling populations of Arctic grayling *Thymallus arcticus* in Alaska and northern Canada exhibit distinct patterns of migration. Adults migrate from overwintering areas after ice break-up in June, to reach small tributaries where they spawn over gravel and thereafter move to feeding areas in lakes and rivers, or remain in the same streams (Craig & Poulin 1975; Bodaly *et al.* 1989; West *et al.* 1992; Hughes & Reynolds 1994; Northcote 1995). On hatching, the young feed in these spawning tributaries until autumn when they migrate to overwintering areas. A similar pattern of refuge migration is displayed by juveniles and adults, which move to overwintering areas, often near springs, which may be more than 80 km from summer feeding habitats. Arctic grayling in the Mackenzie system feed and overwinter in Great Bear River and Great Bear Lake some 280 km from the uppermost spawning tributaries (McCart 1986; Bodaly *et al.* 1989). The juvenile and subadult arctic grayling leave these areas in early June after ice break-up and make extensive feeding migrations within rivers and lakes. Zakharenchko (1973) reported downstream migrations of European grayling *Thymallus thymallus* in the Pechora River in northern Russia to deep pools, up to 37 km downstream of summer feeding areas. These fish made diel migrations of 2 km to rapids for feeding.

In autumn, most lake-dwelling fishes are thought to move out of littoral habitat and into deeper water, although there has been little detailed study this far north. However, this behaviour is strongly influenced by patterns of oxygen availability during the period of winter ice cover. Arctic and subarctic lentic freshwaters may have substantial organic loads, often resulting from incomplete decomposition of vegetable matter at low temperatures. In lakes and ponds, winter freeze-up prevents oxygen diffusion from the air, but uptake of oxygen by decomposition and oxidation processes in sediments continues. Oxygen depletion is greatest nearest to the lakebed, but with prolonged ice cover this may result in extensive oxygen depletion and 'winter fish-kill' conditions. These problems tend to be most apparent in shallow lakes with heavy aquatic vegetation or where substantial deposition of organic matter occurs. These habitats are most commonly inhabited by cyprinids, esocids, umbrids, percids and hiodontids. Although lake-dwelling fish species may not be subjected to the energy demands of swimming in a flow, they may be subject to low levels of oxygen. Responses of fish to these conditions are variable. When oxygen levels decline to very low levels (<0.5 mg per litre) species such as mudminnows (Umbridae), northern pike *Esox lucius* and yellow perch *Perca flavescens* move short distances into shallow water or just below the ice where oxygen levels remain highest (Petrosky & Magnuson 1973; Klinger *et al.* 1982; Magnuson *et al.* 1985). Other species such as walleye *Stizostedion vitreum* usually remain in deeper water or may congregate near to inflows where oxygen levels usually may remain elevated, a behaviour pattern that occurs in the other species also. In truth, most of the fieldwork by Magnuson and colleagues has been carried out on temperate lakes which suffer winter freezing and fish kills, but the processes and behaviour patterns are similar to those occurring further north.

Species such as the goldeye *Hiodon alosoides*, a hiodontid, which occurs in the Mackenzie system of Canada, undergo refuge migrations from shallow lakes to overwinter in rivers. Within the Mackenzie system, this species is probably most abundant in the Peace-Athabasca delta region, and undergoes extensive migrations between wintering and spawning areas (Donald & Kooyman 1977a). The Peace-Athabasca delta population of goldeye, estimated to number about 100 000, overwinters primarily in the Peace River, Alberta, from Vermillion Falls to the Slave River. In May, goldeye migrate from the lower reaches of the Peace River into Mamawi Lake (168 km^2) and then to Lake Claire (1445 km^2) of the Peace-Athabasca delta. They spawn and feed in the delta before returning to the Peace River in late summer and fall (Donald & Kooyman 1977a, b). The fish are repeat spawners and long-lived, maturing at 7–8 years and with a maximum age of 24. Lakes Claire and Mamawi, with a maximum depth of 3 m, are the nursery habitat for the young. Freeze-up occurs in October and lasts for 6 months, so that both Mamawi and Claire lakes are devoid of oxygen by late winter. In order to escape this, both young and adults move into the Peace River to overwinter. Migration to the lakes for spawning in May provides a good egg development and nursery area, but recruitment is strongly influenced by weather, with warm, calm conditions enhancing survival (Donald 1997). Winter refuge migration appears effective for YOY fish since no effect of winter river discharge on year class strength was found.

While most spawning migrations of anadromous species occur in early autumn, those of the potamodromous freshwater species occur in spring, following ice break-up, and during the thaw. The walleye *Stizostedion vitreum* spawns on rocky or sandy shoals in lakes or may migrate into rivers to spawn over riffles, a similar behaviour pattern to that seen further south in this species' range (McPhail & Lindsey 1970; Scott & Crossman 1973). After spawning, adults move to feeding areas, principally in lentic habitat, before moving to wintering areas. Annual circuits as long as 600 km have been recorded for walleye travelling between the Peace-Athabasca delta and Lake Athabasca (Dietz 1973). Longnose suckers *Catostomus catostomus* journey 260 km from Lake Athabasca to spawn in the Slave River at the base of Mountain and Cascade Rapids (McCart 1986). The burbot *Lota lota*, a freshwater gadid, is a benthic species that is typically regarded as rather sedentary (e.g. Bergersen *et al.* 1993) but burbot radio-tracked in the Tanana River, a glacial tributary of the Yukon River, Alaska, moved distances of up to 125 km (Breeser *et al.* 1988). Breeser *et al.* (1988) interpreted these movements, which were greatest in autumn and winter, prior to spawning, as being associated with the harsh environmental conditions. Extensive migrations appear to be common amongst freshwater fishes at arctic and subarctic latitudes, even amongst many species which elsewhere are reported only to exhibit limited migratory behaviour.

As the spring thaw occurs, northern pike *Esox lucius* exhibit relatively local spawning migrations, usually of no more than a few kilometres, into areas of submerged vegetation to lay their eggs. This may involve movements into shallow bays, upriver into tributaries or onto inundated floodplain and reflect a common pattern of movement shown by this species across its range. The young utilise weeded habitats for refuge, entering permanent pools and backwaters as the water recedes. A similar but more restricted pattern of movement is displayed by mudminnows, in particular the Alaska blackfish *Dallia pectoralis* which occurs in Alaska and northeastern Siberia (Scott & Crossman 1973). Dietary studies have also provided indirect evidence of the dynamic nature of fish communities at high latitudes. Little *et al.* (1998) examined seasonal variations in diet of several species within the Slave River. They considered that changes in diets of northern pike and walleye *Stizostedion vitreum*, which they classed as 'resident' species, reflected the patterns of movements by other species. For example diets in spring were dominated by Arctic lamprey *Lampetra japonica* and flathead chub *Platygobio gracilis*, which were aggregating to spawn. Nevertheless it is clear from the information provided above that species such as walleye can also be highly mobile in these systems.

Documentation of anadromous and freshwater migrations in the Churchill system, northern Canada, is less thorough than in the Mackenzie system; this may be because they are not so extensive. Many fish appear to migrate away from deltas in the Mackenzie system for spawning, probably to avoid high sedimentation (Bodaly *et al.* 1989). The relative lack of such deltas in the Churchill system probably means there is less selection for migratory behaviour. Freshwater migrations of walleye in the Churchill system have been documented in Lac la Ronge and Southern India Lake (Rawson 1957; Bodaly 1980). Movements of up to 160 km for individual fish have been recorded, but most were recaptured within 10 km of spawning areas.

Examples of extreme adaptations to very narrow migratory periods at extremely high latitudes have been provided for the Arctic char *Salvelinus alpinus*, the geographical distribution of which ranges up to 82°N in Canada (Johnson 1980) and 80°N in the Svalbard Archipelago in northern Norway, where it is the only freshwater fish species. Rivers and streams in these northernmost territories are frozen 10 months per year, and lakes have worldwide records of low productivity. Gulseth and Nilssen (2000) demonstrated that sea migrants, both smolts and repeat-migrants, left the ultra-oligotrophic lakes where they overwintered, and entered the Dieset River on Spitsbergen island (79°N) within the first 48 h following ice break-up. Most of the downstream run in the high arctic Dieset River was completed within 2 or 3 weeks, without any size precedence, contrary to the situation in subarctic or middle arctic ecosystems (e.g. Johnson 1989). Gulseth and Nilssen (1999, 2000) suggested this might be due to extremely strong time constraints, with the necessity for fish of all sizes to reach rich foraging places in the sea as soon as possible. Polar seas are known to be much richer than polar fresh waters, and marine areas around the Svalbard Archipelago are no exception. In spite of this, the residence time in marine areas was found to average 5 weeks only, and migrants entered fresh water at a time when food availability was still high (Berg & Berg 1993). The upstream migration of Arctic char in the Svalbard Archipelago was found to be size-structured, as observed in more southern locations (e.g. Johnson 1989) with large fish migrating first, and spawning in the autumn of the same year, before returning to the lakes (Gulseth & Nilssen 2000). The proportionally longer residence time of young fish at sea can be interpreted as a need for prolonging the period of fast growth (immature char can double their body weight during this very short period; Johnson 1980) in view of the very limited food availability in freshwater environments. However, there are obvious risks of predation in the surrounding marine environment, particularly by the ringed seal *Phoca hispida* (numerous seal bites are observed on ascending char), and shown by the size-dependent return rate of migrants (Gulseth & Nilssen 2000). Although this remains to be demonstrated, long-life, late maturity, repeat spawning and long freshwater residence may thus have been selected in northernmost populations of Arctic char as an alternative to the lifestyle adopted by other salmonids with year-round marine residence, such as *Salmo* and *Oncorhynchus* spp.

Variations in the size structure of the migration of Arctic char *Salvelinus alpinus* are thought to reflect latitudinal differences in time constraints. There are other, somewhat clearer, examples of how important these latitudinal differences are in shaping the migratory lifestyle of northern species. In the Mackenzie River system, which flows from subarctic (60°N) regions into the middle arctic Beaufort Sea (71°N), northern and southern populations of inconnu *Stenodus leucichthys* show contrasting migrations, as regards distance, timing and habitat use. In the southern, most upstream part of this river system, populations complete their entire lifecycle in fresh water. They overwinter and feed in early summer in Great Slave Lake, a lake with rather high productivity with respect to its geographical situation, and they migrate upstream over 40–300 km into the Taltson, Buffalo or Slave Rivers to spawn in autumn, a few weeks before ice forms (Fuller 1955; Howland *et al.* 2000). Populations spawning in the Arctic Red River (66–67°N) have no chance of overwintering in lakes unless they move to Great Bear Lake, on another branch of the Mackenzie River system, and more than 500 km further upstream from their spawning site. They have adopted an anadromous lifestyle, including overwintering and feeding in coastal areas or the outer Mackenzie delta, then undertaking a long (600–1,500 km) upstream spawning migration in summer (see Howland *et al.* 2000). Because ice forms sooner, and distance travelled is much longer in northern than in southern rivers, northern anadromous populations migrate about 75 days in advance compared to southern populations. This illustrates how latitudinal differences influence life-history strategy and migratory lifestyle, including within a single river system. Similar latitudinal differences have been suggested for the inconnu in other places (e.g. Alt 1977) and for other species living in arctic and subarctic ecosystems (e.g. broad whitefish *Coregonus nasus*; Reist & Bond 1988; Arctic cisco *C. autumnalis*, Dillinger *et al.* 1992). Stable isotope analysis (see chapter 6) of Arctic char *Salvelinus alpinus* from a southern relict population in Quebec showed that a small proportion of individuals continued to adopt anadromy as a life-history strategy, but that the majority did not, presumably due to the reduced fitness advantage of anadromy in this southern population (Doucett *et al.* 1999).

The degree of anadromy, marine residence time, extent and timing of migrations in arctic environments are thus clearly influenced by latitudinal differences in time constraints, as they apply to water temperature, timing of ice break-up, length of river system, predation pressures and availability of suitable overwintering and feeding places. Where rivers are scarce but offer some overwintering place, homing is prevalent and return rates are high (e.g. Spitsbergen Island, where the river nearest to the Dieset River is 150 km away). When river systems offer few or no overwintering places, anadromy prevails, as in the Lower Mackenzie River. When neighbouring rivers offer contrasting overwintering sites, there may be seasonal migration between river systems. Beddow *et al.* (1998) showed from radio-tracking data that populations of Arctic char *Salvelinus alpinus* spawned in all rivers flowing into Voisey's Bay, Northern Labrador (56°N), but they overwintered in the larger system only (Fig. 4.1). This implied downstream migration, coastal movements then upstream migration into the overwintering river, a strategy that had been postulated for other char species, but never proven prior to this study (except for the dolly varden, *Salvelinus malma*; see section 5.23).

4.3 Temperate regions

The extent of discussion in this section is limited, since many of the characteristics have been dealt with in chapter 3 and are examined taxonomically in chapter 5. Many of the migratory movements displayed by freshwater fishes in temperate environments are closely related to seasonal variations in the environment. Even where the interval between links of a migratory cycle for an individual fish is greater than one year the timing of migration is usually closely synchronised with seasonal phenomena. This is most commonly observed in diadromous fish species such as the Atlantic salmon *Salmo salar* in which most of the young, which remain in fresh water for 1–8 years, smoltify and emigrate to sea in late spring (Mills 1989; Shearer 1992). Although the precise timing of this emigration varies with latitude and according to local circumstances, the year-to-year timing is remarkably consistent. Larvae of the European eel *Anguilla anguilla* accumulate in coastal waters along the northeast European coast in their second winter after hatching, and river entry occurs primarily in spring and early summer (Tesch 1977), although this pattern may vary in Mediterranean areas, where the timing of arrival, dictated by ocean currents, is up to a year later and water temperatures are warmer. Temperate climates are characterised by strong seasonal variations in the magnitude of solar radiation, increasing with latitude, with associated changes in temperature and patterns and types (rain or snow) of precipitation. Nevertheless temperate regions comprise a wide range of climatic systems, from continental climates with strong seasonal variations in temperature to coastal 'oceanic' environments with small seasonal fluctuations.

In freshwater systems, the most obvious feature of these conditions is the low temperature in winter which limits metabolic capacity for energy-intensive processes such as feeding,

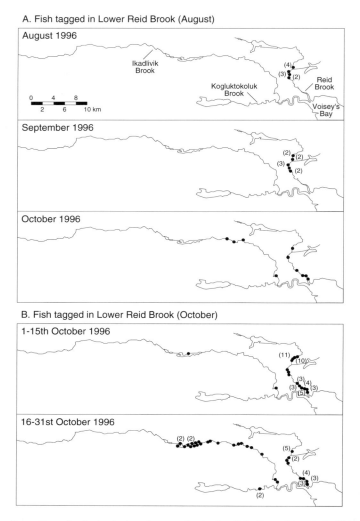

Fig. 4.1 Migration pattern of radio-tracked anadromous Arctic char *Salvelinus alpinus* in the Voisey's Bay region of northern Labrador. Char spawn in several rivers flowing into the bay but appear to overwinter only in the main river system. Redrawn from Beddow *et al.* (1998).

growth and activity. Additionally there is increased river discharge and water level in mild, oceanic climates or freezing of water in lacustrine environments in continental climates, resulting in changes to the fish's physical environment. Under conditions of poor swimming performance resulting from low environmental temperatures fishes may not be able to maintain positions in fast water and may actively seek slower-flowing areas or may be displaced downstream until they arrive in deposition habitats. This behaviour has been found for European cyprinids such as barbel *Barbus barbus* and chub *Leuciscus cephalus* in northern English rivers which are susceptible to flood pulses (Lucas & Batley 1996; Lucas *et al.*, 1998; Lucas, 2000). Although there is evidence of a general trend of downstream movement in winter for a wide variety of species, lateral migration across the floodplain to temporarily or permanently connected lentic environments may occur (Knights *et al.* 1995). Many examples exist of large aggregations of cyprinids that occur in lentic habitats connected to the river channel, but which are absent in summer, although there are considerable variations between fish species (see below).

Although the most common patterns of winter movement in temperate regions would appear to be to deep, slow-moving backwater or lacustrine habitats, the direction and extent of movements is probably strongly influenced by the local geomorphological and physical conditions occurring in a particular river and floodplain system. There are a substantial number of situations where winter migration occurs from lentic to riverine environments. Migratory populations or population components of roach *Rutilus rutilus*, common bream *Abramis brama*, pikeperch *Stizostedion lucioperca* and other species leave feeding areas in the Caspian and Black Seas (and formerly in the Aral Sea) during autumn and migrate to overwintering areas in rivers and lakes (Nikolsky 1963), although these patterns of movement have been disrupted by increasing impoundment of the rivers entering these seas. A similar pattern occurs for ide *Leuciscus idus* in the River Kävlingeån, Sweden, which having fed in the Baltic Sea during the summer return to the river in autumn (Cala 1970). This is not strictly a prespawning migration, since the youngest, juvenile age groups return first, and by November all age groups are present in the lower reaches where they remain until the following spring when spawning occurs. However, in other places, ide are known to overwinter in coastal waters and enter rivers for spawning in spring (e.g. River Angeran, Finland; Müller 1986).

Where freezing occurs and most winter precipitation is in the form of snow, river flows decrease, especially in the upper reaches, reducing water depth and exposing fish to mammalian and avian predators. Susceptibility to homeothermic predators may be increased by reduced 'fast-start' escape responses (see chapter 2), and movement to slow, deep areas may partly serve to limit exposure to such predation. In lakes and ponds with substantial organic loads winter deoxygenation may occur during periods of ice cover. Oxygen depletion is greatest nearest to the lakebed but with prolonged ice cover may result in extensive oxygen depletion and 'winter-kill' conditions, as described in the previous section. As a result of these environmental fluctuations migration to winter refuge habitats is a common migration pattern in temperate freshwater fish species. Where accessible inlet or outlet streams occur, fishes may leave the lake and return in spring, to spawn and feed (e.g. He & Kitchell 1990). Nikolsky (1963) reports the downstream movement of grass carp *Ctenopharyngodon idella* in winter from summer lake habitats to overwinter in the lower reaches of rivers. In other cases, where fish remain in lakes, there is local migration to areas with highest oxygen content (Magnuson *et al.* 1985). Some species such as the common carp *Cyprinus carpio* and the crucian carp *Carassius carassius*, are rather tolerant to these extreme conditions, as they can turn to glycolytic metabolism when the oxygen concentration falls close to zero, accounting for their successful overwintering in harsh conditions, including half-buried in mud substratum for months (Johnsen & Hasler 1977). Similar movements away from hypoxic areas can also occur during summer in thermally stratified lakes suffering oxygen depletion of the hypolimnion. In the reservoirs of the Russian Volga River system, for example, common bream *Abramis brama* move out from and avoid the deep bottom waters as they become hypoxic (Malinin *et al.* 1992).

Spring increases in photoperiod and temperature result in increased primary productivity in temperate climates. Shallow water warms most quickly and generally provides the best environment for feeding and reproduction. In warm, wet climates water levels in spring are still high and the floodplain remains inundated, with access available for fish from the main river channels via tributaries and side channels (Welcomme 1985; Kwak 1988; Masse *et al.* 1991). Shallow backwater and marshland habitats provide important spawning and nursery areas. As spring progresses and evapotranspiration increases, inundated areas dry out and fish remain in isolated ponds or move back into river channels. In areas with cold winters, ice melt may increase river discharge through spring and into early summer, resulting in later inundation of floodplain habitat and providing a different hydrographic response. Under these conditions timing of movement to spawning areas may be delayed by comparison to more oceanic environments, so that spawning occurs once peak levels of discharge have passed. This situation is found for many fish species in the Colorado River in the US where the spring increase in discharge occurs due to snowmelt in the Rocky Mountains. Peak discharge occurs in May when flows are three to ten times higher than normal over a prolonged period (Tyus 1990), despite substantial regulation from several impoundments. The Colorado pikeminnow *Ptychocheilus lucius*, a large riverine cyprinid, commences its spawning migration from wintering areas once peak flows have passed and spawns in cobble riffle areas (Tyus 1986, 1991). On hatching, larvae drift downstream and utilise inundated and deposition areas for early development. Movement by other cyprinids such as humpback chub *Gila cypha* and catostomids such as flannelmouth sucker *Catostomus latipinnis* into smaller tributaries for spawning is common in the upper Colorado system. Although adults of these species have body forms which enable them to maintain a benthic mode of life in fast-flowing water without being displaced, larvae and young are easily washed downstream. Discharges normally decline first in the tributaries, the lower reaches of which become ponded by high flows in the main river, thus retaining young fish which have drifted down the tributaries (Douglas & Marsh 1996; McKinney *et al.* 1999). Small tributaries also warm more quickly than the main river, enabling rapid development. In the context of the Colorado system it has been argued that these processes were important for the survival and recruitment of a variety of endemic riverine cyprinid and catostomid species, but that recent changes to hydrographic regimes through impoundment have upset these natural processes and resulted in lower larval and juvenile survival.

A large amount of data concerning upstream-directed migrations of riverine fishes is available in a number of published and unpublished studies of counts of fish passing upstream through fishways in continental western Europe. Similar information is widely available in other geographic regions. Some western European data are summarised in Fig. 4.2 which provides composite histograms derived from a meta-analysis of data available on the numbers of fish counted in fish passes on the Garonne and Dordogne, France (Travade *et al.* 1998), the Meuse, Belgium (Philippart *et al.* [unpubl.]); Prignon *et al.* 1998) and Netherlands (Lanters 1993, 1995), the Mehaigne stream, Belgium (Philippart [unpubl.]) and the Mosel, Germany (Pelz 1985). It immediately becomes apparent that there is wide variation between fish species in the seasonal patterns of upstream-directed movement. Spawning by all of the species illustrated occurs between early spring and early summer. Some of these peaks of movement by various species clearly seem related to the occurrence of prespawning migrations; for example by European grayling *Thymallus thymallus*, dace *Leuciscus leuciscus*, barbel *Barbus barbus* and white bream *Blicca bjoerkna*. The timing of these peaks varies in relation to the relative timing of reproduction in these species (Fig. 4.2).

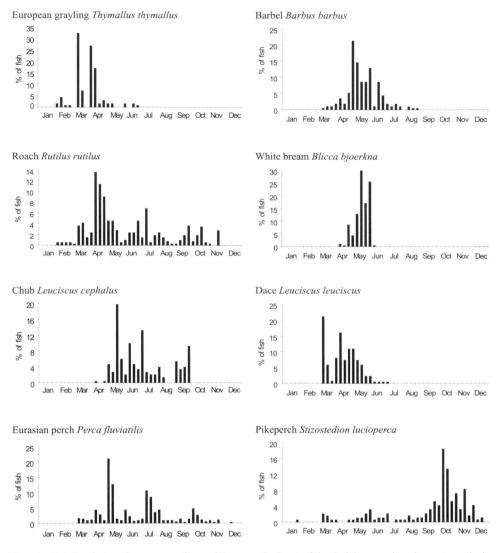

Fig. 4.2 Relative timing of occurrence of several European freshwater fishes in fish passes, based on meta-analysis of data from several western European rivers (see text for further details). It is suggested that seasonal peaks in occurrence are likely to be indicative of upstream-directed migration.

There is a peak in the occurrence of European grayling *Thymallus thymallus* in these fish passes in the early spring which precedes the normal spawning period of this species and is suggestive of a spawning migration, which has also been demonstrated by radio-tracking studies (Parkinson *et al.* 1999). Patterns of movement by adult barbel, dace and white bream, linked with spawning, have also been recorded in other studies of the spatial behaviour of these species (Baras 1992; Lucas & Frear 1997; Molls 1999; Lucas *et al.* 2000). Other species such as roach *Rutilus rutilus* and Eurasian perch *Perca fluviatilis* exhibit upstream-directed activity over a much wider period of the year than around the spawning season, while that of pikeperch *Stizostedion lucioperca* (in the current analysis, exclusively from the Dordogne River) occurs mostly in autumn (Fig. 4.2). These migratory movements are presumably for purposes other than spawning. Moreover a significant proportion of fishes entering and passing upstream through fishways, even during spawning periods, may be immature fishes (Lucas *et al.* 1999; Lucas 2000).

Welcomme (1979, 1985) has emphasised the importance of floodplain habitats and lateral connectivity to the lifecycles of many freshwater fish species in both temperate and tropical climates when seasonal inundation of land occurs. Despite this, remarkably few studies have examined the dynamics of fish behaviour in functional floodplain habitats within temperate regions. Kwak (1988) examined the movements of fish between river and two floodplain habitats on the Kankakee, a lowland river in Indiana, US. Fish were trapped moving between the river and an ephemeral ditch, and between the river and a backwater pool (oxbow) using one-way traps with 6-mm mesh set in the outlets of the ditch and oxbow. Fishes were only trapped during intermediate and reduced flooding during spring–summer, when fish movements between river and floodplain were spatially restricted to depressions in the floodplain that formed connections to the river channel (i.e. via ditch and pool outlet). The very low number of recaptures from fish moving out of ditch and pool showed that use of the floodplain was very seasonal – that fish were moving on or off the floodplain, but not shuttling back and forth on a shorter time scale as might occur with diel movements. An exponential relationship between discharge and CPUE during the flood decline indicated that most fish left as floodwaters started to recede. All fish left the ditch during this period, but at the oxbow some fish entered and remained there during summer. Twenty-five fish species were recorded; mostly grass pickerel *Esox americanus* (28.3% by number), green sunfish *Lepomis cyanellus* (16.0%), pirate perch *Aphredoderus sayanus* (12.7%) and orange-spotted sunfish *L. humilis* (10.6%). In all, 54.7% of fish were juveniles. It would appear that those fish which utilise floodplain habitat continue to seek favourable backwater habitat when forced off the floodplain. Fish often leave floodplain rivers in a distinct sequence (Welcomme 1979), with large fish leaving before small species and juveniles, but in Kwak's study no correlation between date and mean size, or species was found.

It is generally agreed that large backwaters and oxbow lakes function as winter refuges, spawning habitats for adults and nursery habitats for larvae and juveniles, but that most fish leave these areas during summertime. This general picture though, needs to be refined depending on whether connections between flowing and non-flowing environments are permanent. Lateral floodplains have been destroyed in many river systems, especially in Europe and North America, through drainage, channelisation and blockage, with much of the damage being complete in many catchments by the early twentieth century. More subtle, but just as severe, damage has been brought about by modifications of lateral connectivity. While investigating the functional connections between the River Rhine and its oxbows, Molls (1999) found that common bream *Abramis brama*, white bream *Blicca bjoerkna* and roach *Rutilus rutilus* all undertook spawning migrations from the River Rhine into oxbows, with their juveniles (0+ and 1+) migrating back to the main river at times when there still was a functional connection. Adult white bream and roach returned systematically to the river after spawning, whereas common bream emigrated from the oxbows, or remained there until the next spawning season. Molls suggested that this oxbow-resident behaviour in adult bream corresponded to an adaptive advantage in a context of temporally reduced lateral connectivity.

More recently, Tans (2000) showed that several predatory species (*Esox lucius*, *Perca fluviatilis*, *Stizostedion lucioperca*) also adopted such a resident behaviour in some backwaters of the River Meuse where many cyprinids, chiefly roach *Rutilus rutilus*, visited to spawn. Very low numbers of juvenile cyprinids and young predators were found to emigrate from these backwaters, probably because the inlets/outlets had been modified, and restricted to deep water, causing these habitats to become actual fish traps for juveniles, and supraoptimal habitats for predators. In these circumstances, year-round residence by piscivorous predators can be viewed as an adaptive behaviour too, as it minimises the energy expenditure for foraging, but it clearly suppresses the role of backwater habitats as nurseries. This provides another example of how a minor man-made change can affect migratory lifestyle, and population dynamics in a river.

Diadromous species in temperate regions usually enter fresh water when temperatures are intermediate between their annual maxima and minima, typically at times of seasonally elevated discharge. In most cases this is normally in spring and/or autumn. Elevated river discharge reduces the salinity of coastal water and provides navigational cues for identifying the river mouth (McKeown 1984). Elevated flows may aid negotiation of natural obstructions during upstream migration, while low temperatures occurring during winter reduce aerobically fuelled and anaerobically fuelled swimming performance. Two main strategies are available to adult anadromous fish at temperate latitudes. They may migrate upriver and spawn during the same period of seasonally elevated flow in autumn or spring or they may migrate during one period of elevated flow and remain in the river until their spawning season (see section 2.3.6). The strategy adopted partly appears to depend on the migration distance and the geomorphology of rivers in which migration occurs.

Migration by adult Atlantic salmon *Salmo salar* into short coastal rivers, with few deep resting areas, normally occurs in autumn, a short time prior to spawning (Mills 1989; Thorstad & Heggberget 1998). In larger rivers, with abundant deep pools, there is often a migratory stock component of multi-seawinter Atlantic salmon which enters the river in late winter or early spring, and moves upstream in several distinct phases typically involving a period of limited movement during summer, before spawning in the upper reaches of the river in autumn (Hawkins & Smith 1986; Shearer 1992). Many rivers contain a combination of different stock components of Atlantic salmon, with variations evident in the timing and extent of migration and the final spawning positions (Laughton & Smith 1992; Shearer 1992). Variations in patterns of migratory behaviour within and between populations are found in temperate climates for a substantial number of spring-spawning fishes such as sturgeon (Box 4.1). Nikolsky (1961) described the occurrence of both autumn/winter and spring return migrations of Atlantic sea lamprey *Petromyzon marinus* in Russia, with spring-run lampreys having more mature gonads than those arriving in late autumn. Those entering the river in autumn tend to remain in the lower reaches and complete their migration in spring.

Migration timing and life-history strategy, for example the degree of iteroparity, may vary with latitude and environmental conditions for widely distributed species such as the American shad *Alosa sapidissima* (Glebe & Leggett 1981a, b; Leonard & McCormick 1999) and brown trout *Salmo trutta* (Jonsson & L'Abée-Lund 1993), and were considered in detail in section 2.3.5. Climate change is thought likely to affect the habitat availability, migration patterns and timing for a range of species with wide latitudinal ranges such as striped bass *Morone saxatilis* (Coutant 1985, 1990a) and effects of environmental changes over several

Box 4.1 Variations on a theme – spawning migrations of sturgeons

All sturgeons spawn in fresh water, but display a range of migratory behaviour which varies between species and populations. The shortnose sturgeon *Acipenser brevirostrum*, a predominantly freshwater species from eastern North America, is one of the smaller sturgeon species (reaching 'only' about 1 m in length) and tends not to migrate long distances (Kynard 1997). Shortnose sturgeon often migrate upriver in autumn to near the spawning site and spawn in the following spring. In the Connecticut River, there is a 'short' spring migration of 23–24 km from overwintering areas to a presumed spawning site or a longer autumn migration of 77–81 km with overwintering at the spawning site (Buckley & Kynard 1985). Similar behaviour has been observed in the Merrimack River, Massachusetts, where spring spawning migrations were 9–19 km upstream of the overwintering area (Kieffer & Kynard 1993). Annual migration and spawning has been recorded in the Merrimack and Connecticut Rivers, in both cases characterised by short-distance migrations (Kieffer & Kynard 1996), while in the Saint John River, New Brunswick, Canada, where longer distance migrations occur, spawning is biennial (Dadswell 1979). Kieffer and Kynard (1993) suggest that male shortnose sturgeons in populations that must make long migratory journeys do not have the energy resources for annual gonad production and a long migration. They suggest that the timing of spawning migration is related to spawning distance, with an autumn migration for distances greater than 50 km and a spring migration for distances less than 25 km. Due to egg production, the energy investments by females are of course likely to be much greater than those of males.

A similar lifestyle has been reported for the Atlantic sturgeon *Acipenser oxyrinchus*, which has one of the southernmost distributions of sturgeons worldwide, as this species is encountered in rivers flowing into the Gulf of Mexico. Fox *et al.* (2000), who radio-tracked Atlantic sturgeon in the Choctawhatchee River system (Alabama–Florida) found that ripe fish, both males and females, entered the river as early as March–April, and migrated 100–150 km upriver to spawn on hard-bottom areas. Atlantic sturgeon eggs are adhesive, and they require a hard-substratum on which to be laid. Fox *et al.* (2000) further argued that greater predation hazards and lesser water quality in the lower reaches also were driving forces behind the rather long upstream migration of these sturgeons. Early spring migration was deemed to enable spawners to start reproducing at times of the year when water temperature can optimise their reproductive effort (15–20°C). Non-ripe fish also made upstream migrations, but these started over a much more protracted period (March–September), and they migrated further upstream. Atlantic sturgeon are known to require prolonged periods of time between spawning events, as long as 3–5 years in females, and 1–5 years in males (Smith 1985). For Atlantic sturgeon spawning in the northern Hudson River, New York, there is now evidence that adults 'reside' in marine waters for 3–5 years between spawning events (Bain 1997). In the Choctawhatchee River system, females do not spawn on an annual basis, but males seemingly do (Fox *et al.* 2000).

decades have been observed for populations of American shad and sockeye salmon *Oncorhynchus nerka* in a single river system (Quinn & Adams 1996).

4.4 Tropical regions

4.4.1 Introduction

Tropical regions host more than 7000 freshwater fish species (Nelson 1994). For example, the Amazon River basin alone hosts more than 1300 freshwater fish species, and the Congo River basin about 800 species, exceeding by far the 450 freshwater species recorded over the entire Palaearctic region. Some species, such as the cyprinid *Tor putitora* in the Brahmaputra River, or the osteoglossid *Arapaima gigas* in the Amazon River, can exceed 3 m in length, whereas other species, such as the Burmese cyprinid *Danionella translucida*, reach a maximum size of less than 20 mm. In spite of this diversity and richness, the movements of tropical freshwater fishes have been far less investigated than those of their European and North American counterparts, and knowledge of numerous tropical freshwater ecosystems and fish communities is limited to information derived from fishermen's experience and catches.

Environmental harshness, ecosystem size, temporally and spatially limited access to sampling sites, richness of fish communities, and diversity of lifestyles in tropical fish communities are partly responsible for this fragmentary information. Also, the scientific priorities have been and still mostly are to determine and map the diversity of fish communities, and to define other key parameters of fish species ecology, such as maximum size, age at maturity and fecundity. The proportionally greater importance of fisheries in local economies (fish food consumption amounts up to 500 g per person per day in the Central Amazon) paradoxically has contributed to this lack of information, as solutions to climatic or anthropogenic changes often have had to be implemented from an economic, rather than ecological, perspective. To a degree, political instability in some regions, and the more recent development of science relative to northern countries, has also limited the acquisition of detailed knowledge on freshwater fish behaviour in tropical regions. For example, telemetry techniques have been applied in no more than in a dozen studies of freshwater fishes in Africa, Latin America and Southeast Asia. Finally, language has often been an obstacle to the dissemination of information, as English is not always the most widespread language in tropical areas.

Entering into a detailed review of the migratory patterns of tropical fishes within families or genera (chapter 5) would have resulted in an unbalanced presentation, laying an exaggerated emphasis on those species which have received some attention, and for which information is accurate enough to be given in such a review. This could result in some misinterpretation, notably for river management, as species not listed in the present chapter nor in the corresponding sections of chapter 5 could just be viewed as non-migratory, and thus be deemed to deserve no attention in river management schemes and mitigation of obstacles to migration that have become increasingly abundant in tropical regions. We do not want to repeat the same mistakes that arose from the confusion brought about by the definitions of 'migratory' and 'non-migratory' as applied to fishes in fresh water. Instead, we prefer to present a comprehensive review of the processes governing diel and seasonal migrations in tropical ecosystems, chiefly by analysing the nature and extent of environmental constraints, and how these interact with the species' biological characteristics. When no information on migratory behaviour is at hand for a particular species, this approach may give the reader the opportunity to take advantage of existing information on the species' lifestyle, morphology and physiology and determine the likelihood that this species has strong migratory or sedentary habits.

4.4.2 Setting the scene: what makes tropical freshwater systems similar or different to those of temperate areas?

Most freshwater tropical ecosystems throughout the world share many common traits, which make it possible to deal with fish migration in tropical regions (almost) as a whole. Among these is the biogeographical distribution of fish families that are typical of tropical areas, as a consequence of a common origin in the former Pangea or Gondwana. For example, cichlids and nandids are found in Africa, Southeast Asia and South America. Dipnoans and osteoglossids occur on these three continents, and also in Australia. Siluriforms are spread throughout the temperate or tropical areas of all continents except Australasia. Among these, there is a strong overlap between the African and Southeast Asian faunas, notably for bagrid, clariid and schilbeid catfishes, in contrast to the South American fauna. Conversely, characins, which are more primitive members of the Ostariophysi, are found both in Africa and South America (and no longer in Asia). Except for recent introductions (e.g. cichlids), most similarities or differences between tropical freshwater fish faunas can be accounted for by continental drift, volcanism and the timings of their appearance over the past 200 million years (see also review by Lévêque 1999). Continental drift was at the origin of notable climatic changes in areas presently located in the intertropical region, but these were proportionally less severe with respect to water temperature than in present temperate ecosystems, where long glaciations caused major restructuring of the fish fauna.

As in temperate regions, the diversity of freshwater fishes per unit area is proportionally greater than that of their marine counterparts. Also, the vast majority of freshwater fishes are riverine in habit, essentially as rivers are old and lakes are quite young. According to Fernando and Holcik (1982, 1989), only five of 160 fish families living in fresh water are exclusively lacustrine, and another 12 families have lacustrine adapted members. However, some old lakes of the Australasian area have no truly lacustrine species. Lacustrine ecosystems hosting truly lacustrine species include Lake Titicaca in Latin America (about 20 species), Lake Lanao in Southeast Asia (about 20 species) and the Rift Valley lakes in East Africa (about 700 species). However, lacustrine fish communities are generally dominated by secondary freshwater species, essentially clupeids and centropomids in the pelagic zone, and cichlids and cyprinodontids in the littoral zone (Welcomme 1979; Lowe-McConnell 1987). Many other species enter lakes at various stages of the lifecycle, but return to rivers to breed. Dominey (1984) suggested that the failure of non-cichlid fishes to adapt thoroughly to lacustrine conditions in Africa originated from their need to find an appropriate breeding site, whereas cichlids became independent from it, essentially through mouthbrooding behaviour. Some species like *Synodontis multipunctatus*, a small catfish endemic to Lake Tanganyika, somehow bypassed the problem of adapting to a lacustrine reproductive style through cuckooing, as it lays its eggs at the time when mouthbrooding cichlids are engaged in courtship. The eggs of the 'cuckoo catfish' are then incubated by the cichlids, and its larvae prey on cichlid embryos inside the parent's mouth (Sato 1986). Cuckooing is also frequent among cichlids (Ribbink 1977), despite the fact that most of these species are truly adapted to a lacustrine lifestyle.

Proportionally, there are more freshwater clupeids in tropical lacustrine ecosystems (about 20 species in Africa alone; Marshall 1984) than in comparable temperate regions (e.g. four species only in Canada; Scott & Crossman 1973). However, in view of the considerable diversity of the tropical fish fauna, there is a general paucity of truly pelagic species in lacustrine ecosystems (Fernando 1994), and even so-called pelagic species frequently forage in littoral areas (e.g. *Limnothrissa miodon* in Lake Kivu: Iongh *et al.* 1983; *Clupeichthys aesarnensis* in Thailand's reservoirs: Sirimongkonthaworn 1992, in Fernando 1994). Reasons for this paucity can be found in the relatively recent origin of most lakes, and in the considerable changes undergone during their recent history, mostly as a consequence of volcanic activity. Huge variations of water level caused the truly lacustrine species to collapse, as they could not accomplish their lifecycle in the lake tributaries, by contrast to the coastal species, which later recolonised the recreated lake from these preserved environments.

Food availability is a key difference between tropical and temperate ecosystems. In tropical lakes, zooplankton comprises more rotifers and less crustaceans than in temperate lakes, and tropical fish have more frequently adapted to an herbivorous, omnivorous or detritivorous lifestyle (Nilssen 1984; Winemiller 1991). The proportionally lower diversity of truly planktonic invertebrates in the tropics is compensated for by a greater diversity of aquatic insects, and these are deemed to provide a large proportion of food for fishes in lakes (Lowe-McConnell 1987). There is also greater niche specialisation among invertebrate feeders and piscivores in the tropics than in temperate areas, and some feeding niches (e.g. scale feeding by the characid *Roeboides dayi*; blood feeding by trychomycterid catfishes) are found in the tropics only. Tropical freshwaters generally have a higher year-round productivity than temperate rivers (Welcomme 1979, 1985). According to Gross *et al.* (1988), the environmental conditions in tropical freshwater which provide fish with an abundant food supply may have favoured the greater occurrence of catadromous life histories in the tropics (e.g. Anguillidae, Mugilidae, milkfish *Chanos chanos*), whereas anadromy prevails in temperate and high latitude areas, where fish leave poorly productive rivers and streams to exploit the richer marine environment.

As a bridge to some terminology frequently used to describe tropical environments, especially as regards Amazon tributaries, those rivers that are rich in nutrients are known as 'whitewater' rivers. These tributaries (e.g. the Ucayali) and the upper Amazon, which drain the Andes, have a near-neutral pH, and carry much sediment which restricts the primary production due to increased turbidity, but calm areas host abundant vegetation and floating meadows. By contrast, 'blackwater' tributaries (e.g. the Rio Negro in Brazil) are acid (pH <5.0), brown-stained rivers, as a result of humic acids being leached from flooded vegetation. They generally are poor in nutrients, as are the so-called 'clearwater' rivers such as the Tocantins or Xingu rivers. 'Rainforest' rivers are slow meandering permanent rivers crossing the forest, not to be confused with the flooded forests, named '*várzea*' or '*igapo*' depending on whether they are flooded by whitewater or blackwater rivers, respectively. Floodplain lakes which are seasonally flooded during high water periods are generally named *várzea* lakes, whereas those places filled mainly by rain water are called '*terra firma*' (all definitions after Lowe-McConnell 1991).

Temperature is a key difference between temperate and tropical ecosystems. In the former, seasonal variations are huge, and daily variations are relatively small, whereas in the latter, the amplitude of daily variations can exceed seasonal differences in average daily temperatures. Also daylength shows little or no seasonal variation in tropical regions. Implicit in these traits is that temperature and daylength, which are the two major stimuli for seasonal migrations in temperate regions, most probably have a lesser importance in the tropics. Implicit too is that the influence of temperature is often greater at the daily than at the seasonal scale. However, the relative stability of temperature and daylength over the year does not imply the absence of seasonality, nor does the relative stability of temperature during the history of intertropical regions imply that the environment has not changed. Historical changes principally involved droughts, variations in water level and probably in oxygen content, which caused tropical fish species to evolve in response to these environmental constraints, which still prevail in present tropical environments.

Tropical regions undergo more intense and seasonal rainfalls than temperate regions, and these produce seasonal patterns of river discharge and habitat availability. For example, the mean variation of water level in the Amazon River is approximately 10 m per year, and variations as high as 15 m per year have been recorded (Junk 1983). Flood pulses reduce the availability of periphyton and benthic food in small tropical streams, which become decreasingly attractive during this season (Pringle & Hamazaki 1997). Conversely, the floodplain becomes increasingly attractive, as organic and inorganic nutrients are leached into floodwaters within a greatly expanded habitat, and increase the primary and secondary production (e.g. Camargo & Esteves 1995). Extended wet periods are further characterised by slightly lower water temperature and conductivity, greater oxygen concentration and water clarity, thus increasing predation risks from visual predators (see sections 4.4.3–4.4.5). Receding waters are accompanied by lower availability of many aquatic habitats, greater conductivity, higher temperature, and a reduction of oxygen concentration, which begins as soon as flooded terrestrial vegetation starts decaying. Deteriorating habitat quality and availability tend to result in increased fish density and biomass per unit area of water surface, increased competition between and within species, as well as enhanced risks of predation by piscivorous fishes and birds. Receding waters also make the environment more unpredictable. There are obvious risks of being stranded in pools, and many fish perish in pools that dry completely (Winemiller & Jepsen 1998) (Fig. 4.3), or following algal blooms in temporarily hypoxic or anoxic low waters (Fernandes 1997).

Forested watersheds such as the Amazonian *várzeas* are the most frequently cited examples of these marked seasonal variations, but these also apply, to a lesser extent, to about 40 000 km^2 of papyrus swamps in Eastern Africa and about 1.4 million km^2 of rice fields throughout tropical regions. Even within a restricted geographical area, there is mosaic of habitats that approximately fit the requirements or preferences of specialised fishes. This extreme diversity makes lateral migrations, onto and from the floodplain, more important in the tropics than anywhere else. The seasonality of flooding also implies that connections between environments are not permanent, and that lateral migrations have a strongly marked seasonality.

Temperate ecosystems also undergo seasonal variations of water levels, but these are of much lower amplitude, and drying-out is far less frequent, except for Mediterranean ecosystems and their counterparts. Similarly, variations of oxygen concentration in temperate

Fig. 4.3 Clariid catfishes stranded in a ditch in Lake Katavi, Tanzania during the dry season. Although catfishes can make terrestrial movements, the steepness of the banks prevents them from doing so here, and they survive through aerial respiration. ©Alan & Joan Root/Oxford Scientific Films.

ecosystems rarely approach those observed in the tropics. Winter is often associated with the so-called harsh and critical period in temperate environments, but water temperature is much lower and fish metabolism is substantially reduced at this time of the year (Northcote 1978). By contrast, in the tropics, the dry period combines with high temperatures, and rather high metabolic needs.

Therefore, tropical riverine environments are generally much more directive than their temperate counterparts in shaping migration patterns among fish assemblages, at the daily, seasonal and lifecycle levels. Central to this overview of migration of tropical freshwater fishes is that such behaviour incorporates many more key ontogenetic and trophic components than in temperate ecosystems. Implicit too is the idea that behavioural and physiological adaptations of tropical fish that have been selected for during the evolution of these species, are shaped as a response to the regularity of seasonal hydrological features, and in relation to food web characteristics (Winemiller & Jepsen 1998). As a corollary, tropical fish communities probably are more sensitive to changes of climatic or anthropogenic nature than their temperate counterparts. This sensitivity makes the study of their biology in general, and migration in particular, a true priority in view of recent changes brought about by the El Niño phenomenon and man-made impoundments for hydroelectric purposes as well as more general degradation of freshwater habitat (Dudgeon 1992).

4.4.3 Influences of predation pressure on fish migrations

As mentioned above, the widespread specialisation of feeding regimes in tropical freshwater fish communities has included the emergence of various top piscivores. As early as 1961, Jackson (in Winemiller & Jepsen 1998) hypothesised that the life-history patterns, and particularly the migration patterns of many African river fishes, represented ecological or evolutionary responses to the threat of predation by tigerfishes *Hydrocynus* spp., the adults of which are mainly restricted to main channels (Daget 1954; Welcomme 1985). A similar statement was proposed by Goulding (1980) for South American assemblages. Winemiller and Jepsen (1998) provided further evidence for Jackson's assumption, as they noted that typical primary channel-dwelling species in Africa were either large (clariid catfishes), possessed anti-predation morphological adaptations (e.g. spines of the mochokid catfish *Synodontis* spp.), or were confined to shallow (e.g. the characid *Micralestes acutidens*) or densely vegetated areas (e.g. the mormyrid *Petrocephalus catostoma*; the anabantid *Ctenopoma intermedium*). Even typical piscivores such as the African pike *Hepsetus odoe*, and the silver catfish *Schilbe intermedius*, were reported to be preyed upon by tigerfishes (Winemiller & Kelso-Winemiller 1994), and this is presumably why they were confined to lagoon habitats during the dry season.

Hence, predation by tigerfishes in Africa, and probably by other top predators in other tropical regions (e.g. erythinids such as *Hoplias* spp. and pimelodid catfishes such as *Brachyplatystoma* and *Pseudoplatystoma* in South American assemblages), may be a sufficient driving force to shape migrations of tropical fishes, chiefly in terms of lateral migrations onto the floodplain, or to shallow tributaries or turbid waters. Analyses of tropical fish communities have revealed substantial proportions of piscivores, representing up to 35–40% of fish biomass, at the entrance of secondary channels and lagoon mouths (Barthem 1981; Bayley 1983, 1988), further supporting the idea that predation plays a key role in shaping tropical fish communities. Conversely, any ecosystem with such a high proportion of piscivores would collapse without a substantial immigration component of prey species, and this supports the idea that most tropical fish species, or at least the most abundant ones, have marked migratory habits.

Some piscivores such as the South American erythrinids *Hoplias* spp., are ambush predators that hide in vegetated areas at mouths of lagoons and creeks, where they can tolerate extremely low oxygen concentrations thanks to their highly vascularised swim bladder, large gill surface (Fernandes *et al.* 1994) and capacity to sustain high levels of glycolysis. However, this mostly applies to juveniles, as adults prefer pool habitats (e.g. *Hoplias malabaricus* in the Orinoco River; Winemiller 1990). Large pimelodids such as *Pseudoplatystoma fasciatum* and *P. tigrinum* occur less frequently in structured habitats (Reid 1983), and essentially inhabit the main channels of South American rivers, where they may undertake trophic migrations, in pursuit of migrating shoals of curimatids (e.g. *Prochilodus mariae* in the Rio Apure, Venezuela; Reid 1983). These migrations, which involve only a part of the population, differ fundamentally from the amazingly long spawning or feeding migrations that characterise these species (see sections 4.4.6 and 5.16). Some other piscivores also make rare restricted migrations outside of the main channel (e.g. tigerfishes *Hydrocynus brevis* and *H. forskalii*; Loubens 1973). Main channels, and especially pools, are therefore risky places in tropical freshwater systems, and are generally avoided by non-predatory fishes. In streams, the high water velocity in shallow riffles and runs may provide protection against predation by piscivorous fishes (Power 1983, 1984). However these are less frequently inhabited by tropical fishes all year round than in temperate stream ecosystems (e.g. Martin-Smith & Laird 1998), as morphological adaptations are frequently needed to counter water currents and minimise energetic investment in swimming. Such traits include small body size (e.g. characids *Creagrutus* spp. in tributaries of the Magdalena River, Colombia; Baras *et al.* 1997), benthic lifestyle of fish hiding or resting between cobbles (e.g. loricariids in South American rivers), expanded lateral fins (e.g. balitorids such as *Parhomaloptera* and *Protomyzon* spp., and the cyprinid *Garra borneensis* in Southeast Asia), thoracic adhesive organs (e.g. the sisorid *Glyptothorax major*) or suction discs formed by fusion of pelvic fins in gobioids and some other species (e.g. homalopterids such as *Gastromyzon* spp.; Roberts 1982). Most other species inhabiting lotic habitats apparently need to forage at riffle–pool interfaces, where they are frequently preyed on by piscivores (e.g. *Hoplias malabaricus* in Venezuela; Winemiller 1990; *Hampala sabana* in Malaysia, Martin-Smith & Laird 1998). As in most aquatic ecosystems, shallow edges may provide a size-limiting refuge against piscivorous fishes, but they increase risks of predation by avian piscivores (e.g. herons and egrets).

Hence, the lateral migration into secondary channels and pools on the floodplain under rising waters meets both trophic and refuge criteria. Except during the early wet season, most main channel predators rarely venture into these habitats, as their predatory behaviour relies on vision and their oxygen requirements are generally higher than those of their prey (Daget 1954; Welcomme 1979; Junk *et al.* 1983; Winemiller & Kelso-Winemiller 1994). Floodplains in general, and flooded forests in particular, have fewer predators of young fishes than rivers, but some are still well adapted to hypoxic environments. For example, Winemiller and Jepsen (1998) reported intense ambush predation by the erythrinid *Hoplias malabaricus* and the cichlid *Caquetaia kraussii* that hide in vegetated areas, and open-water predation by the piranha *Pygocentrus cariba* in Caño Maraca, Venezuela. They also highlighted that vegetation-dwelling species were less frequently preyed on by these piscivores, even by those employing ambush from vegetated areas. This further supports the role of vegetation as a refuge against predation, and emphasises the role of oxygen tolerance in determining the capacity to escape this threat (see section 4.4.4).

Sooner or later, receding waters and decreasing oxygen levels in floodplain waters during the dry season force most fish to move into the main river channel. Even the vast majority of hypoxia-tolerant species must return to the main channel at some stage of their life as they require access to riverine environments for spawning (Welcomme 1985; Lowe-McConnell 1987; Fernando 1994). As they return to the river, fish move through a series of habitats (Johnson *et al.* 1995) associated with rather characteristic predatory pressures. For example, in the Zambezi River floodplain, the African pike *Hepsetus odoe* and the silver catfish *Schilbe intermedius* prey on floodplain residents, and on migrants in secondary and primary channels. In swamps, shoals of clariid catfishes (*Clarias gariepinus* and *C. ngamensis*) exert intense predation pressure on small or young fishes along the vegetated shoreline (Merron 1993). Juveniles of the tigerfish *Hydrocynus vittatus* prey in side channels, whereas adults essentially forage in the main Zambezi River channel (Winemiller & Kelso-Winemiller 1994; Winemiller & Jepsen 1998). Similar niches in South American ecosystems are occupied by the piranhas *Pygocentrus* (side channels) and *Serrasalmus* spp. (main channel), erythrinids *Hoplias* spp. and pimelodids *Brachyplatystoma* or *Pseudoplatystoma* spp. (main channel) (Goulding 1980; Reid 1983; Barthem & Goulding 1997; Winemiller & Jepsen 1998).

Not all species leave or enter the floodplain at the same time, depending on their tolerance towards low oxygen levels, feeding niche (Fernandes & Mérona 1988) or opportunity of finding a drifting shelter (Henderson & Hamilton 1995; Box 4.2). This enables piscivores to intercept prey fish almost all year round, as they move onto or from the floodplain. Implicit in this is that many piscivorous fishes do not need to undertake long trophic migrations, as prey abound at short range almost all year round. Implicit too is that most predators do not need to make extensive spawning migrations, provided that appropriate spawning substratum and

Box 4.2 Escaping predation at confluences... take a taxi!

During the high-water season, lakes within the floodplain of inundation forests (locally called *várzea* in Brazil) are connected to the main channel through flowing channels of variable length and width, depending on floodplain geomorphology and forest. Lakes frequently support numerous aquatic plants (chiefly *Paspalum repens* and *Echinochloa polystachya*) that form floating mats with submerged rhizomes and roots. These produce extremely high quality shelters and feeding areas for small fish species and larvae or juveniles of larger species (Goulding & Carvalho 1982). Considering that food is abundant in open waters, shelter is probably the primary function of these habitats. Many of the infaunal species inhabiting these floating meadows are reluctant to enter open water or flooded forest, probably because of predation risk, and these are ideal conditions for genetic isolation and speciation.

However, wind or flow can break such 'meadows' into pieces, sometimes as large as a hectare, and force them into the main channel. Anchored meadows are usually inhabited by juveniles of many fish species (anostomids, cichlids, erythrinids, synbranchids, and siluriforms), including piranhas. Most fish species usually leave meadows soon after they have become detached from the main body, thus avoiding being swept out of the lake (Henderson & Hamilton 1995). This is most understandable for gymnotiforms (knifefishes) or synbranchids (swamp eels) which are highly specialised fishes, which are particularly well adapted to exploit *várzea* environments irrespectively of their intrinsic drawbacks (low dissolved oxygen, restricted space), but poorly adapted to openwater environments. Erythrinids and serrasalmines can also leave *várzea* lakes by migrating through the forest, and cichlids are adapted to a truly lacustrine lifestyle. By contrast, Henderson and Hamilton (1995) found siluriforms in greater numbers in drifting than in anchored meadows. This suggests that catfishes, and possibly other species, moving from the floodplain to the main river channel may take advantage of these drifting shelters, especially for crossing risky areas such as river confluences, where the biomass of predators is extremely high (Barthem 1981; Bayley 1988).

nursery habitat is available at close range. Although there are exceptions, such as the long-range migrations of piscivorous pimelodids in South America, most predatory fishes exhibit very restricted migratory behaviour (e.g. Nile perch *Lates niloticus*, Loubens 1974; the tigerfishes *Hydrocynus brevis* and *H. forskalii*, Loubens 1973; the sciaenid *Plagioscion squamosissimus*, Agostinho *et al.* 1994; snakehead *Channa striata*, Halls *et al.* 1998; erythrinids such as *Hoplias* spp., de Vazzoler & Menezes 1992; piranhas such as *Pygocentrus* and *Serrasalmus* spp., Myers 1972; Sazima & Zamprogno 1985; and the cichlid *Cichla temensis*, Winemiller *et al.* 1997). Detritivores migrating onto the floodplain as larvae, then from the floodplain as juveniles, permit the efficient transport of energy and production from the eutrophic floodplain to oligotrophic tributaries (Winemiller & Jepsen 1998), where intense piscivory by resident predatory fishes may contribute to stabilise it in the longer term. An additional type of energy transfer occurs in frugivorous species, as seeds may pass through the fish gut intact and viable, and are dispersed upstream by moving fish (e.g. characiforms *Brycon* and *Colossoma* spp.; Araujo-Lima & Goulding 1997; Horn 1997).

4.4.4 Influence of dissolved oxygen on fish migrations

Tropical regions are characterised by marked seasonal cycles of oxygen concentration, as richly oxygenated waters in the early wet season become progressively more hypoxic through the late wet season and dry season, under conditions of receding water and increased turbidity that restrict photosynthesis (Kramer *et al.* 1978; Welcomme 1979; Junk *et al.* 1983). Daily variations in oxygen concentration superimpose on this general pattern, making some environments particularly unfavourable. For example, *várzeas* have oxygen concentration reaching daytime maximal values of only 0.5 mg per litre, and become almost anoxic at night (Bayley 1983; Junk *et al.* 1983). Also, most deep-water layers in *várzeas* are anoxic day and night (Saint-Paul & Soares 1987). These conditions also prevail in vegetated areas where fish most frequently hide from avian and piscine predators, such as papyrus *Cyperus papyrus* swamps in Eastern and Central Africa (Beadle 1981).

Anoxic or severely hypoxic waters can play a similar role to physical obstacles and absence of water in creating a barrier to the movements of fishes (Kramer 1983a, 1987; Saint-Paul & Soares 1987). Conversely, these adverse conditions can provide temporary or permanent shelters from predation to species or life stages that have adapted to low oxygen concentration (Chapman & Liem 1995), or act as barriers preventing oxyphilic predators from accessing more favourable habitats, such as marginal pools or lagoons (Chapman *et al.* 1995; Kaufman *et al.* 1997). Probably the best example for the predation refugium hypothesis is provided by studies on habitat use by native fish species in lakes where the Nile perch *Lates niloticus* was introduced. The Nile perch is relatively intolerant to oxygen concentrations lower than 5 mg per litre, as its blood has a low affinity for oxygen (Fish 1956). Marginal wetland ecotones or valley swamps create a gradient of habitats with decreasing oxygen content that limit the dispersal of Nile perch and its access to satellite lagoons or lakes. Surveys in these 'protected' habitats around Lake Nabugabo (Uganda), where the papyrus swamp is particularly dense, revealed that they harboured many fish species reported as extinct or rare in previous surveys in the lakes, but all of these species showed adaptations to oxygen deficiency (Chapman *et al.* 1996).

Any thorough understanding of how oxygen concentrations shape daily and seasonal migrations of tropical freshwater fish species intimately relies on knowledge of their capacities to adapt to hypoxia. During their evolution, fish species have (almost) independently adopted a wide variety of morphological, physiological and behavioural adaptations for coping with deficiency in oxygen, and these are proportionally more abundant in the tropics than anywhere else (Junk *et al.* 1983; Kramer & McClure 1982; Kramer 1983a). A case of particular interest is provided by the gymnotiforms, as oxygen tolerance and residency may have evolved from the diversification of electric signals (Crampton 1998; Box 4.3).

Box 4.3   Oxygen-related habitat selection and migration in gymnotiform electric fishes

Possibly because *Electrophorus electricus* is an obligate air-breather, it had been incorrectly assumed that most gymnotiform fishes were tolerant to hypoxia or anoxia. While investigating the oxygen tolerance of 64 gymnotiform species from the Upper Amazon basin, Crampton (1998) found that 9 species were endemic to well-oxygenated, low-conductivity habitats, as they could not tolerate anoxic or hypoxic conditions. Another group of 38 species, exhibiting a benthic lifestyle and essentially found in deep whitewaters, were strict water breathers, with relatively high oxygen demands. They only entered the floodplain under conditions of rising, well-oxygenated water to breed and feed, returned to the river as waters receded, and avoided crossing downward gradients in oxygen concentration. Only 17 species inhabited poorly oxygenated habitats of the *várzea* floodplain, and were *várzea* 'resident' species. Yet, not all species of the oxygen-resistant group have adapted to air breathing (e.g. the hypopomid *Brachyhypopomus* spp., the rhamphichthyid *Rhamphichthys* sp., the eigenmanniid *Eigenmannia* gr. *virescens*), and they turn to aquatic surface respiration (ASR) under hypoxic conditions, just like the species restricted to better-oxygenated waters. Crampton (1998) postulated that this separation between resident oxygen-tolerant and migratory oxyphilic species was secondary to their mode of electric discharge, which had been the true driving force behind habitat selection, as depicted below.

All gymnotiforms generate dipole-like electric fields through electric organ discharges (EODs), these being used for electrolocation, object detection and/or electrocommunication. Electric organ discharges commence soon after hatching, and continue day and night over the entire duration of the fish's life. Even if the voltage of EOD is weak, its continuous nature implies a rather high energy investment which can be affected by fish metabolism and oxygen concentration. Two distinct types of EODs have been identified in gymnotiforms: discrete pulses separated by silences (pulse-type EOD in the families Electrophoridae, Gymnotidae, Hypopomidae and Rhamphichthyidae), and continuous wave-type EOD (tone-type EOD in the families Apteronotidae, Eigannmannidae and Sternopygidae). Tone-type signals give greater resolution of information about the substratum, object or fish examined (Heiligenberg 1991). However, this strategy requires the fish to alternate forward and backward

> movements almost all day long (Lannoo & Lannoo 1993), and this is more energy-demanding than the alternation of foraging and rest in pulse-type species. Also, tone-type species generally show higher repetition rates of EODs than pulse-type species, and they are unable to substantially reduce the rate or the amplitude of their EODs, whereas some pulse-type species seem to be able to accomplish this. Hence there is little doubt that tone-type EOD is more energy-demanding than pulse-type EOD, for reasons relating to the generation of the electric signal and/or to the associated behaviour patterns. Interestingly, Crampton (1998) found that among the 17 *várzea* resident gymnotiform species, there was a single tone-type species, whereas 72% of gymnotiforms in well-oxygenated habitats were tone-type species. Relying on the arguments above, he proposed that the evolution of EOD from pulse to tone type confined tone-type species to well-oxygenated habitats, where they evolved means of avoiding low oxygen environments. Conversely, it can be hypothesised that the competition between tone-type and pulse-type gymnotiforms promoted the migration of the latter group on the floodplain, and its adaptation to protracted hypoxia.

When the oxygen content of the water declines, almost all fish species increase their ventilation frequency to cope with this deficit, although this is an energetically expensive and merely temporary solution. Kramer (1987) estimated that up to 10% of the metabolic rate supports gill ventilation, even in water saturated with oxygen, and this contribution soars under hypoxia. For example, Saint-Paul & Soares (1987) documented increases in ventilation rate of up to 600% in *Schizodon fasciatus* when oxygen concentration dropped from 8 to 1 mg O_2 per litre, and rates as high as 270 opercular beats min^{-1} were observed in *Piaractus brachypomus* (Baras [unpubl.]). Below a certain threshold (usually between 0.5 and 1.5 mg O_2 per litre for tropical species), most fish move to the surface of the water and exhibit so-called aquatic surface respiration (ASR), by which they exploit the oxygen-rich surface layer of the water (Lewis 1970; Gee *et al.* 1978; Kramer & Mehegan 1981; Kramer & McClure 1982). Kramer and McClure (1982) demonstrated the capacity for this behaviour in about 94% of the tropical aquarium fishes they tested versus about 85% in fish species from temperate areas (Gee *et al.* 1978). When the oxygen concentration in the water approaches zero, fishes spend virtually all their time at the surface (Kramer 1987). Continuous swimming at the air–water interface while employing ASR can improve its efficiency, as more oxygenated water passes across the gills (so-called ram-assisted ASR; Chapman *et al.*, 1994). The shift to ASR is almost always accompanied by a substantial decrease in ventilation rate (Kramer 1983b; Saint-Paul & Soares 1987), and this reduces the cost of respiration, and enables fish to maintain feeding and growth under hypoxic conditions (Weber & Kramer 1983). However, implicit in ASR is that fish need to move to the surface layer and swim in it, resulting in extra energy costs, and increased predation risks, since ASR can only be performed in relatively open environments. Sonar observation in *várzeas* indeed showed there was a measurable increase in the fish density in open waters at night, when an oxygen deficit occurs (Saint-Paul & Soares 1987).

While the vast majority of tropical species can effectively use ASR to cope with temporary oxygen deficiency, the threshold below which they turn to ASR varies substantially between species, reflecting their variable physiological or morphological adaptations to hypoxia. Among these is the relative gill surface, which substantially varies between lifestyles and ontogenetic stages (Pauly 1981; Hughes 1984; Severi *et al.* 1997), and determines the cost for respiration and the threshold below which ASR should be employed. Morphological adaptations also contribute to increase the efficiency of ASR (Kramer 1983a), such as small body size, flattened head, upturned mouth or vascularised extensions of the lips (e.g. *Brycon*, *Colossoma* and *Osteoglossum* spp.; Braum & Junk 1982; Saint-Paul & Bernardino 1988). Hence, there is little doubt that diel migration between macrophyte shelters, where fish hide from predation during the day, and open waters, where they can utilise ASR, is a common trend in tropical ecosystems, but its occurrence and timing are intimately dependent on the fish's intrinsic capacity to cope with oxygen deficiency. Some fish which extract their oxygen from the water, including *Schizodon fasciatus*, and *Colossoma macropomum*, do not take part in these horizontal and vertical diel migrations (Saint-Paul & Soares 1987). The haemoglobin of *Colossoma macropomum* has an extremely high affinity for oxygen. Saint-Paul and Soares (1987) further suggested that this physiological feature could enable *C. macropomum* to take advantage of the minute discharge of oxygen into the water exhibited by some macrophytes with a rhizoid system, such as the water hyacinth *Eichhornia crassipes*, and remain in macrophyte beds by day and night. Species with lesser oxygen-extracting capacities were forced to leave this shelter and turned to ASR in open environments at night. *C. macropomum* has no particular capacity of sustaining high levels of glycolysis in its tissues, unlike the erythrinid *Hoplias malabaricus*, which can remain in its ambush site, even during protracted hypoxic or anoxic periods.

Long periods of hypoxia or anoxia in tropical environments have also promoted the evolution of morphological adaptations to air breathing in many fish species. Junk *et al.* (1983) found that about 30% of neotropical fish species could survive long periods of hypoxia, 27% of which showed morphological adaptations. Air breathing capacities have evolved almost independently in various groups of teleosts (e.g. see Moreau 1988, for African species). In some species, air breathing can be performed by the gills, as an air bubble becomes trapped in the gill chamber (e.g. *Hypomus* spp., cited in Wootton 1990). Suprabranchial organs, with secondary lamellae carried on structures that do not collapse in the air, are found in many species (clariid catfishes, and *Amphipnous*, *Anabas*, *Betta*, *Heteropneustes*, *Osphrenomus* and *Macropodes* spp.). The swimbladder of other species has become more vascularised and enables the rapid absorption of oxygen (*Arapaima*, *Erythrinus*, *Gymnotus*, *Lepidosiren*, *Polypterus* and *Protopterus* sp.). Different parts of the alimentary tract have also evolved as air-breathing organs in many species that gulp a bubble of air, which passes through the tract until it reaches this particular area (Kramer 1983b). These include the mouth cavity of *Electrophorus* and *Synbranchus* spp., the stomach of loricariids *Hypostomus* and *Pterygoplichthys*, and the intestine of callichthyids *Brochis* and *Hoplosternum* sp. With few exceptions (e.g. the pirarucu *Arapaima gigas*, the electric eel *Electrophorus electricus*) most air-breathing species are facultative air breathers, and use aquatic respiration under normoxic conditions. The threshold between aquatic and aerial respiration varies substantially between species (Kramer 1983a, 1987). As pointed out for ASR, air breathing requires that the fish accesses the surface of the water, and it is thus more frequent in open water than in vegetated areas, and energetically more advantageous in shallow than in deep habitats. As a corollary, air breathers too may be subjected to greater predatory pressure than species such as *Colossoma macropomum*, which remain in vegetated areas by day and night due to physiological adaptations.

Tolerance of freshwater fishes to hypoxia and the nature of their adaptive response thus intimately condition their diel migration pattern in the tropics. Species able to survive hypoxic conditions prevailing at night in vegetation beds decrease their night-time activity, whereas others make lateral migrations to the surface of open-water environments, where the predation risk is greater. Predation pressure and daily variations in dissolved oxygen thus restrict the feeding times of fish to dawn and dusk, causing greater competition for food resources (Zaret 1984). This competition may have proved a sufficient evolutionary pressure for the extreme specialisation of trophic niches in the tropics and may represent an additional driving force, complementing oxygen, predation and population density pressures in shaping seasonal migration patterns. For example, Fernandes (1997) postulated that decreasing oxygen levels at the end of the wet season can serve as a stimulus for many species (e.g. the characins *Curimata kneri*, *Potamorhina latior*, *Psectrogaster amazoniana*, *P. rutiloides*) to leave Lago do Rei, in the Amazon floodplain, and enter the main river as early as August or September, at a time when they are not ready to spawn. However, not all individuals of these species leave the lake at that time, and others enter the Amazon as late as December–January, just before spawning (see section 4.4.6). Other species, with greater tolerance of hypoxia, such as the iliophagous curimatids *Prochilodus nigricans* and *Semaprochilodus taeniurus*, remain in Lago do Rei until December, when they undertake their spawning migration into the river. Odinetz Collart and Moreira (1989) and Mérona and Bittencourt (1993) provided further evidence that this seasonal pattern indeed was dependent on variations in water level and oxygen, as its intensity varied between consecutive years with contrasting rainfall and drought regimes. However, the coincidence between decreasing water levels, increased turbidity, reduced oxygen, lower primary and secondary production, and increased fish biomass, makes it difficult to discriminate which is the actual driving force, if there is a single one. Comparable statements relating to broadly similar seasonal migration patterns have been put forward for other ecosystems (e.g. Chari River deltaic zone near Lake Chad: Loubens 1973; Gebel Aulia Reservoir on the White Nile in Sudan: Hanna & Schiemer 1993). In any case, the migration of a fraction of the population may favour the survival of the remaining fraction, which probably has not stored enough fat resources for undertaking spawning migrations or to build up gonadal tissues (Fernandes 1997; see also section 2.3.5).

Oxygen concentration and water level also shape shorter-term migration patterns resulting from short-term tropical showers in foothill tributaries. For example, in the Andean Piedmont, night-time rainfalls can raise the water level by 5–6 m, forcing most species to leave riffles and glides, and to enter backwater and irrigation canals, locally called *caños* (Baras *et al.* 1997). The decay of recently flooded terrestrial vegetation and the local increase of fish biomass cause the oxygen level to decline very rapidly, usually within a few days. When the oxygen concentration drops below 2 mg per litre, fish undertake a reverse lateral migration to the main stream, and the use of *caños* becomes restricted to the most oxygen-tolerant species. Some predators that are tolerant to low oxygen concentrations, such as *Hoplias malabaricus*, heavily colonise the mouths of the *caños*, and take advantage of the repeated passage of their prey to employ ambush predation.

4.4.5 Other environmental factors shaping habitat use and seasonal migrations

Marked decreases in water level can cause shallow habitats to dry completely, forcing most fish species to migrate to avoid desiccation. Exceptions to this rule are some species that have adapted to temporary (e.g. up to 24 h in *Arapaima gigas*; Neves 1995) or protracted drought periods. Most clariid catfishes with suprabranchial respiratory organs can survive in temporary pools, and they avoid desiccation through burrowing in damp mud or wet sand (Bruton 1979a). Still, this does not compare to the formidable adaptation of lungfishes *Protopterus* spp. that burrow into the bed of drying pools and secrete a cocoon of hardened slime, in which they can aestivate until the next flood. Annual species do not survive desiccation, but their diapaused eggs do (see also opportunistic strategists, section 4.4.6.1).

Seasonal variations in water levels generally are concomitant with variations in water clarity. Tropical freshwater assemblages are highly structured by predation pressure, which in turn is dependent on water clarity for visually-oriented species. Rodríguez and Lewis (1997) proposed a model of fish community organisation, the piscivory–transparency–morphometry (PTM) model, which emphasised the interaction between water clarity and the sensory and foraging abilities of predatory species of the Orinoco River basin. This model was later evaluated in floodplain lakes of the Araguaia River (Amazon River basin, Brazil) under non-limiting oxygen conditions (Tejerina-Garro *et al.* 1998). Characins, which are visually-orientated diurnal fishes dwelling in well-lit surface waters (Lowe-McConnell 1975), decrease in abundance as turbidity increases during the dry season. Conversely, there is an increase in the relative abundance of gymnotiforms, which use electric sensors for foraging (see Box 4.3), and siluriforms (chiefly pimelodid catfishes) which have chemical and tactile receptors and can forage at reduced light levels (Hara 1971). Implicit in these studies was that predatory characins were emigrating from floodplain lakes sooner and/or in greater numbers than other predators as a result of increased turbidity. Although transposition of the PTM model to other ecosystems seems valid (e.g. Vinces River floodplain in Ecuador; Landívar, 1996 in Tejerina-Garro *et al.* 1998), special attention should be paid to species with eyes possessing tapeta lucida that enable more efficient foraging under dim-light conditions (e.g. Nile perch *Lates niloticus*).

In shallow tropical lakes, seasonal variations of water levels may induce marked changes in conductivity and salinity, which are variously tolerated by fish, depending on their osmoregulatory capacities. A similar situation applies to coastal lakes and lagoons with permanent connections to the sea, and estuaries, where salinity rises during the dry season, as the input of fresh water from rivers declines. Some species restricted to lacustrine habitats have adapted to extremely high salinities (e.g. 40‰ in *Cyprinodon fasciatus* in North African lakes; Beadle 1943; *Oreochromis alcalicus grahami* of Lake Magadi, and *O. alcalicus alcalicus* in Lake Natron, Eastern Africa; Coe 1966, 1969). Some tropical freshwater fish species are truly euryhaline. For example, *Oreochromis amphimelas* is encountered in waters with conductivities as low as 200 μS cm^{-1}, and in waters with exceptionally high salinities (58‰ in Lake Manyara; Fryer & Iles 1972; Lévêque & Quensière 1988), and other *Oreochromis* spp. are found over a salinity range of about 30–35‰ (Fryer & Iles 1972; Philippart & Ruwet 1982; Trewavas 1983; Lowe-McConnell 1991). Many other species can tolerate moderate changes in salinity, and can survive in brackish waters (e.g. up to 8‰ for *Tilapia rendalli* in Lake Poelela in Mozambique; Whitfield & Blaber 1976; up to 15‰ for *Clarias gariepinus* in Lake Chilwa; Clay 1977).

Variations of salinity beyond the osmoregulatory capacities of fish species are a major driving force behind the seasonal migrations of fish from coastal lagoons or shallow lakes into tributaries. For example, *Oreochromis macrochir* only inhabits Lake Mweru (Zambia) at salinities lower than 7‰, and is exclusively found in its tributaries at other times of the year. During the protracted drying of Lake Chilwa in the late 1960s, *Oreochromis shiranus chilwae* entered the lake tributaries as soon as the salinity in the lake exceeded 5‰ (Furse *et al.* 1979). Similarly, the gradual increase of salinity (11‰ in 1920 and 22‰ in 1932) in Lake Quarum (Egypt) caused *Oreochromis aureus* and *O. niloticus* to permanently leave the lake, where they were replaced by the more tolerant *Tilapia zilli* (Fryer & Iles 1972). This emigration may seem surprising at first sight, since *O. niloticus* can tolerate higher salinity (up to 30‰). However, it should be remembered that osmoregulation has an energy cost that decreases the scope for activity and imposes a large environmental constraint on foraging or reproduction by fishes. The cost of osmoregulation in *O. niloticus* at 22‰ is about 20% of the oxygen consumed (Farmer & Beamish 1969). In lagoons with a permanent connection to the sea, effects of changes in salinity include the seasonal immigration of marine piscine predators that impose additional pressures on habitat use and migration. For example, in Southeast African coastal lakes, carangids and sphyraenids occur in clear-water lagoons, while elopids and sciaenids occur in turbid waters (Blaber 1988) and there are parallels for West African lagoons (Welcomme 1979).

4.4.6 *Life history, breeding systems and migration patterns of tropical fish*

Slobodkin and Rapoport (1974) proposed an interesting metaphor for nature and evolution, as animals playing a game of chess against nature where the winners were those playing longer than others. Short-term environmental changes usually elicit behavioural responses only, whereas protracted or chronic perturbations are frequently accompanied by deeper changes of the animal's physiology, biochemistry and anatomy. Staying alive is thus the key to the game, and the previous sections illustrated how tropical fish have morphologically, physiologically and behaviourally adapted to cope with predation pressure, oxygen-deficient waters, variations in water clarity and salinity, and the need to forage.

However, because a fish's life is finite, playing for longer can only be achieved through relaying its genes to the next generation, and minimising the cost of environmental variation can thus be viewed as reducing the loss of fitness. As in a chess game, there may be several different solutions for postponing checkmate more or less indefinitely, and animals have adopted a wide range of life-history traits of phenotypic or genotypic nature, that reflect adaptations to environmental conditions (Breder & Rosen 1966; Wootton 1990). These include variation in the age at first reproduction, fecundity, egg size and buoyancy, seasonality of reproduction, breeding system or reproductive guild, and spawning substrate choice, if any. Based on these traits, three major types of life-history strategies, listed below, have been identified for tropical freshwater fishes (Albaret 1982; Lowe-McConnell 1987; Winemiller 1989, 1992; see also Welcomme 1979 and Régier *et al.* 1989 for categorisation into 'white, black and grey fish species').

(1) **Opportunistic strategists**, with a small size at maturity, short lifespan, low degree of iteroparity, batch spawning over the year and no parental behaviour.
(2) **Equilibrium strategists**, with a protracted spawning season, rather independent from flood regime, low fecundity, large eggs, sophisticated courting and mating behaviour, and exhibiting parental care.
(3) **Seasonal strategists**, with synchronic, total spawning occurring in a restricted period of the year, usually during the flood season and in well-oxygenated waters. These fish reach sexual maturity at a relatively large size, have high fecundity with small eggs and do not exhibit any parental care.

This delimitation relies on the way the energy is partitioned, on the variable capacities of individuals to adapt to environmental harshness, and on the environmental requirements of eggs and larvae. It also closely overlaps the delimitation between species migrating over long distances, and others showing less mobility (Table 4.1). More detailed information on the reproductive traits of Amazonian and African species can be found in Ruffino and Isaac (1995), and Paugy and Lévêque (1999), respectively.

4.4.6.1 Opportunistic and equilibrium strategists

Opportunistic strategists do not adapt to environmental variations at the individual fish level. As pointed out by Hirshfield (1980), they give a strict priority to reproduction, and proportionally reduce the allocation of energy to maintenance, even if it leads to an increase in individual fish mortality. This is not an unbearable challenge since they have evolved towards early sexual maturity, small size, short lifespan and no parental care. Eggs are laid in drying pools, where they survive in the muddy substratum through one or more diapauses. However, there always is the risk that eggs emerge too early from the diapause, and hatch at inappropriate times. Some cyprinodonts (e.g. *Nothobranchius* spp.) have adapted to this environmental unpredictability by adding a more or less unpredictable component to the development of their progeny (Wourms 1972). Their eggs may enter one or more diapauses of variable length that cause them to hatch at different times of the year, almost removing altogether the risk that the entire progeny dies from hatching at an inappropriate season. Opportunitistic fishes, the small size of which increases predation risk, exhibit reduced mobility (except for spreading or passive drift during periods of rising water levels), and energy that might otherwise be used for migration can be invested in reproduction.

Fishes of the equilibrium group comprise very large species with late sexual maturity (e.g. the osteoglossid *Arapaima gigas*) and small species reaching sexual maturity at an early age and small size (e.g. many cichlids). In seasonally flooded environments, the functional delimitation between opportunistic and small-sized equilibrium strategists may not be conspicuous, as the populations of both groups decline during the dry season, and recover during the flood season. However, compared to opportunistic species, equilibrium strategists exhibit larger egg size, lower fecundity and partial spawning, with the breeders producing less eggs at a time, but more frequently during a spawning season, which is usually protracted, or takes place all year round in some environments. Most equilibrium strategists can tolerate relatively low oxygen concentrations, through behavioural, physiological or anatomical adaptations that enable them to live in floodplain habitats or river edges that are optimal or at least

Table 4.1 Relationship between demographic traits, diet, spawning season and migratory habits in some South American fishes (modified from Ruffino & Isaac 1995). Equilibrium, opportunistic and seasonal strategists after Albaret (1982), Lowe-McConnell (1987) and Winemiller (1989).

Strategy and species name	Season	Habitat	Spawning	Behaviour	Fecundity	Diet	Migration	References
Equilibrium								
Arapaima gigas	Late dry	Lentic-bottom	Batch	N, PC	Low	Piscivorous	Home range	Lowe-McConnell 1975
Astronotus crassipinis	Flood	Lentic, stones	?	T, PC	Low	Omnivorous	Home range	Fontenele 1951
Auchenipterus nuchalis	Late dry	Lotic	Batch	IF	Low	Omnivorous	?	Vazzoler et al. 1993
Cichla spp.	Flood	Woods, lentic	Batch	N, PC	Low	Piscivorous	Home range	Winemiller et al. 1997
Electrophorus electricus	Late dry	Residual lentic pools	Batch	N, PC	Low	Piscivorous	Home range	Assunçao & Schwassmann 1992
Hoplias malabaricus	Late dry	Shallow waters	Batch	N, PC	Medium	Piscivorous	Home range	Vazzoler & Menezes 1992
Osteoglossum bicirrhosum	Early flood	Lake, lentic	Batch	PC	Very low	Omnivorous	Home range	Goulding 1980
Opportunistic								
Plagioscion spp.	Protracted	Lake, lentic	Partial	?	Medium	Omnivorous	No	Worthmann 1982
Serrasalmus spp.	Flood	Lentic, aquatic vegetation	?	?	?	Omnivorous, Piscivorous	No	Sazima & Zamprogno 1985
Seasonal								
Brachyplatystoma spp.	Late dry, early flood	Lotic	Total	NI	Very high	Piscivorous	Very long	Barthem et al. 1991
Brycon spp.	Late dry, early flood	Water fronts, lotic	Total	NI	Very high	Omnivorous	Long	Junk 1985
Colossoma macropomum	Early flood	Lotic	Total	NI	Very high	Fruits, plankton	Medium to long	Goulding & Carvalho 1982; Loubens & Panfilli 1997
Metynnis, Mylossoma spp.	Early flood	Water fronts, lotic	Total	NI	High	Herbivorous	Yes	Junk 1985
Potamorhina latior	Early flood	Water fronts, lotic	Total	NI	High	Iliophagous	Yes	Junk 1985
Prochilodus and *Semaprochilodus* spp.	Early flood	Water fronts	Total	NI	Very high	Iliophagous	Long	many, see section 5.15
Pseudoplatystoma spp.	Early flood	?	Total	NI	Very high	Piscivorous	Very long	Reid 1983
Schizodon fasciatus	Early flood	Water fronts, lotic	Total	NI	High	Herbivorous	Yes	Santos 1982
Triportheus elongatus	Early flood	Water fronts	Total	NI	?	Omnivorous	Yes	Junk 1985

IF internal fertilisation; N nest building; NI no investment; PC parental care.

suitable to host their eggs and young. Another key feature of most equilibrium strategists is that they proportionally invest more into territoriality, spawning site preparation and parental care (see reproductive guilds in Balon 1975, 1981), which may also contribute to preserve their eggs from deoxygenation (Roberts 1975), and reduce environmental unpredictability (Krebs & Davies 1981).

For example, the anabantids *Ctenopoma* spp., which have a labyrinthine accessory respiratory organ in a suprabranchial cavity, lay their eggs in a bubble nest at the water surface (Lowe-McConnell 1988). Egg ventilation is performed through fanning by males, females or both sexes in nesting cichlids (Keenleyside 1991). In mouthbrooding cichlids, the parent(s) incubate (and thus ventilate) the eggs and embryos in their buccal cavity until the young have resorbed their yolk sac and start exogenous feeding (e.g. the tilapiines *Oreochromis* and *Sarotherodon* spp.). Some tilapiine species even prolong parental care in accompanying their young for a couple of weeks, providing shelter from predation in their buccal cavity (Fryer & Iles 1972). Livebearers (poeciliids) provide de facto a live shelter to their young until they start exogenous feeding (Wourms 1972). Because their reproductive style and parental care made most of these species independent of substrate availability, and because they show partial spawning, these species can spawn all year round, or at least during protracted seasons. However, territorial defence, courtship and parental care are time-consuming and have a high energy cost (Hirshfield 1980; Feldmeth 1983; Townshend & Wootton 1985; Baerends 1986; Turner 1986; Keenleyside 1991).

Because of this considerable energy investment, repeated spawning in the same habitat, and correspondence (or proximity) between the habitats of young and adults, it is deemed that most equilibrium strategists have strong sedentary habits. Evidence of home-ranging behaviour is available for large species such as the osteoglossid *Arapaima gigas*, which breathes air, tolerates temporary periods of desiccation, builds a nest, guards its eggs and young over periods of up to 6 months, and remains within a home range (about 20 km) which is small relative to the size of the mature adults (about 1.7–1.8 m) (Fontenele 1948, 1952; Lüling 1964; Neves 1995).

Further evidence of the strong sedentary habits of equilibrium strategists has been indirectly provided by the haplochromine cichlid species flocks in Rift Valley lakes (Malawi, Tanganyika and Victoria), which host no less than 700 different cichlid species, most of them being endemic (Witte 1984, Ribbink 1991). African Great Lakes cichlids are considered to have evolved from distinct but closely related ancestors (Greenwood 1979; Ribbink 1988, 1991), and similarities between the faunas of the different lakes are now regarded as the results of parallel evolution of fishes with a common heritage that were submitted to rather similar selection pressures (Ribbink 1988). It is nowadays proposed that the high rate of speciation in all three African Great Lakes originated from the combination of variable preferences for microhabitat, marked (about 100 m) changes in lake levels and sophisticated courtship ('specific mate recognition system'; Paterson 1978). These acted together to produce a typical pattern of punctuated evolution (Mayr 1963; Eldredge & Gould 1972). Implicit in this interpretation is that cichlids remained highly sedentary during periods of environmental stasis to become evolutionary stable units. As pointed out by Lowe-McConnell (1975), parental care makes it possible for cichlids to grow, breed and rear their offspring within a single habitat, which is presumably extremely conducive to ecological specialisation and speciation. The most restricted geographical distribution of most African Great Lakes cichlids, with some species being found nowhere else than on the shoreline of a small island, and the extinction of some cichlid species formerly inhabiting habitats accessible to the introduced Nile perch *Lates niloticus*, both support the hypothesis of restricted migration during the evolution of most Great Lakes cichlids. Apart from these 'coastal' cichlid species, small pelagic, shoaling planktivorous species such as *Cyrtocara* or *Cyprichromis* spp. inhabit deep water as adults, but require appropriate substratum, shelter and shallow water for their young. These species show little or no geographical variation, supporting the idea that they were, and probably still are, too mobile to develop geographic isolates (Eccles 1986).

However, not all equilibrium species are strictly 'resident'. Species showing home range or sedentary behaviour in a particular ecosystem may exhibit seasonal migrations in another, depending on habitat structure and seasonality of flow regimes. For example, tilapias are usually relatively sedentary in lakes, whereas they undertake seasonal migrations in savanna ecosystems, entering the floodplain as it becomes flooded, and leaving it before it dries out or becomes isolated from the main channel, as in the central delta of the Niger River in Mali (Bénech *et al.* 1983). The African pike *Hepsetus odoe* becomes mature at a relatively small size (14–15 cm), spawns several times per reproductive season and lays its eggs in a floating bubble nest where the larvae will remain attached, this strategy having been interpreted as an adaptation to oxygen-deficient waters (Merron *et al.* 1990). However, in the Kafue River, Williams (1971) found that the same species could migrate over 180 km.

4.4.6.2 Seasonal strategists

Seasons, what seasons?
From a cursory examination over a wide latitudinal range, seasonality in the tropics appears to be very limited, especially for biologists of northern temperate areas, where seasons are clearly defined by strong variations in daylength, reflected in the calendar. However, seasonality (in terms of precipitation) in the tropics is strongly influenced by latitude. For example, peak water levels in the Sanaga River (Cameroon) takes place in July. In the Chadian Ba Tha River, and central delta of the Niger River (Mali), corresponding peaks take place in late August–early September, and mid-September, respectively. In the Central African Republic, the Oubangui reaches its highest level in mid-October, and the Gabonese Ogôoué River does so in late November. Finally, the highest flows of the River Congo in Brazzaville on the Equator, are recorded in mid or late December but flows are high almost all year round (information on all six rivers as compiled by Lévêque, 1999). The greater the distance to the equator, the greater the predictability of the moment of the peak water levels in terms of time of the year. This variability, together with the low amplitude of fluctuations in temperature and daylength throughout the year, contributed to spread the false idea that 'there were no seasons in the tropics'. Yet, regarding a single locality or region, seasonality is extremely marked, and similar, both in nature and level of constraints, to that taking place sooner or later in another place.

On a local or regional basis, inland tropical fisheries in Africa, Southeast Asia and South America have a strong seasonal component that reflects both the poor efficiency of most traditional fishing gear under high or rising waters, and the migratory patterns of most exploited fish species, which belong to the seasonal group. In some cases, the mobility of the targeted species is so high, and their economic importance so great (e.g. Amazon River), that they have produced a class of itinerant full-time fishermen. In other places with no itinerant fishermen, the community comprises a class of seasonal fishermen or appointees contributing to a more efficient (and sometimes too efficient) exploitation of the very intense migration of fishes over a short period, which is known as *'piracema'* (fish time or fish swarms) in Brazil, *'ribazon'* in Venezuela or *'subienda'* in Colombia. These aggregations mostly comprise spawners of fish species that are seasonal strategists. To some extent, this compares to the exploitation of diadromous salmonids by man and other mammals in cold and temperate northern regions, except that, here, the migration is potamodromous, and proportionally more synchronised than in temperate rivers of similar size.

Why are seasonal strategists so seasonal?
Seasonal strategists have moderate or long life expectancies, exceptionally as long as 70 years in the characid *Colossoma macropomum* (Loubens & Panfilli 1997), and they reach sexual maturity much later than opportunistic or equilibrium strategists. They have high fecundity, small eggs, no sophisticated mating behaviour, and show no parental care. Embryos hatch at a small size, and show altricial development, thus having greater exigencies for feeding, lesser swimming capacities, and greater risks of being preyed upon by a vast range of predators (including omnivorous fish species), than the young of equilibrium strategists. Probably the best nursery habitat for these embryos is the floodplain, which provides a widespread mosaic of calm microhabitats, including vegetation shelters and refuges against predation (Winemiller & Jepsen 1998). High fecundity may compensate for high initial mortality. In any case, as these species with total spawning generally reproduce no more than once a year, the time of the year when they spawn should also be selected to minimise early losses.

As suggested by Lowe-McConnell (1991), there are opposing selection pressures for breeding at low or high water. At low water, the energetic investment required by migrants is reduced as the water velocity is low, and the eggs and young suffer little turbidity and turbulence. On the other hand, in the rainy season, there are greater swimming costs for migrating adults, but food for the young in the floodplain is plentiful, and abundant plant growth provides cover from the numerous predators. Also, the dispersal of the young is greater at high water, reducing the risk that the entire progeny will be eliminated by piscivores. Regarding fast-growing species with larvae showing intense and early piscivory, such as the characid *Brycon moorei*, it has been suggested further that the dispersal of eggs or larvae minimises sibling cannibalism that can eliminate up to 99% of a clumped population within the first week of exogenous feeding (Baras *et al.* 2000b). Obviously, constraints upon the most critical stage of the species lifecycle (exogenous feeding of larvae) have been traded off positively against less critical stages (energy invested by spawners), as spawning of most seasonal fish species in tropical regions takes place under rising waters. In this context, spawning at the onset of the rainy season probably is the most advantageous solution, as both parents and eggs are exposed to rather moderate water velocity and turbulence, and larvae arrive into the floodplain at the time when phytoplankton then zooplankton blooms occur.

Fundamental to this approach is that the time window for spawning is rather narrow, plausibly as narrow as a few weeks, especially in view of the exigencies of embryos and larvae under high temperatures. Most seasonal species, such as characids, have small eggs that hatch rapidly (from a dozen hours to a couple of days) and the embryos have small yolk reserves and starve rapidly if appropriate food is unavailable. Araujo-Lima (1994) found that in larvae of most characin species, age at starvation at 29°C was 140–250 h post-hatch, almost independent of egg size and yolk content within the size range evaluated, probably because having less tissue to maintain is less energy-demanding. Early spawners might thus incur a loss of fitness as their larvae would forage less efficiently, or die of starvation within less than 2 weeks. Late spawners would also incur a loss of fitness, as their offspring would arrive on the floodplain after the plankton blooms, and in any case at times when larvae spawned earlier would already have gained a growth advantage that might prove decisive in the context of intracohort competition in general, and cannibalism in particular. These arguments may account for why most seasonal strategists spawn within a restricted period of the year.

This view is slightly oversimplistic, as reproduction by all seasonal strategists over a very restricted period of time would result in unsustainable competition for food. Also, it overlooks the intrinsic characteristics of eggs (essentially their buoyancy and adhesive properties), as well as the irregularity of flow regimes. Loubens *et al.* (1992) reported that the water level of the Mamoré River, in Bolivian Amazon, could rise then fall by several metres during the early rainy season, causing recently flooded habitats to shrink, become hypoxic and eventually dry out. Subsequently, Loubens and Panfilli (1995) proposed there were two groups of seasonal strategists, which can be viewed as 'risk-takers' and 'care-takers' (Fig. 4.4). Group A (risk-takers) seemingly is the least represented group. Species belonging to this group (e.g. *Brycinus leuciscus*, *Marcusenius senegalensis*, *Schilbe mystus*, *Synodontis schall*; Paugy & Lévêque 1999) start spawning at the beginning of the rainy season, just before the floodplain becomes inundated. Some species such as the African characin *Alestes baremoze* spawn in streams, ensuring the dispersal of their young into downstream reaches, then onto the recently inundated floodplain (Durand & Loubens 1970), but risking higher mortality of the eggs and young if the water recedes relatively early. Other species like the characid *Colossoma macropomum* lay their eggs in sheltered bays, with little early dispersal followed by a typically riverine stage before they move onto the floodplain (Loubens & Panfilli 1997). Species of group B (care-takers; e.g. most characids and curimatids) spawn when the water level, and availability of floodplain habitat, are maximal, with reduced risks that eggs or larvae become stranded, but implying further lateral and downstream dispersal and greater energetic constraints for adults migrating in faster currents (see also section 2.3). Some species, such as the characid *Brycinus nurse*, seemingly take advantage of both rising waters and maximum flood to migrate and spawn (Paugy & Lévêque 1999), but it is unclear whether these traits reflect different subpopulations or adaptive behaviour.

In truly equatorial areas with little variation in water levels, species which are total spawners may have a protracted spawning season, even though each fish spawns once (e.g. *Brycinus longipinnis*; Paugy 1982). The characid *Brycinus imberi* spawns throughout the year in the equatorial Ivory Coast (Paugy 1980), whereas it shows synchronous spawning under high flood in the Zambezi River (Marshall & van der Heyden 1977). Similar changes in spawning seasonality have been brought about by man-made changes, notably by impoundments. In the Volta River, the clupeid *Pellonula afzeluisi* and the schilbeid catfish *Physalia pellucida* exhibit seasonal spawning, under conditions of low and high water levels, respectively. In Lake Volta, both species spawn all year round (Reynolds 1974), indicating that seasonality in riverine environments is an adaptation to the cycle of flowing and receding waters.

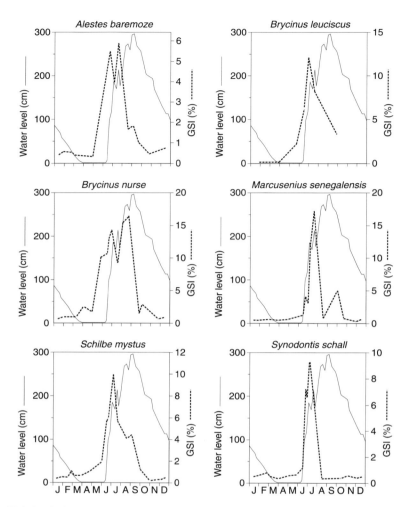

Fig. 4.4 Variation in the spawning periodicity of seasonal strategist freshwater fishes in Africa. Plain and dotted lines stand for water level (cm) and gonadosomatic index female fish (%), respectively (left and right axes of each graph). Redrawn from Paugy and Lévêque (1999).

Functional implications of seasonal spawning.
Implications of spawning at the onset of the rainy season, and early dispersal of larvae onto the floodplain, relate to the choice of spawning habitats and daily periodicity of spawning behaviour. Examples of spawning habitat selection can be found in Kramer's (1978) study on the characid *Brycon petrosus* in the Chagres River basin, Panama. This species spawns in very shallow water, on damp gravel, among dead leaves and in recently flooded herbaceous vegetation. Kramer (1978) hypothesised that the risk of egg desiccation, resulting from spawning in these sites, was traded off positively against the risk of egg predation by small egg-eating fish species if *B. petrosus* spawned elsewhere. This may be especially so, since spawning takes place at the onset of the rainy season and as characins have slightly pelagic eggs (Pavlov *et al.* 1995). The same applies to many species moving inland to spawn, including clariid catfishes that lay slightly sticky eggs on recently flooded vegetation in African and Asian rivers (Bruton

1979b). Most characins and many other seasonal strategists have small eggs that hatch rapidly (about 16 h; Araujo-Lima 1994), and this may account for the diel periodicity of spawning, as observed in the wild (Kramer 1978) or as inferred from studies upon drift. While investigating the drift of eggs in the rivers of Amazonian Peru, Pavlov *et al.* (1995) noticed that drifting eggs were abundant at night, and scarce during daytime except for the early morning (Fig. 4.5). Since eggs are non-motile, it has been suggested that this seemingly paradoxical periodicity originated from a marked diel periodicity of spawning in the late afternoon, which was associated with a relatively short incubation period, so that only hatched embryos or larvae were found by day. Embryos of characins (e.g. *Brycon* spp.) may show an early phototactic response while still in the egg (Baras [unpubl.]), and, provided that embryonic development is sufficiently complete, exposing eggs to light causes them to hatch earlier, as the embryo shows greater activity and eventually breaks the egg membrane. In view of the nature and periodicity of predatory pressures upon pelagic organisms, there obviously are fewer risks for eggs drifting at night, and this may have represented a sufficient evolutionary pressure upon the diel periodicity of spawning in seasonal strategists with fast embryonic development.

In the previous paragraphs, we have argued, from a conceptual viewpoint, for the importance of egg dispersal in seasonal strategists by reference to foraging needs, predation risks and cannibalism. Thorough studies on the drift of eggs and larvae in tropical ecosystems are scarce, and lack detailed information, with few exceptions (Amazon River basin: Araujo-Lima *et al.* 1994; Pavlov *et al.* 1995; Araujo-Lima & Oliveira 1998; Jamuna River basin, Bangladesh: de Graaf *et al.* 1999). Pavlov *et al.* (1995) observed that drift contained much greater numbers of eggs and early larvae relative to late larvae and juveniles, suggesting that passive drift is the main mechanism behind the downstream movement of young fishes, contrary to the situation in temperate ecosystems, where there is both active and passive migration of fry, underyearlings and yearlings (Pavlov 1994). As depicted in the paragraphs dedicated to environment and climate, these differences between tropical and temperate ecosystems can be interpreted within the context of season-dependent environmental constraints that shape migrations onto and from the floodplain, as they apply to seasonal strategists. In the rivers of Amazonian Peru where Pavlov *et al.* (1995) conducted their study, the most frequently encountered taxa, especially during the main drifting peak at the onset of the rainy season, were characins (chiefly characids) and siluriforms (chiefly pimelodids), both having typical demographic traits of seasonal strategists. This supports the idea that the early drift of seasonal strategists' eggs and larvae is a common trend in many tropical ecosystems. Further evidence on the early drift of seasonal strategists' eggs and larvae has been provided by Kramer (1978), who reported that not a single egg or embryo could be found on the spawning sites of *Brycon petrosus* in the Chagres River in Panama, as early as 3 days after spawning.

How far do eggs and larvae drift and disperse?
Early drift of seasonal strategists, as a consequence of pelagic eggs being laid during high or rising waters, increases drifting time, and the distance over which the progeny is dispersed. Other factors governing the drifting distance are the topography of the spawning site, flow regime, and behaviour of fish embryos and larvae. In lowland, turbid rivers and streams (e.g. Amazon, Maranon, Ucayali) early life stages of seasonal strategists drift throughout the day and night, whereas in clear water rivers (e.g. Mamon, Nanay, Samiriya, all of them being part

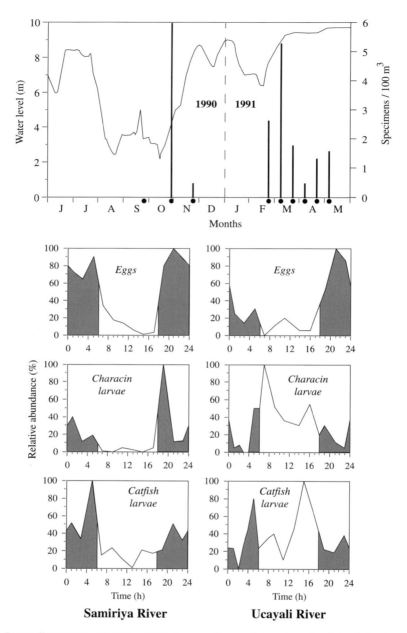

Fig. 4.5 Seasonality and diel periodicity of drift of eggs and larvae of characins and catfishes in the rivers of Amazonian Peru. The upper graph shows seasonal variations in the flow regime of the Amazon River, and in the numbers of drifting larvae at different sampling dates in 1990–1991 (solid circles on the horizontal axis). The six lower graphs show diel drift patterns of eggs and larvae of characins and catfishes in two major tributaries with contrasting water transparencies (40–240 cm versus 3–40 cm in the Samiriya and Ucayali River, respectively). The diel periodicity of egg drift owes to rapid embryonic development (12–18 h) and diel periodicity of spawning (mid–late afternoon). Redrawn after Pavlov *et al.* (1995).

of the Amazon River basin), most of the drift takes place at night (Fig. 4.5; Pavlov *et al.* 1995). Larvae of siluroid species drift primarily near to the bottom, whereas characin larvae occur in the upper layer of the water column, suggesting they may drift faster too, in view of the positive correlation between water velocity and distance from the substratum. Drift at night in clear water presumably limits predation risk upon larvae, in the same way as described for the eggs. In turbid water, predation by visually-orientated predators is hardly influenced by daylight, and drift can take place day and night (Fig. 4.5). However, because clear-water rivers are generally flowing faster than turbid rivers, distance drifted and dispersal probably are just as great, despite drift occurring only at night. In fast-flowing upper tributaries in the Andean foothills, the drift of larvae occurs throughout the day (Pavlov et al. 1995), probably due to the physical impossibility of larvae moving to inshore shelters during daytime in fast currents.

At present, there are no accurate data on how far eggs and larvae drift in tropical riverine ecosystems. However, rough estimates can be produced by combining the duration of embryonic period, age at starvation, diel periodicity of drift in larvae depending on environments, and estimates of mean water velocity. For example, eggs spawned in the late afternoon, carried by currents and drifting in foothill streams at 1 m s^{-1} over 12 h at night would have moved more than 40 km downstream by the time they hatched next morning. Eggs spawned in the Andean Piedmont, where the current is faster, and temperature cooler, would take longer before hatching (about twice as long at 23–24°C than at 28°C) and they could drift over more than 150 km prior to hatching. In view of their reduced swimming capacities relative to water current, drifting fish larvae have very little chance of capturing passing prey. Hence, from a functional viewpoint, their drifting period cannot extend beyond the age at starvation, which ranges from 6 to 10 days in characids (exceptionally as long as 15 days; Araujo-Lima 1994). Assuming that water velocity decreases along the downstream migration route as stream order decreases, distances as long as 500–1,300 km could be covered by eggs and larvae prior to starvation (Araujo-Lima & Oliveira 1998). Some species such as the sorubims (*Brachyplatystoma* and *Pseudoplatystoma* spp.) spawn during the period of high water in mountain tributaries, where the cool temperature presumably postpones hatching, and high water velocity causes eggs to drift faster. For these species, distances as long as 2500–3500 km could possibly be travelled within less than 2 weeks. This may account for why larvae are found in the estuary of the Amazon River whereas adults spawn in the Andean Piedmont (Barthem & Goulding 1997). Such estimates may seem exaggerated at first sight but they remain perfectly sound in view of the upstream migrations of spawners (section 5.16).

Spawning migrations.
Implicit in drift and dispersal of eggs and larvae in floodplain nurseries, is that equivalent distances should be travelled upstream at older life stages. Although this does not exclusively occur, the upstream migration of a few adults, with greater swimming capacities and greater chances of escaping predation, is less energy-demanding and less hazardous than when numerous young fish move the same distance. This general principle, originally put forward by Margalef (1963), applies perfectly to seasonal strategists in tropical freshwater ecosystems, where high predation pressure in the main channel, and huge variations in water level and velocity restrict considerably the possibility of upstream compensatory migration at the larval or juvenile life stages, especially for distances of several hundred kilometres.

Long potamodromous upstream migrations are most frequent in tropical ecosystems, both in river corridors, and in lake tributaries (Table 4.2). In the Chari River, which flows from

Table 4.2 Examples of the extent of potamodromous migrations by some tropical freshwater fish species.

Fish species	Family	Ecosystem	Migration (km)	References
Alestes baremoze	Alestiidae	Lake Chad tributaries	650	Blache & Milton 1962
	Alestiidae	Senegal River	400	Reizer *et al.* 1972
Alestes dentex	Alestiidae	Lake Chad tributaries	650	Blache & Milton 1962
Barbus altianalis	Cyprinidae	Lake Victoria tributary	80	Whitehead 1959
Brachyplatystoma flavicans	Pimelodidae	Amazon River	*c.* 3500	Barthem & Goulding 1997
Brachyplatystoma vaillanti	Pimelodidae	Amazon River	*c.* 3500	Barthem & Goulding 1997
Brachysynodontis batensoda	Mochokidae	Lake Chad tributary, Chari River	*c.* 150–200	Bénech *et al.* 1983; Bénech & Quensière 1989
Brycinus leuciscus	Alestiidae	Niger River	400	Daget 1952
Colossoma mitrei	Characidae	Paraguay River	*c.* 400	Bayley 1973
Distichodus rostratus	Citharinidae	Lake Chad tributary, Chari River	hundreds	Bénech *et al.* 1983
Hemisynodontis membranaceus	Mochokidae	Lake Chad tributary, Chari River	250–300	Bénech & Quensière 1983
Hepsetus odoe	Hepsetiidae	Kafue River	180	Williams 1971
Hydrocynus brevis	Alestiidae	Lake Chad tributary, Chari River	hundreds	Bénech *et al.* 1983
Hyperopisus bebe	Mormyridae	Lake Chad tributary, Chari River	hundreds	Bénech *et al.* 1983
Labeo altivelis	Cyprinidae	Lupuzlz River, Lake Mweru	150	Welcomme & Mérona 1988
Labeo senegalensis	Cyprinidae	Lake Chad tributary, Chari River	250–300	Bénech & Quensière 1983
Leporinus obtusidens	Anostomidae	Middle Paraná River	hundreds	Bonetto *et al.* 1981
		Paraguay River	*c.* 400	Bayley 1973

Marcusenius cyprinoides	Mormyridae	Lake Chad tributary, Chari River	hundreds	Bénech *et al.* 1983
Prochilodus lineatus (also named *P. scrofa*)	Curimatidae	Upper Paraná River Basin	600–700	Godoy 1972; Toledo *et al.* 1986; Agostinho *et al.* 1994
Prochilodus mariae	Curimatidae	Orinoco River Basin	hundreds	Saldaña & Venables 1983
Prochilodus nigricans	Curimatidae	Rio Mamoré, Bolivia	hundreds	Loubens & Panfilli 1995
Prochilodus platensis	Curimatidae	Middle Paraná River	700	Bonetto *et al.* 1981
		Paraguai River	*c.* 400	Bayley 1973
Prochilodus spp	Curimatidae	Orinoco River Basin	hundreds	Lilyestrom 1983
Pseudoplatystoma coruscans	Pimelodidae	Paraguai River	*c.* 400	Bayley 1973
Pseudoplatystoma fasciatum	Pimelodidae	Magdalena River Basin	*c.* 500–700	Baras [pers. comm.]
Pterodoras granulosus	Doradidae	Paraná River Basin	200	Agostinho *et al.* 1994
Rhinelepis aspera	Loricariidae	Paraná River Basin	60	Agostinho *et al.* 1994
Salminus maxillosus	Characidae	Middle Paraná River	850	Bonetto *et al.* 1981
		Paraguay River	*c.* 400	Bayley 1973
Schilbe mystus	Schilbeidae	Lake Victoria tributary	25	Whitehead 1959
		Lake Chad tributary, Chari River	*c.* 150–200	Bénech *et al.* 1983; Bénech & Quensière 1983
Schilbe uranoscopus	Schilbeidae	Lake Chad tributary, Chari River	hundreds	Bénech *et al.* 1983
Schizodon fasciatus	Anostomidae	Paraguay River	*ca.* 400	Bayley 1973
Synodontis schall	Mochokidae	Lake Chad tributary, Chari River	*c.* 150–200	Bénech *et al.* 1983; Bénech & Quensière 1983

North Cameroon into Lake Chad, the distance moved (150–300 km) is closely associated with the geographical distribution of floodplain in North Cameroon. Species which travel over longer distances (up to 650 km), like *Alestes baremoze*, frequently lay their eggs in midstream, from where they drift to downstream reaches. *Prochilodus lineatus* (also named *P. scrofa*) also makes upstream spawning migrations over hundreds of kilometres (Table 4.2, and see section 5.15). For this particular species, there is evidence that some spawners can migrate repeatedly, year after year, at very regular times, as exemplified by one of the rare long-term mark recapture programmes undertaken in tropical rivers; that on the Mogi-Guaçu River, a tributary of the upper Paraná River (Godoy 1959, 1967, 1972). Probably the best-known example from this investigation is a 58-cm long *P. lineatus*, tagged on 6 November 1962 and recaptured at the same site on 5 November 1963 and 16 October 1964. Analyses of the seasonality of recaptures at different sites upstream and downstream of the tagging site from 1954 to 1965 indicated that *P. lineatus* travelled roundtrip distances of about 1300 km per year (Toledo *et al.* 1986; Fig. 4.6). Much longer distances are covered by the large pimelodid catfishes of the Amazon, which descend the Amazon as eggs, larvae and juveniles, and ascend it as large juveniles or adults (a lifetime journey of 7000–8000 km; Barthem & Goulding 1997), although in this particular case the lifetime migration occurs over several years (see also section 5.16). Other examples of upstream spawning migrations of tropical fishes are provided in chapter 5 (especially in sections 5.15 and 5.16 on characins and catfishes respectively).

Most seasonal strategists migrate upstream at relatively slow speeds, but they are capable of fast runs, especially in the final stages of migration (Table 4.3). The vast majority of

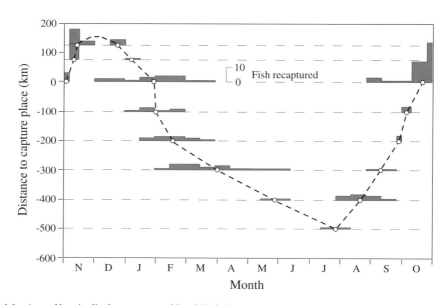

Fig. 4.6 Annual longitudinal movements of *Prochilodus lineatus* (also named *P. scrofa*) in the Mogi-Guaçu River, Brazil (Upper Paraná River Basin), as determined by mark recapture. Positive and negative values on the vertical axis stand for places upstream and downstream of the capture place (sited at zero). Grey bars indicate the number and periodicity of recaptures in different river stretches, the midpoint of which is shown by dotted horizontal lines. Redrawn after Toledo *et al.* (1986).

Table 4.3 Examples of migration speeds during the upstream runs of tropical seasonal strategists

Fish species	Ground speed	Reference
Brachyplatystoma spp.	av. 11 km d^{-1}	Barthem & Goulding 1997
Brycinus leuciscus	up to 1.0–1.5 km h^{-1}	Daget 1952
Hypophthalmus marginatus	av. 3–4 km d^{-1}	Carvalho & Mérona 1986
Leporinus copelandii	av. 3.0 km d^{-1}	Godoy 1975
Leporinus elongatus	av. 3.5 km d^{-1}	Godoy 1975
Prochilodus lineatus	av. 5–8 km d^{-1}	Toledo *et al.* 1987
	up to 43 km d^{-1}	Godoy 1975
Prochilodus platensis	av. 7 km d^{-1}	Bonetto *et al.* 1981
Pterodoras granulosus	up to 27 km d^{-1}	Agostinho *et al.* 1994
Rhinelepis aspera	av. 10 km d^{-1}	Agostinho *et al.* 1994
Salminus maxillosus	up to 21 km d^{-1}	Bonetto *et al.* 1971; Godoy 1975

seasonal spawners undertake downstream migrations soon after spawning, a few days or weeks of rest in flooded vegetation being frequent in these species. Factors responsible for the downstream migration of adults may include environmental harshness and excessive water velocity for feeding efficiently and sufficiently, especially for those species having little or no adaptation to fast current. This argument may also apply to several *Prochilodus* spp., in South American environments where appropriate feeding grounds are upstream of spawning grounds. During high floods, spawners first descend tributaries then enter the main channel to spawn, with the reverse movement during receding waters. Other complex migration patterns have been described for these species, and they are detailed in section 5.15. Winemiller and Jepsen (1998) suggested these migratory strategies enabled minimisation of intraspecific competition for food, which is perfectly sound for iliophagous species.

4.4.7 Conclusion

At first sight, most of the previous section concerning tropical regions might seem more appropriate to a book on the biology of tropical fish and not to one dedicated to fish migration. This point would certainly be valid for well-documented fish assemblages, but for 90% of the tropical fish species, knowledge of migration is restricted to scanty empirical data or information from fishermen's catches, the reliability of which might be questioned in countries where subsistence fishing may distort official data. For example, the fisheries statistics in Benin during the early 1970s to the mid 1980s showed only minute variations between annual catches, until there was a change in the administration and catches surprisingly soared by 200% the next year.

We hope that the way this section on tropical regions was structured, and particularly the emphasis laid on the functional environmental and biological mechanisms underlying the migration processes, might serve as a basis for a better understanding and management of tropical fish assemblages. We are most aware that mapping biodiversity is a definite priority, but combining these surveys with some targeted measurements of key variables such as the demographic traits as determined from growth patterns and histological structure of gonads might be just as important for species or populations, the migratory habits of which are unknown. As pointed out in chapter 2, basic studies of proximate composition and proportion of muscle mass might contribute to a clearer picture of the migratory capacities of these species. Finally, there are many techniques and methods for studying fish migration, which may appear excessively costly, but the benefits of which exceed by far the investment required (chapter 6). Assessing whether a species may migrate before knowing much else about its biology may seem inappropriate, but may turn out to be a reality for sustaining fisheries or preserving biodiversity when landscape or climate becomes modified, which is precisely the case with the El Niño phenomenon that impacts on the regularity of flow regimes. These considerations are valid for any fish assemblage, but they probably are more relevant for tropical assemblages, which have produced a huge diversity of species as a response to more or less predictable environmental conditions, and for which man-made environmental changes might be more far-reaching than elsewhere.

Chapter 5
Taxonomic Analysis of Migration in Freshwater Fishes

5.1 Introduction

Migration takes place in a wide range of taxa occurring in fresh water, although the most conspicuous migrations are principally associated with anadromous and catadromous species. Migrations between fresh water and salt water occur in a wide variety of fishes, although constituting less than 1% of species (McDowall 1988). For these species, estuaries and river channels form 'highways' for migration (Novoa 1989; Roy 1989). For a variety of species including anadromous salmonids and clupeids large shoals may aggregate within river channels, providing important fisheries. However, most families of stenohaline freshwater fishes contain species for which clear migratory behaviour occurs irrespective of spatial scale (Northcote 1978, 1984; Welcomme 1979, 1985). Additionally large population components of euryhaline marine fishes may enter low-salinity areas of estuaries, creeks and marshes at key lifecycle stages (Marotz *et al*. 1990). Although these fishes do not fulfil McDowall's (1988) definition of diadromy, these are movements between key habitats, of regular occurrence and often involving large proportions of populations.

In many cases, especially for tropical regions, it is at present impossible to assess the relative occurrence of migratory behaviour across whole fish communities and geographic regions. However, to provide some perspective of the occurrence of migration, we reviewed the biological details of Canadian species of fish provided by Scott and Crossman (1973), and examined these in detail for clear evidence of movement between distinct habitats. This analysis indicates that about 55% of Canadian freshwater fishes are migratory or contain significant migratory elements (Table 5.1), according to the definition we provided in chapter 1. The families with the highest proportions of migratory species are familiar groups such as the Salmonidae, which includes the Coregoninae and Thymallinae, and the Petromyzontidae. Groups such as the Cyprinidae have relatively low proportions of migratory fishes, although this is almost certainly an underestimate relating to the comparative paucity of spatial ecological information for this and similar families (Smith 1991). However, it is clear that a diverse range of taxa substantially contributes to the occurrence of migration and that consideration of diadromous fishes alone, especially anadromous salmonids, gives a very limited view of the occurrence of migration in freshwater fishes. This is undoubtedly even more apparent in many of the large tropical river systems where well over 95% of migratory fish species are potamodromous, based on existing published information (Welcomme 1985; McDowall 1988; Nelson 1994; section 4.4).

Table 5.1 Incidence of migratory behaviour (by our definition, see section 1.2.5) within families of Canadian freshwater fishes, determined according to biological information presented for each species in Scott & Crossman (1973).

Family and common name	Number of species	Number of diadromous species	Number of potamodromous species	% migratory species
Petromyzontidae (lampreys)	9	4	2	67
Acipenseridae (sturgeons)	5	4	1	100
Lepisosteidae (gars)	2	0	1	50
Amiidae (bowfin)	1	0	0	0
Clupeidae (herrings, shads)	4	3	1	100
Salmonidae (salmons, trouts, chars, whitefishes, grayling)	32	16	10	81
Osmeridae (smelts)	4	3	0	75
Hiodontidae (mooneyes)	2	0	2	100
Umbridae (mudminnows)	2	0	2	100
Esocidae (pikes)	4	0	4	100
Cyprinidae (minnows, chubs etc.)	44	0	13	30
Catostomidae (suckers, redhorses)	17	0	9	53
Ictaluridae (catfishes)	7	0	1	14
Anguillidae (eels)	1	1	0	100
Cyprinodontidae (killifishes)	2	0	0	0
Gadidae (cods)	2	1	1	100
Atherinidae (silversides)	1	0	0	0
Gasterosteidae (sticklebacks)	5	2	1	60
Percopsidae (trout-perches)	1	0	1	100
Moronidae (temperate basses)	3	3	0	100
Centrarchidae (sunfishes)	10	0	5	50
Percidae (perches)	14	0	8	57
Sciaenidae (drums)	1	0	0	0
Cottidae (sculpins)	8	1	0	13
Total	181	38	62	–
% migratory	–	21	34	55

The remainder of this large section is dedicated to a group-by-group discussion of the migration and lifecycle characteristics, where appropriate, of species representative of particular families exhibiting migration in fresh and brackish water environments. The division of the chapter is principally made in terms of subclasses, orders and families, for ease of reference by readers seeking information on particular groups, although in some cases we group several closely related families within the same section for convenience of discussion. The taxonomic groupings used mostly follow those of Nelson (1994), with some modifications using the ICLARM FishBase database (Froese & Pauly 2000). This taxonomic review is necessarily limited for some groups of fishes, notably the salmonids and anguillid eels for which a massive literature already exists and for which excellent reviews and sources of information are available elsewhere (e.g. Harden Jones 1968; Tesch 1977; McDowall 1988; Groot & Margolis 1991). The extent of information provided within this section is also limited for some groups of freshwater fishes mostly found in tropical freshwater regions. This is partly because of a paucity of information concerning spatial ecology at the species level for these groups, partly because some of these groups (Cichlidae, Characiformes and Siluriformes) are so speciose, totalling over 5000 species and representing nearly 50% of all fish species utilising freshwater (Nelson 1994), and partly because excellent syntheses of material for these taxa are available elsewhere (Lowe-McConnell 1975, 1987; Welcomme 1979, 1985). Additionally, we have already provided an integrated account of migratory behaviour of these groups within the context of fish migrations in tropical freshwater systems (section 4.4). Lastly, given the space constraints of this book and the limited amount of information at our disposal we do not provide information in this chapter on migratory behaviour of a number of fresh or brackish water groups including the Polypteridae (bichirs), Pantodontidae (butterflyfish), Kneriidae and Phractolaemidae (mudfishes), Gyrinocheilidae (algae eaters), Amblyopsidae (cavefishes), Synbranchidae (swamp eels), Mastacembelidae (spiny eels), Ambassidae (Asiatic glassfishes), Haemulidae (grunts or croakers), Polynemidae (threadfins), Monodactylidae (moonfishes), Toxotidae (archerfishes), Elassomatidae (pygmy sunfishes), Blenniidae (blennies), Kurtidae (nurseryfishes), and Anabantoidea (pikehead, gouramies and snakeheads). Nevertheless, reference is made to several of these groups (e.g. Haemulidae, Toxotidae) in other sections. We apologise for this lengthy list, and our undoubted omissions of migratory behaviour in these and other taxa, for which we would be pleased to receive information.

5.2 Lampreys (Petromyzontidae)

Lampreys from the northern hemisphere (Petromyzontinae) and from the southern hemisphere (Geotriinae and Mordaciinae) display a range of lifecycle strategies. Although they are jawless primitive vertebrates most lamprey species exhibit clear migratory patterns and a substantial number are anadromous, with the attendant osmoregulatory advances needed for such behaviour. Of the 41 species 18 are parasitic in the adult growth phase (Nelson 1994). In many continental European countries populations of lampreys have been dramatically reduced by river regulation and pollution, and they now receive increased protection through the EC Directive on the conservation of natural habitats and flora and fauna. This represents something of a paradox given that one of those species, the Atlantic sea lamprey *Petromyzon marinus*, is a major pest in the North American Great Lakes where it has caused serious mortality to native fishes, resulting in declines of several fisheries and dramatic changes to the fish communities (Smith 1968). Invasion of the upper lakes occurred as a result of the opening of the Welland Canal linking Lake Ontario with the Upper Great Lakes of Erie, Huron, Michigan and Superior (see also section 7.2.2). Of the 41 species of petromyzontids nine are anadromous, including two of the three species of *Mordacia* and the single species of *Geotria* (McDowall 1988; Nelson 1994). They are widely distributed but most common in temperate climates.

The Atlantic sea lamprey *Petromyzon marinus* is the most well-known of the anadromous lamprey species and is characteristic of this group. Adults may approach 1 m in length and are the largest of the lampreys. They are poor swimmers and, although they may migrate 300 km up rivers, spawning usually takes place in the middle to lower reaches in fresh water (Bigelow & Schroeder 1953). The rate of upstream progress by sea lampreys has been estimated as about 0.18 km h^{-1}, although this may vary with the strength of the downstream current opposing this movement (Hardisty 1979). Although poor swimmers they are capable of passing many obstructions, even in fast-flowing water, by swimming in the boundary layer and using the sucker to attach to firm substrate during periods of rest. Nikolsky (1961) described the occurrence of both winter and spring return migrations in this species, with spring-run lampreys having more mature gonads than winter-run lampreys. Movement into shallow spawning areas occurs in late spring and shallow riffles are selected for spawning. In the sea lamprey and the smaller European river lamprey *Lampetra fluviatilis*, males have a tendency to reach the spawning grounds first and begin preliminary nest building (Hardisty 1979; Maitland 1980), by displacing stones using the sucker. After spawning, adults may drift or swim some distance downstream, but it seems that all eventually die after a single spawning, a feature that appears to be common to most and perhaps all lamprey species (Hardisty 1979; Maitland 1980) (Fig. 5.1).

On hatching, larval ammocoetes of the Atlantic sea lamprey *P. marinus* and the European river lamprey *L. fluviatilis* burrow into mud and silt along sluggish stream margins and live for several years, filter-feeding in fresh water (Fig. 5.1). A metamorphosis takes place during the summer and autumn (Hardisty & Potter 1971b) and the small subadults migrate downstream during the autumn to lakes or the sea and parasitise fish during a second growth phase (Fig. 5.1). Metamorphosis to the parasitic adult form is dependent on body size and temperature, with increasing spring temperatures above 9–13°C acting as an important cue for metamorphosis of *P. marinus* (Holmes *et al.* 1994). Hardisty (1979) argued that, within a river system, the distribution of larval lamprey populations results from the interaction of the passive downstream drift of the larva and the rheotactic upstream migration of the spawning adult. Thus, throughout the larval period the larval population will tend to move downstream towards the middle and lower reaches of the river but this is counteracted each year by the ascent of spawning adults to higher reaches (Hardisty & Potter 1971a). The subadults migrate downriver, normally to the sea, but in the Great Lakes *P. marinus* grows to adulthood within the lake environment. In general, diadromous lampreys do not feed until they reach the sea

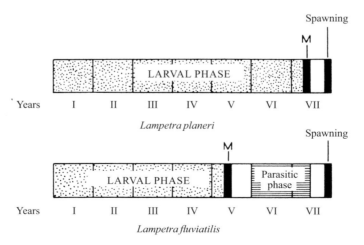

Fig. 5.1 Schematic illustration of the lifecycles of the river lamprey *Lampetra fluviatilis* (parasitic) and the brook lamprey *L. planeri* (non-parasitic). M represents metamorphosis, while unshaded areas represent periods when the animals are not feeding. Redrawn from Hardisty (1979).

although adult lampreys sometimes feed in fresh water (Davis 1967; Maitland 1980). In this adult growth phase they live as ectoparasites of fish for 2–3 years. As they mature they return to coastal areas and migrate back up river systems. The gonads develop as they migrate upriver, during which time they do not usually feed. Some species such as the brook lamprey *Lampetra planeri* lack a parasitic adult stage and metamorphose directly from the ammocoete to the mature adult form, and are therefore characteristically small in size when mature (Hardisty 1979; Fig. 5.1).

The Great Lakes of North America provide important feeding habitats for landlocked Atlantic sea lamprey and these are well-known to have caused catastrophic effects on the fish communities of the lakes (Smith 1968; McDowall 1988). Downstream movement of juveniles into the lakes occurs in two peaks in spring and autumn, associated with high flows in the tributaries (Applegate & Brynildson 1952). Using fyke netting and the proportion of released and recaptured ammocoetes, Hanson and Swink (1989) estimated the average annual emigration of juveniles from the 42-km long Ocqueoc River in Michigan to be 62 000 lampreys. An autumn peak in November–December and a spring peak in April occurred during periods of high flow at temperatures of 5°C. There is good evidence from experiments of emigrating juveniles tagged with coded wire tags and subsequent recapture of adults, that the returning adult sea lampreys do not home to their natal streams (Bergstedt & Seelye 1995) but appear to rely on innate attraction to abiotic stream odour cues, or more likely to pheromones from larvae in streams (Teeter 1980). Recent studies show that several components of bile products in juvenile sea lampreys stimulate dramatic olfactory bulb responses in electrophysiological studies of adult conspecifics, strongly supporting a pheromone-mediated orientation response (Li *et al.* 1995; see also section 2.4.5). The rapid spread of sea lampreys across the Great Lakes (Smith & Tibbles 1980) also suggests that a high degree of homing is unlikely. Given the rapid geographical spread, it is surprising that only about 8% of Great Lakes tributaries have supported populations of sea lamprey larvae (Morman *et al.* 1980). This may reflect the unsuitability of some streams for hatching and development of larvae, and the selection of only certain streams by adults. Following chemical treatment with lampricides, the number of Great Lakes streams with lamprey larvae has decreased, although regular treatment remains necessary for effective control (Torblaa & Westman 1980).

Other anadromous members of the Petromzyontinae, the Arctic lamprey *Lampetra japonica* of the North Pacific and Arctic Ocean, *Lampetra tridentata* and *L. ayresii* of western North America, *L. fluviatilis* of western Europe and the Mediterranean, and *Caspiomyzon wagneri* of the Caspian Sea, have lifecycles and patterns of migration similar to that of the Atlantic sea lamprey *P. marinus*. Similar patterns are also observed for the anadromous members of the southern hemisphere lampreys of the families Geotriinae and Mordaciinae, and more information is presented by McDowall (1988). *Mordacia mordax* has been reported to exhibit diurnal burrowing during its upstream migration, and to travel exclusively under low light intensity, this trait being enabled by its retinal structure, which consists only of rods (in contrast to the two other subfamilies, which possess at least one type of cone cell) and possesses a tapetum lucidum (Collin & Potter 2000).

European river lampreys *L. fluviatilis* are typically 30–40 cm long as adults and exist as anadromous and potamodromous forms (Maitland 1980). Potamodromous 'landlocked' river lampreys grow to adulthood in lakes, for example in Loch Lomond, Scotland (Maitland 1980), where they mainly parasitise coregonines. The River Endrick is the largest feeder stream of Loch Lomond and Maitland *et al*. (1994) found, by trapping, that adult river lampreys started to appear in the river in late September with the main spawning runs in October to December. Spring and autumn spawning runs of *L .fluviatilis* have been recorded (Nikolsky 1961), although in many cases the autumn migration represents an initial period of activity which is halted by low winter temperatures and is resumed during early spring (M. Lucas [unpubl.]). In western Europe, river lampreys normally reach the spawning areas a month or so before sea lampreys. Malmqvist (1980) pointed out that spawning in the freshwater-resident brook lamprey *Lampetra planeri* is also preceded by upstream migration, although this generally involves limited distances of up to a few kilometres.

5.3 Sharks and rays (Elasmobranchii)

Sharks and rays are primarily marine fishes, but a number of species regularly occur in fresh water including several taxa that are known only from fresh and brackish water (Nelson 1994). The spatial behaviour of most of these sharks and rays in fresh water is extremely poorly known, but well-defined migrations do not seem to occur. The best-known example of such freshwater movements is the bull shark *Carcharhinus leucas* (Carcharhinidae), which has been captured more than 4200 km up the Amazon River and regularly traverses the 175-km long Rio San Juan between Lake Nicaragua and the Caribbean Sea (Thorson 1971, 1972). Another six carcharinid species may be found in fresh water and one, the Ganges shark *Glyphis gangeticus*, may be confined to fresh- and brackish water (Nelson 1994). A few triakid sharks such as the smoothound *Mustelus canis* are reported to enter fresh water for short periods, but such movements are probably rare (Nelson 1994). Several species of sawfishes (Pristidae) regularly enter fresh water, including *Pristis microdon* which ascends some rivers in Australia and New Guinea (Nelson 1994) and *P. perotteti* which moves between lakes and the ocean in South and Central America and has established genetically distinct, reproducing populations in fresh water (Thorson 1982). Some guitarfishes (Rhinobatidae) may also enter fresh water.

Over 20 species of stingrays (Dasyatidae) rarely, if ever, enter marine environments, but their behaviour in fresh water is largely unknown. These include three genera, *Potamotrygon*, *Paratrygon* and *Plesiotrygon*, which comprise the river stingrays of South American Atlantic drainages, and several species of *Dasyatis* and *Himantura* which inhabit rivers in Africa, Southeast Asia and New Guinea (Seret 1988; Nelson 1994). When in fresh water, euryhaline elasmobranchs such as the bull shark reduce their plasma salt concentration by up to 20% and their plasma urea concentration by up to 50%. By contrast, potamotrygonine stingrays lack functional rectal glands, contain negligible urea, cannot concentrate it and die in water more saline than about 15‰ (Thorson *et al*. 1967).

5.4 Sturgeons (Acipenseridae)

The sturgeons represent an ancient group of large, bony-plated chondrostean fishes. Some species such as the American Atlantic sturgeon *Acipenser oxyrinchus* grow to in excess of 3 m in length and 1000 kg in weight. Sturgeons have provided important fisheries in many countries, especially for caviar, the eggs of gravid females. The family contains 27 species and most are currently viewed as extinct, endangered or threatened (Bemis & Kynard 1997; Birstein *et al.* 1997). Overharvesting, barriers to migration, habitat damage and pollution have been important factors for population declines. However, blockage of migratory routes may, in many cases, be the greatest factor in preventing recovery of populations for a number of sturgeon species (Auer 1996a). Most if not all sturgeon species exhibit migratory behaviour, though to varying degrees. They are generally long-lived and all are repeat spawners. Sturgeons exhibit three general patterns in habitat selection and migratory strategy. Some outmigrate to brackish or estuarine environments after spawning and juveniles follow later; some outmigrate to the sea immediately, young and juveniles too; and a few species spend their whole lives in fresh water, especially where there are large lakes (Rochard *et al.* 1990). Adult sturgeon usually spawn in fast-flowing main channel river habitat, usually in the middle and upper reaches of large river systems, although spawning often occurs further downstream in heavily impounded rivers.

The most well-studied sturgeon species are probably those of the Atlantic seaboard of North America; in particular the American Atlantic sturgeon *Acipenser oxyrinchus* and the smaller shortnose sturgeon *Acipenser brevirostrum*. The American Atlantic sturgeon is found from the Gulf of St Lawrence to Florida and the Gulf of Mexico and perhaps a little further south. Adults spawn over cobbles in fresh water. The eggs are adhesive and the tiny larvae hatch after about a week and move to slacker water in the lower river and estuary for 3–7 years before moving to sea. The species is regarded as anadromous and may make extensive coastal migrations, but in many cases, especially in more southerly regions, subadults remain in brackish waters and may not spend significant periods of time at sea (Moser & Ross 1995). In a study of the behaviour of subadult *A. oxyrinchus* in the lower Merrimack River, Massachusetts, Kieffer and Kynard (1993) used radio and acoustic telemetry to track the movements of 23 fish. These entered the river from coastal areas by mid to late May when increasing river temperatures reached 14.8–19.0°C, and river discharge was decreasing to 303–675 m^3 s^{-1}. They occupied the saline reach, experiencing salinities of 0–27.5‰ and used one discrete area during river residence. Residence areas were often associated with sediment deposition and appear to relate to areas of high food availability (especially bivalves). They migrated from river in autumn when temperatures were 13–18.4°C. In the Cape Fear River, North Carolina, juvenile *A. oxyrinchus* remained in the estuary throughout the year. Moser and Ross (1995) tracked 14 fish varying between 69 and 122 cm throughout the year. They preferred deep areas, greater than 10 m, in the vicinity of the saltwater–freshwater interface at river kilometre 46 (rkm 46). In summer they held position and apparently fasted, but were more active and ranged over greater areas in cooler conditions of autumn, winter and spring, with average rates of movement of 1.3 km d^{-1}.

Foster and Clugston (1997) tracked the movements of 67 Gulf sturgeon *Acipenser oxyrinchus desotoi*, the southern subspecies of the American Atlantic sturgeon (75–212 cm, 2.5–74.8 kg) in the Suwannee River, Florida from March 1989 to August 1992. Fish entered the river from mid-February to the end of April, moved upstream at an average speed of 3.5 km d^{-1} to areas 50–200 km upstream and they spawned after the March new moon, at water temperatures >17°C but lower than 21–22°C. From April to October, adult sturgeon congregated in deep holes serving as summer-autumn holding areas, and displayed restricted movements, of no more than an average of 0.6 km upstream or downstream of their established summer home area. Summer ranges comprised areas around major cool-water springs, but there was no evidence of behavioural thermoregulation within the spring's thermal plume, or of any differences in oxygen levels. Sturgeons lose weight during this period of the year (Clugston *et al.* 1995), which is compensated for by winter feeding in marine habitats of the Gulf of Mexico. They begin leaving the area in October, moving downstream at an average speed of 6.2 km d^{-1}, and after a period of remaining near the river mouth they enter nearshore mesohaline waters, then move further offshore into deeper water as the temperature drops. Water temperatures associated with spring and autumn migrations averaged 22.1°C (range 16–28°C) and 21.3°C (16.9–26.8°C) respectively. Many of the Gulf sturgeon that left the Suwannee returned the next year (at least 76% of tagged fish), but most of these are thought to have been immature. This population exhibits much higher river fidelity than many other populations of American Atlantic sturgeon *A. oxyrinchus*, such as those in the Hudson River, New York (Dovel & Berggren 1983). Fox *et al.* (2000) provide further information of the migratory behaviour of Gulf sturgeon in the Choctawhatchee River (Alabama-Florida; see Box 4.1, chapter 4).

The shortnose sturgeon *A. brevirostrum* has a similar range to *A. oxyrinchus* and is found as far south as Florida. Shortnose sturgeon are found primarily in riverine and estuarine areas, but unlike *A. oxyrinchus*, they do not make extensive coastal migrations. In the north of their range shortnose sturgeon tend to spawn and spend much of their life in fresh water, but make periodic movements, usually during spring, into the salty estuary (Dadswell 1979). In southern rivers such as the Savannah River, Florida, they spend most of their time in estuarine waters and make excursions into freshwater to spawn (Hall *et al.* 1991). In the north it seems that shortnose sturgeon and Atlantic sturgeon are spatially separated, with shortnose sturgeon occurring mostly in fresh water and Atlantic sturgeon in saline water, but not in the south where both spend most time in saline estuaries. The reason for this difference is unknown but may relate to water quality or food availability for growth, especially during the summer.

Shortnose sturgeon *Acipenser brevirostrum* make use of very discrete areas for spawning, summering and wintering, with well-defined movements between (Kieffer & Kynard 1993; Kynard 1997). Across the latitudinal range, spawning adults normally move upstream to about rkm 200 or further to spawn, although there is significant variation in the actual distances moved (Kynard 1997). In rivers where spawning takes place in relatively upstream reaches, most shortnose sturgeon live and move wholly within fresh water (Buckley & Kynard 1985). If suitable spawning areas are in the lower reaches, then movement into saline waters is common (Kieffer & Kynard 1993; Moser & Ross 1995; Kynard 1997). Where the spawning migration distance for shortnose sturgeon is relatively long they will often migrate upriver in autumn to near the spawning site and spawn in the following spring. In the Connecticut River, there is a long autumn migration of 77–81 km with overwintering near the spawning site or a shorter spring migration of 23–24 km from an overwintering area to a spawning site (Buckley & Kynard 1985). Similar behaviour has been observed in the Merrimack River, Massachusetts, where spring spawning migrations from overwintering areas were as short as 9–19 km (Kieffer & Kynard 1993). Male shortnose sturgeon have been observed to spawn every 2 years in the St John River, New Brunswick, Canada (Dadswell 1979) where wintering areas are 50–120 km from the spawning areas. Kieffer and Kynard (1993) suggest that males in populations that must make long migratory journeys do not have the energetic resources for annual gonad production and a long migration. They suggest that the seasonal periodicity of the spawning migration is related to spawning migration distance, with an autumn migration for distances greater than 50 km whereas a spring migration seems to suffice for distances of less than 25 km.

The energetic losses by spawning female sturgeon, which invest much energy into egg production, are of course likely to be much greater than those of males. Mature *Acipenser oxyrinchus* may spawn only once per 2–5 years (Smith 1985) and in unobstructed rivers typically migrate several hundred kilometres. Annual migration and spawning has been recorded in long-term telemetry studies of shortnose sturgeon in the Merrimack and Connecticut Rivers, in both cases characterised by short-distance migrations (Kieffer & Kynard 1996). Net-capture and telemetry studies on the Merrimack have also provided evidence that males could show repeated fidelity to the same spawning ground up to 4 years in a row (Kieffer & Kynard 1993, 1996). Shortnose sturgeon embryos vigorously seek cover in laboratory tests (Richmond & Kynard 1995). They would normally be concealed within the substrate at the spawning site so this is a behaviour that ensures they remain there until development to the larval stage has been completed. Larvae prefer deep water and a silt substrate. They are most active at night, leaving cover and actively entering the water column, suggesting that these actively migrate downstream from the spawning site.

The white sturgeon *A. transmontanus* of the Pacific coast of North America from Alaska to California is another species which migrates extensively in rivers but rarely moves into truly marine environments and may carry out its full lifecycle in fresh water (Miller 1972). It normally inhabited rivers of the Pacific coast of North America from Alaska to California, although it is now very rare in the south of its range, possibly because of negative effects of high temperatures, which may result from water regulation, on the ovulatory response and egg quality in this species (Webb *et al.* 1999). The western green sturgeon *A. medirostris*, another Pacific sturgeon, occurs in Siberia and western North America and is regarded as an anadromous species (McDowall 1988). The European Atlantic sturgeon *A. sturio* may remain in fresh water for 3 years before going to sea (Wheeler 1969). It is now extinct through much of its former range and remaining populations are threatened (Lelek 1987). The story of its extinction as a result of human activity has been traced in several rivers, including the River Meuse in the Netherlands and Belgium (Philippart & Vranken 1983) and the Guadalquivir, Spain (Fernandez-Pasquier 1999). In the latter case, a dam built in 1932 close to the spawning area strongly interfered with the upstream migration, as gravid females were unable to clear this obstacle under spring flows lower than $100 \, m^3 \, s^{-1}$. Frequent droughts and the increasing use of water for irrigation led to repeated reproductive failures, and gathering of spawners in the lower estuary, where fishing pressure was highest, resulting in the extinction of this population.

The Arctic Lena sturgeon *A. baeri* is another anadromous species and is recorded as migrating 1300 km up the Yenisei River in Russia, with the young spending 5 or 6 years in fresh water before entering the marine environment (Nikolsky 1961). The beluga *Huso huso* is perhaps the largest sturgeon species, and is reported to reach a weight of 1500 kg (Berg 1962). It occurs in the Black Sea and Sea of Azov and is more numerous in the Caspian Sea. This species also has spring and autumn migrating population components, with fish migrating in autumn tending to move longer distances and spawn further upstream (Nikolsky 1961). Upstream movements as long as 2500 km were historically reported in the Danube River prior to the construction of a large dam at the Iron Gate corridor. The young tend to move relatively rapidly towards the sea and subadult beluga are uncommon in rivers. The kaluga *H. dauricus*, another very large species, occurs in eastern Siberia and is anadromous, but the adults do not seem to move far from the shore. They may spawn in rivers just a few kilometres upstream of the saltwater limit, although upstream migrations of up to 700 km have been recorded (Berg 1962).

Several other anadromous species of sturgeon are found in eastern Europe and Russia. The ship sturgeon *A. nudiventris* occurs in the Sea of Azov, the Caspian and Black seas, and the Aral Sea prior to its degradation. It spawns in rivers, often migrating long distances to do so. It is recorded as migrating beyond Budapest on the Danube but upstream migration is now severely hampered by the increasing impoundment of this river. Spawning appears to occur at no more than once per 2 years and is associated with periods of freshwater residence of up to 1 year (Berg 1962). The stellate sturgeon *A. stellatus* and Danube sturgeon *A. gueldenstaedtii* occupy similar ranges but neither occurs in the Aral Sea, while stellate sturgeon also occur in the Adriatic (Svetovidov 1984). *A. persicus* is very difficult to differentiate from *A. gueldenstaedtii* and occurs in the Caspian Sea, migrating up the Kura, Volga and Ural rivers to spawn (Lelek 1987). Mature males first enter the Kura River at about 8 years of age and females at 12 years. Adriatic sturgeon *A. naccarii* are restricted to spawning populations in the Po and Adige rivers and are very rare. As for *A. oxyrinchus* and *A. brevirostrum*, autumn and spring river entry are common for several of these species especially for *A. nudiventris*, and although Berg (1962) reports separate spring and autumn spawning stocks, it seems likely that autumn-migrating fish overwinter and spawn in spring.

The genus *Scaphirhynchus*, as well as several species of *Acipenser* such as the lake sturgeon *A. fulvescens* of North America, are wholly freshwater in habit. Other species such as the sterlet *A. ruthenus* of eastern Europe and Asia exhibit primarily potamodromous life histories. The sterlet is a relatively small species, rarely exceeding 80 cm in length. They do occur in less saline areas of the Caspian and Black Seas but tend to be more common in fresh water (Lelek 1987). They tend to overwinter in deep areas of the lower reaches of rivers and make conspicuous spawning migrations, moving upstream in shoals to spawn over gravel. The shovelnose sturgeon *Scaphirhynchus platorynchus* appears to be relatively sedentary in its behaviour (Hurley *et al.* 1987). Their radio telemetry studies of 22 shovelnose sturgeon in pool 13 of the Mississippi River, between April and September 1982, showed that they mostly occupied limited areas, but sometimes moved up to 17 km at speeds of up to 11.7 km d^{-1}, mostly during May and June. They also found that eight of the tracked fish returned to discrete areas that had formerly been heavily utilised.

The lake sturgeon *Acipenser fulvescens* is principally associated with the Great Lakes region of North America. This species feeds and grows in lake environments but migrates up rivers to spawn. Most remaining populations in the US are restricted in movement by obstructions (dams, locks and weirs). An exception in terms of lakeward movement is the Sturgeon River population. These migrate 69 km up the river to spawn in spring below Prickett hydroelectric facility which blocks further upstream progress. The young/juveniles can move down to Portage Lake, Michigan (Auer 1996b). Until and including 1988, the Prickett power station used a hydropeaking regime of water use. In 1990 and 1991 there was a transitional pattern of flow and in 1991 and 1992 this was altered to a flow regime mimicking natural discharge patterns measured further upstream, as a result of relicensing of the hydroelectric facility. Under near natural flows adult lake sturgeons spent 4–6 weeks less at spawning sites. This is interpreted as an increase in spawning efficiency, with fish returning to recovery and growth activities earlier. Whether this actually results in increased egg deposition and/or recruitment remains to be seen. However, 74% more fish were observed at spawning sites under the more natural regime. There were many more females, giving a higher mean weight per fish, and fish showed increased reproductive readiness (with 79% of males and 39% of females in ripe-running condition compared to less than 5% during hydropeaking years). Also in hydropeaking years, many lake sturgeon were found near the spillway and powerhouse, seeking to move further upstream, perhaps for more appropriate spawning conditions. During the years of near-natural flow 95% of fish were captured in the lower rapids downstream of the dam, and only 10 of 224 were captured from close to the dam. These results seem to suggest a change in conditions which have met the demands of the spawning migration of these sturgeon, and provide appropriate conditions for spawning to occur.

There is a strong positive correlation between recorded maximum river migration distance and the adult body size for species of sturgeon (Fig. 5.2), although the value for lake sturgeon *Acipenser fulvescens* appears to be an outlier from this linear relationship, with relatively low migration distances for its body size (Auer 1996a). There are only a few relatively unrestricted populations of this species. Probably a barrier-free 250–300 km combined river and lake distance is needed for maintenance of a viable population of lake sturgeon. Barrier removal, rather than habitat improvement, should be considered the primary management task for sturgeons (Auer 1996a). There are several proven examples of sturgeon homing to historic spawning grounds (Dadswell 1979; Lyons & Kempinger 1992) but not many, although some netting and telemetry studies have provided evidence of repeat spawning by male shortnose sturgeon *A. brevirostrum* in successive years (Kieffer & Kynard 1993, 1996). Lyons and Kempinger (1992) found that only 1.5% of lake sturgeon tagged at spawning in the Wolf River, Wisconsin, moved to another river to spawn the following year. It is possible that in many natural systems inhabited by sturgeons, fidelity to natal rivers and spawning sites may be high.

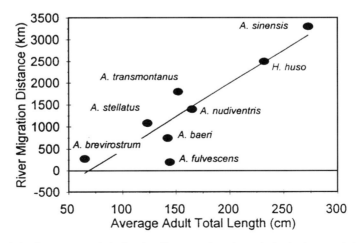

Fig. 5.2 Correlation between recorded migration distance and maximum body size in several sturgeon species. Reproduced from Auer (1996a).

5.5 Paddlefishes (Polyodontidae)

The paddlefishes are bizarre though impressive chondrostean fishes with elongate, flattened snouts. One species, the American paddlefish *Polyodon spathula*, occurs in North America, living in the Mississippi drainage and some adjacent Gulf of Mexico rivers, while the Chinese paddlefish *Psephurus gladius* occurs only in China in the Yangtze River and the lower reaches of adjacent rivers. The American paddlefish is a filter feeder while the Chinese paddlefish has a protrusible mouth and is a piscivore. Both species are large, often growing to over 1.5 m in length. Both species make migrations, although these may not be conspicuous due to the turbid water which they often inhabit, and to their nocturnal habits (Paukert & Fisher 2000). Increasing concern as to the habitat requirements of these fishes has lead to increased study of their behaviour.

Southall and Hubert (1984) radio-tracked 17 adult paddlefish in pools 12 and 13 of the upper Mississippi in spring and summer. They found high levels of movement in spring pre-spawning and spawning periods, leading to congregations of fish below dams. Paddlefish are not normally able to pass the dam gates which, when in the normal mode of partial closure, have water velocities of 7 m s^{-1}. In 1981, gate 12 was opened for maintenance and paddlefish successfully passed through, travelling upstream as far as gate 11. This represents a distance of approximately 70 km, measured from the middle of pool 13. It seems clear that in unobstructed systems these fish would seek to make extensive movements. Upstream movements were associated with spring high water periods when dam gates were partially opened and flows increased. After spawning adult paddlefish mainly used channel and backwater habitat where zooplankton was abundant.

The Alabama River carries a more diverse range of habitats than the Mississippi in the area described above and has been the subject of several recent paddlefish studies (Hoxmeier & DeVries 1997; Lein & DeVries 1998). Where appropriate habitat is available juvenile paddlefish use and may remain in oxbow lakes until adulthood. Significantly higher levels of zooplankton remain in the oxbows in winter than in channel or backwater habitat. Rising temperatures and flows stimulate the spawning migration. Spawning occurs in March–April at temperatures approaching 18°C. Immature fish that have left the oxbows migrate upstream in spring with adults to spawning areas and tend to congregate at Claiborne Dam 80 km upstream of the main area of oxbows. Males appear to arrive on the spawning grounds first and leave last. Lein and DeVries (1998) found some evidence of spawning site fidelity, and of little mixing between Tallapoosa and Cahaba tributary populations. After spawning, adult paddlefish mainly use channel and backwater habitat where zooplankton is abundant, and they seemingly avoid the highest temperatures available (Paukert & Fisher 2000). Feeding in this species is greatly assisted by their possession of passive electroreceptors that can detect planktonic prey producing electrical noise (e.g. swarms of cladocerans), and this capacity is enhanced by stochastic resonance (Russell *et al*. 1999). Electroreception is a definite advantage for foraging on plankton, especially at night, but it also causes the avoidance of metallic obstacles such as dams and locks, which may thus interfere with the migration of the species to a greater extent than for species with lesser sensorial capacities (Gurgens *et al*. 2000).

Larvae of paddlefish drift downstream and may enter oxbows or backwaters during the 1–2 months when they are connected to the river channel. Juveniles reaching maturity may leave these habitats in pulses, and thereafter appear to remain in channel and backwater habitat. However, paddlefish larvae are known to exhibit positive rheotactic responses, and are frequently regarded as main- or side-channel fishes. As a result of this rheotactic behaviour, they might be subjected to greater risks of being stranded when subjected to navigation-induced drawdowns, since they tend to swim towards the shoreline as the water recedes (Adams *et al.* 1999), although this risk is intimately dependent on habitat structure, and distance between their typical habitats and littoral areas.

The Chinese paddlefish *Psephurus gladius* is regarded as the most endangered fish in China because of overfishing, habitat destruction and dam construction in the Yangtze River which blocks the upstream spawning migrations and other movements of this species (Wei *et al.* 1997). The Three Gorges Dam project on the Yangtze is likely to exacerbate these problems, yet little is known of the spatial ecology or other biology of this species (Liu & Zeng 1988; Wei *et al.* 1997). Given our knowledge of the migratory behaviour of the American paddlefish and of sturgeons, and the arguments invoked behind their declines, it seems likely that Chinese paddlefish exhibit extensive migrations and that they are probably potamodromous.

5.6 Gars (Lepisosteidae) and bowfins (Amiidae)

The gars are elongated predatory fish of slow-flowing rivers and pools in North and Central America. They are principally freshwater species, although they may occur in brackish and salt water (Hildebrand & Schroeder 1927). Species such as the longnose gar *Lepisosteus osseus* and the spotted gar *Lepisosteus oculatus* normally remain in vegetated areas of lakes, backwaters and submerged floodplain habitat, lying in wait to ambush prey. There are few recorded long-distance migrations, but short-distance migrations in spring to inflowing streams appear quite common (Scott & Crossman 1973; Johnson & Noltie 1997). These often involve a large proportion of the adult population and are likely to be important in population maintenance. On the basis of radio telemetry investigations in the Atchafalaya River basin, Louisiana, Snedden *et al.* (1999) described substantial movement of adult spotted gars onto inundated floodplain during the spring flood pulse. During this period, median home ranges (265 ha) were 25 times larger than in summer and 43 times larger than in autumn/winter. In all seasons, home range and rate of movement were greater at night than at any other time of the day. The inundated floodplain provides important spawning and nursery areas for this species, and although adults utilised a variety of habitats during spring, these movements may be regarded as incorporating lateral spawning migrations.

There is one species of bowfin, *Amia calva*, which occurs in fresh water in eastern North America. It lives in shallow, vegetated lakes and backwaters and is not considered migratory, although adults may move into shallow water in lakeside margins in spring for spawning (Scott & Crossman 1973).

5.7 Bonytongues, mooneyes, featherfin knifefishes, elephant fishes (Osteoglossiformes)

The osteoglossiforms are a group of teleosts of primitive origin, comprising six families of freshwater fishes, of which several groups will be considered here. The bonytongues (Osteoglossidae) comprise seven species, together exhibiting a circumtropical distribution. Although they are generally large fish (the pirarucu *Arapaima gigas* of South America may exceed 3 m in length), they do not appear to exhibit strong migratory habits but home ranges may be extensive (see section 4.4.6).

There are two species of mooneyes (Hiodontidae), herring-like fishes which occur in freshwater systems of northern North America, principally the Mackenzie, Saskatchewan, Mississippi and St Lawrence river systems. Both mooneye *Hiodon tergisus* and goldeye *Hiodon alosoides* exhibit migratory movements and are mostly active by night (Scott & Crossman 1973). Mooneye are characteristic of large, clear freshwater systems, and in spring they undergo conspicuous migrations up clear streams to spawn (Scott & Crossman 1973). Goldeye more usually spawn in sheltered, slow-flowing areas and are much more common in turbid habitats. Because spawning is pelagic and the eggs are semi-buoyant, large numbers of eggs and larvae may drift downstream, especially in channelised rivers such as the middle Missouri (Hergenrader *et al.* 1982). Within the Peace-Athabasca delta and adjacent rivers of Canada, goldeye undergo extensive migrations between wintering and spawning areas (Donald & Kooyman 1977a). The Peace-Athabasca delta population of goldeye winters primarily in the Peace River, Alberta, from Vermillion Falls to the Slave River. In May, goldeye migrate from the lower reaches of the Peace River into Mamawi Lake and then to Lake Claire of the Peace-Athabasca delta. The adults, which may live for 24 years, spawn and feed in the delta before returning to the Peace River in late summer and autumn (Donald & Kooyman 1977a, b). The lakes are the rearing habitat for the young and are shallow with a maximum depth of 3 m. Freeze-up occurs in October and lasts for 6 months, so that both lakes are devoid of oxygen by late winter. In order to escape this, both young and adults move into the Peace River to overwinter. Migration to the lakes for spawning in May provides a good egg development and nursery area, but recruitment is strongly influenced by environmental conditions, with warm, calm conditions enhancing survival (Donald 1997).

Eight species of featherfin (Old World) knifefishes (Notopteridae) occur in Africa and Asia and although some species such as *Chitala lopis* reach 1.5 m in length they appear mostly sedentary, tending to remain in floodplain habitat during wet and dry seasons (Welcomme 1985). They are air breathers and so are well adapted to stagnant pool habitats. Many of the elephantfishes (Mormyroidea, about 200 species in Africa) are also relatively sedentary (Welcomme 1985) and this probably also applies to *Gymnarchus niloticus*, the single member of the closely related Gymnarchidae. These families have well-developed electrosensory organs (Heiligenberg 1991) and are predominantly nocturnal in habit, exhibiting local diel movements between refuge and foraging areas (e.g. Friedman & Hopkins 1996). However, some species such as *Brienomyrus niger* and *Mormyrus rume* do exhibit clear patterns of movement related to changes in water level (Carmouze *et al.* 1983; Welcomme 1985), while *Hyperopisus bebe, M. rume, Mormyrops deliciosus* and *Marcusenius senegalensis* exhibit upstream movements in the Senegal River to avoid saltwater intrusion that occurs during the dry season (Reizer 1974 in Welcomme 1985).

5.8 Tenpounders and tarpons (Elopiformes)

The elopiforms comprise about eight species of large, streamlined, fast-swimming, herring-like fishes, normally found in tropical and subtropical seas worldwide. The Atlantic tarpon *Megalops atlanticus* may reach 2.5 m in length. Both species of tarpon (Megalopidae), *M. atlanticus* from the Atlantic, and *M. cyprinoides* from the Indo-Pacific, may enter fresh water but this behaviour appears to be somewhat facultative. Movement into fresh water may be a more significant component of the lifecycle for some populations of *M. cyprinoides* (McDowall 1988), which has been recorded 900 km up the Fly River in Papua New Guinea (Roberts 1978). The latter author found that substantial numbers of subadults were always present in the upper reaches of the Fly River, while juveniles occurred in the middle and lower river and larvae were collected in the lower river and in coastal shallows. It is thought that adult *M. cyprinoides* may move downstream to spawn in estuarine or coastal areas and further consideration of the lifecycle of this species is given in McDowall (1988). Several species of tenpounders (Elopidae) *Elops* are also known to enter fresh water, but in most rivers such occurrences appear to be relatively infrequent.

5.9 Freshwater eels (Anguillidae)

The freshwater eels are the classical group of fishes associated with catadromous migration patterns. There are 15 species, all with an elongated body form, no pelvic fins, and many ecological traits in common, among which are generalist feeding habits, old age at maturity, semelparity, wide geographic distribution and catadromy. They are found in tropical and temperate waters except those associated with the eastern Pacific and southern Atlantic. The majority of species are tropical and subtropical, but most knowledge concerns the two North Atlantic species, the North American eel *Anguilla rostrata* and the European eel *A. anguilla*, and the Japanese eel *A. japonica*. Although much is known of the coastal, estuarine and freshwater life histories of these three species, there is still a paucity of knowledge concerning the marine spawning migration. The migration and lifecycles of anguillid species are reviewed in detail in Tesch (1977) and a useful review of the four species found in South Africa is given in Bruton *et al.* (1987).

All species of anguillid eels are regarded as obligately catadromous. Indeed in textbooks of fish migration they are usually portrayed as the 'type' lifecycle for catadromy. This is understandable since the life history of anguillids typically involves a freshwater growth phase to maturity and a long-distance migration to a marine spawning site. Nevertheless, evidence is growing that the migratory cycle of some anguillids may be less rigid than previously thought. Microchemistry studies of the otoliths from maturing (silver) European eels has provided good evidence that a very substantial proportion of North Sea or Baltic Sea European eels have never ventured into fresh water (Tsukamoto *et al.* 1998; Tzeng *et al.* 2000) (Fig. 5.3). A similar 'sea-locking' phenomenon has been put forward for Japanese eels in the China Sea (Nakai *et al.* 1999). The extent to which this oceanodromous migratory component exists over the full geographical ranges of these species is still unclear. Whatever the case, it remains that catadromy probably represents the dominant life-cycle strategy for anguillids.

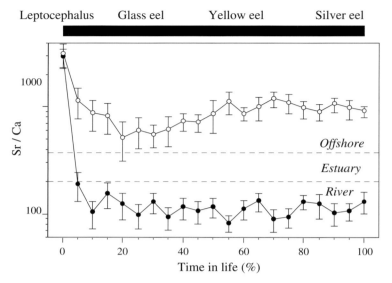

Fig. 5.3 Diadromous and oceanodromous life-styles in the European eel *Anguilla anguilla*, as determined from otolith microchemistry of specimens captured in the North Sea (40–61 cm in length, open symbols) and in the River Elbe, Germany (40–78 cm in length, closed symbols). Redrawn after Tsukamoto *et al.*, 1998.

Spawning of anguillid species typically occurs at a depth of about 500 m over deep tropical, oceanic waters, distant from land. Both species of Atlantic anguillid eels spawn in the Sargasso Sea (western Atlantic), Japanese eels spawn in the northwestern Pacific, those from the Australasian region spawn in the mid-western Pacific and those from Africa spawn in the western Indian Ocean. It is generally thought that all adult eels die after spawning, but strict semelparity is speculation since no one has located an adult anguillid eel at the presumed spawning grounds (Avise *et al.* 1990). The eggs are pelagic and hatch into laterally flattened leptocephalus larvae, which grow and develop as they move back on currents, often across large areas of ocean, to land. This migration is largely passive but may also involve an active component (Jellyman 1987). As they approach coastal areas leptocephali metamorphose into transparent glass eels of adult body form, about 50 mm in length. The mean age at metamorphosis varies between species, being as short as 120–125 days in *Anguilla celebensis* and *A. marmorata* (Arai *et al.* 1999a), 186 days in *A. australis* (Arai *et al.* 1999b), 200 days in *A. rostrata* and 350 days in *A. anguilla* (Wang & Tzeng 2000). It also varies, to a lesser extent, between populations of a single species, as for example, *A. australis* found in Australia metamorphose earlier and enter fresh water 3 weeks earlier than those found in New Zealand, whereas the duration of metamorphosis, about 27 days, is similar in the two stocks (Arai *et al.* 1999b).

There is increasing evidence that leptocephalus larvae that metamorphose earlier also enter fresh water earlier, this factor thus governing partly the partitioning of eels among river basins. Based on the capture locations of the smallest leptocephali, American eels and European eels spawn in greatly overlapping areas of the Sargasso Sea (McCleave *et al.* 1987). Both species drift northwards, but American eels metamorphose earlier (Wang & Tzeng 2000) and then move westwards, perhaps utilising an element of active locomotion. Sex is also a governing factor, as shown in *A. rostrata*, with the sex ratio of riverine populations being skewed to the detriment of males in northern latitudes. Male American eels are more abundant at southern latitudes and in estuaries and grow no larger than 45 cm, maturing between 3 and 10 years of age. By contrast the females migrate as far as northern Canada and Iceland (Avise *et al.* 1990) and to the upper limits of the longest river systems, occurring in an extremely wide range of habitats. They mature at 4–13 years of age in the south, but may be as old as 43 years in Nova Scotia (Jessop 1987). Alternatively to the influence of sex on migration, it is also possible that migration to northern latitudes influences eel sex through a thermolabile sex determinism in this species, with cold temperatures promoting the differentiation of female gonads, although this remains to be demonstrated.

Attraction to fresh water at this stage appears to be due to dissolved organic material in outflowing freshwaters (Sorenson 1986; Tosi & Sola 1993). Glass eel movements in the estuary are influenced by water flow and rainfall, light, lunar cycle and water clarity, which affect the vertical distribution of glass eels, and their chances of taking advantage of tidal transport. As they move upstream glass eels become pigmented and are termed elvers and may move inland substantial distances, negotiating major obstacles. These develop into 'yellow' eels, the juvenile stage which grows for several years (and up to 60 years) before returning to the sea. This seaward migration is also associated with a change in colour to a silvery bronze, deposition of fat stores and enlargement of the eyes, especially in males. The non-feeding silver eels purportedly migrate back to the spawning grounds, using orientation cues that are poorly understood. Swimming depths during the spawning migration are shallower than thought initially, as exemplified by a recent study of Aoyama *et al.* (1999) who tracked acoustically tagged Japanese eels released from a submersible, and found that they migrated rather leisurely (<0.4 m s^{-1}) in warm waters less than 200 m deep.

Although it represents most of the freshwater life of *A. anguilla*, migration at the yellow eel stage is the least extensively studied part of its lifecycle. Migrations of yellow European eels deeper into river catchments can continue in successive years after entry as elvers until eels reach sizes as large as 40–45 cm and 10+ years of age (Moriarty 1990). The elver is capable of migrating 150 km upstream before it is fully pigmented (Tesch 1965). Once it is fully pigmented it can travel considerably further in its first year although this may be less if hindered by obstructions. Upstream migration of young eels is slow with some individuals still found in the lower reaches of rivers after 2 or more years (Tesch 1977). Pigmented eels do not exhibit strong rheotaxis to the main current for migration. They continue to swim even if the current is reduced or ceases completely. As a result they often end up in backwaters and only relocate the current after some delay (Tesch 1977).

Yellow eels migrate mostly at night, which also is the time of the day when they normally forage (e.g. Helfman *et al.* 1983; LaBar *et al.* 1987; Baras *et al.* 1994b), but diurnal movements can take place on cloudy days or in turbid waters (McGovern & McCarthy 1992; Parker 1995; Baras *et al.* 1998). In temperate latitudes, the upriver migration of yellow eels has a strong seasonal component, as eels rarely migrate at temperatures less than 15°C (Sörensen 1951), and the number of migrants increases with increasing temperature in mid- or late spring (Mann 1963; Baras *et al.* 1996b). In the River Meuse (rkm 227), the upriver migration extends over a short period of time, and ends in mid- or late July (Baras *et al.* 1996b), whereas eels in the Irish Shannon River or in the Norwegian Imsa River migrate until early autumn (Moriarty 1986; Vøllestad & Jonsson 1988). Moriarty (1986) showed that the size of eels in the River Shannon decreased throughout the season due to a later and shorter migration period of small eels. Baras *et al.* (1996b) also showed a marked variation in yellow eel size, but exclusively during the first migratory wave in spring. Later in the season, eels migrated in waves, among which the number and size of eels could not readily be accounted for by environmental factors, which led Baras *et al.* (1996b) to suggest these migratory waves might result from increasing population density in downstream reaches as a result of immigration. Upstream migration rates vary substantially between rivers, averaging 8 km yr^{-1} in the Tadnoll Brook, England (Mann & Blackburn, 1991), 15 km yr^{-1} in the Shannon (Moriarty 1986), 10–15 km yr^{-1} in the River Dee, England, 20–30 km yr^{-1} in the River Severn, England (Aprahamian 1988), 45–46 km yr^{-1} in the Rivers Tweed, England (Hussein 1981) and Meuse, Belgium (Baras *et al.* 1996b). Gradient steepness had been invoked as a key factor behind slow migration rate (Aprahamian 1988), but water clarity may be invoked too, as it restricts the period of the day during which eels make their ascent, especially at high latitudes. Eels can clear many physical obstacles, including under fast currents, provided the substratum is rough (irregular) enough to enable their reptant behaviour (Legault 1992). Conventional fish pass design may impede or slow down their migration, and clearing a 5-m difference in height through a Denil fish pass with slow current may require more than one night (Baras *et al.* 1994b). Where navigation locks are available, and provided these are open late at dusk or early at dawn, they may serve as alternative pathways for yellow eel migration (Baras *et al.* 1996b) (see also section 7.4).

Part of the difficulty when studying yellow eel migration is that migrants and non-migrants may be of similar size (see comparisons between the distributions shown by Baras *et al.* (1996b) and those by Philippart and Vranken (1983), Aprahamian (1988) or Vøllestad and Jonsson (1988)). Nevertheless, upstream migrants rarely exceed 40 cm in length, a size above which most eels become relatively sedentary and migrate only as a result of meteorological, hydrological or seasonal factors. During this period home ranges are very small. Mann (1965) showed that on the River Elbe, 16 out of 47 eels were recaptured where they were originally caught and 21 had moved only 10–60 m. Baras *et al.* (1998) radio-tracked seven yellow eels in the Awirs stream, a small tributary of the Meuse, demonstrating a low level of movement, and fidelity to discrete refuges, as also found by McGovern and McCarthy (1992). Net journeys were higher in May and June which corresponded to the immigration of migratory yellow eels from the Meuse. Baras *et al.* (1998) argued as a result of this that eels adopt a sedentary lifestyle in fast-flowing streams when eels in the main river were usually migratory. If eels do change habitats during freshwater residence, movement takes place during the transition phases between summer and winter. McGovern and McCarthy (1992) used acoustic tracking to show that yellow eels in the Clare River, Ireland, were relatively sedentary. Movements did however increase in the autumn and were attributed to eels moving to overwintering habitats.

At a certain age and size, anguillid eels start their downstream migration to the sea, at the yellow or silver stage. Conversely, not all eels regarded as having started silvering move to the sea within this season. In French Brittany, only about 20% of the silver eel emigration candidates did so (Feunteun *et al.* 2000). In *A. rostrata*, the size of females at downstream migration is positively correlated with latitude, while male size is not (Oliveira 1999). Downstream movements of *A. anguilla* take place at temperatures between 18°C and 4°C, with no apparent threshold temperature, but a peak at 9–12°C (Vøllestad *et al.* 1986). Within this thermal range, migration is encouraged by increased discharge, and downstream movement is faster too (see also sections 2.2.2.2 and 2.2.2.3). Silver eels mostly migrate at night, at light intensities of less than 0.06 lux, rendering the timing of the downstream migration dependent on the lunar cycle (review for *A. anguilla* in Jonsson 1991).

5.10 Anchovies, shads, herrings and menhaden (Clupeiformes)

There are about 360 species of clupeiforms, of which about 80% are marine. They tend to be small, silver-bodied, shoaling fishes, most of which are pelagic and planktivorous. The single species of denticle herring *Denticeps clupeoides* (Denticipidae) is a freshwater species, occurring in coastal rivers of West Africa, but there appears to be no clear evidence of migratory behaviour. Similarly, relatively little information is available for those members of the Engraulidae (anchovies) that occur in fresh water, these being most widespread in tropical and subtropical waters. The manjuba *Anchoviella lepidentostole* is known to migrate into rivers along the Brazilian coast during spring and autumn to spawn, adults remaining in the rivers for several months (Giamas *et al*. 1983). The Atlantic sabretooth anchovy *Lycengraulis grossidens* occurs in the coastal waters, estuaries and rivers of the Atlantic coast of South America. It shows anadromous migrations, although some populations or forms seemingly complete their lifecycle in fresh water (e.g. form limnichthys in Lake Maracaibo). The anchovy *Thryssa scratchleyi* is a freshwater species occurring in the Indo-Pacific region which appears to migrate to the lower reaches of estuaries to spawn (Roberts 1978). Diadromous behaviour may well occur for several other anchovy species (McDowall 1988), but is poorly documented.

There are about 180 species of clupeids, among which anadromy is exhibited by shads including various species of *Alosa*, *Tenualosa* and *Hilsa* which are usually 30–50 cm long at adulthood. Anadromy is common in several North Atlantic clupeids, represented by various species of *Alosa* and *Dorosoma* and commonly known as shads and gizzard shads respectively. The alewife *Alosa pseudoharengus*, blueback herring *Alosa aestivalis* and American shad *Alosa sapidissima* of eastern North America are perhaps the most important and well-studied species. Detailed reviews of the biology of alewife and blueback herring are provided by Loesch (1987). The alewife occurs from Labrador in the north to Carolina in the south, while blueback herring are found from New Brunswick in the north to Florida in the south. Study is complicated by the fact that these two species occur sympatrically over much of their range and that they are morphologically similar. Collectively they are known by commercial fishermen as alewife or gaspereau in Canada, and river herring in the US. Extremely large numbers migrate in some river systems such as the St John River, New Brunswick. Upstream migration of both species tends to be limited to the lower reaches of rivers but in the larger systems this may be well over 100 km inland. Both species occur at the Mactaquac Dam, 148 km from the mouth of the St John River (Messieh 1977) and some fish which pass upstream of the dam migrate another 100 km upstream.

In general, blueback herring tend to migrate further upstream than alewife (Loesch 1987) despite previous contradictory statements which are reflected in McDowall (1988). This makes sense since alewife tend to use lentic sites for spawning whereas, within their sympatric range, blueback herring use lotic sites. In allopatry blueback herring also use lentic sites. Thus the differential migration and selection of spawning sites by blueback herring may serve to reduce interspecies competition for spawning grounds (Loesch 1987). Blueback herring tend to be more abundant in high river gradient systems than alewife and vice versa. Adult alewife move into rivers from March in the south to June in the north and may continue for several months in the larger river systems. In the St John River alewives begin running upriver in large numbers at the end of April and blueback herring in the middle of May and both species begin moving downstream from Mactaquac Dam in late June (Jessop 1994). By late August and through September they move out of the river and along the New Brunswick coast. Results of mark recapture using flag and spaghetti Floy tags showed that adults migrate upriver at rates of 8–21 km d^{-1}. Recapture data of repeat spawning fish between 1973 and 1983 showed that fish not only returned to their home river but to previous spawning sites. Tagged repeat spawning fish of both species have been recaptured at original spawning sites at rates of 63–97%, suggesting high site fidelity (Jessop 1994). This may also reflect homing to natal sites but this has been technically difficult to prove due to the difficulty of tagging underyearling fish. Young shad remain within the river system during the first summer but move downstream progressively and out of estuaries in autumn. Outward migration of alewife, blueback herring and American shad in the Annapolis River, Nova Scotia, was associated with increased precipitation, a sharp decline in temperature and the occurrence of a new moon (Stokesbury & Dadswell 1989).

The American shad occurs in North American Atlantic coast rivers from Labrador southwards to Florida and has a similar life history to alewife and blueback herring. This species ranges widely at sea and ascends rivers all along the coastline, and may migrate to spawn just above the head of tide or as far as the upper tributaries of rivers (Bigelow & Schroeder 1953). Upriver migration as far as 600 km inland is known (McDowall 1988). Estimated rates of upstream riverine migration for American shad are 1.6–3.1 km d^{-1} (Leggett 1976). Postspawning survival is negligible in rivers south of Chesapeake Bay, but the proportion of repeat spawners increases northwards (Leggett 1973; Leggett & Carscadden 1978; see also section 2.3.5). Although American shad display homing to natal rivers, sufficient straying occurs to permit only marginal population differentiation as determined from microsatellite DNA and mitochondrial DNA differences between widely separated spawning sites (Waters *et al.* 2000). The eggs are demersal and non-adhesive, and they and the newly-hatched larvae are carried downstream. The young grow in fresh or brackish water for the first summer before moving to sea and mature after 3 years or more (Leggett 1973). Broadly similar patterns of anadromous migratory behaviour are displayed by the gulf shad *A. alabamae* and the skipjack herrring *A. chrysochloris* in rivers draining to the Gulf of Mexico, and hickory shad *A. mediocris* in temperate western Atlantic drainages (McDowall 1988).

Several species of shad ascend rivers on the northeast Atlantic coast to spawn. The allis shad *A. alosa*, distributed from Norway south to the Bay of Biscay and the Mediterranean, enters rivers during spring and spawns in swiftly flowing fresh waters (Wheeler 1969). The adults return to sea, while the young spend 1–2 years in fresh water before migrating to the sea. In the Loire River, France, migration of allis shad usually starts in April but may begin as early as February (Boisneau *et al.* 1985). They found that migrating shad used the main channel where water velocities were highest and that daily and hourly activity of migration was strongly influenced by temperature and light. No migration was observed below 11°C, while migratory activity was greater at night. Migration was limited by obstacles (weirs and dams of nuclear power stations), although some individuals managed to penetrate as far as

500 km upstream of the estuary. The twaite shad *A. fallax* occurs from Norway and Iceland south through the Faeroes to the Bay of Biscay and the Mediterranean. The adults migrate into rivers in late spring and spawn in the lower reaches of rivers, above the tidal limit (Wheeler 1969). The fry move downstream to the sea within the first summer. Both allis shad and twaite shad have lacustrine 'landlocked' populations (Whitehead 1984). Pontic shad *A. pontica* occur in the Black Sea while Caspian shad *A. caspia* occurs in the Black and Caspian Seas. Pontic shad are strongly anadromous, spending much of their adult life in pelagic marine environments and moving in shoals as much as 1000 km upriver between mid-May and August in rivers such as the Don and Danube (Whitehead 1984). Caspian shad tend to remain in estuaries and although they may enter fresh water to spawn in Black Sea rivers, the Caspian form does not enter fresh water.

The most well-known of the tropical shads are the anadromous hilsa shads which comprise several species of *Hilsa* and *Tenualosa* from Africa, India and Southeast Asia. The Indian hilsa *Tenualosa ilisha* of India, Bangladesh and Burma is the largest of the clupeids, reaching 60 cm in length, and this and other anadromous hilsa shads such as the more widespread *Hilsa kelee* provide important fisheries (Whitehead 1985). The Indian hilsa has been widely researched and there are records of spawning migrations of at least 1200 km up some river systems in India, although most migration distances are 50–100 km and movements are increasingly restricted by impoundments (Jones 1957; Pillay & Rosa 1963; Islam & Talbot 1968; Whitehead, 1985; Miah *et al.* 1999). River entrance and upstream migration by mature Indian hilsa occurs during the period of high water in the monsoon season, when rivers are swollen, during which period they do not feed. Spawning activity appears to be synchronised to the full moon and new moon phases (Miah *et al.* 1999). The eggs are deposited in fresh water and they and the hatched larvae drift downstream. The young may remain in fresh water for several months, growing before entering the marine environment, where they reach maturity at an age of one or two. Surviving postspawners return to sea and repeat spawning is relatively common. Some Indian shad populations remain in fresh water and may exhibit limited movement or potamodromous behaviour, occurring in the same river systems as anadromous population components (Pillay & Rosa 1963; Whitehead 1985). *Tenualosa macrura*, *T. reveesi* and *T. toli* have broadly similar life histories (Whitehead 1985).

The gizzard shads include a variety of genera occurring in fresh water, of which the best known and studied is *Dorosoma* which occurs in North and Central America. *Dorosoma cepedianum* is found along eastern North America from New York south to Mexico. The species spends most of its time in brackish water in estuaries but ascends rivers to spawn in fresh water in spring and returns to the estuary. The young spend several years in fresh water before entering estuarine areas. They may live for 10 years and are repeat spawners (Miller 1960). This and several other *Dorosoma* species also occur in fresh water and have been introduced into many lakes and reservoirs as 'forage' fish for predatory species. In Grand Lake (Oklahoma) large numbers of juvenile gizzard shad are entrained in the dam outflow. This is interpreted as being due to their wintering behaviour at low temperatures (Sorenson *et al.* 1998) but could reflect a natural tendency for outmigration from the lake.

Many temperate sprats and herrings are marine but some exhibit regular diadromous patterns, though these seem somewhat facultative. *Clupeonella cultriventris* is a brackish water species, tolerating salinities up to 13‰, but with semi-anadromous and purely freshwater forms in rivers and lakes in Eastern Europe. Some Black Sea populations spawn in the lower reaches of rivers such as the Dnieper in May. The adults then migrate to the Black Sea for the summer before returning to the rivers in August to November (Whitehead 1984). The subfamily Pellonulinae comprises about 44 species of exclusively freshwater herrings, but because these schooling clupeids range widely, feeding on plankton, structured migratory behaviour is not conspicuous or of great importance. The tropical freshwater clupeid *Limnothrissa miodon* is a truly lacustrine species, endemic to Lake Tanganyika, but it has been introduced into many African lakes and water bodies, including Lake Kivu and Lake Kariba, where it supports important fisheries. Adults are generally found in the pelagic zone of the lakes, and may spawn at a small size (>3.5–8 cm) all year round (Iongh *et al.* 1983). Larvae are found in littoral waters exclusively, where they make no or little horizontal daily migrations. At the juvenile stage (>20 mm TL), *L. miodon* makes more extensive daily movements into littoral areas during the day, and offshore at night (Isumbisho, Descy, Baras [unpubl.]).

Migratory clupeids are most common in the Indo-Pacific region and especially in Southeast Asia. *Ilisha megaloptera* ascends and spawns in rivers, and similar behaviour may be shown by *I. novacula*, although the latter species may be entirely riverine (Whitehead 1985; McDowall 1988). Several *Pellona* (Asia) and *Pristigaster* (South America) species may also exhibit migratory behaviour although details are sketchy (Whitehead 1985). Several species of clupeids breed in marine environments but have young which move into brackish water and may penetrate fresh water. Such behaviour appears common in tropical and subtropical environments. Young *Escualosa thorocata* from the Indian Ocean region enter rivers but later return to sea (Whitehead 1985). Juvenile bonga shad *Ethmalosa fimbriata* and the herrings *Sardinella maderensis*, *S. aurita* and *Ilisha africana* exhibit tidal and seasonal patterns of movement into weakly brackish parts of the Sine-Saloum estuary, Senegal (Guillard 1998). Bonga shad aged 1 and 2 years are reported to migrate 200–300 km up the Gambia River in March and January respectively (Whitehead 1985).

Perhaps the best examples of such amphidromous movements with regard to clupeids are for the Atlantic menhaden *Brevoortia tyrannus* and the Gulf menhaden *B. patronus*. Although McDowall (1988) classed menhaden as 'marginally amphidromous', for several populations of these species, entry to brackish water refuge environments is predictable, forms a key part of the lifecycle and appears important to survival and recruitment processes (Weinstein 1979; Marotz *et al.* 1990). Incursion into fresh water does occur, but is not typical. Gulf menhaden *Brevoortia patronus* in the Gulf of Mexico rely on the extensive marshlands on the coast of Louisiana as nursery habitats, but these are increasingly being regulated, using canals with low weirs which allow water movement during only part of the tidal cycle (Marotz *et al.* 1990). These authors used traps to study movements of Gulf menhaden through three routes between marsh and the Gulf of Mexico to determine ways to assist passage through water control structures. They found that the young are born offshore (August–April), move to coastal waters over a period of 6–10 weeks and then move inland through brackish marsh waters (November–April). There is an exit of fish mostly over the same period, but extending over May–October. Fish of up to 3 years of age utilise the marsh habitat. Fish entry and exit to the marshes is a tidal phenomenon. Marotz *et al.* found that seaward migrations of Gulf menhaden were especially high during periods of low oxygen in the marshes, which occurred frequently and unpredictably over the summer period, with an apparent critical level of 2 mg O_2 per litre.

Further discussion concerning the lifecycles and possible migratory behaviour of a range of clupeids including *Nematalosa, Clupanodon, Anodontostoma, Herkotslichthys* and *Pellonula* is available elsewhere (Whitehead 1985; McDowall 1988).

5.11 Milkfish (Chanidae)

The milkfish *Chanos chanos*, the only species in the Chanidae, is a large (up to 180 cm SL) long-lived (15 years), principally marine and brackish water species occurring in tropical and subtropical Indo-Pacific regions. McDowall (1988) regards this fish as a euryhaline wanderer, and it frequently enters coastal lakes where it competes with other iliophagous species (Blaber 1988). Milkfish are highly fecund (several millions of eggs per female) and are egg scatterers, spawning exclusively at night in shallow marine water, above sand or coral at no more than 30 km from the shore. Pelagic eggs hatch within 24 h and the pelagic larvae migrate to the shore, seeking water warmer than 23°C. Metamorphosis from the ribbonlike larval stage occurs at a body weight of 25 mg in brackish water, and juveniles settle in the estuaries or mangroves, but frequently enter freshwater lakes (Nelson 1994). The lifecycle of the milkfish may, therefore, be somewhat more reliant on migratory behaviour between seawater and brackish environments than is sometimes assumed.

5.12 Carps and minnows (Cyprinidae)

The cyprinids are principally a freshwater group, widely distributed in North America, Africa and Eurasia, and represented by over 1700 species. This group is found in a wide variety of habitats from swift-flowing streams to deep, slow-flowing lowland rivers and backwaters, although they are perhaps most characteristic of the latter. Reports of migratory behaviour by cyprinids have, until recent years, been few and were often anecdotal (Mills 1991; Smith 1991). The paucity of confirmed migration by comparison with the large number of cyprinid species has been suggested as being due to little research (Smith 1991) but may also be associated with smaller scale migration than for some other taxa and difficulty of observation in some habitats. More recently a greater amount of research has been carried out on the movements of cyprinids, especially in Europe and North America, and clear migratory patterns have been demonstrated in an increasing number of studies. The decline in some cyprinid populations, especially of rheophilic species, in many river systems has been linked to interruption of cyclical movements by dams and weirs (see chapter 7).

Perhaps the most celebrated example of migration in cyprinids has been the Colorado pikeminnow *Ptychocheilus lucius*, a large, endangered cyprinid endemic to the Colorado River basin of western North America (Box 5.1 and Fig. 5.4). Adult Colorado pikeminnow can attain a large size, often exceeding 0.6 m and 2 kg, and are top predators, mostly eating fish. This species was once widely distributed in the upper and lower Colorado River basins, but has been eradicated from 80% of its former range (Tyus 1986). The decline in distribution appears to have resulted from building of dams, river regulation, loss of habitat and expansion of an exotic fish community in many areas. The pikeminnows appear to have evolved a large body size and mobile habit during the late Cenozoic era, enabling alternation between feed-

Box 5.1 Shooting the rapids – migration and homing of the Colorado pikeminnow

Natural populations of the Colorado pikeminnow *Ptychocheilus lucius* currently exist only in the Colorado, Green, Yampa, White, Gunnison and San Juan Rivers of the Upper Colorado River basin. Adult fish are known to use restricted spawning sites in the Yampa (Yampa Canyon) and Green Rivers (Desolation Canyon) but migrate to these to spawn and subsequently return to their 'home' feeding areas, showing high site fidelity for these (Fig. 5.4). The rivers run through arid scrublands and have strong seasonal temperature variation: water temperature can approach 0°C in winter and 25°C in summer. They tend to exhibit extremes in discharge and frequently carry a high silt load. The high gradient river reaches are dominated by riffles, runs and rapids that run through canyon areas with sheer cliffs and boulder-strewn banks. In lower gradient areas there are deep eddies and pools and runs that meander through slower areas with partially vegetated banks. Radio-tracking has shown that adult Colorado pikeminnows in the Green River system migrate upstream or downstream to spawn in highly oxygenated whitewater rapids that may be well over 100 km from their preferred feeding habitat of deep pools and runs (Tyus 1985, 1990; Irving & Modde 2000). The spawning migration commences at or slightly after the spring elevation in discharge in May during which flows are three to ten times higher than normal over a prolonged period, and homing occurs between discrete feeding and spawning areas.

Twelve adult Colorado pikeminnows radio-tagged by Irving & Modde (2000) in the White River moved downstream out of the White River, to spawn in the Green and Yampa rivers in June and July 1993. From the confluence, seven fish moved upstream into the Green and Yampa Rivers to use spawning areas above the confluence of the White and Green Rivers, while five moved downstream to spawn at Desolation Canyon (Green River). All fish returned to the White River after spawning to their home feeding areas (Fig. 5.4). In July 1994 all 12 fish were again relocated in these spawning areas. The fish travelled 231–624 km to reach spawning areas in 1993 and the average distance moved during the 1993 migratory period was 644 km (range 438–745 km) over an average period of 97 days (range 46–167 days) between May and October 1993. Between the start of tracking for each fish (1 October 1992 to 28 April 1993) and its end (13 July 1994) the total minimum distance moved was 763–1405 km per fish (average 1060 km). However, seasonal movements of radio-tagged adult Colorado pikeminnows in the upper Colorado River were much smaller than those in the Green River system (McAda & Keading 1991), with movements mostly less than 50 km. Wide distribution of larvae and radio-tracked adults during the spawning season, together with occurrence of areas of cobble substrate throughout the 350 km study reach, strongly suggested that spawning occurred at widely scattered locations. If spawning areas are interspersed with appropriate feeding areas, the apparent wide distribution of spawning habitat could explain the difference in migratory patterns. At the end of the flood period, eggs hatch and larvae drift downstream to occupy warm, shallow habitats where rapid growth is possible (Tyus 1991). This strategy may aid the dispersal of the young from what is a low carrying capacity environment. Although Tyus suggested that Colorado pikeminnow larvae may drift well over 100 km before settling in appropriate conditions, where suitable YOY habitat is widespread, it appears drifting may be more restricted (McAda & Keading 1991).

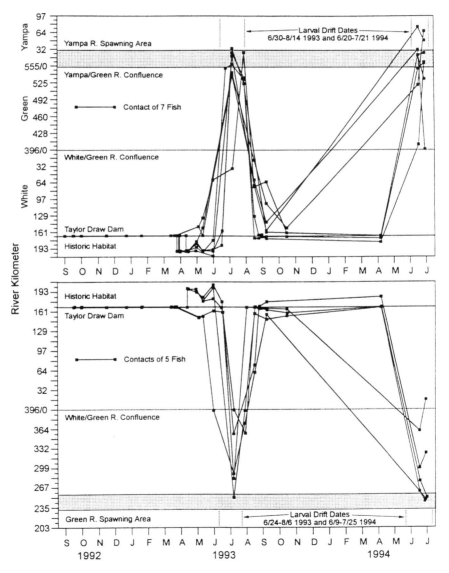

Fig. 5.4 Migration and homing of radio-tracked adult Colorado pikeminnow *Ptychocheilus lucius* from the White River to the Green and Yampa Rivers, in the Rocky Mountains, US. Reproduced from Irving and Modde (2000). Fish moved downstream, out of the White River, to spawn at two historic spawning sites, and later returned to their 'feeding and growth' home ranges.

ing in rich, eutrophic areas and reproduction in more favourable upriver areas (Tyus 1986). As the southwestern United States became more arid, these migratory habits may have assisted the survival of this species in a harsh environment. However, blockage of migration routes would be expected to have a serious effect in an ecosystem where spawning, nursery and adult areas are widely separated.

Another cyprinid which is endemic to the Colorado system is the humpback chub *Gila cypha*, with a body form which reflects a highly specialised mode of life. It has a massive protrusion on the dorsal surface behind the head which acts as a compression keel, so that the force of the water allows it to maintain a benthic existence and move in the boulder-strewn rapids of the Colorado with relatively little energy expenditure. Its range has greatly declined due to impoundment and related factors, and it is most abundant at the confluence of the Colorado and Little Colorado rivers in northern Grand Canyon, Arizona. Monthly sampling including mark recapture of humpback chub in the Little Colorado River confluence area over the period 1991–1992 indicated an upstream migration into and up the Little Colorado River by some fish in early spring, followed by a slow downstream post-reproductive movement (Douglas & Marsh 1996). There was a tendency for continuous presence by many fish around the confluence area which provided good habitat in a system which has been substantially altered by the effects of impoundment. Current limited migration may be due to the reduction in the number of patches of good juvenile and adult habitat. Spawning and egg hatching for humpback chub and other tributary-spawning cyprinids such as speckled dace *Rhinichthys osculus* occurs at the end of the spring high flow period in these tributaries. Prior to significant impoundment, annual peak Colorado River flows between April and July typically coincided with decreasing flows in the tributaries and timing of reproduction. High main river flows ponded canyon-bound tributary mouths which probably helped to retain larvae, preventing them from being forced out into the fast canyon waters until more fully developed. Larvae of humpback chub exhibit passive drift and resultant settling in depositional areas, providing sheltered nearshore habitat until greater swimming capacity is achieved (Brown & Armstrong 1985; Robinson *et al.* 1998). Transport of eggs and larvae out of tributaries into main river environments reduces survival through physical damage, lack of juvenile habitat or low growth due to coldwater release.

The northern pikeminnow *Ptychocheilus oregonensis* is a piscivorous relative of the Colorado pikeminnow, but is much more characteristic of lakes and slow-flowing rivers. They spawn on gravel in lakes and rivers but do not appear to exhibit the large-scale migrations of the Colorado pikeminnow (Scott & Crossman 1973). The occurrence of large numbers of northern pikeminnow in summer below Lower Granite Dam, on the Columbia River, Washington, is a regular occurrence and it has been suggested that they aggregate to feed on juvenile salmonids migrating downstream past the dam. They may be responsible for an estimated 75% of predation in John Day Dam on the lower Columbia River. Radio-tracking of 140 adult northern pikeminnow tagged 0–7 km below Lower Granite Dam showed that they moved upstream in summer and accumulated below the dam, reaching peak numbers in July as discharge was decreasing (Isaak & Bjornn 1996), and after most juvenile salmonids had passed the dam. It seems that upstream movement was not directly related to a coordinated feeding migration to take maximum advantage of the bottleneck for salmonid migration, but probably reflected an upstream spawning migration by northern pikeminnow.

A large number of rheophilic cyprinid species of temperate river systems exhibit spring spawning migrations, particular where gravel or stony habitats are required for spawning. The adults may occupy a range of habitats and in winter often move to deeper, slower areas of the main river. Movement to swift-flowing spawning areas may involve short movements into the mouths of tributaries but may require longer migrations in the main river channel, typically in an upstream direction, to find appropriate spawning habitat. The adults often remain in this habitat to feed during the summer. This behaviour is characteristic of many of the species of *Aspius*, *Barbus*, *Chondrostoma*, *Leuciscus* and *Vimba* in Eurasia (Lelek 1987; Povz 1988; Baras 1992) and of at least some species of *Rhinichthys*, *Notropis*, *Semotilus* and probably also *Acrocheilus alutaceus* in North America (Scott & Crossman 1973).

Barbel *Barbus barbus* are large (up to 1 m TL), long-lived (up to 35 years) iteroparous fishes inhabiting rivers of central and western Europe, and they are highly mobile in the spawning season (Baras & Cherry 1990; Baras 1992, 1993a; Baras *et al.* 1994a; Lucas & Batley, 1996). Barbel spawning migrations show strong seasonal periodicity with peaks in May in the Rivers Meuse and Ourthe, Belgium (Baras 1992, 1993a; Baras *et al.* 1994a). Both males and females migrate in spring to spawning grounds (Lucas & Batley 1996; Lucas & Frear 1997), but females may move greater distances than males (Baras 1992), owing to their larger size and preference for more lentic habitats outside of the spawning season. Baras *et al.* (1994a) showed that the first fish in the migration are males and immature individuals on their way to spawning grounds. Males usually gather around the spawning grounds at least 1 week before the beginning of spawning (Hancock *et al.* 1976; Baras 1992, 1994). Females remain less than 48 h on the spawning grounds, then either move downstream or rest for several days in habitats nearby the spawning area before undertaking a downstream homing movement to the summer area. Males remain on the spawning grounds for longer, apparently searching for receptive females (Baras 1994; Lucas & Batley 1996). During the summer, barbel movements become much more stable, reflecting fidelity to a defined activity area with very high local activity (Pelz & Kästle 1989; Baras 1993b; Philippart & Baras 1996). *Barbus meridionalis* may exhibit similar migratory behaviour to *B. barbus*, although studies by Chenuil *et al.* (2000) using microsatellite DNA mapping and mark recapture indicate that movements of both species in the River Lergue, southwestern France, were mostly limited to just a few kilometres, resulting in a stable hybrid zone.

The nase *Chondrostoma nasus* is a moderately large cyprinid which forages on the bottom and often attains a length of over 40 cm. It lives in shoals in swiftly flowing streams and rivers throughout much of Europe but is absent south of the Alps and in Denmark, Scandinavia and the British Isles (Lelek 1987). The species has dramatically declined in abundance and distribution in recent years (Penáz 1996). Where healthy populations occur the spawning migration is conspicuous, involves large numbers of fish and often occurs over long distances, some exceeding 100 km (Povz 1988). In spring they migrate upriver and into tributaries to spawn in swiftly flowing shallow water, less than 30 cm deep on a gravel or cobble substrate. The migrating shoals may be large, with many documented reports of several thousand fish in a shoal although these are rare today (Zbinden & Maier 1996). Migrating (and spawning) shoals are also typically segregated into fish of varying size and age classes, so that a single spawning shoal typically exhibits a single mode and narrow distribution of size (Penáz 1996). In the summer, shoals of nase can also show high mobility, as their grazing habits cause them to utilise new patches of periphyton, whereas movements are most limited during autumn and winter (Huber & Kirchhofer 1998). Although the eggs are negatively buoyant and have adhesive properties, a high proportion may drift downstream and fail to hatch. The young exhibit a nocturnal pattern of near-surface drift downstream and rely on finding sheltered bays and slow-flowing areas in which to feed and grow (Hofer & Kirchhofer 1996). Although juvenile nase expand their range of habitat use as they grow and swimming capacity increases, they still require adequate lentic habitat in which to take refuge during periods of high flow (Baras & Nindaba 1999b).

Broadly similar behaviour is shown by asp *Aspius aspius* which occur in running water throughout much of eastern and central Europe. They move into swift water and spawn over a clean, sandy or pebbly substrate in April and May (Lelek 1987). The young drift downstream into more lentic sections and form shoals of juveniles. Radio-tracked chub *Leuciscus cephalus* in the River Spree, Germany, exhibited an upstream spawning migration in May of up to 13 km. After spawning they homed back to their original location (Fredrich 1996). Migratory behaviour of chub may be somewhat variable with annual habitat use and range of movement exhibiting substantial variability between individual fish and rivers (Lucas *et al.* 1998, Allouche *et al.* 1999). Adult chub commonly occur in fish pass catches between May and September, which may reflect an extended spawning period that appears to occur in at least some rivers (M. Lucas [unpubl.]) or merely widespread upstream directed activity during summer. Migrating adult chub passing through a Denil fish pass on the Yorkshire Derwent, England, on their spawning migration, principally in May and June, did so almost exclusively at night (Lucas *et al.* 1999; Lucas 2000).

Although eurytopic cyprinids such as common bream *Abramis brama* may not have highly specific spawning habitat requirements, they often exhibit distinct migrations between feeding, refuge and historical spawning areas. Whelan (1983) showed that in the River Suck, Ireland, some individual common bream exhibited exceptional movements of up to 59 km. Most fish, however, remained in shoals which displayed regular spawning migrations of up to 10 km. Observations of common bream at the Derrycahill spawning site on the River Suck over several years showed that there was a Derrycahill shoal, a shoal which moved upstream to spawn and two shoals which moved downstream to a spawning site, producing altogether a spawning shoal of over 4000 fish. After spawning the aggregation broke down into separate shoals which returned to their respective feeding grounds. However, much larger migrations of common bream occur in the River Don, Russia (Federov *et al.* 1966), and from rivers that drain into the Black Sea and Caspian Sea and vice versa (Nikolsky 1963). Caffrey *et al.* (1996) found that the movements of radio-tracked bream in the Barrow, Grand and Royal Canals in Ireland became erratic during the spawning season and shoals moved considerable distances from their home ranges. However, other fish did not move from their home range and the extent of movements during the spawning season was similar to movements at other times of the year. This led Caffrey *et al.* (1996) to conclude that these movements could not necessarily be attributed to spawning migration. Data from European fish passes show that there is a peak in the occurrence of white bream *Blicca bjoerkna* in the spring which would appear to precede the main spawning period (see Fig. 4.2), and this species may also exhibit well-defined movement patterns (Molls 1999). The spring peak in occurrence of common bream *Abramis brama* in European fish passes is less clearly defined than for barbel *Barbus barbus* or white bream and occurs over a more extended period (Baras *et al.* 1994a; Prignon *et al.* 1998).

Champion and Swain (1974) recorded counts of roach *Rutilus rutilus* and dace *Leuciscus leuciscus* passing through a fish trap on the River Axe, Devon, at monthly intervals from 1960–1969 inclusive. They showed that the main downstream movements of both roach and dace occurred regularly during March, April and May and were probably associated with spawning movements. This is notable in that most recorded potamodromous spawning migrations are in an upstream direction. Diamond (1985) showed that spawning shoals of roach migrate each year to utilise the same spawning grounds in a variety of different environments. Data from fish pass catches (Fig. 4.2) show that there is a peak in roach occurrence in the spring, which may be associated with prespawning movements for this species although upstream-directed activity occurs over a significant proportion of the year and may also relate to movements for other purposes, such as feeding or refuge. Radio-tracking of adult dace, in the River Nidd, northern England, demonstrated that in some rivers adults may migrate significant distances (Lucas *et al.* 2000). Dace are rheophilic and spawn on sand/gravel riffles in April in northeast England. The dace moved from the Yorkshire Ouse into the lower reaches of the Nidd, where they were tagged, and some subsequently moved further upstream to areas with suitable spawning habitat 10–21 km upstream of the Ouse confluence, after passing a flow-gauging weir (Lucas *et al.* 2000). Radio-tagged dace in the River Frome, southern England, also exhibited movements to discrete spawning areas (Clough & Beaumont 1998) and displayed a postspawning migration into small, slow-flowing side channels (Clough *et al.* 1998). This postspawning movement was interpreted as a refuge migration enabling recovery and minimal energy expenditure in habitat which was characterised by reduced predation risk from northern pike *Esox lucius*. Starkie (1975) showed that marked dace in the River Tweed, Scotland, moved average distances of 6.3 km. Pitcher (1971) observed that European minnows *Phoxinus phoxinus* undertake a spawning migration in May in which they move 250 m to 1 km upstream to gravel beds in open shallow water. Kennedy (1977) showed that tagged minnows homed back to their non-spawning area after about a month on the spawning grounds.

A substantial number of temperate cyprinids occupy lentic or slow-flowing water throughout much of the year, but move into faster-flowing water, often by ascending lake tributaries to spawn. Movements may be no more than a kilometre or two but tend to be associated with location of gravel habitat by lithophilous spawners. Such behaviour occurs for *Richardsonius*, *Couesius* and to a lesser extent for some species of *Notropis* in North America (Scott & Crossman 1973; Reebs *et al.* 1995), as well as for some populations of species such as roach *Rutilus rutilus*, in Europe (L'Abée Lund & Vøllestad 1985, 1987). In British Columbia, redside shiners *Richardsonius balteatus* move from lakes into streams to spawn (Lindsey & Northcote 1963). The migration commences in the spring when maximum daily temperature reaches 10°C. Lindsey and Northcote (1963) found that the daily number of migrating adults was correlated with increasing water temperature rather than river flow. The latter authors also found that adults tend to migrate upstream during daylight and marked fish tend to return to spawn in the same stream each year, though it is not known whether they return to their natal streams. Males arrive at spawning areas before females and spawning occurs over gravel in water as little as 10 cm deep. They are batch spawners and individuals, especially males, may return to the spawning grounds several times in one season (Weisel & Newman 1951). Downstream movement of fry occurs at night and results in the young being redistributed into the lake to grow through the juvenile stage. Scott and Crossman (1973) report that in lakes redside shiners form shoals and that there is a vertical and horizontal stratification of their sizes with smallest fish highest in the water column and closest inshore. In summer during the day adults feed over vegetated, shallow-water shoals but disperse and move out to the centre of the lake in the surface waters at night. In autumn redside shiner shoals move into deeper water.

The lake chub *Couesius plumbeus* occurs widely in lakes and rivers in northern North America and exhibit upstream spawning migrations, often moving out of lakes into shallow streams between April and June (Scott & Crossman 1973). Use of a counting fence with upstream and downstream traps was employed by Reebs *et al.* (1995) who showed that most upstream movement occurred in June and most downstream movement occurred in June, July and October. The peak in upstream movement was at dusk and downstream movement occurred at night, with negligible amounts of migratory movement during the day. Although lake chub are not large (adults are about 10 cm in length), in shallow streams they are relatively immune from predation by fishes but susceptible to avian predators (Power 1984; Schlosser 1988). Reebs *et al.* (1995) speculate that adult lake chub adopt a nocturnal strategy for movement because they are moving through shallow water, and twilight or night-time movement minimises the risk from avian predation, the principal source of predation for this species.

In streams with Mediterranean climates, seasonal movements of fish may be related to physical and biotic factors. Cyprinids are an important component of the endemic communities occurring in Mediterranean streams, the upper reaches of which often partially dry out in summer, leaving a series of isolated pools. Species such as Iberian chub *Leuciscus pyrenaicus* commonly penetrate into the upstream reaches in spring to spawn, but as water depth decreases in summer, adults retreat to deeper areas in the permanent flowing areas further downstream (Magalhães 1993). Juveniles or small species (e.g. *Barbus haasi*, Aparicio & De Sostoa 1999; *Rutilus alburnoides*, Pires *et al.* 1999) remain in the shallow disconnected areas throughout the summer and are joined by larger fishes when flows increase in autumn. Large fish in the shallow, intermittent stretches create greater demand for oxygen than small fish in these environments and may be subject to respiratory stress, thus limiting their persistence in these habitats. Size-dependent predation risk may also influence movement, with shallow water acting as a refuge for small fish from predation by other fishes and making larger chub susceptible to predation by mammalian and avian predators.

Similar hot, dry conditions, with infrequent and variable rainfall patterns, occur throughout much of the Middle East and arid parts of Africa. Although cyprinids are abundant in these regions there is a relative paucity of information concerning their behaviour. Nevertheless, under these conditions there seems little doubt that river discharge acts as an important trigger for migration of a wide variety of juvenile (Cambray 1990) and adult (Fishelson *et al.* 1996) cyprinids. In the River Jordan catchment *Barbus canis*, *B. longiceps* and *Capoeta damascina*, which are all relatively large cyprinids (adults often exceeding 30 cm in length), aggregate and move upstream at the start of the rainy season in November–December (Fishelson *et al.* 1996). This period is characterised by a reduction in water temperature (temperatures normally of 16–18°C in November–December) and flood activation of perennial tributary streams. The spawning season of *C. damascina* is principally December–February and large groups of this species migrate to the highest reaches of tributaries (25 km, 400–900 m ASL) through strong rapids. The females dig shallow nests in sand and gravel in which to deposit adhesive eggs and after spawning return to the main river. Similar migratory habits are exhibited by *B. canis* and *B. longiceps*, although migration tends to begin later and is more facultative, with some individuals spawning in lakes or river mouths.

In the Elands and Letaba rivers, South Africa, Meyer (1974; cited in Cambray 1990) found that several small *Barbus* species, mostly *B. trimaculatus* and *B. unitaeniatus*, exhibited major upstream migrations. Cambray (1990) found substantial numbers of smallscale redfin minnow *Pseudobarbus aspus* and chubbyhead barbs *Barbus anoplus* congregating below a weir on the Groot River, South Africa. These had moved upstream from isolated pools following pool reconnection during a major flood. However, the numbers of redfin minnow and chubbyhead barb were dwarfed by those of moggel *Labeo umbratus*, which from simple density packing measurements were estimated to number several million across the 160-m wide river, immediately below the weir. Most of the moggel were juveniles and Cambray (1990) interpreted the movements of this and the other species as being an explorative migration for conditions of improved feeding and reduced competition. As adults, a variety of *Labeo* species are known to exhibit longitudinal (Whitehead 1959; Van Someren 1962; Cadwalladr 1965; Bowmaker 1973; Balirwa & Bugenyi 1980) and lateral (Jackson & Coetzee 1982) spawning migrations, but migrations of juveniles also seems important, with 31% and 78% of catches of migrating *L. molybdinus* and *L. cylindricus* respectively comprising 4–10 cm juveniles (Meyer 1974, cited in Cambray 1990). Adult *Labeo cylindricus* from Lake Chicamba, above a hydroelectric dam in Mozambique, concentrated at river mouths before migrating up flowing rivers in January to spawn (Weyl & Booth 1999).

In their native geographical range of East Asia various species of indigenous carp including black carp *Mylopharyngodon piceus*, grass carp *Ctenopharyngodon idella*, silver carp *Hypophthalmicthys molitrix* and bighead carp *Aristichthys nobilis* typically exhibit regular movements between lake and river environments (Liu & Yu 1992). Upstream spawning migrations of several hundred kilometres appear common in unobstructed river systems and the semi-buoyant eggs of these species drift downstream into lentic areas. With damming of the Hanjiang River in China to form Danjiang Reservoir, and construction of sluices in lakes connected to the river, the migratory paths of these fishes have been blocked and they have tended to utilise lotic environments. However, the reservoir populations of these and at least another ten cyprinid species continue to engage in large spawning migrations from the reservoir to the upper reaches of the Hanjiang. The resultant numbers of eggs and fry produced from these spawning areas are distinctly higher than in the river below the dam. The cyprinids *Coreius heterodon* and *Rhinogobius typus* originally inhabited the upper reaches of the Hanjiang system and Liu and Yu (1992) report that the young fish habitually migrated upstream, presumably due to being displaced as eggs or larvae. The population of *C. heterodon* restricted to below the dam overwinter in the lower reaches of the Hanjiang and begin to move upstream in mid-April, reaching the Danjiangkou district below the dam (a distance of about 150 km) by mid-June.

The cyprinids are essentially a freshwater group in which diadromy between fresh water and full-strength seawater has not been reported. However, a number of species undergo migration between freshwater and brackish water environments, some of which appear to be obligatory processes for successful population maintenance and meet the criteria used by McDowall (1988) for discussion of other fishes as slightly diadromous species. A much larger range of cyprinid species enter brackish water facultatively, but in locations where they do so, movement often occurs at very specific times and may involve certain age groups.

Migrations into estuarine reaches of rivers are common in winter and may involve large numbers of cyprinids of several species and a wide size range. Such annual migrations involving species such as bleak *Alburnus alburnus* and roach *Rutilus rutilus* appear to occur on the River Gudenå in Denmark (Koed *et al.* 2000). The Southeast Asian cyprinids *Catla catla*, *Labeo rohita* and *Cirrhinus mrigala* move into the estuary of the Halda River, Bangladesh, for reproduction, but spawning takes place during the rainy season when the estuary has low salinity (Tsai *et al.* 1981). The peamouth *Mylocheilus caurinus* principally occurs in lakes and rivers of northwestern North America, but is described by McPhail & Lindsey (1970) as one of the few species of cyprinids in that region that can tolerate saltwater. They report its capture from marine waters near Vancouver, and suggest this tolerance to be the reason that this is the only cyprinid native to Vancouver Island. Takeshita and Kimura (1991) found that in spring the small cyprinid *Hemibarbus barbus* exhibited an upstream spawning migration in the Chikugo River of Japan. Juveniles of this species are normally found in tidal, but low salinity (2‰), reaches and when mature at 3–4 years migrate 20–40 km to spawn in the river's middle reaches. Soon after hatching, the young begin to move downstream and are common in tidal waters by mid-summer.

The splittail *Pogonichthys macrolepidotus* is a moderately large cyprinid (up to 40 cm in length), endemic to the Central Valley of California where it was once widely distributed. It is now largely confined to the Sacramento-San Joaquin basin, especially the delta. They are found in fresh and brackish water environments of Suisun Bay, Suisun Marsh and the Sacramento-San Joaquin Delta, the latter comprising over 1600 km of channels. Adult splittails undertake an annual upstream spawning migration from the estuary in late autumn and winter, when delta inflow increases from the seasonal rains (Meng & Moyle 1995). Spawning occurs in winter and spring on flooded vegetation in the inundated floodplain. The young spend a period of a few weeks to a year or more before moving to tidal fresh and brackish waters. Historically, the splittail ranged from Redding on the Sacramento River to Millerton on the San Joaquin River and this may therefore have involved migrations of over 400 km on each river. Spawning runs that ascended tributaries of these rivers have largely disappeared after dams were built. Splittail migration and early juvenile rearing are primarily limited to 250 km upriver on the Sacramento and San Joaquin Rivers. Year class strength is positively correlated with annual flow of the Sacramento River (Daniels & Moyle 1983), relating to the extent and duration of inundation of the floodplain. Increased water diversion and regulation, reduction of freshwater flows by a half due to irrigation and urban use, and a succession of dry years, have caused concern regarding the ability of this species to sustain populations. However, 1995 was a very wet year and large numbers of splittail were caught migrating and recruitment appears to have been high (Sommer *et al.* 1997). Like other Californian cyprinids, particularly those found in larger rivers and lakes, the splittail seems resilient to the wet and dry cycles typical of the Mediterranean climate of California. They have a relatively high fecundity which helps population recovery after several low flow years. Its long life (5–7 years) and high fecundity allow the splittail to delay spawning during droughts or at other times when conditions are unsuitable. Use of ephemeral aquatic areas may reduce loss of eggs, larvae and juveniles to aquatic predators.

There are distinct forms of several lacustrine cyprinids such as roach *Rutilus rutilus* and common bream *Abramis brama* which have summer feeding grounds in the brackish environments of the Black, Caspian and Azov seas (and formerly the Aral Sea, prior to its desiccation – see section 7.2.1) but which may migrate hundreds of kilometres up inflowing rivers to spawn (Mills 1991). Vobla *Rutilus rutilus caspicus* are more fusiform in shape than typical roach and together with vimba *Vimba vimba* exhibit highly conspicuous and synchronised spawning migrations (Omarov & Popova 1985; Orlova & Popova 1987). The kutum *Rutilus frisii* is reported to have migrated over 1000 km in Russian rivers running into the Caspian Sea (Nikolsky 1961). Impoundment of rivers such as the Volga and Terek means that migration patterns of vobla, vimba, kutum and common bream are almost certainly less extensive than they once were. Prior to barrage development, vimba also used to migrate into the Vistula River, Poland, from the Baltic Sea, but have now almost become extinct (Backiel 1985).

5.13 Suckers (Catostomidae)

The suckers are a freshwater group of cypriniforms occurring in China, northeastern Siberia and North America. There are about 68 species, typically occurring in riffle–pool sequences of rivers, but many species also occur in lakes, at least as adults. Most information concerning patterns of migration in this group come from studies in North America. The degree of regular movement between habitats varies between species and river systems and seems to depend, at least in part, on the spatial distribution of habitats. Many species are dependent on riffle–pool sequences and if these are intact then movements may be small. Gerking (1953) reported home ranges of 91–122 m for adult northern hog sucker *Hypentelium nigricans* in small Indiana streams. Matheney and Rabeni (1995) radio-tracked adult northern hog sucker in the Current River, Missouri, where northern hog suckers generally required at least one pool and one riffle. Fish normally used slow, deep water with fine substrate in winter, and fast shallow water and large substrate in summer. Fish typically moved about 500 m upstream or downstream into faster water in February and March before spawning. In winter they moved back down into pools but remained at home sites during floods and used flooded riparian areas. As well as seasonal habitat separation, northern hog sucker demonstrated diel migrations, feeding in run and pool habitat respectively in summer and winter and resting in riffle/edge habitat at night in both seasons.

Late spawning (in October) may occur in several catostomids (*Catostomus discobolus* and *C. latipinnis*) in the Arizona Grand Canyon, as a result of off-season patterns of tributary outflow (Douglas & Douglas 2000). However, spring spawning migrations are characteristic of this group and often involve movement out of lakes into tributaries, although spawning may also occur in inlets (Geen *et al*. 1966). Quillback *Carpiodes cyprinus* from Dauphin Lake, Manitoba, exhibit clear spawning migrations into the Ochre River in spring at 5–7°C (Parker & Franzin 1991). Quillback utilise the meandering last 30 km of river where gradients are 1.8 m km^{-1}, with riffles and mud/sand bottomed pools. The Ochre is prone to flooding during snowmelt or rain, but is often near zero discharge in July–August. Upstream migratory activity was closely associated with high discharge due to snowmelt or flooding and during these conditions upstream migrations of at least 32 km by quillback were achieved. White sucker *Catostomus commersoni* were observed to begin migrating in early April before quillback, and shorthead redhorse *Moxostoma macrolepidotum* (another sucker) started migrating after quillback in late April (Parker & Franzin 1991). During low water conditions migration and spawning of fish could be as little as 2–3 km upstream of Dauphin Lake in the first few spawning areas. Since discharge tended to decline throughout spring and summer, early migrants tended to penetrate and spawn furthest upstream, and late migrants spawned furthest downstream. Downstream migration of quillback (following spawning) was mostly between early May and the end of June. In some cases low discharge inhibited downstream movement and fish remained in deep pools until a high discharge event.

Discharge during spring explained greater variation in the number of upstream-migrating white suckers and longnose suckers *Catostomus catostomus* in an Alberta tributary, than temperature, although the greatest variation was explained by the product of two variables (Barton 1980). Upstream spawning migration to gravel riffle areas has also been reported for the greater redhorse *Moxostoma valenciennesi* (Cooke & Bunt 1999) in the Grand River Ontario, and for *Cycleptus meridionalis* in the Pearl and Pascagoula rivers, Mississippi (Peterson *et al.* 2000). Kennen *et al.* (1994) reported downstream migration of white sucker *Catostomus commersoni* to Lake Ontario between April and August. Bigmouth buffalo *Ictiobus cyprinellus*, smallmouth buffalo *Ictiobus bubalis*, river carpsucker *Carpiodes carpio*, largescale sucker *Catostomus macrocheilus*, golden redhorse *Moxostoma erythurum*, silver redhorse *Moxostoma anisurum* and black redhorse *Moxostoma duquesnii* are also regarded as relatively mobile species of sucker (Scott & Crossman 1973; Hesse *et al.* 1982; Bulow *et al.* 1988). Hesse *et al.* recorded maximum upstream movements of 405 km and 234 km respectively for tagged smallmouth buffalo and bigmouth buffalo in the middle Missouri River.

Considerable information has also been gained on the life history and migratory behaviour of one of the rarest fishes of the Colorado system, the razorback sucker *Xyrauchen texanus* which is endemic to the Colorado River basin (Tyus 1987; Modde & Irving 1998). Razorback suckers declined in numbers following completion of impoundments, almost disappeared from the lower basin, and only a small highly endangered population exists in the upper basin (Modde *et al.* 1996). They are long-lived and may reach 40 years of age like some other catostomids such as the cui-ui *Chasmistes cujus*. The decline has been attributed to lack of successful recruitment due to loss of floodplain nursery habitat resulting from substantially reduced peak flows (Tyus & Karp 1990) and interaction with exotic species that have become established in altered river environments. They are mostly restricted to calm water areas, but may move substantial distances. Tyus found that 24 of 52 razorback sucker at liberty for 2 weeks to 8 years moved less than 10 km, while the remaining 28 fish were captured an average of 59.3 km away from the release site (range 13–206 km). There are two principal spawning sites: the Escalante spawning area on the Green river at rkm 492–501, and the Yampa area 0.3 km upstream of the Yampa confluence at rkm 555. This area includes the longest low gradient (0.1%) length of the Green River downstream of Flaming Gorge Dam and provides the largest area of floodplain habitat in the Green River basin. Spring high flows here are largely determined by the unregulated Yampa River. Because razorback suckers migrate with spring increases in discharge (Tyus & Karp 1990), changes to natural hydrographic patterns could affect migration of fish to form spawning aggregations. Timing of reproduction is clearly important if floodplain inundation optimises survival of the early life stages of this species.

Radio telemetry of male razorback suckers through three successive spawning years together with long-term mark recapture data for 15 years suggested the repeated use of historical spawning areas, but individual fishes switched between sites in different years (Modde & Irving 1998). Migration to spawning areas occurred in April and May and was followed by a downstream movement of up to 112.7 km, but typically about 20 km, usually to the mouths of tributaries, during the high-flow period. Within 3 months fish returned to their prespawning residence areas. Downstream movements of razorback sucker during the high flow season may be associated with postspawning recovery or foraging (Tyus & Karp 1990). Another catostomid indigenous to the Colorado system, the flannelmouth sucker *Catostomus latipinnis*, also appears to be migratory, on the basis of seasonal variations in CPUE and mark recapture information (McKinney *et al.* 1999).

5.14 Loaches (Cobitidae) and river loaches (Balitoridae)

The loaches (Cobitidae, about 110 species) are small benthic freshwater fishes found throughout Eurasia and North Africa while the river loaches (Balitoridae, about 470 species) are similar in form and occupy similar habitats throughout Eurasia, but are especially diverse in Southeast Asia. They are mostly small and cryptic, with an elongated body form and occur in a range of habitats from swift-flowing streams to stagnant backwaters. Most species remain buried or partially buried in bottom sediments during the day and are active during twilight or at night. Little is known of the spatial behaviour of most species, especially the many Southeast Asian balitorid loaches. Loaches are not generally assumed to demonstrate significant migratory behaviour, and are frequently used as models for population genetics, although the stone loach *Barbatula barbatula*, found in Europe and in Asia, is listed as migratory within rivers (Cowx & Welcomme 1998). Stone loach have been observed moving upstream over small weirs in the River Sheaf, northern England, in spring (S. Axford [pers. comm.]). Captured fish had enlarged gonads, suggesting that at least some populations may migrate to spawn.

Slavík and Rab (1995, 1996) studied an isolated population of spined loach *Cobitis taenia* in the Pšovka Creek, Bohemia, Czech Republic. Downstream movements started in March (mainly males) and April (the rest of the males followed by females) to the stream's lower limit where it ended in a boggy area. Spawning occurred in June followed by an upstream migration in July. The youngest and oldest reproductively inactive females remained in overwintering sites and did not migrate to spawn. Juveniles steadily migrated upstream from the spawning area over the summer period, reaching wintering areas by October. Distances of 200–800 m were travelled. Lelek (1987) reports local movements of about 500 m and an exceptional record of 10 km from the site of capture for this small species, which rarely exceeds 12 cm. He suggests that populations contain mobile elements not exceeding 25% of the population. Most species of loaches are phytolithophous spawners and these habitats are similar to those utilised for refuge and feeding, resulting in limited movement.

5.15 Characins (Characiformes)

The characins are one of the world's largest groups of fishes, and as for the vast majority of ostariophysan fishes, they occur in fresh water only. They are widespread in Africa (about 250 species), and especially in Central and South America (over 1100 species) as well as in the southern part of the United States, although several species have been introduced more or less involuntarily in several water bodies, notably small characins used in ornamental aquaculture. Some larger escapees from fish farms are locally found in warm water effluents of power plants (e.g. *Piaractus brachypomus* at 50°N in Belgium), but there is no evidence for natural reproduction. Most characins have well-developed teeth, including at early life stages (<24 h after hatching in *Brycon moorei*; Baras *et al.* 2000b), but they are omnivorous, although some exceptions have earned characins the reputation of being highly predatory species. These include the South American piranhas with their sharp teeth and the African tigerfishes (*Hydrocynus* spp., Alestiidae) with their fang-like teeth, especially the giant tigerfish *H. goliath* that can attain 130 cm and weigh over 50 kg. However, many characins are small species, sometimes not exceeding 5–10 cm.

Like many predatory fishes, the large tigerfishes show limited migratory activity. They mainly occur in open waters of lakes and large river channels, where they exert intense predation on almost any other fish species (Daget 1954; Welcomme 1985), including other predators, such as the African pike *Hepsetus odoe* (Hepsetidae), another large African characin. Nevertheless, these species exhibit lateral ontogenetic migrations (Niaré & Bénech 1998), as the juveniles live and forage in lateral channels, probably minimising the risk of being preyed on by their larger conspecifics in the main river channel (Winemiller & Kelso-Winemiller 1994). Lacustrine populations of these predators may undertake potamodromous migrations, as occurs for *Hydrocynus forskalii* from Lake Chad into Chari River (Bénech & Quensière 1989), although they range over much shorter distances than many other characins in the same river (Loubens 1973). The African pike *Hepsetus odoe* has mainly been reported to show restricted migration, mainly laterally, although long potamodromous migrations have been reported from the Kafue River (Williams 1971). Several piranha species (mainly genera *Pygocentrus* and *Serrasalmus*) seemingly undertake ontogenetic lateral migrations onto and from the inundated floodplain or forest, as juveniles can be found in places where adults are absent (Henderson & Hamilton 1995). The erythrinid *Hoplias malabaricus* seemingly exhibits most restricted migrations, but ontogenetic movements from inshore vegetation to deeper pools may take place (Winemiller 1990; see section 4.4.3 for further details). The giant traira *Hoplias aimara* are large (up to 120 cm) erythrinids inhabiting the rivers and lakes of South America, and they also show strong home range habits, as illustrated by one of the rare radio-tracking studies conducted on characins. In 1994, 19 large (65–100 cm) traira were radio-tracked in the Sinnamary River, French Guyana (Tito de Morais & Raffray 1996), and most (15 of 19) of them returned to their capture place after they had been displaced. Except for this homing behaviour, traira showed no sizeable movements under natural flowing conditions. When studied over 24-h cycles, they showed no distinct activity patterns, and daily activity areas were restricted to a few tens of metres, although greater distances were travelled in river stretches at low flow, when competition for food was greater. Traira frequently returned to formerly occupied places, especially during low flows, when they exhibited territorial behaviour. A few months later, while the fish were still being radio-tracked, the filling of the Petit-Saut Reservoir took place. With one exception, fish located further upstream from the newly created reservoir consistently showed home range behaviour, whereas most of those closer to the reservoir undertook rather long (up to 46 km) upstream migrations as the water rose by 8 or 9 m. This, together with the post-displacement homing behaviour observed after tagging, clearly indicates that traira have the capacity to migrate, but that they rarely do so under normal flows, for reasons that may be similar to those invoked for the limited migratory behaviour of *H. malabaricus*.

In subsaharan Africa, many alestiids (also named silversides, but we will avoid this name here, as a confusion might arise with atherinids; section 5.27) undertake long seasonal potamodromous migrations from lacustrine environments or lower reaches of river channels to upstream spawning grounds. *Alestes baremoze* and *A. dentex* are mainly found in reservoirs (e.g. Gebel Aulia reservoir in Sudan, Hanna & Schiemer 1993) or lakes (e.g. Lake Chad, Carmouze *et al.* 1983) from October through April, when water level is high, plankton is abundant and temperatures are mild. At the onset of the rainy season, in late April or early May, *A. baremoze* undertakes spawning migrations under low flows. Migrations of 150–200 km are frequent (Bénech & Quensière 1989), but they have been found to travel over 400 km in the Senegal River (Reizer *et al.* 1972), and as far as 650 km to upstream swamps of the Chari River in Cameroon (Blache & Milton 1962; Carmouze *et al.* 1983). They reproduce in June–July, as water levels increase (Paugy & Lévêque 1999), and spawn in midstream, from which their eggs and larvae drift onto the inundated floodplain (Durand & Loubens 1970). In the Chari River, the juveniles were found to enter Lake Chad from November to mid-December, when waters were receding (Bénech & Quensière 1983). Except for the spawning site, a similar life-history pattern within this fish assemblage has been depicted by Loubens (1973) and Bénech & Quensière (1989) for several mormyrid species (*Hyperopisus bebe, Marcusenius cyprinoides, Mormyrus rume;* Osteoglossiformes), catfishes (mainly schilbeids and mochokids, see section 5.16) and for other characins (e.g. *Distichodus rostratus*), although the last-mentioned species seemingly enters the Chari River during the floods in late summer.

In Eastern African lacustrine systems, such as Lake Turkana, a similar pattern involving long potamodromous migrations into large lake tributaries such as the River Omo has been reported for *A. baremoze* and some other large characins (*Citharinus citharus* and *Distichodus niloticus*), whereas smaller characins such as *Brycinus nurse* move over shorter distances into temporary streams (Hopson 1982). *Alestes dentex* shows a similar life history, although it seemingly starts moving slightly later than *A. baremoze*, and most migrants leaving the lake and entering the Chari River are immature individuals. Loubens (1973) suggested that *A. dentex*, which may also ascend rivers over several hundreds of kilometres, carry out their migration in a two-step mode, with large juveniles entering rivers, achieving growth and maturation in riverine habitats, then continuing their upstream migration as adults in the following years. A somewhat similar situation has been reported for shortnose sturgeon *Acipenser brevisrostrum* travelling over long distances (section 5.3). Whether all mature individuals or large juveniles of the aforementioned species undertake long potamodromous migrations is unknown, except for *A. baremoze*. Loubens (1970) reported that *A. baremoze* also spawned on the shores of dune islands in Lake Chad, where immersed macrophytes and food are abundant all year round, but that not all lacustrine fish became mature each year. This indicates that the migration between riverine and lacustrine environments is facultative. Probably numerous individuals undertake long migrations because of restricted spawning and nursery habitat availability in Lake Chad compared to the extensive Chari River basin which extends over Chad, Cameroon and the Central African Republic.

As depicted for anguillid eels and some other species, long-distance migration is not restricted to large fish, and it is certainly not the case for the long-range migrations of the small (<12 cm SL and <10 g) characin *Brycinus leuciscus*, which is widespread across Western Africa, and better known as dabé in Senegal or tinéni in Niger and Mali. Both adult (5–8 cm) and small juvenile tinéni enter the floodplain during the flood period (Niaré & Bénech 1998), but under receding waters they return to the Niger River, where they undertake long potamodromous migrations. These are structured into several waves of *B. leuciscus*, with fish spaced just a few centimetres apart so that they pave the entire riverbed (Daget 1952). Such waves can last for hours. Although the ground speeds are rather slow (about 0.2–0.3 m s^{-1}), they require very fast swimming by these small fish (about 3–5 BL s^{-1}). The start of the upstream movements of tinéni schools coincides with full moon days. At the end of the lunar cycle, the fish stop and disperse, reforming the school and resuming their upstream progression when the moon is waxing. According to Daget (1952) some schools might travel over distances of about 400 km. The building of the Markala Dam in the Inner Delta of Niger River, Mali, interfered with this migration, causing rapid dispersal and downstream movements of individuals or small groups. From Daget's observations, it can be suggested that grouping matters for upstream migration, possibly because schooling might provide a hydrodynamic advantage to migrants, although this is uncertain. Moonlight might serve as a stimulus, possibly because sight is important in coordinating individuals within the school. Conversely, migrating in school under higher light levels might also be interpreted as an anti-predator tactic, especially in view of the fishes' small size and the abundance of large predators in the river channel. All three functions might be equally important, but there is no major evidence for determining which, if any, function is prevalent.

Apart from the large predatory *Hoplias* described above, South American characins comprise many small species, which have been depicted as resident (e.g. the characids *Characidium* spp., *Roeboides dayi*; Winemiller 1990) or making restricted lateral migrations onto the floodplain during rising waters, and off the floodplain under receding waters (e.g. the cynodontid *Cynodon gibbosus* and the characid *Roeboides myersi*; Fernandes 1997). Many small *Astyanax* and *Creagrutus* spp. (Characidae) are riverine fishes inhabiting rapids, riffles or plunge pools, and most of them are usually regarded as non-migratory (Roman-Valencia 1998; Winemiller & Jepsen 1998). However, it is possible that small-scale migrations might occur between stream margins and deeper habitats, possibly as there are obvious risks of predation in the two major types of habitats, sometimes by different life stages of a single species (e.g. adults and juveniles of *Hoplias malabaricus*, respectively). *Astyanax caucanus* and *A. fasciatus* are seemingly quite eurytopic species, being encountered in all habitats of the Magdalena River, Colombia, from backwaters to rapids, but they occur in greater number in side channels during high floods, suggesting they are capable of lateral migration too (Baras et al. 1997). However, as for most small species, there is a dearth of knowledge on whether they are migratory or not, notably because these species are of little importance to local fisheries, in contrast to the medium- or large-sized characins belonging to the 'seasonal strategist' group, and without which most South American fisheries would collapse. They essentially include the anostomids *Leporinus*, *Rhytiodus* and *Schizodon* spp., the characids *Brycon*, *Colossoma*, *Mylossoma*, *Piaractus*, *Salminus* and *Triportheus* spp., the curimatids *Curima-*

ta, *Potamorhina*, *Prochilodus*, *Psectrogaster* and *Semaprochilodus* spp., the cynodontid *Rhaphiodon vulpinus*, and the hemiodontids *Anodus* and *Hemiodus* spp. For example, in the Magdalena River basin, Colombia, more than half of the landings relate to the curimatid *Prochilodus magdalenae* and the anostomid *Leporinus muyscorum*, and about half of these are captured from mid-January to mid-March, during the so-called *subienda*, which is the Colombian counterpart for the Brazilian *piracema*. The functional bases of these migrations have been considered through the example of *Prochilodus* spp. in section 2.3 and comparative information in section 4.4. Here, we will focus on some specific examples of characins, the migratory lifecycles of which have been documented to a large or reasonable extent (by comparison with less documented species).

The tambaqui *Colossoma macropomum* is a large (over 90 cm and 30 kg), long-lived (up to 70 years) fish with a high fecundity (1–2 million 1 mm eggs in a 15 kg female) and large size at sexual maturity (>60 cm in the Mamoré River, Bolivia; Loubens & Panfilli 1997). It naturally occurs throughout the Amazon and Orinoco river basins, and is locally on the decline in some parts of its original range. It has been introduced in many places over South and Central America, and in many locations with warm climates such as Hawaii, Taiwan, Indonesia and China, as this and some other closely related species (*Colossoma* or *Piaractus* spp.) are most valuable species for tropical or subtropical aquaculture (Araujo-Lima & Goulding 1997). In contrast to *Piaractus brachypomus* which occurs in the upper reaches of many South American rivers (e.g. Bolivian Amazon, Lauzanne *et al.* 1991), *C. macropomum* mainly inhabits the lowland rivers and floodplains. The following description of its migratory life cycle (see also Fig. 5.5) relies essentially on descriptions from the Madeira (Goulding & Carvalho 1982), Orinoco (Novoa 1982) and Mamoré (Loubens & Panfilli 1997) river systems. At the start of the rainy season, mature individuals move upstream during low flows to spawn along the levees of main whitewater river channels, most frequently near river confluences. The extent of the potamodromous migration in the main river channel is variable, and in some environments there may be no major upstream movement (e.g. Orinoco River; Novoa 1982). As Loubens and Panfilli (1997) suggested, spawning in the main channel might be a less risky strategy than spawning on the floodplain for species reproducing during the early rainy season, as it is frequent that the water level drops by a substantial margin (up to 2 m) during this season. Spawning behaviour is poorly known, but there is some evidence that males occupy the spawning grounds prior to, and over longer periods than females. Spent females, then males, enter the floodplain, inundated forest or blackwater tributaries, not necessarily the places from which they came (Lowe-McConnell 1987). During the periods of high flood, they spread over tens of kilometres and feed vigorously, mainly on fruits which they are able to crush with their massive teeth. Seeds pass through the fish's gut and are redistributed along the fish's daily range, as occurs in other frugivorous characins and notably *Brycon* spp. (Horn 1997). As waters recede, tambaqui move out of the inundated areas or descend blackwater tributaries. They enter main river channels and are generally found near shore, close to the wooded riparian areas.

As in most other seasonal characins, eggs of tambaqui hatch rapidly (<16–18 h). Although there is some evidence that eggs and hatched embryos can be transported downstream over distances of about 100 km, they seemingly drift over shorter distances downstream before they enter the floodplain. Young tambaqui have long and thin branchiospines that permit planktonophagous feeding, although they possess teeth at an early age, which may also en-

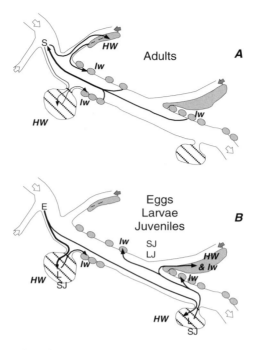

Fig. 5.5 Schematic drawing of the lifecycle of the tambaqui, *Colossoma macropomum* (inspired by Lowe-McConnell, 1987, and based on additional information from Goulding & Carvalho 1982, Araujo-Lima & Goulding 1997 and Loubens & Panfilli 1997). HW and LW stand for high and low waters, respectively. S stands for spawning, E for eggs, L for larvae, and SJ and LJ stand for small and large juveniles, respectively. Hatched areas represent the floodplain, and grey areas are the inundated forest or nearshore trees. White and grey broad arrows indicate the direction of flow in white and blackwaters, respectively. Plain black arrows indicate the direction of migration, and variable thickness relates to the probability that this movement occurs when alternative behaviours can be exhibited.

able them to handle larger prey, including fishes. Tambaqui are fast growers, and they may attain sizes of up to 16 cm (about 100 g) within a few months in the floodplain, and fish eaters might just do so within 8 weeks (by reference to observations in aquaculture environments). As waters recede, they leave the floodplain and enter the main river channel, where they occur in nearshore areas, or permanently inundated forests. Older juveniles move between these habitats and recently inundated forests, where they feed on fruits. Araujo-Lima and Goulding (1997) reported that the noise of tambaqui crushing nuts is so loud and distinct that it can attract conspecifics and predators, including fishermen. Tambaqui are preyed on by several predatory fish species inhabiting floodplains and inundated forests (the pirarucu *Arapaima gigas*, the electric eel *Electrophorus electricus*, the erythrinids *Hoplias* spp. and piranhas *Serrasalmus* spp.), but they have several traits enabling them to evade or deter predation. Juvenile tambaqui are silvery, and they mimic young piranhas, although this mimicry is not as complete as in *Piaractus brachypomus*. As described in section 4.4.4 tambaqui have evolved means of coping with protracted hypoxia, including during night-time, resulting in little or no oxygen-dependent daily migration, in contrast to many species that utilise aquatic surface respiration. This general pattern (except for the use of floodplain during the first months of life) probably applies to the first 7–10 years of life of the tambaqui, when males and females reach sexual maturity. Loubens and Panfilli (1997) reported that in the Mamoré River large juveniles joined adults in the main channel nearshore habitats, and they did not exclude the possibility that these large juveniles might undertake potamodromous movements together with or following mature adults during the spawning season. Similar observations have been made by Godoy (1959) for *Prochilodus lineatus*, with large juveniles joining the adults in their spawning runs in the Paraná River. This might partly address the issue raised in section 2.4.7 on how species with drifting eggs might be able to home or find scattered spawning places, although clear reproductive homing is not apparent in tambaqui.

Curimatids of the genus *Prochilodus* and related genera (e.g. *Semaprochilodus*) are among the most important fishes comprising the fish swarms ascending South American rivers (Welcomme 1985), where their potamodromous migrations can extend over several hundred kilometres (Table 4.2). These iliophagous species live up to 10 years, they attain smaller size and body weight (60 cm and 3 kg) than tambaqui, but they reach sexual maturity at a younger age (e.g. 2 years in *P. nigricans* in the Mamoré River, Bolivia; Loubens & Panfilli 1995). There is strong evidence that most *Prochilodus* spp. exhibit rather similar migratory histories, which involve a combination of longitudinal and lateral spawning and feeding migrations onto and off the floodplain or tributaries, although the extent of the migration and degree of complexity vary between species and environments (Fig. 5.6). The common trend is that *Prochilodus* spp. spawn at the high water stage, with their semi-buoyant eggs and small larvae being swept downstream, over variable distances depending on whether the adults spawn in flooded areas or in the main river channel, and how far the spawning grounds are from the banks. High water levels and relatively low water clarity during the reproductive season have limited the knowledge on *Prochilodus* spawning behaviour. However, there have been several reports from different species that prespawning and courting males on the spawning grounds emit low-pitched drumming sounds that seemingly attract the ripe females and cause them to move from nearshore vegetated habitats onto the spawning grounds (Godoy 1959; Lilyestrom 1983).

Prochilodus lineatus (also named *P. scrofa*, a junior synonym) is probably the best documented species after long-term studies were conducted in the Paraná River system (Godoy 1959, 1972; Toledo *et al.* 1986, 1987; Agostinho *et al.* 1993). During the late dry season, when water level is lowest and water temperature is cool (<25°C), adults undertake long upstream migrations which they pursue at a faster pace in the early rainy season, at warmer temperatures (27–29°C) and under conditions of rising water levels. Sexual maturation is completed during the migration. Shoals of migrants, which may comprise several tens of thousands fishes, mainly mature individuals, aged 3–7 years, but also some large juveniles aged 2 years, reach the upstream spawning grounds in the main channel or tributaries and spawn slightly before or during the peak of high waters (Fig. 5.6A). After spawning, they move down the main river or tributaries, back to the downstream feeding grounds. Halts or sojourns in intermediate locations where flooded areas are available seemingly are frequent, causing the downstream migration to take place at a much more leisurely pace than its upstream counterpart (see Fig. 4.6). Eggs spawned under rising or high water move into recently flooded areas, where warm temperature and abundant food permit fast growth. Young *Prochilodus* are rather tolerant to low oxygen concentration, and iliophagous feeding enables them to remain in floodplain lakes even under low water levels, when the lakes are no longer connected to the main river. A similar migration pattern was exhibited by *P. nigricans* in the

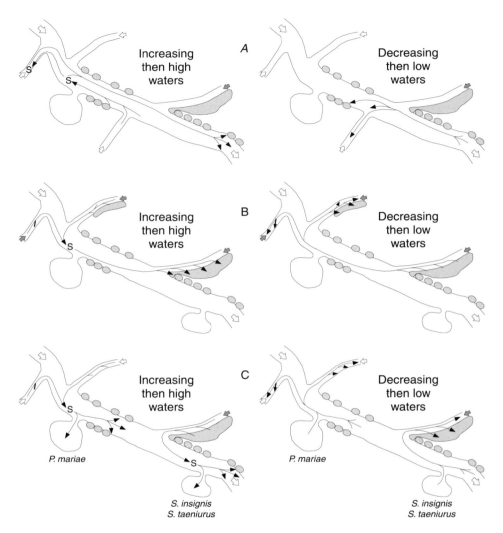

Fig. 5.6 Schematic representation of the annual migrations by adults of some *Prochilodus* and *Semaprochilodus* spp. in different environments. (a) *P. lineatus* in the Paraná River system (also applies to *P. nigricans* in the Tocantins River system). (b) *P. nigricans* in the Madeira River basin. (c) *S. insignis* and *S. taeniurus* in the Central Amazon (right part of each map), and *P. mariae* in the Orinoco River system. S stands for spawning. Hatched areas represent the floodplain, and grey areas are the inundated forest or nearshore trees. White and grey broad arrows indicate the direction of flow in white and blackwaters, respectively. Plain black arrows indicate the sense of migrations, and variable thickness relates to the probability this movement occurs when alternative behaviours can be exhibited. Based on descriptions by Goulding (1980), Lilyestrom (1983), Carvalho & Mérona (1986), Toledo *et al.* (1986), Ribeiro & Petrere (1990), Loubens & Panfilli (1995), Winemiller *et al.* (1997) and Winemiller & Jepsen (1998).

Tocantins River before a large dam was built at Tucurui (Carvalho & Mérona 1986), but with some differences. Adults moved shorter distances both downstream and upstream, and larvae and juveniles moved further downstream from the spawning grounds, grew in downstream reaches and finally started migrating upstream to the middle reaches. Both strategies observed in the Tocantins and Paraná rivers are consistent with the idea that the migratory history of *Prochilodus* spp. limits the competition between life stages through habitat compartmentalisation on a very large scale (Winemiller & Jepsen 1998).

Possibly because these are among the best-known examples of *Prochilodus* life histories, they might have contributed to spreading the false idea that all upstream migrations of large fish during the *piracema* were spawning migrations. In environments where floodplain nurseries are found in downstream reaches, the migration pattern might just be reversed or involve no long migration within the main river channel. In the Mamoré River, Bolivia, adult *P. nigricans* exhibit no marked upstream movements during the spawning season. They spawn in the floodplain where larvae and small juveniles spend their ongrowing period, and they return to the main channel under receding waters, ascending the river to upstream reaches (Loubens & Panfilli 1995). In the Madeira River, *P. nigricans*, as other *Prochilodus* and *Semaprochilodus* spp., exhibit a distinct and more complex migration pattern (Goulding 1980; Fig. 5.6B). During the periods of rising water, adults move down blackwater tributaries to spawn in the Madeira River, and then enter the flooded forests on the margins of large tributaries, this being a typically feeding-orientated migration. As waters recede, they leave the inundated forest, enter the main river channel again, ascend the river and disperse in upstream tributaries. A slightly different pattern has been described for *P. mariae* in the Orinoco River, with adults descending the whitewater tributaries draining the Andes, and entering the main channel for spawning, then feeding in the lowland floodplain or river reaches (Lilyestrom 1983; Fig. 5.6C). As in the Madeira River, upstream movements of large shoals in the main river and ascents of tributaries during the dry season are food and refuge-orientated migrations. An even more downstream orientated pattern occurs in the Central Amazon, where *Semaprochilodus insignis* and *S. taeniurus* descend blackwater tributaries during flood periods, move downstream and spawn in lower reaches of the main river, then return to the flooded forest (Ribeiro & Petrere 1990). Winemiller *et al.* (1997) and Winemiller and Jepsen (1998) have reported a similar pattern for *Semaprochilodus kneri* and *S. laticeps* moving in between the Orinoco River and blackwater tributaries such as the Cinaruco stream. Finally, it is worth noting that not all mature adults in a population of *Prochilodus* enter this large-scale annual migration, and that maturation and spawning may take place in the floodplain, sometimes with a much greater energetic investment into reproductive products than in migrants, as described for *P. mariae* in the Orinoco River (Saldana & Venables 1983; see also section 2.3.5).

In a remarkable paper, Winemiller and Jepsen (1998) synthesised how the lifecycle of these highly migratory species, which involves movements between the floodplain, the inundated forest, whitewater and blackwater rivers, permits the longitudinal and lateral transport of energy from the nutrient-rich areas to poorer locations where it is stabilised through predation by piscivorous species exhibiting more restricted migrations (see also section 4.4.3). Many tributaries or river channels in Africa and South America have become blocked by huge dams, notably for hydropower purposes, and these have interfered considerably with the migrations of tropical characins and the distributions of their populations (e.g. *Prochilodus lineatus* in the Paraná River system; Toledo *et al.* 1987). With respect to the arguments above, the impact on their predators and energy transport in the entire ecosystem should also be considered.

5.16 Catfishes (Siluriformes)

The Siluriformes are an extremely broad group of species widespread over Europe, Africa, Southeast Asia, South, Central and North America. In 1984, Nelson listed no fewer than 2211 species belonging to this group. Ten years later, he had added another 200 species, and Teugels (1996) who has provided one of the most recent comprehensive reviews of their taxonomy, listed 2584 species. While searching the recent literature on these species, we found an average of about 30 species being discovered each year, and many others being renamed. Here, we rely on Teugels (1996) who divided the Siluriformes into 33 families and 416 genera.

Catfishes are primarily benthic, thigmotactic freshwater ostariophysan fishes. They have adopted a broad range of lifestyles, notably as regards reproductive styles, starting from the open-substrate spawning of madtoms (ictalurids *Noturus* spp.) to mouthbrooding in ariids, internal fertilisation in auchenipterid woodcats and cuckooing mouthbrooding cichlids in some *Synodontis* (Mochokidae; review in Bruton 1996). Maximum sizes are as small as 14 mm in some scoloplacids, as large as 2 or 3 m in some clariids (e.g. the African *Heterobranchus longifilis*), pangasiids (e.g. *Pangasianodon gigas* from the Mekong River, Southeast Asia) and pimelodids (South American *Brachyplatystoma* spp.), and up to 5 m in the wels *Silurus glanis* (Siluridae) occurring in the Palaearctic. The electric catfish *Malapterurus electricus* can produce discharges of up to 600 V, and many species (ariids, aspredinids, bagrids, callichthyids, chacids, clariids and doradids) produce sound (review in Heyd & Pfeiffer 2000). Catfishes are found from open waters to caves, and they have adopted a wide range of trophic niches and habits, including scale eating, wood eating and blood sucking (trichomycterids, with *Vandellia cirrhosa* being the only fish known as a parasite of humans). These traits make the Siluriformes one of the most diverse and interesting groups of fishes. Paradoxically, knowledge of their migratory habits is generally limited, except for the two families occurring in the northern temperate region (see the end of this section), the two families occurring in marine waters and some large tropical catfishes of interest for fisheries.

Although many catfish species are found in coastal lakes and lagoons, at least over several months of the year (e.g. clariids and claroteids in South Benin; Welcomme 1979, 1985; Blaber 1988), only the coral catfishes (Plotosidae) and some forktailed catfishes (Ariidae) have marked osmoregulatory capacities and have penetrated marine environments. However, several *Arius* spp. are truly freshwater species, such as the large *A. gigas*, which have been found as far upstream as the Gauthiot falls in the River Niger basin, and are now endangered. Several ariid species (*A. felis*, *A. graeffei*, *A. heudeloti*, *A. latiscutatus*, *A. madagascariensis* and *A. melanopsis*) are thought to be anadromous, but definite evidence is lacking for most of them, except possibly for *A. madagascariensis*. The issue of anadromy in this family has been reviewed by McDowall (1988).

In tropical environments, many catfishes have evolved means of coping with protracted hypoxia, anoxia and low water, notably the clariids, which have developed suprabranchial organs enabling air breathing (see section 4.4), and many of these species are thought to exhibit little migratory activity, either laterally or longitudinally. This is partly supported by the relatively low abundance of catfishes in the catches in Parana do Rei, a channel connecting a large floodplain lake to the Amazon River (<10% of all catches; Fernandez 1997). Ageneiosids (2.4%) and hypophthalmids (1.5%) outnumbered other catfish families which abound in the River, especially the loricariids (1.2%) and the pimelodids (1.3%), but all of them fell second to the characiforms, which comprised more than 75% of the catches. Similarly, Niaré and Bénech (1998) reported that catfishes contributed to less than 13% of the fishes making lateral migrations in the River Niger Inner Delta, the most abundant families being the bagrids, claroteids and schilbeids.

However, there is evidence from sampling at different stations along South American rivers and streams indicating that some small loricariid catfishes may exhibit a migratory pattern, although the extent of the migration seems limited. For example, de Menezes and Caramaschi (2000) found that juvenile *Hypostomus punctatus* were found exclusively in inshore habitats of the Ubatiba stream (Brazil), whereas adults occurred in other reaches of the stream, either in fast-flowing sections with rocky bottom or in pools with submerged branches. Such patterns of habitat use, involving lateral and longitudinal short-range ontogenetic migrations, seem common among South American loricariids (E. Baras [unpubl.]). Lateral migrations onto the floodplain may be more rare, as suggested by Fernandez's study, probably because loricariids have adapted to high floods through their benthic lifestyle, and to low oxygen concentration through air breathing (see section 4.4).

Nevertheless, there are many examples showing that some tropical catfishes exhibit migratory habits, and this especially applies to non-guarding egg scatterers, like several clariids, pimelodids and schilbeids. In African rivers, the sharptooth catfish *Clarias gariepinus*, and the vundu catfish *Heterobranchus longifilis* both undertake upstream spawning migrations during high floods and lay large numbers of eggs on recently flooded vegetation (Bruton 1979b; Welcomme 1979, 1985; Bénech & Quensière 1989; Lung'Ayia 1994). Both adults and juveniles remain on the floodplain, with adults sometimes remaining in shallow pools or localised ditches known as fish holes, and used for rural aquacultural practices. For populations living in coastal lagoons such as the Lake Nokoué–Port Novo Lagoon in southern Benin, additional upriver migrations may occur during the dry season, as salinity increases. Hocutt (1989) radio-tracked *C. gariepinus* in Lake Ngezi, Zimbabwe, and found that they moved quite leisurely (about 100 m h^{-1}), essentially at night. More recently, Hérissé and Bénech (in press) observed that mudfish, *C. anguillaris* spawners could move at a much faster pace (about 3.5 km h^{-1}) while making lateral migrations onto the floodplain. However, such fast swimming speeds were sustained for short periods of time only (about 15 min), and mudfishes travelled no more than 7 km d^{-1}, with females travelling on average greater distances than males. Many other African catfishes undertake potamodromous migrations to upstream spawning grounds in lake tributaries. *Schilbe mystus* ascend Lake Victoria tributaries over 25 km (Whitehead, 1959), but they can move as far as 150–200 km along the Chari River (Lake Chad) in July to September, as other catfishes do (the schilbeids *S. niloticus* and *S. uranoscopus*, and the mochokids *Brachysynodontis batensoda* and *Synodontis schall*; Bénech & Quensière 1989). Their eggs, larvae and juveniles remain for a few months in the floodplain, then move down to the lake with increasing floods in February (Bénech & Quensière 1983). In the same river system, another mochokid, *Hemisynodontis membranaceus*, was found to undertake potamodromous migrations as far as 300 km from the lake (Bénech & Quensière 1989).

In the Amazon River basin, many eggs and larvae of siluriforms (mainly pimelodid catfishes) have been found drifting in currents, both in white- and blackrivers, in clear or turbid waters (Pavlov *et al.* 1995; Araujo-Lima & Oliveira 1998). Catfish larvae mainly drift near the bottom and are presumed to drift over shorter distances than larvae of characins. However, drifting distances of several thousand kilometres, and fast downstream movement (averaging 250 km d^{-1}) have been suggested for the early life stages of some pimelodids drifting on rain-swollen currents. These traits apply to the large (L_∞; c. 1.5 m) predatory catfishes (sorubims *Brachyplatystoma* and *Pseudoplatystoma* spp.) which have been suspected to undertake the longest freshwater migrations ever documented (Barthem *et al.* 1991; Barthem & Goulding 1997). In the Amazon River basin, it has been estimated that the five species of large catfishes can consume as much as 100 000 tons of food fishes (about half the amount consumed by humans), making their exploitation by fisheries not only a matter of predation but also a matter of competition. Examining the seasonality of captures and sizes of dorado *Brachyplatystoma flavicans* and piramutuba *B. vaillanti* at the major fish-landing sites of the Amazon River basin, Ruffino and Barthem (1996) noticed that the longer the distance to the estuary, the greater the fish size and the less marked the seasonality of landings. In Leticia (Colombian part of the Amazon River, in the Andean Piedmont), only fish greater than the size at first sexual maturity (L_m; c. 80 cm) were found, but they were captured almost all year round. Fish captured in the lower Amazon River (from Manaus down to Belem, in the estuary) were smaller than L_m, and catches showed a strong seasonal periodicity (e.g. landings in Santarem peak in August–September, during the period of receding waters; see also Ruffino & Isaac 1995). It has been proposed that larvae (and possibly juveniles) drift downstream, as far as the Amazon estuary, grow in these warm and rich areas, then undertake upstream migrations over several thousands of kilometres to upstream reaches of the Alto Solimoes, Madeira or Japurá tributaries where they spawn in places undiscovered to date (Goulding 1979, 1980; Barthem *et al.* 1991; Barthem & Goulding 1997). Young fish are found in the estuary about 3 months later. The presence of large fish all year round in the upper reaches may suggest either that a part of the population makes more restricted movements, or that younger fish migrate progressively during their life up to these upstream reaches or tributaries. The many factors which may have shaped the lifestyle of these species in such a spectacular way have been reviewed and discussed by Barthem and Goulding (1997). Among the most exciting hypotheses is the possibility that this is a heritage from the Miocene era, when the Amazon was flowing westward before the Andean uplift, and spawning and nursery areas were much closer together than they are now.

In the northern temperate region, only two main families of catfishes occur, with the ictalurid catfishes (Ictaluridae) occurring throughout North America from southern Canada to Guatemala and the sheatfishes (Siluridae) occurring mostly in Asia, with two species in Europe. Many of the 45 species or so of ictalurid catfishes are regarded as sedentary but there is substantial evidence to show that channel catfish *Ictalurus punctatus* are migratory. A thorough study by Pellett *et al.* (1998) involved the tagging of more than 10 000 fish in the lower Wisconsin River and adjacent waters of the upper Mississippi captured by the study authors, commercial fishermen and anglers. Channel catfish occupied small home ranges in summer, migrated down to the upper Mississippi in autumn and homed back up the Wisconsin in spring to spawn and to occupy the same summer home range as in previous years. Larger fish showed greater summer home site fidelity than small fish. In a substantial number of cases the distances moved by channel catfish were at least the length of the Wisconsin River which could be negotiated before being limited by a dam, a distance of 126 km. Patterns of upstream movement in spring and downstream movement in autumn by channel catfish seem typical

(Dames *et al.* 1989; Newcomb 1989; Smith & Hubert 1989). Newcomb (1989) found that 53% of all recaptures of channel catfish in the Missouri River in winter were from tributary streams, and that at least some of these were tagged in summer. As migrants, channel catfish can exploit spawning and feeding habitats in shallow rivers in the summer and retreat to the safety of deepwater habitats in winter. Kennen *et al.* (1994) reported downstream migratory movements of brown bullhead *Ameiurus nebulosus* towards Lake Ontario between April and August.

The sheatfishes (Siluridae) are found in freshwater and brackish waters and seas. Most of the 100 species occur in Asia and only two species occur in Europe. The largest species, the European wels *Silurus glanis*, can reach a length of 5 m and weight of 330 kg. It is found in river systems from the Baltic to the Caspian, Azov and Aral Sea regions. It also occurs in brackish water seas such as the Baltic Sea. The adults are solitary nocturnal predators and tend to occupy deeper holes during the day. Short-distance migrations into shallow water occur for spawning, after which wels return to deeper water (Lelek 1987). It is unclear whether wels in brackish water migrate into freshwater to spawn. Overall, migratory behaviour may be of limited occurrence in this family, but there appear to have been no studies of its spatial behaviour.

5.17 Knifefishes (Gymnotiformes)

Knifefishes comprise approximately 120 nominal species of ostariophysan fishes with a compressed or rounded eel-like body, which are restricted to neotropical freshwaters, although occurrence in brackish waters has been reported (Nelson 1994). Their systematics undergo frequent revision and some nominal species with broad distributions, such as *Gymnotus carasco*, tend to be deconstructed. Gymnotiforms are far better known with respect to their electrogenic and electrosensory capacities that rely on the use of electric organ discharges (EOD) and are the functional bases of electrolocation and electrocommunication, than as regards habitat use, movement or migratory habits. Araujo-Lima and Oliveira (1998) reported that larvae of gymnotiforms represented an almost negligible proportion of drifting larvae in the Amazon River, possibly because larvae possess cephalic attachment organs, as recently demonstrated in some apteronotids and rhamphichthyids (Britz *et al.* 2000). Fernandes (1997), who studied the lateral migrations between the Amazon River and a floodplain lake (Lago do Rei), found that gymnotiforms represented no more than 0.4% of the fishes entering or leaving this floodplain lake. Similarly, Henderson and Hamilton (1995) found that gymnotiforms, and especially *Hypopomus* spp., represent a major component of attached meadows in *varzea* habitats, but that they move to a neighbouring residence once the meadow has become detached and starts drifting, suggesting they have strong residential habits.

Combining these rare pieces of information at hand suggest these species exhibit restricted movements. However, comparisons between lifestyle, EOD signals and tolerance to oxygen (Crampton 1998, see Box 4.3) suggest that the rather sedentary lifestyle depicted above fits mainly those species with a high tolerance to low oxygen and which produce pulse-type EODs. These comprise the hypopomids, the gymnotids and the electric eel, *Electrophorus electricus*, which are found almost permanently in *varzea* habitats or in *terra firme* swamps, and the rhamphichthyids, which mainly occur in low conductivity blackwater systems. Species with tone-type electrical signals (apteronotids, eigenmanniids and some sternopygids) are far less tolerant to protracted hypoxia, are mainly found in whitewater river channels under low flow and enter the floodplain, inundated forest and blackwater lakes under high flow, which also corresponds to the spawning season. Although seasonal movements appear frequent in these species, there is no evidence for long-range potamodromous migration, although, as for most tropical fish species, information is scarce.

5.18 Pikes and mudminnows (Esociformes)

The Esociformes comprises the five species of pike (Esocidae) and six species of mudminnows (Umbridae) found from the north temperate zone to the arctic (Nelson 1994). Several species of pike may exceed 1 m, measuring up to 1.6 m for the muskellunge *Esox masquinongy* from North America, while the mudminnows rarely exceed 20 cm. One species, the northern pike *Esox lucius*, has a circumpolar distribution.

Esocids tend to display limited migration, although local migration may be of key significance for population maintenance. They principally occur in fresh water and typically spawn in seasonally inundated areas or over aquatic vegetation in shallow water. During the main period of summer feeding and growth, adult northern pike *Esox lucius* may range widely through appropriate habitat without following any clear route (Diana *et al*. 1977; Diana 1980; Lucas *et al*. 1991). Where spawning occurs within lake habitats, although overall activity increases dramatically during the spawning season in March–April, spawning areas may not be distinct from feeding and nursery areas, particularly in small, shallow lakes (Lucas 1992). However, in other systems northern pike exhibit clearly defined spawning migrations into lake tributaries, drainage ditches and marsh areas (Carbine 1942). Clark (1950) showed that pike migrated from Lake Erie, Ohio, into feeder streams. Any stream or ditch was utilised provided that some vegetation or debris, with enough water to partially cover the fish, was available. Males predominated in the early upstream movement and females in the later part of the run. Franklin and Smith (1963) also showed that northern pike moved out of Lake George, Minnesota, to spawn in a feeder stream. They found no difference in sex ratios as the spawning run progressed nor did they find any changes in the size of pike over the time of the run. Adult fish began leaving the breeding grounds shortly after spawning. Some individuals remained for considerable periods but 62–64% of fish had left within 40–60 days of spawning. Miller (1948) observed that individual pike were not faithful to a single spawning ground but would move around visiting several spawning grounds.

Increasing discharge and temperature are shown to be key factors in stimulating spawning migrations by northern pike (Masse *et al*. 1991). In many cases the maintenance of these migrations and of appropriate habitat are shown to be of key significance in maintaining northern pike populations. Young northern pike may migrate out of inundated areas within 2 days of hatching, seeking refuge in vegetated areas of lakes and backwaters. Franklin and Smith (1963) showed that northern pike alevins began to emigrate from their nursery stream into Lake George, Minnesota, at 16–24 days after hatching. Juvenile fish left the nursery stream in mid-May to early June and in two out of three years 98% of juvenile fish left the stream within 20 days of the start of emigration. Studies of feeder streams like this show that the availability of spawning and nursery areas in small tributaries can be important for the maintenance of northern pike populations in some lake systems.

Although esocids are principally freshwater species they will move into brackish water environments, exhibiting clear migratory patterns, particularly during the spawning run into freshwater streams. Most of the studied populations occur in the Baltic Sea. Northern pike in the coastal areas of the Baltic Sea ascend coastal rivers in spring to spawn. In the oligohaline coastal zone of the Gulf of Bothnia they migrate up to 6 km in the Angeran stream and mostly spawn in a shallow lake in areas which are usually inundated during the spring thaw (Johnson & Müller 1978b; Müller & Berg 1982, Müller 1986). After spawning the pike leave the stream and most migrate back to the sea. The majority of young emigrate in their first summer. There is low productivity in the stream system and the rapid downstream migration appears to be due to relatively high populations which may be associated with elevated competition and cannibalism pressure. For adults, emigration therefore represents a feeding migration, while for juveniles it is probably a combined feeding and refuge migration. Temperatures reach 10°C in the stream about a month earlier than in the coastal sea. The aforementioned authors term this migration anadromous, although it is somewhat marginal and McDowall (1988) does not consider esocids to exhibit any degree of diadromy.

Muskellunge *Esox masquinongy* generally exhibit clear migratory patterns and seasonal use of specific areas of habitat. Radio-tracking studies have shown that in Canada muskellunge occupy well-defined home ranges during summer and winter, but not during spring and autumn when rates of movement are greatest (Minor & Crossman 1978; Dombeck 1979; Miller & Menzel 1986). High levels of spring movement with mean maximum rates of 408 m d^{-1} and 562 m d^{-1} by males and females respectively in the period 16–30 April were recorded (Minor & Crossman 1978). In these studies spawning migrations resulted in aggregations of fish, especially around tributary entrances, with distances of up to 4.5 km travelled from winter home ranges. Dombeck (1979) suggested that increased levels of movement in autumn were associated with foraging activity. Displaced muskellunge exhibit clear homing ability (Miller & Menzel 1986) and spawners may exhibit repeated fidelity to spawning grounds, although this does not occur systematically (Crossman 1990). Migratory behaviour has not been widely described for the other species of esocids. However, localised longitudinal and lateral spawning migrations of grass pickerel *Esox americanus vermiculatus* to weedy streams and inundated areas are common (Scott & Crossman 1973; Kwak 1988).

The mudminnows are small freshwater fish, less than 20 cm in length, found in parts of the northern hemisphere from temperate areas in Europe and North America to northeastern Siberia and Alaska. They are most common in floodplain pools and oxbows and can withstand low oxygen levels. They are not known to exhibit large seasonal migrations within river systems, although it is likely that movement is increased during inundation of the floodplain. Small upstream lateral spawning movements have been reported for Alaska blackfish *Dallia pectoralis* after ice breaks up in May and for central mudminnow *Umbra limi* (Scott & Crossman 1973). Due to their tolerance of hypoxia, mudminnows are often an important component of the fish fauna in pools which suffer periodic deoxygenation. Rahel and Nutzman (1994) demonstrated that central mudminnow regularly entered hypoxic water in a Wisconsin pond to feed on *Chaoborus* larvae. Winter ice cover over ponds results in oxygen depletion near the bottom. In arctic and continental temperate climates, many of the small ponds inhabited by mudminnows become icebound during winter, often resulting in hypoxic conditions. Mudminnows display localised but regular movements in response to ice cover, with fishes moving to positions immediately underneath the ice where oxygen concentrations remain highest (Magnuson *et al.* 1985).

5.19 Smelts (Osmeridae)

The smelts are small, silvery shoaling fishes that are widely distributed in the cool areas of the Northern Hemisphere. There are about 12 species and these exhibit a variety of migratory behaviour patterns. Detailed descriptions of the biology of diadromous smelts are provided by McDowall (1988). The rainbow smelt *Osmerus mordax* is one of the most well-studied species and is widespread in eastern North America, but also occurs on the Pacific coast of North America as far south as British Columbia and on the west Pacific coast as far south as Korea. Both anadromous and freshwater populations occur. Anadromous rainbow smelt move into brackish water in early spring before moving upstream to spawn in early summer (May–June). McAllister (1984) stated that this species may migrate as far as 1000 km upstream; but in most cases spawning occurs no more than 20 km upstream of the head of tide. A substantial proportion of fish survive spawning and these move to estuarine and coastal waters (McKenzie 1964). The eggs hatch in about 2 weeks and the young are carried down into the estuary where they feed and grow. Although this species is often described as anadromous, many populations do not move out of brackish, estuarine waters where conditions remain favourable. At the northern and southern limits of their range greater movements offshore appear to occur, reflecting movements to avoid ice and high temperatures respectively. The eulachon *Thaleichthys pacificus* grows to adulthood in marine environments along the Northwest Atlantic coast and enters rivers in spring to spawn in their lower reaches (Scott & Crossman 1973). Most eulachon die after their first spawning.

The European smelt *Osmerus eperlanus* is mainly found along the eastern Atlantic coast, south through the Baltic to the Bay of Biscay. It has a similar life history to *O. mordax*, normally being anadromous and moving into rivers between February and April to spawn just above the head of tide (Wheeler 1969; McAllister 1984). After hatching, the young remain in the river and estuary for the remainder of the summer. Wholly freshwater populations occur in Jutland, southern Norway, Sweden, Finland and Poland, and a population in Rostherne Mere, Cheshire, England became extinct in the 1920s (Lelek 1987). These may spawn in the lakes or ascend rivers to spawn. Anadromous and freshwater populations of a subspecies also occur in rivers entering the White Sea. The longfin smelt *Spirinchus thaleichthys* of North America and the delta smelt *Hypomesus transpacificus* which occurs on East and West Pacific coasts comprise fully marine, freshwater and anadromous populations. Both species occur in the extensive Sacramento-San Joaquin basin of California, in which the brackish water delta systems provide extensive nursery habitat (Stevens & Miller 1983). This system originally had large migratory runs of both longfin smelt and delta smelt. However, these species have been impacted by changes in the water management, especially through increases in water diversion that reduce spawning and rearing areas and reduce the area over which migration can occur. Reduced freshwater flow has been correlated with low year class strength for longfin smelt and to a lesser degree for delta smelt (Stevens & Miller 1983).

Ayu *Plecoglossus altivelis* occur in Japan, Korea, Taiwan and China, rarely exceed 20 cm in length, and have diadromous and freshwater forms. The diadromous form is often reported as being anadromous (e.g. Kusuda 1963; Nelson 1994) but McDowall (1988) describes it as amphidromous, and this description seems most appropriate. Most live for only 1 year, spawning in the autumn and then dying. Amphidromous ayu dig small pits in gravel in which to spawn, in the lower reaches of rivers, and the newly hatched young drift into the ocean, with greater survival when arriving in coastal areas with water temperatures less than 20°C. The young overwinter in coastal areas and migrate upriver in spring where they grow over the summer, feeding on algae attached to rocks (Kusuda 1963). They mature in autumn and move downstream to the lower reaches of the river to spawn. Landlocked populations of ayu occur in many lakes such as Lake Biwa, Japan (Azuma 1973; Tsukamoto *et al.* 1987). Larvae of landlocked populations may also enter seaward migrations. In the Japanese Shou River, Tago (1999) estimated that 5–15% from the 300 million to 2.9 billion larvae drifting each year originated from the upstream landlocked population in Lake Bi Na. In Lake Biwa, landlocked ayu ascend inlet streams to spawn in autumn, but there is much greater variation in migratory behaviour. Some fish migrate upstream in spring and grow large (about 15 cm), some migrate in summer and remain small and others migrate in autumn and remain small (Tsukamoto *et al.* 1987). The major determinant of growth and migratory behaviour is the date of hatching. Since the small, late-migrating fish spawn earlier than the large spring migrants, Tsukamoto *et al.* suggest that their offspring will hatch first, grow rapidly and become spring migrants. This could result in an alternation of growth and migratory characteristics between generations, but with all types present at any one time. Coexisting diadromous and freshwater-resident forms of pond smelt *Hypomesus nipponensis* in Lake Ogawara, Japan, have also been demonstrated through examination of strontium to calcium ratio profiles in otoliths (Katayama *et al.* 2000).

5.20 Noodlefishes (Salangidae and Sundsalangidae)

Salangid noodlefishes are small slender fishes of the cool temperate northwestern Pacific and East Asia. Various species appear to be marine, freshwater or anadromous (Roberts 1984), but relatively little is known of their lifecycles and migratory behaviour. *Salangichthys microdon* migrates into rivers to spawn near the head of tide in Japan, and the eggs develop there despite substantial fluctuations in salinity (Senta 1973). A substantial number of species appear to show some degree of migration between marine and brackish-freshwater environments. We are not aware of any information regarding migratory behaviour of holobiotic freshwater noodlefishes, including the sundsalangid noodlefishes of Southeast Asia. McDowall (1988) provides a more detailed review of known diadromous migratory habits for this group.

5.21 Southern smelts and southern graylings (Retropinnidae)

The southern smelts are small fishes, reaching about 15 cm in length, which occur in New Zealand and the southwest of Australia. Several members of this family exhibit anadromous behaviour. The most well-described species is the New Zealand common smelt *Retropinna*

retropinna. Sexually mature adults of this species migrate from the sea into rivers during spring and summer to spawn near the tidal limit, sometimes 30–50 km from the sea, but usually much closer (McDowall 1988; Fig. 5.7). The adults spend several months in fresh water during which time they feed, and are in some cases accompanied by immature fish. The larvae, which are less than 10 mm long, are carried into the sea on hatching where they grow to maturity. The species is mainly annual, with adults dying after spawning (McDowall 1979). Non-diadromous populations of this species frequently occur in coastal and inland lake environments where they breed without undergoing significant migrations (McDowall 1979). Other species such as *R. tasmanica* from Tasmania have a similar lifecycle. *Retropinna semoni* from Southwest Australia occurs widely in fresh and brackish water throughout its lifecycle and exhibits potamodromous migratory behaviour (Harris *et al.* 1998). Further information on the behaviour and life histories of diadromous southern smelts is provided by McDowall (1979, 1988).

The Australian grayling *Prototroctes maraena* is an amphidromous smelt-like schooling fish rarely larger than 30 cm and occurs in coastal rivers of southeast Australia and Tasmania (Berra 1987). Another species of southern grayling, *P. oxyrhynchus* from New Zealand, also appears to have been amphidromous but is now extinct (McDowall 1988). Most information

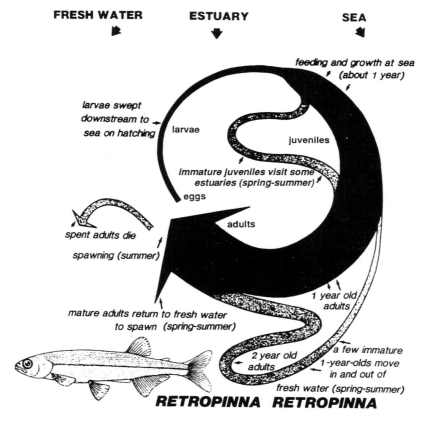

Fig. 5.7 Lifecycle of the southern smelt *Retropinna retropinna*, an amphidromous fish from New Zealand and southern Australia. Reproduced from McDowall (1988).

on Australian grayling life history and migration is based on a study in the Tambo River, Victoria (Berra 1982, 1987). Australian grayling spawn in fresh water in late April to early May (autumn), at least as far as 100 km from the sea. The small eggs settle in gravel interstices and on hatching it is likely that the young are carried downstream to brackish water and coastal environments where they grow for some time before migrating back into rivers, typically in spring. These fish tend to move upstream within the first year or two of life, but without obvious seasonal movements. They mature at 1–2 years of age and are repeat spawners, living for up to 5 years.

5.22 Galaxiids, southern whitebaits and peladillos (Galaxiidae)

The family Galaxiidae (about 40 species) is widely distributed in the cool temperate southern hemisphere, occurring in New Zealand, southern Australia, Patagonian South America and the southern tip of Africa. They do not normally exceed 30 cm in adulthood. Most species of the subfamily Galaxiinae of Nelson (1994) occur in fresh water, principally in Australia and New Zealand. They may exhibit limited migratory behaviour (e.g. *Galaxias depressiceps*, *G. gollumoides*) although this has not been examined in detail. However, diadromous behaviour in this group is extensive, since McDowall (1988) describes at least eight species of the genus *Galaxias* as being clearly amphidromous or catadromous. The juveniles of these species often aggregate in coastal areas, and are known as 'whitebait' to the fishermen who catch them.

Most of these species are amphidromous. *Galaxias brevipinnis* spawns in fresh water as much as 200 km from the sea, although migrations associated with spawning itself appear to be minor (McDowall 1988). The larvae are washed downstream and early growth takes place in the sea, and is followed by migration of shoals of juveniles, about 50 mm long, into fresh water. These shoals often comprise several species. These fish, especially *G. brevipinnis*, demonstrate impressive migratory tendencies, moving upriver and negotiating rapids and falls. The juveniles of *Galaxias* can move up damp rock faces and dams. After growing for several years the fish mature and may spawn for several years. There are many examples of landlocked populations of galaxiids such as *G. brevipinnis* (in New Zealand and Tasmania) in which the adults tend to live in lake tributaries and the young spend the first few months in lakes. Further information on the lifecycles and migration of these species is provided in McDowall and Eldon (1980), McDowall (1988) and McDowall (1990).

One species of *Galaxias*, the inanga *G. maculatus* of Australia and New Zealand, exhibits somewhat catadromous behaviour. As an adult it migrates from lowland freshwater habitats to spawn in tidal estuaries. Egg deposition and development occurs over a wide range of salinities from fresh to almost full-strength seawater (McDowall 1990). McDowall (1988) presented evidence of a clear lunar rhythm of migration and spawning for this species, with maximum activity occurring on both new and full moons. The eggs are deposited in marginal vegetation at the spring tide highwater mark and develop out of water before being submerged on a subsequent spring tide. At low temperatures, development is slowed and the eggs can survive for at least 6 weeks out of water. The larvae (<10 mm) are washed into the sea and grow there for about 6 months before returning as juveniles of about 5 cm in length. The young typically spend another 6 months feeding and growing in fresh water before migrating downstream to spawn. Several landlocked stocks of inanga occur, including an Australian population in which the pre-adults grow in the lower reaches of tributaries and in lake margins, and the adults migrate upstream during floods to spawn in marginal vegetation (Pollard 1971).

The peladillos *Aplochiton* spp. occur on the southern fringes of South America and the Falklands. Some species, for example *A. taeniatus*, spawn in fresh water and larval *Aplochiton* have been collected from the sea, but the life histories and movements of *Aplochiton* species are not generally clear. They are most likely amphidromous but may be anadromous. A helpful discussion of migration and lifecycles within this group is presented by McDowall (1988). The Tasmanian whitebait *Lovettia seali*, a close relative of the peladillos, is known to display clear anadromous behaviour. Adults mature at 1 year old when they are about 7 cm long, and migrate into lowland Tasmanian rivers where they spawn and die. The newly hatched larvae drift straight out to sea where they grow to adulthood (Blackburn 1950).

5.23 Salmons, trouts, chars, graylings and whitefishes (Salmonidae)

The salmonids are the family of fishes most synonymous with migratory behaviour to most people, particularly the species of *Oncorhynchus* and *Salmo* that display large-scale anadromous migrations. Providing a detailed review of their migratory behaviour in proportion to the colossal knowledge collated over the past decades would make this section twice as thick as the rest of this volume. Also, several sections of chapters 2, 3 and 4 have been illustrated with various examples on salmonids. Additionally, there are many detailed references dealing specifically or generally with this family elsewhere (Harden Jones 1968; Banks 1969; Foerster 1968; Scott & Crossman 1973; Northcote 1978, 1995, 1997; Balon 1980; McKeown 1984; Dadswell *et al.* 1987; Holcik *et al.* 1988; McDowall 1988, Mills 1989, 2000; Groot & Margolis 1991; Shearer 1992; Moore 1996). Northcote (1997) specifically reviews potamodromy in salmonids. Therefore, we provide a relatively brief overview of the migrations of salmonids, focusing on groups or genera which have received less attention than others, or for which integrated information is less readily available.

The salmonids are represented by about 66 species in 11 genera, and are characteristic of cool and cold climates of all continents of the northern hemisphere as far south as the Mediterranean and northern Africa on the eastern Atlantic coast, and as far south as Mexico on the eastern Pacific coast (Nelson 1994). Some anadromous populations of salmonids such as the chinook salmon *Oncorhynchus tshawytscha* and the Atlantic salmon *Salmo salar* have been recorded at lengths approaching 1.5 m, while some huchens *Hucho* spp. may reach 2 m, though such large fish are extremely rare today. Because of their economic importance and their migratory behaviour, several species of *Salmo*, *Oncorhynchus* and *Salvelinus* have been well studied, although behavioural information tends to be less complete for the whitefishes and ciscoes (Coregoninae), the graylings (Thymallinae) and several *Hucho* species.

Anadromy in salmonids is exemplified by the Atlantic salmon *Salmo salar* which spawns in fresh water in autumn or winter and in most cases migrates to sea where most growth occurs. River entry may occur in any month of the year and a few fish may remain in fresh water for up to a year before spawning (Shearer 1992). Adult Atlantic salmon may migrate up to 1000 km in some large river systems, but in short coastal streams upstream migration distances may be as short as a few kilometres. The females dig spawning pits or 'redds' in gravel in swift-flowing water in which the eggs are laid and then covered. The eggs hatch in the gravel and the young alevins emerge in late spring where they feed and grow, adopting and defending territories. Young 'parr' tend to remain in fresh water for 1–8 years, the duration of freshwater residence depending on environmental conditions and growth potential (Mills 1989). Some young, particularly males, do not migrate to sea and become sexually mature 'precocious parr', which are capable of and successful in 'sneak' matings of eggs from sea-run adult female Atlantic salmon. Two main periods of downstream movement occur, one in autumn and one in spring, but most salmon emigrate from rivers in spring when they go through a number of physiological changes and become silvery 'smolts', with an osmoregulatory competence to adaptation in marine environments (Høgåsen 1998). Early smolts descend river at night, orientating along the strongest flows, which causes many problems at water abstractions by hydropower turbines (see section 7.3.1 and 7.4.6). Later runners descend faster, as movements tend to occur throughout the day and night. Upon arrival in the estuary, smolts can show some residency or immediate seaward movement, depending on tide height, and there is strong evidence for a selective ebb transport (Moore *et al.* 1995; see also section 2.4.3). As is common with many of the salmonid species that are typically anadromous, 'landlocked' populations of Atlantic salmon exist, which complete their lifecycles in fresh water, and display potamodromy (e.g. Trépanier *et al.* 1996).

At sea, Atlantic salmon from North America and Europe converge on rich feeding grounds off Greenland, many hundreds of kilometres distant from their 'home' rivers, although other less-distant waters such as around the Faeroe Islands (from the perspective of European stocks) may also be used. Some fish stay at sea for only one winter ('grilse'), while others may remain at sea for 2 or more years. Most of their behaviour at sea remains to be discovered, essentially due to current limitations of remote sensing systems for tracking submerged animals in the open sea (see section 6.3.6), and archival tags certainly offer great promise for elucidating these (see section 6.3.11; Moore *et al.* 2000). Coastal migrations of returning adults have been more fully documented, and salmon have been shown to possess the capacity of following a particular compass heading over large distances irrespective of tide direction. Returning adults of both *Salmo* and *Oncorhynchus* spp. frequently exhibit an alternation of deep dives and sojourns near the surface, which have been interpreted with respect to thermoregulation and search for directional cues, respectively (Westerberg 1982; Døving *et al.* 1985; Tanaka *et al.* 2000). A long acclimation period from saltwater to fresh water is seemingly not needed and Atlantic salmon may rapidly enter fresh water (Priede 1992), the timing of entry in the estuary being essentially dependent on its topography, and on river discharge, but many factors including wind, light level, and temperature may be important too. The influence of the tidal cycle is less clear, as salmon enter some rivers at all stages of the tidal cycle, whereas in other rivers, river entries happen during flood tides, and in some others during ebb tides (e.g. Smith & Smith 1997). During the upstream migration adult Atlantic salmon do not feed and energy reserves are used to fuel body maintenance, gonad growth and migration. Males remain on the spawning grounds for longer than females and suffer higher mortality than females, a few percent of which may survive as 'kelts' and regain condition at sea, to spawn again (Mills 1989, 2000).

Broadly similar lifecycles and patterns of migration are exhibited by other anadromous *Salmo* and *Oncorhynchus* species, and McDowall (1988) provides a comprehensive review of the anadromous species or populations. Some species such as the sockeye salmon *O. nerka* exhibit complete semelparity, with all adults dying after spawning, and this has been attributed to the considerable investment into migration and reproductive activity (see section 2.3.5). There is great plasticity in the migratory behaviour of stocks of a wide variety of salmonids, influenced both by genetic and environmental factors (see chapter 2), with many populations existing as freshwater residents and others having anadromous and/or holobiotic freshwater components (Northcote 1997).

Such plasticity is exemplified in the brown trout *S. trutta* with environmental conditions having a major influence on whether the population is holobiotic or anadromous, thereby exhibiting a lifestyle similar to that of salmons, except that sea trout eat during their upstream migration, and that they exhibit a much greater degree of iteroparity than species of salmon (review in Baglinière & Maisse 1991). The progeny of a single pairing may develop into forms which remain in fresh water, migrate to sea or move into estuarine waters to feed (see also section 7.3.4). Brown trout *S. trutta m. fario* has frequently been reported as a territorial species making restricted migrations, although the advent of telemetry techniques has questioned their residency (Gowan *et al.* 1994). During summertime, under low flow, brown trout generally show home range or territorial behaviour, and move short distances only (Clapp *et al.* 1990; Meyers *et al.* 1992; Young 1994; Ovidio 1999), notably as they feed on drifting insects, and may adapt to variations in prey abundance through time-budgeting (Giroux *et al.* 2000), as has also been reported in the cutthroat trout *Oncorhynchus clarki* (Young *et al.* 1997). Spawning migrations during spring (cutthroat trout) or autumn (brown trout) can be stimulated by increased flows (Allan 1978; Baglinière *et al.* 1987; Brown & Mackay 1995) or varying temperatures (Clapp *et al.* 1990) or by the combination of the two factors (Ovidio *et al.* 1998), probably depending on how predictive and reliable these cues or combinations of cues are in the rivers or streams considered. Spawning runs have variable lengths, with some trout migrating dozens of kilometres away from their summer home range, and others using nearby spawning grounds sited in the main stream or in small tributaries (e.g. Ovidio 1999). Although upstream migration of young salmonids soon after emergence is relatively rare it can and does occur, particularly in lake outlet stream-spawning populations (Godin 1982; Roussel & Bardonnet 1999).

Similar patterns are exhibited by a variety of *Oncorhynchus* species, notably by *O. mykiss*, of which anadromous and freshwater forms are known as steelhead and rainbow trout, respectively. McDowall (1988) provides a comprehensive view of how the life history of steelhead trout changes with latitude and climate, with longer periods of residence in rivers or sea, and greater size of spawners in the northern populations that show a lower repeat spawning frequency than southern populations. *Salmo trutta* and *O. mykiss* have been introduced in a wide range of river systems throughout the world, with variable success and variable impacts on indigenous fauna, which are beyond the scope of this book. However, it is worth pointing out that the environmental conditions that rule the migratory patterns and life history of these species in their natural distribution range, might be different in other climates. For example, Dedual and Jowett (1999), who radio-tracked rainbow trout in the Tongariro River, New Zealand, found that the seasonality, extent and speed of upstream migrations were highly variable and could not be predicted from environmental factors. This is possibly because the species flourishes there and the climate is so favourable to them that the constraints on migrations are less conspicuous than in other river systems. The only noticeable environmental cue that represented a consistent incentive to trout movement, a downstream movement in this case, was the eruption of Mount Ruapehu, and this, to our knowledge, is the only study where fish movements were considered in the course of a volcanic eruption, although studies on the subsequent frequency of homing by anadromous salmonids are available elsewhere (Whitman *et al.* 1982).

The genus *Hucho* comprises five large-sized species, including a single anadromous species, the Sakhalin taimen *Hucho perryi* (Holcik *et al.* 1988). Adults migrate from the Sea of Japan or the Sea of Okhotsk, move over short distances inland in the rivers and lakes of the islands of Honshu, Hokkaido, Kunashir and Sakhalin, and construct spawning redds, as do other *Hucho* spp. (Fukushima 1994). Gritsenko and Churikov (1977, in Holcik *et al.* 1988) reported that smoltification in the Sakhalin taimen takes place at an older age (5–7 years) and greater size (40–50 cm) than in other anadromous salmonids, possibly because this species is relatively recent, and has a shorter history of diadromy (Holcik *et al.* 1988). *Hucho bleekeri* occurs exclusively in mountain gorges of the Yangtze River in China, and the Korean taimen (or chachi) *H. ishikawae* is endemic to the upper reaches of the Yalu River, in Korea. Both species have a most restricted geographical distribution, and very little is known on their biology in general and migration history in particular (Holcik *et al.* 1988).

The taimen *Hucho taimen* is the most widespread *Hucho*, occurring throughout the Siberian and Ural regions, whereas the huchen *Hucho hucho* is endemic to the Danube River system, although it has been introduced in many rivers from Morocco to Norway, and in North America. Huchen and taimen are large (up to 2 m and over 100 kg for huchen) and long-lived (up to 55 years for taimen) species that share a similar lifestyle, with shoals of spawners exhibiting nocturnal potamodromous migrations, ascending rivers and entering tributaries, where they construct redds and spawn in pairs (or with a slight predominance of males). The two species spawn in early spring, and many authors have reported that the onset of their spawning migration coincides with snow melting. Depending on climate, spawning takes place 2–6 weeks later, when temperature ranges from 6 to 10°C (Witkowski 1988), but there are records of taimen reaching their spawning grounds at 3°C (Misharin & Shutilo 1971 in Holcik *et al.* 1988). The young remain in tributaries for variable periods, depending on climate, but downstream movements at the onset of the first winter are frequent in foothill or mountain tributaries. Both the huchen and the taimen, by virtue of their large size, have been suspected to undertake extremely long potamodromous migrations, over several hundreds of kilometres. The decline of the huchen after the Danube River became dammed has reinforced this suspicion. Although there are some historical reports of migrations as long as 190 km for huchen (Ivaska 1946, in Holcik *et al.* 1988), it is now assumed that spawning migrations are a matter of kilometres or tens of kilometres, and that the decline of huchen in the Danube is due to changes in physical habitat more than long-range migrations being blocked by dams or weirs (Waidbacher & Haidvogl 1998). Holcik *et al.* (1988) suggested from the regular occurrence of huchen in identical places, that homing was frequent in huchen, although evidence from tagging studies is lacking. Similarly, in view of the paucity of tagging studies on huchen and taimen, their 'resident' habits outside of the spawning season must be considered with caution, especially if these fish exhibit strong homing behaviour. Also, both species are reputedly piscivorous, and trophic migrations may be more or less extensive as they chase their fish prey, as happens in the Lena River, where the taimen seemingly tracks shoals of *Coregonus* spp. (Pirozhnikov 1955 in Holcik *et al.* 1988).

The genus *Thymallus* (graylings) comprises five species, although *Thymallus jaluensis* (found in Korea, Siberia and Northern Europe) is probably identical to *T. arcticus grubei*. *Thymallus nigrescens* is endemic to Lake Kosogol and its tributaries (Central Asia), and the Mongolian grayling *T. brevirostris* is only found in landlocked basins of northwestern Mongolia, and possibly in Russia. Virtually nothing is known of their migratory habits. The biology of the two most widespread species of grayling, the Arctic grayling *T. arcticus* and the European grayling *T. thymallus* is reviewed in detail by Northcote (1995). Both species are spring spawners, with females laying small eggs (by salmonid standards) in spawning pits dug in gravel substratum defended by a territorial male (Jankovic 1964). Although movement through estuarine regions has been demonstrated for the Arctic grayling (West *et al.* 1992), this behaviour might be restricted to the harshest part of its geographical range, where riverine habitat is dramatically restricted by ice during winter and no lacustrine habitat is available. Otherwise, graylings are typically potamodromous and exhibit migrations between overwintering, spawning and summer feeding areas. Migrations of Arctic grayling can range from a few kilometres to over 100 km in northernmost latitudes (Craig & Poulin 1975; West *et al.* 1992; Hughes & Reynolds 1994; Northcote 1995), frequently depending on how close or distant overwintering habitat can be found, and thus on the harshness of the climate, a trend which also applies to other salmonid species making potamodromous or diadromous migrations (see section 4.2; Northcote 1997). In similarly harsh Eurasian environments, as occur in Russia, seasonal migrations of European grayling over several tens of kilometres have also been reported (Zakharchenko 1973). However, most migrations of European grayling are seemingly much more restricted. Gustafson (1949) followed spawning migrations of European grayling after ice thaw, from Storsjö Lake to the small brook, Svärtbacken, Sweden, and he found that spawning took place no further than 3 km from the lake. Parkinson *et al.* (1999), who radio-tracked European grayling in the Aisne Stream, Southern Belgium, also found that neither males nor females migrated over more than 5 km upstream during their upstream spawning runs. Each male grayling utilised a single spawning ground, where it showed territorial defence, whereas females could visit several spawning grounds within the same season.

The environmental factors triggering the spawning runs of Arctic and European grayling (*Thymallus arcticus* and *T. thymallus*) vary between environments and climates, with ice thaw or snow melt being most frequently invoked in northern latitudes (e.g. Gustafson 1949; Witkowski & Kowalewski 1988), and temperature increases and/or changes in flow regime in temperate regions (e.g. Fabricius & Gustafson 1955; Parkinson *et al.* 1999). The role of water clarity is less obvious as Northcote (1995) reports that in some rivers, the spawning of Arctic grayling coincides with decreasing loads of suspended sediments, whereas in other rivers, spawning occurs in highly turbid conditions. The latter phenomenon is more surprising as grayling are visually orientated fishes, courtship is seemingly based on vision, and European grayling were found to cease spawning during a period of high turbidity following a freshet, and resumed it when turbidity decreased (Parkinson *et al.* 1999). Homing behaviour is frequent in Arctic and European grayling, with adults returning to home ranges previously occupied (Parkinson *et al.* 1999; Buzby & Deegan 2000), or to spawning grounds occupied during previous spawning seasons (Kristiansen & Døving 1996), and/or to their birthplace (Witkowski & Kowalewski 1988). While reviewing the current knowledge on the population genetics of graylings, Northcote (1995) reported that small differences were frequent, though not systematic, between Arctic graylings found spawning in different tributaries, supporting the idea that reproductive homing was frequent in this species. Grayling hatch at a small size relative to *Salmo* and *Oncorhynchus* spp., and they are vulnerable to high stream discharge. Avoiding immediate downstream drift may be the reason why grayling larvae emerge during the day (Bardonnet & Gaudin 1990), unlike the aforementioned genera. Young-of-the-year grayling generally show fast growth, and they exhibit marked ontogenetic migrations away from spawning and then nursery areas (Scott 1985; Bardonnet *et al.* 1991), usually downstream to swiftly flowing areas of larger rivers or into lakes.

The genus *Salvelinus* comprises 38 species of the temperate and arctic regions (Balon 1980), of which several are threatened (e.g. *S. japonicus*) and one is seemingly extinct (*S. inframundus*). Anadromy is frequent in four species (*S. alpinus*, *S. fontinalis*, *S. leucomaenis* and *S. malma*), rare in the North American bull trout *S. confluentus*, and strongly suspected in the Russian *S. albus*, although detailed studies on this species are lacking. The lake trout *Salvelinus namaycush* can be found in brackish water (Scott & Crossman 1973), but its occurrence in full-strength seawater is most occasional, suggesting it is not anadromous (McDowall 1988). All other species are encountered in fresh waters only, and a vast number of them are endemic to a single river drainage or lake system (e.g. *S. boganidae* and *S. drjagini* in the Taimyr Peninsula, *S. elgyticus* from Lake El'gygytgyn in the Anadyr River drainage, *S. jacuticus* in small mountain lakes in the Lena River Delta, *S. neiva* in the Okhota River, *S. tolmachoffi* in the Esei Lake, Khatanga River). Information on their biology is generally lacking, and they are assumed to behave similarly to riverine or landlocked populations of other *Salvelinus* species, migrating between tributaries for spawning and lakes or rivers for feeding, with life-history traits being dependent on latitude and climate, as illustrated for potamodromous populations of the brook char *Salvelinus fontinalis* (Power 1980; see also Northcote 1997). In all four *Salvelinus* species that may exhibit an anadromous lifestyle, the degree of anadromy varies depending on latitude and climate, but also on lake and river morphology. In the Arctic char *Salvelinus alpinus*, deep lakes and long river watercourses are correlated with a low degree of anadromy, whereas in drainage systems combining shallow lakes and short watercourses, anadromy is frequent (Kristoffersen 1994; Kristoffersen *et al.* 1994). Reasons behind these differences include limited swimming capacity (char being reputedly poor swimmers compared to other salmonids; Beamish 1980) or capacity of passing natural migration barriers, predation risks at sea (Jonsson 1991) and the growth potential in lacustrine environments. McDowall (1988) gives further accounts of anadromy versus potamodromy in this species, and a more complete picture of how Arctic char adapt to extreme latitudes can be found in section 4.2. Regarding landlocked populations, Näslund (1990, 1992) has demonstrated that Arctic char in northwest Sweden could undertake seasonal potamodromous migrations, shifting between small eutrophic lakes for feeding during summer and a deep oligotrophic lake for spawning and overwintering. Johnson (1980) provides further examples of homing in Arctic char, although he points out that it is probably not as rigid as in other salmonids. In a recent radio-tracking study, McCubbing *et al.* (1998) demonstrated that both male and female landlocked Arctic char made repeated migrations between Ennerdale Lake (English lake district) and two tributaries during the spawning season (males and females made an average of 6.1 and 2.4 migrations per season, respectively). They interpreted this behaviour as males attempting to spread their genetic material with a broader proportion of the population. This straying capacity, which might provide better chances of adapting to less predictable environments in harsh climates, might also account for the considerable invading abilities of chars (see also Balon 1984), and there are numerous examples of Arctic chars travelling over considerable distances between river drainages. For example, a char tagged in the Vardnes River, northern Norway, was recovered in the Tuloma River, Russia, at a distance of about of 950 km (Jensen & Berg 1977), and distances of 400–500 km have been reported to be travelled by char in the Canadian archipelago (Gyselman 1984).

Such long movements are also exhibited by the Dolly Varden *Salvelinus malma*, which is a frequent interdrainage mover (Armstrong & Morrow 1980; DeCicco 1989, 1992). In northwestern Alaska (67°N), radio-tracked Dolly Varden have been found to exhibit very complex movement patterns between the Noatak River (a 640-km long river, draining over 30 000 km^2), and the Kivalina and Wulik Rivers (two shorter, about 100-km long, neighbouring rivers with a combined drainage of about 4000 km^2) (DeCicco 1989). The simplest pattern involves overwintering in the lower Noatak River, summer growth in the Chukchi Sea, upstream spawning runs during August and spawning in September–October in the same river (Fig. 5.8). Other fish exhibit similar behaviour, except that they enter the Kivalina or Wulik River, at about 100 km distance from the mouth of the River Noatak, and spawn in these. Other fish which overwintered in the lower Noatak River do not make any seaward run during early summer, and migrate straight to upstream spawning grounds for summer spawning. Spawners that have overwintered elsewhere also ascend the Noatak River at the same period of the year. The rest of the sequence for all these summer spawners seemingly depends on whether they spawn in July or August. Fish spawning in August descend during September to overwintering sites in the Lower Noatak, whereas those

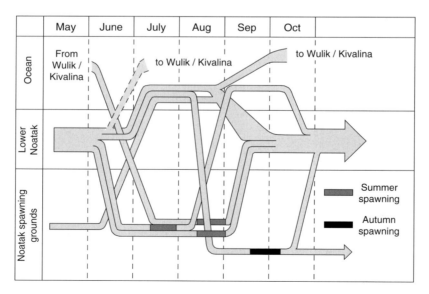

Fig. 5.8 Migratory patterns of the Dolly Varden *Salvelinus malma* between the Chukchi Sea, the Noatak River and other drainages (Wulik/Kivalina Rivers), northwestern Alaska. Dashed lines indicate probable but undocumented movements. Redrawn from DeCicco (1989).

spawning in July undertake a seaward migration, remain in the Chukchi Sea until mid-October, then ascend the lower River Noatak, and overwinter there. Hence, seaward movements in this species concern either postspawners or prespawners, depending on whether the Dolly Varden spawn in summer or autumn. DeCicco (1989) indicates that the summer spawning scheme dominates in this region, suggesting it is an adaptation of chars to local environmental conditions. To our knowledge, it is still uncertain whether populations reproducing at different seasons are genetically distinct, notably as DeCicco points out there is a real possibility for a gene flow between them.

Whether or not *Salvelinus* spp. frequently move between river drainages, they seemingly remain in coastal areas during their marine growth period, and rarely venture over long distances in the sea (Johnson 1980; Power 1980), although Moore (1975b) reports distances of up to 40 km for Arctic char at sea. The Dolly Varden tells another story, as several fish tagged in the Wulik River were found to cross the Bering Strait (DeCicco 1992). One of them was recaptured on St Lawrence Island, 530 km away from its tagging site and 180 km off the Alaskan Coast. Two other fish were recovered over 500 km upstream in the Anadyr River, Siberia, one of them having covered this distance (1560 km) within approximately 2 months during summertime. This gives an average ground speed of 26 km d^{-1}, and a much faster swimming speed when considering that marine water in the Bering Strait flows in a northern direction during summer, and that the fish ascended the Anadyr River over 540 km. The three fish mentioned here were among 118 tag recoveries in the study of DeCicco: another fish was recovered in Norton Sound, Alaska, about 700 km from its tagging site, and the 114 other fish all were recovered in drainages of the Chukchi Sea. However, they clearly illustrate that Dolly Varden are adapted to long marine migrations, and that not all migrations of chars are coastal in nature. The advent of archival tags provides great opportunities for elucidating further the migratory patterns of this and other char species.

Many of the 32 species of whitefishes and ciscoes (*Coregonus, Stenodus and Prosopium*) also exhibit substantial migratory behaviour (Northcote 1997). While in lakes, coregonines generally range in pelagic habitats, and they show seasonal vertical migrations, descending to the deep and 'warm' layers in winter, and progressively moving up to the surface layers during spring and early summer (e.g. vendace *Coregonus albula*; Jurvelius & Heikkinen 1988). These fishes also frequently exhibit distinct diel movements in relation to foraging. Seasonal migration of lake-dwelling coregonines occurs to local spawning areas on shallow-water gravel or into tributaries (Scott & Crossman 1973; Lelek 1987; Stanford & Hauer 1992; Northcote 1997; Howland *et al.* 2000). Significant riverine migrations have also been reported for a number of holobiotic, freshwater forms of several *Coregonus* and *Prosopium* species (Nikolsky 1963; Scott & Crossman 1973; Pettit & Wallace 1975; Reist & Bond 1988; Bodaly *et al.* 1989). *Stenodus leucichthys* in the Mackenzie River system, Canada may include predominantly lacustrine, fluvial and anadromous forms (Howland *et al.* 2000), although differentiation of the migratory behaviour of the first two forms for this and other coregonine species can be problematic.

Many of the riverine coregonine populations of northern Canada and Alaska, especially for inconnu *Stenodus leucichthys*, are anadromous, maturing and feeding at sea and entering fresh water to spawn in a similar manner to salmons (Alt 1977; Howland *et al.* 2000; see also review in McDowall 1988). Anadromous and holobiotic freshwater forms of broad whitefish *Coregonus nasus*, lake whitefish *C. clupeaformis*, lake cisco *C. artedii*, least cisco

C. sardinella, Arctic cisco *C. autumnalis* and whitefish *C. lavaretus*, are also known or strongly suspected, and probably occur for many other coregonines (McPhail & Lindsey 1970; Scott & Crossman 1973; Dodson *et al.* 1985; McDowall 1988; Reist & Bond 1988; Bodaly *et al.* 1989; Dillinger *et al.* 1992; Laine *et al.* 1998b). However, their movements at sea tend to be much more limited to coastal regions than for salmon species, coregonines often remaining in areas of freshwater influence (McPhail & Lindsey 1970; Reist & Bond 1988). River-spawning stocks of inconnu often mix within the same estuarine or coastal areas but appear to exhibit homing to their respective spawning rivers, without intermixing, as occurs for the Kobuk and Selawik Rivers, Alaska (Underwood 2000). Similarly, anadromous populations of the vendace *Coregonus albula* mix in the Baltic Sea during summer, and home to their spawning sites in autumn (Enderlein 1989). In contrast to the general picture for coregonines, vendace in the Baltic Sea can be found in offshore areas, although Enderlein suggests this spreading was restricted to surface waters, where salinity was lowest. Coregonine populations further inland and to the south tend to be holobiotic, either completing their lifecycle in upland lake environments or moving between rivers and lakes (Scott & Crossman 1973; Northcote 1997). Howland *et al.* (2000) provide a detailed and informative comparison of the migratory behaviour and lifecycles of holobiotic inconnu *S. leucichthys* from the Great Slave Lake area and anadromous inconnu from the lower Mackenzie River, Canada.

5.24 Trout-perches (Percopsidae) and pirate perch (Aphredoderidae)

There are two species of trout-perches and one species of pirate perch, all occurring in fresh water in North America. *Percopsis omiscomaycus*, a trout-perch, may spawn in lakes over inshore sand and gravel where this occurs, but is frequently observed to migrate into streams in about May to spawn and then move back to lakes after spawning (Scott & Crossman 1973). In lakes it exhibits clear patterns of diel migration, moving into shallow water at night and back into deeper water at dawn (McPhail & Lindsey 1970). Little is known of the pirate perch *Aphredoderus sayanus*, but a recent study in the Atchafalaya River basin, Louisiana, revealed they spawn under conditions of increasing temperatures (12–22°C) in February–March, and lay adhesive eggs on leaf litter and woody debris. With increasing size, larvae leave nearshore habitats and no larvae larger than 14 mm are found in limnetic areas (Fontenot & Rutherford 1999).

5.25 Cods (Gadidae)

The majority of the cod family are exclusively marine species and only two species are commonly encountered in brackish and fresh water: the burbot *Lota lota* and the tomcod *Microgadus tomcod*. Adult tomcod are coastal marine species along the east coast of North America from Newfoundland in the north to Virginia in the south. Mature adults are repeat spawners and migrate into estuaries in winter, often under ice cover in the north of their range (Peterson *et al.* 1980). It had been suggested that tomcod entering estuaries avoided the greater turbulence and turbidity associated with rising tides, but Fortin *et al.* (1990) reported that adult tomcod migrated upstream during both rising and falling tides. More recently, Bergeron *et al.*

(1998) provided evidence that the use of rising or falling tides could depend on the morphology of the estuary and ambient water velocity. In the Sainte-Anne River (Québec), increasing impoundment of the estuary by sand bars resulted in excessive water velocity during falling tides, and video recordings revealed that tomcod favoured the short period of flow reversal, when the tide is maximum for migrating upstream. Tomcod may penetrate fresh water and it appears that the eggs need low salinity water to develop properly. Adults do not remain in low-salinity water for more than about a month, and after spawning they return downstream to recover. The eggs hatch during the elevated flows of the spring thaw and the larvae are carried down to more saline waters. Although it has been suggested that high salinities are necessary for larval development, landlocked populations of tomcod are known, suggesting this may not be the case (McDowall 1988).

Immature and adult burbot *Lota lota* are found in fresh and brackish water environments (Morin *et al.* 1980; Hudd & Lehtonen 1987) but spawning is in fresh water. Burbot tagged with Carlin tags in the Gulf of Bothnia and Gulf of Finland were shown to exhibit home range and homing behaviour (Hudd & Lehtonen 1987). After spawning, burbot remain for a short time near the spawning areas, then they migrate to the outer parts of archipelagos but they rarely wander further than 20 km from the tagging sites. In the late autumn, they return to coastal areas, and show consistent homing behaviour, suggesting there is little or no intermixing between stocks. In lacustrine populations young burbot larvae are pelagic (Fischer 1999) and presumably drift in lotic environments. Based on evidence from otolith microstructure, Fischer (1999) suggested that older larval burbot in Lake Constance descend to the cold hypolimnion and carry out their inshore migration from the profundal zone. Early telemetry studies of the behaviour of adult burbot suggested them to be sedentary by day and night (Malinin 1971). However, more recent telemetry studies have produced varying results. Adult burbot in Bull Lake, an impoundment in Wyoming, moved little through winter and early spring (Bergersen *et al.* 1993). In Lake Opeongo, Ontario, acoustically-tracked adult burbot displayed nocturnal activity and local diel and seasonal migrations, tending to remain offshore, but utilising deeper water in summer (Carl 1995). Although a cold-water species, during summer in Carl's study some burbot spent extended periods of nocturnal activity in shallow water warmer than 20°C, returning to cool deep water during the day.

The long movements of up to 125 km displayed by adult burbot in the Tanana River, Alaska, may be related to the harsh conditions occurring there (Breeser *et al.* 1988). The authors point out that burbot remained in the main channel in summer when peak flows associated with high turbidity and channel scouring resulted in an apparently inhospitable environment, and that the longest and most rapid movements occurred in autumn and early winter, prior to the spawning season. Burbot are often regarded as poor swimmers but one burbot moved 101 km upstream in 13 days or less. Upstream-directed spring migrations of burbot are also known to occur and substantial numbers of burbot are captured from a fish pass operating on the River Ohre, a tributary of the River Elbe, Czech Republic (Slavík [unpubl.]; Fig. 5.9). A dam 18 km upstream releases hypolimnial water, resulting in relatively low spring–summer temperatures and elevated autumn temperatures. The spring peak in burbot moving through the fish pass appears to be related to reduced discharges stimulating burbot to leave overwintering refuges in the river banks. Mitochondrial DNA analysis shows that, in large river systems, natural barriers such as waterfalls provide effective barriers to

Fig. 5.9 Adult burbot *Lota lota* (a) captured by O. Slavík, from a pool and weir fish pass (b) on a tributary of the Elbe River, Czech Republic. In early spring, substantial numbers of burbot enter this and similar fish passes. See text for further information.

burbot dispersal and migration, resulting in genetically divergent stocks, but that in the short term at least, some impoundments have not caused similar effects (Paragamian *et al.* 1999).

5.26 Mullets (Mugilidae)

The mullets or grey mullets are widely found in coastal and brackish water of all temperate and tropical seas, but about a dozen of the 70 species also frequently occur in fresh water. One species *Liza abu* occurs only in fresh and brackish water in southeastern Asia (Nelson 1994). In a number of cases these movements into fresh water follow predictable migratory patterns, which are usually described as catadromous in so far as that most species reproduce at sea. Nevertheless, the mullets are distinctly euryhaline and most species are able to move freely, taking advantage of feeding opportunities. Some mullet species grow to about 90 cm and most species graze algae and sift sediments.

The most well-studied species is the thick-lipped grey mullet (striped mullet) *Mugil cephalus*, which appears to be facultatively diadromous, frequently but not always entering fresh and brackish water systems. It is a very widespread species, occurring across all oceans in tropical to warm temperate regions. It frequently occurs in rivers and may be found long distances upstream, with records of thick-lipped grey mullet penetrating 190 km up the Colorado River in the US (Johnson & McLendron 1970). Migratory behaviour of thick-lipped grey mullet is extremely variable. In some cases only the juvenile stages occur in fresh and brackish water, as in South Africa (Bok 1979), in others the young grow to adulthood in this environment (Shireman 1975), and in yet others movement into fresh and brackish water is principally a summer phenomenon engaged in by a wide range of age groups, as occurs in many English rivers. Shireman (1975) found that in Louisiana adult thick-lipped grey mullet would develop gonads but that these regressed if mullet were prevented from going to sea, and landlocked populations are not known to reproduce (McDowall 1988). More recently, some evidence has been provided that thick-lipped grey mullet may spawn in fresh water in the Lower Colorado (Bettaso & Young 1999). The hypothesis of freshwater-resident populations has also been put forward after Chang *et al.* (2000) found that the spawning period of thick-lipped grey mullet in the Tanshui River in northwest Taiwan extended over much a longer period than for migratory populations. However, they did not reject the alternative hypothesis that multiple cohorts might also correspond to early and late catadromous spawners.

A variety of other mullet species spend significant periods of time in fresh water. *Liza ramada*, *L. aurata*, *L. saliens* and *Chelon labrosus* regularly occur in freshwater riverine habitats but spawn at sea (Hickling 1970; Torricelli *et al.* 1982). *Chelon labrosus* tends to utilise fresh- and brackish water as nursery habitat, but has less osmoregulatory competence in fresh water than the thin-lipped grey mullet *L. ramada*. Adult thin-lipped grey mullet have been reported 200–300 km up rivers in Morocco, and they ascend European rivers to feed for several months at a time (Hickling 1970). Where they occur with other species, they are always found to penetrate further upstream than the other species, including thick-lipped grey mullet. Whereas juvenile *L. ramada* remain in brackish water in the Loire River, France, adults reach freshwater feeding grounds as far as 300–350 km upstream of Saint-Nazaire (Sauriau *et al.* 1994). Fish enter the Loire in spring and remain there until late summer when they make a seaward migration, typically moving through the estuary at 10–15 km d^{-1} based on mark recapture information. During summer residence in tidal stretches of the Mira estuary, Portugal, tracked fish mainly moved back and forth with the tide, covering a median distance of 6.3 km in a complete tidal cycle (Almeida 1996). In the River Frome, England thin-lipped grey mullet migrate upriver into fresh water and past a fish-counting facility at predictable times of the year in order to graze diatom blooms (W. Beaumont [pers. comm.]).

Myxus capensis, a South African species of mullet, lives in freshwater up to 120 km from the sea for a substantial part of its lifecycle as a juvenile and subadult (Bok 1979; Bruton *et al.* 1987). This seems to be an obligatory component of the lifecycle. The young enter fresh water in late winter and early spring, soon after hatching, and remain there for 2–5 years until nearly mature. They then cease feeding, migrate downstream to estuaries to complete gonad development and probably spawn in coastal areas. The mountain mullet *Agonostomus monticola* of Central and North America occurs in rivers for long periods of its lifecycle and moves well upstream of tidal limits. It is thought that this species may spawn in fresh or brackish water, and that the eggs drift seawards, with the young subsequently returning to fresh water (Loftus *et al.* 1984). A similar life-history cycle may occur for *Liza abu* in Southeast Asia. A substantial number of other species have been recorded moving into freshwater environments including *Valamugil robustus* (Madagascar), *V. cunnesius* and *Liza parsia* (India), *Agonostomus telfairii* (Africa and western Indian Ocean), *Joturus pichardi* (Central America), *Myxus petardi* (Australia), *Liza falcipinnis* and *L. dumerilii* (North Africa). Others may enter brackish environments, but not necessarily fresh water. An overview of the extent of diadromy in mullet species is provided by McDowall (1988).

5.27 Silversides and their relatives (Atheriniformes)

The silversides (Atherinidae) are principally a marine group of small, silvery, shoaling fishes, with over 150 species. However, about 50 species are found only in fresh water and others enter fresh water, mostly in warm temperate to tropical regions (Nelson 1994). True diadromy in this group is rare or non-existent (McDowall 1988) and migratory behaviour in the freshwater species (mostly from Mexico, Central America and Australasia) has not been widely reported. The brook silverside *Labidesthes sicculus* occurs in lakes and rivers of the eastern US and aggregates in shallow water before spawning. On hatching the young become pelagic and move offshore over deep water, perhaps to minimise predation from centrarchids and percids (Hubbs 1921). They form large schools during daylight but disperse at night. Older fish roam more widely and accumulate in shallow water in autumn. River estuaries are commonly used as nursery areas by Australian species such as *Leptatherina presbyteroides* and *Atherinomorus ogilbyi* (Prince & Potter 1983). The sand smelt *Atherina boyeri* is common in shallow, brackish water and principally occurs along the Atlantic coast of southern Europe and in the Mediterranean, Black, Aral and Caspian seas. It is a euryhaline species but regular migratory patterns do not appear to occur in many coastal lagoon populations, although movements between the sea and lagoons do occur. Part of a population of sand smelt living in the Camargue wetlands of southern France exhibits a regular migration, with many fish moving from the brackish lagoon, the Étang du Vaccares, to freshwater marshes via a drainage canal (Rosecchi & Crivelli 1992, 1995). The upstream migration occurs in autumn and winter and spawning occurs in the marshes between April and June. From June onwards, the adults and young leave fresh water and return to the brackish lagoon where they remain until the next migration. The authors note that this behaviour results in low mortality from predation by other fish species, since the marshes are only seasonally inundated. They also utilise the large zooplankton that are characteristic of the marshes due to low and unsustained predation pressure.

The pejerrey *Odontesthes bonariensis* is found in estuaries of Argentina, Uruguay and southeastern Brazil, where it enters fresh water as far as several hundred kilometres inland. They can attain 50 cm in length, represent an important species for commercial fisheries and are highly adaptive, which has promoted their introduction into many freshwater reservoirs and lakes throughout Argentina, Brazil, Chile, as well as in the Lake Titicaca system, having been recorded in Lake Titicaca from 1955 (review in Loubens & Osorio 1988). Spawners reproduce in 3–10 m deep water, where they lay eggs that adhere to *Chara* or *Potamogeton* spp. Larvae and small (<10 cm) juveniles are found close to inshore emergent vegetation, at depths generally less than 1 m. Juveniles of increasing size utilise increasingly deeper waters (4 m deep for 10-cm fish, 7–9 m deep for fish larger than 15 cm), but they rarely go beyond 9–10 m. This marks the limit of aquatic submerged vegetation, which can extend as far as 10 km offshore. Juveniles larger than 20 cm and adults move further offshore where they exert intense piscivory, mainly on *Orestias* spp. They are rarely encountered in water layers deeper than 50 m (up to 40 km offshore in the northwestern and southeastern bays of the lake), possibly due to interactions with another introduced predatory species, the rainbow trout *Oncorhynchus mykiss*. Competition between young pejerrey and young rainbow trout is virtually non-existent, as young rainbow trout essentially live in the lake tributaries. Pejerrey may spawn all year round, but as Lake Titicaca has a cold climate (10–14°C in surface layers), spawning is restricted to the warmest periods of the year (January to early March), when large fish enter shallow bays more frequently.

The Melanotaeniidae (rainbowfishes) comprise 53 species of small-size freshwater fishes occurring in the Australia-New Guinea region, mainly in lakes and quiet backwaters of major rivers (Allen & Cross 1982). Batch-spawning is the norm in rainbowfishes, and they produce relatively large eggs, but not *Glossolepis multisquamatus*, which is endemic to the Sepik and Ramu river systems in Papua New Guinea, and to the Taritaru system in northern Irian Jaya. This species, which is the only rainbow fish to inhabit river floodplains, undertakes migrations onto and off the floodplain in response to changing flood conditions. Coates (1990) related the shift in the reproductive strategy of this species to its migratory habits and greater environmental unpredictability compared to other rainbowfishes living in more stable freshwater habitats. In the Alligator Rivers region (Northern Territory of Australia), *Melanotaenia splendida* has been reported to make presumably long (the extent is unknown) migrations at a fast pace relative to its size (about 5 km d^{-1}), and exclusively during daytime between March and May (Bishop *et al.* 1995). Bishop *et al.* found that similar seasonal and diel patterns were shared by several (and probably many) species inhabiting this tropical floodplain river, which comprises lowland floodplains, river corridors, billabongs (oxbow lakes) and escarpments, which are common features of many northern Australian rivers (see also Bishop & Forbes 1991). Among these species are members of fish families which are not dealt with in detail in this book (e.g. the spotted archerfish *Toxotes chatareus* (a toxotid), the ambassids *Ambassis* spp.).

Several other families of atheriniforms, the Bedotiidae, Pseudomugilidae, Telmatherinidae and Phallostethidae, are common in fresh or brackish water, mainly in the Indo-West Pacific and Southeast Asia. However, there is little information available concerning the spatial behaviour of these fishes.

5.28 Needlefishes, half-beaks and medakas (Beloniformes)

About 51 of the 191 beloniform species are confined to brackish or fresh water (Nelson 1994), while several other species may move between marine and freshwater environments (McDowall 1988). Most of the 11 freshwater needlefish (Belonidae) species occur in the neotropical region, most of the freshwater half-beaks (Hemiramphidae) in the Indo-Australian region and most of the medakas (Adrianichthyidae) in Southeast Asia. However, there is a paucity of information concerning the existence of migratory behaviour or otherwise in this group. The pelagic habits of the half-beaks and needlefish suggest that they might be quite mobile. However, information collected in the course of flood control schemes in Bangladesh suggested that the needlefish *Xenentodon cancila* showed limited migratory activity in the floodplain (Halls *et al.* 1998, see also section 7.2.3).

5.29 Killifishes, livebearers, pupfishes and their relatives (Cyprinodontiformes)

The Cyprinodontiformes comprises over 800 species and eight families (Aplocheilidae, Profundulidae, Fundulidae, Valencidae, Anablepidae, Poeciliidae, Goodeidae and Cyprinodontidae) of small (usually <15 cm long), mostly omnivorous fishes (Nelson 1994). They are widespread in North America, South America, Africa, southern Europe and southern Asia, in fresh and brackish water, and many species are tolerant to extreme conditions of high temperature and salinity, and poor food supply. Most species are opportunistic, with rapid rates and specialised modes of reproduction. Although they may disperse effectively, enhanced by the euryhaline osmoregulatory capacities of many species, well-characterised patterns of migration are largely unknown.

5.30 Sticklebacks and their relatives (Gasterosteiformes)

Sticklebacks are small fishes, usually 4–7 cm long when adult. They occur throughout the cool temperate zone of the northern hemisphere. There are relatively few species (about eight) but some of these, notably the three-spined stickleback *Gasterosteus aculeatus* and the nine-spined stickleback *Pungitius pungitius*, are extremely widely distributed. Both of these species occur in freshwater and marine environments and include anadromous and potamodromous forms.

The three-spined stickleback is found in freshwater and marine habitats of all northern continents between 35–70°N (Wootton 1976, 1984). This distribution results from repeated colonisations of freshwater habitat by marine-dwelling individuals. The species has a number of different morphological forms which appear to be associated to some degree with life-history patterns. The form '*trachurus*' has well-developed lateral skutes and is marine or anadromous; '*leirus*' has weakly developed skutes and is found in fresh or brackish water, as is '*semi-armatus*' which has moderate development of the skutes (Wootton 1976). The '*trachurus*' form tends not to be found as far south as the other forms. Furthermore it should be pointed out that these associations between the forms and behaviour and distribution are typical and by no means universal.

Anadromous three-spined sticklebacks (usually of the '*trachurus*' form) typically overwinter in coastal marine or estuarine environments and migrate upriver in spring, the timing of entry increasing with latitude. Many of the populations migrate into small coastal streams or tributaries from estuaries and tend not to move more than a few kilometres upstream (Craig-Bennett 1931; Kedney *et al.* 1987). The males build nests which they defend and to which they attract females for egg-laying, after which they protect and ventilate the eggs with their pectoral fins until they hatch. The young and surviving adults normally remain in fresh water for the summer and return to the sea in autumn. These populations are mostly annual and the majority of adults die before returning to the sea. The young tend to remain in brackish or coastal waters but some may disperse over hundreds of kilometres. There is no evidence of homing behaviour of '*trachurus*' three-spined sticklebacks, and it seems likely that there is substantial population mixing. Three-spined sticklebacks of the '*trachurus*' form exhibit osmoregulatory adaptation to fresh and salt water during very specific time windows. This physiological adaptation appears to occur in response to environmental stimuli such as changing temperature and daylength and is mediated by hormonal control (Wootton 1976). This is a similar physiological response to that demonstrated during smoltification by anadromous salmonids.

Some populations of three-spined stickleback reproduce in marine habitats such as marine rockpools. Kedney *et al.* (1987) found that three-spined sticklebacks in the St Lawrence estuary comprise a component which is strictly anadromous and another which remains and spawns in a near-marine environment. They found no evidence that these different components came from different gene pools. After hatching, the young of anadromous parents feed and grow in fresh water before returning, together with adults that have survived spawning, to the sea in the summer and autumn. Kedney *et al.* argued that the energy requirements for upstream migration to spawn in fresh water were relatively low and, although hatching takes longer in fresh water, there is lower predation in rivers. The two modes of reproductive behaviour may reflect different strategies for minimising loss of reproductive fitness.

Anadromous three-spined sticklebacks are larger and more fecund than resident fish in the Little Campbell River, British Columbia (Hagen 1967); this may be a trade-off against the extra costs of migration. In this river the sticklebacks migrate 2.4 km while in other rivers such as the Fraser River, British Columbia, migration distances appear to be much greater (Taylor & McPhail 1986). Freshwater migrations of 15 km are confirmed for freshwater-resident three-spined sticklebacks in the Chignik catchment, Alaska (Harvey *et al.* 1997). Several studies have now produced evidence of genetically-based variation in growth rate and body size among populations of differing migratory behaviour (Snyder & Dingle 1989; Snyder 1991), and that these differences are consistent with the hypothesis that adaptation to different life histories has occurred. Snyder and Dingle reared estuarine-migratory, freshwater-migratory and inland non-migratory populations from the Navarro River, California, under the same conditions but separately in the laboratory, and found significantly lower growth rates for non-migratory fish, with some evidence of heritability.

Three-spined sticklebacks of marine origin appear to have greater swimming performance than freshwater conspecifics and can sustain speeds of up to 5 BL s^{-1} for hours (Taylor & McPhail 1986). This may be of significance for migration to and in rivers, as well as for dispersal. Quinn and Light (1989) reported the occurrence of substantial numbers of three-spined sticklebacks of the '*trachurus*' form in the open Pacific Ocean, hundreds of kilometres from the nearest land. These may have been swept there by currents, but Quinn and Light argue that a 7-cm stickleback swimming in a straight line at 3 BL s^{-1} could swim 800 km in less than 2 months. A stickleback of this size might indeed travel over one or several hundreds of kilometres at U_{opt} in still water if it had accumulated large energy reserves (see section 2.3.5). However, the assumption above is most unlikely as swimming at the swimming speed invoked is faster than U_{opt}, requiring about twice the amount of energy per unit distance travelled. Also it would require sticklebacks to swim about 18 h per day, leaving very little time for feeding and replenishing the energy reserves depleted by swimming. Nevertheless, it is clear that whether by passive or active dispersal there is a high potential for colonisation of new habitats.

The nine-spined stickleback *Pungitius pungitius* is very widespread throughout the cool temperate fresh waters and coastal areas of Eurasia and North America. Unlike the three-spined stickleback, it rarely strays far from coastal or brackish environments. This species spawns in brackish and fresh water, and frequently overwinters in brackish or marine environments, entering fresh water to spawn (Scott & Crossman 1973). McDowall (1988) reports the existence of marine and freshwater forms, but it is unclear whether these form discrete populations. Harvey *et al.* (1997) showed that three- and nine-spined stickleback underwent simultaneous spring migrations in the Chignik catchment, Alaska. In the summer and autumn 1-year-old and YOY fish emigrated upstream from Black Lake towards Chignik Lake, an estimated distance of 15 km. All migrating fish had enlarged, mature gonads and had developed spawning coloration. Upstream migration ceased at the end of June and returning 2-year-old fish were found in poor condition, suggesting that spawning mortality was high.

The brook stickleback *Culaea inconstans* of northern North America is principally a freshwater species and rarely occurs in brackish or salt water. In a small Ontario stream which flows into Georgian Bay, Lake Huron, spawning in this species was followed by a striking downstream migration in June or July (Lamsa 1963). In one week in June 1958, 2851 brook sticklebacks were caught in a downstream trap. In successive years, almost all activity was also concentrated into 1 week.

About 17 species of pipefish (subfamily Syngnathinae, family Syngnathidae) occur in fresh water and another 35 species in brackish water (Nelson 1994). McDowall (1988) reviews the very limited evidence for diadromy in this group, and although little is known of their spatial behaviour, given their cryptic form and behaviour, large-scale migratory behaviour seems unlikely. This view is also likely to be applicable to the single species of indostomid *Indostomus paradoxus*, which occurs in Southeast Asian swamps.

5.31 Scorpionfishes (Scorpaenidae and Tetrarogidae)

The scorpionfishes are a large group of principally marine fishes and few enter fresh water. However, the Australian bullrout *Notesthes robusta* (Tetrarogidae), a venomous species of coastal eastern Australia, primarily appears to be a freshwater fish, but very young fish often occur in estuarine environments and this is suggestive of catadromous behaviour. Accumulations of fish often appear below dams, suggesting the existence of upstream migration. Harris (1984) regards the Australian bullrout as catadromous and McDowall (1988) reviews the evidence for diadromy in this species.

5.32 Sculpins (Cottidae)

Sculpins are small benthic fishes (mostly less than 30 cm) with a squat body form, principally occurring in the subarctic and cool temperate regions of the northern hemisphere. Migratory movements of many freshwater sculpin species appear to be of limited importance, since most species are cryptic and carry out their lifecycles within the same stony, benthic habitat. Mills and Mann (1983) described the bullhead *Cottus gobio* as a solitary animal, driving off other individuals from its territory (especially when it guards eggs) to which it showed strong homing behaviour. However, they also suggested that bullhead migrate to deeper water to spawn although presented little evidence to support this.

Crisp *et al.* (1984) and Crisp and Mann (1991) showed that the numbers of bullheads in many streams above Cow Green Reservoir on the River Tees in Northeast England showed some seasonality after impoundment. Peak numbers occurred in mid-summer and numbers diminished rapidly during autumn and winter and increased again in spring or early summer. They argued that the best explanation for this was that these bullhead formed part of the reservoir breeding population, overwintering in the reservoir and returning to the streams for spawning. Bless (1990) recorded upstream movement of bullhead in German rivers, which was pronounced in May and June. Bullheads are small fish, the upstream movements of which can be blocked by obstacles with a height of 18–20 cm (Utzinger *et al.* 1998), and it is most likely that such obstructions have led to major population declines of *Cottus gobio* in the upper parts of some rivers (see also section 7.2.1), whether the upstream movements correspond to true upstream migration or recovery of initial territory after displacement by high floods, or emigration from habitats that become unsuitable during receding waters. Fischer and Kummer (2000) demonstrated an increased activity of bullhead during receding waters in an alpine stream, with most fishes withdrawing into deep pools. As suggested by the poor condition of fish remaining in transitional stretches, abandoning a territory probably represents a loss of fitness lower than that incurred by maintaining territoriality under unsuitable conditions.

It has been thought for a while that early life stages of bullhead were riverine, or at least connected with some form of structured habitat, but recent investigations in the Austrian Hallstattersee provided evidence that some late larvae could have pelagic habits (Wanzenbock *et al.* 2000). Other species, such as the Bear Lake sculpin *Cottus extensus* have lacustrine habits, and they exhibit an ontogenetic migration during their first year of life (Ruzycki & Wurtsbaugh 1999). After dispersing during an initial pelagic stage, late larvae settle either directly in the warm littoral areas, or in poorly productive deep habitats, from which they undertake an inshore migration.

Several species of sculpin exhibit varying degrees of catadromous and amphidromous behaviour patterns (McDowall 1988). The prickly sculpin *Cottus asper* occurs in Pacific coastal and inland streams of North America and typically spawns in fresh water, but coastal populations may spawn in brackish water (McAllister & Lindsey 1959; Scott & Crossman 1973). Coastal populations often exhibit a prespawning migration downstream, with males preceding females. The females move back upstream while the males guard the nests in which eggs were laid, returning upstream later. The larvae are pelagic and metamorphose after about a month; for coastal populations this may occur in the estuaries that are then used as nursery areas. The migration distances are reported to be up to about 16 km. The coast-range sculpin *Cottus aleuticus* of the Pacific coast of North America is described as spawning in a range of habitats from streams to estuaries. The young are described as occurring in estuaries, but it is not clear that this applies for young which hatch in fresh water also (McAllister & Lindsey 1959). The fourhorn sculpin *Triglopsis quadricornis*, which has a circumpolar distribution, has marine and freshwater forms but migratory behaviour has not been described. The Japanese river sculpin *Cottus kazika* moves downriver to spawn in estuaries, and embryos move to rocky shore habitats in coastal waters. In the Shimanto estuary, western Japan, larvae and small juveniles stay in a stagnant layer (12°C, 20‰ salinity) in January–February, and juveniles migrate upstream in March when the stagnant layer disappears from the estuary (Kinoshita *et al.* 1999).

Another species of Japanese river sculpin (named kankyo kajita), *Cottus hangiongensis*, is amphidromous, spawning in the lower reaches of rivers in spring (Goto 1987). It appears that there is a spatial separation of the sexes in fresh water, with males tending to predominate in upstream streams, and migrating downstream to spawn with females. The life history of this species is also characterised by the size of males becoming greater with increasing distance upstream, as also occurs in the bullhead *Cottus gobio*. The pelagic larvae of *C. hangionensis* are carried to sea and return about a month later. Some cottids such as the northern Far East belligerent sculpin *Megalocottus platycephalus* are principally marine but exhibit regular seasonal feeding migrations into estuaries, adults of this species moving up to 12 km from the mouth of the Bol'shaya River (eastern Russia) into low salinity water (Tokranov 1994).

The related cottid-like fish of the Abyssocottidae and Comephoridae (Baikal oilfishes) are characteristic of cold, deepwater Asian lakes, principally Lake Baikal, in which many are endemic. They have been little studied and their migratory habits, if any, are unknown.

5.33 Snooks (Centropomidae)

The centropomids are a small group (22 species) of relatively large, spiny-finned predators occurring mostly in temperate to tropical marine environments. A few species occur solely in fresh water, among which four species of Nile perches, *Lates angustifrons*, *L. mariae*, *L. microlepis* and *L. stappersii*, are endemic to Lake Tanganyika, *L. longispinis* to Lake Turkana and *L. macrophthalmus* to Lake Mobutu Sese Seko. The young are found in vegetated inshore habitats, whereas adults are almost exclusively found in epilimnetic waters (<30 m deep) of offshore areas, where they prey on small freshwater clupeids (*Limnothrissa miodon* and *Stolothrissa tanganicae*) and shrimps. In addition to probable lateral movements during the reproductive season, species like *L. stappersii* have been reported to undertake seasonal northwards movements during the dry season (Phiri & Shirakihara 1999). Captures of adults have also been reported in inshore areas, notably near the estuary of the Malagarazi River, Tanzania. However, genetic analyses using RAPD markers strongly suggested this population was distinct from those captured offshore, suggesting limited migratory activity, or at least little intermixing, in this species (Kuusipalo 1999). *Lates niloticus*, one of the largest piscivorous fish species (up to 180 cm long and 160 kg in weight) have been introduced into many reservoirs and lakes, where they have generally spread rapidly. Beyond dispersal, they do not show long-range migrations, and rarely enter lake tributaries (Loubens 1973), but movements resulting from changes in oxygen concentration are supposedly frequent as this species is rather intolerant to oxygen levels lower than 5 mg per litre (see also section 4.4).

The barramundi *L. calcarifer* is widely distributed in the Indo-West Pacific Ocean and is diadromous. Barramundi grow to a large size, utilise freshwater and saltwater habitats and provide important commercial and recreational fisheries. The lifecycle of the barramundi is generally known and principally reflects a catadromous pattern of migration (Roberts 1978; Moore 1982; Moore & Reynolds 1982; Davis 1985, 1986; Griffin 1987; McDowall 1988). Spawning occurs in shallow marine or estuarine areas, and after 3–6 months in coastal nursery swamps, juvenile fish migrate upstream to fresh water, moving long distances up river systems as they grow. Barramundi have been recorded 800 km up the Fly River in Papua New Guinea. After 3–4 years, maturing fish move back downstream to estuarine areas to spawn (Moore 1982; Moore & Reynolds 1982) and then remain in the estuarine or upper tidal sections of rivers (Davis 1986). Fish moving downstream are males, while all females appear to result from sex reversal of males that remain in tidal waters after spawning (Moore 1979; McDowall 1988). Larger mature barramundi are found in estuarine and coastal waters and immature fish mostly occur in upper freshwater reaches (Griffin 1987). However, there are suggestions that some populations of barramundi may complete their lifecycle without using fresh water. Using inductively coupled plasma atomic emission spectroscopy, Pender and Griffin (1996) examined barium and strontium levels in *L. calcarifer* otoliths as markers of freshwater and seawater history respectively. Cluster analysis showed that there were probably marine, mixed and freshwater groups. They conclude that barramundi found remote from Mary River, northern Australia, probably have no freshwater phase, but spend a short initial period in coastal brackish water environments. Homing behaviour in barramundi has seemingly not been investigated, but recent evidence from genetic studies indicates that the genetic differences between most barramundi populations are extremely small, suggesting that substantial migration and hybridisation has occurred naturally between eastern and western Australian populations, which had been isolated for at least 110 000 years due to ice-age effects and variations in sea level (Keenan 2000).

The Japanese lates, *L. japonicus*, a large subtropical centropomid and a popular gamefish in Japan, seemingly is catadromous too, with adults being encountered in fresh, brackish and marine waters, and juveniles ascending rivers.

5.34 Temperate basses (Moronidae)

The temperate basses are large, perch-like predators, with some species exceeding 1 m in length and 20 kg in weight. They are widely distributed, principally through the northern hemisphere. The best-known and probably economically most important species is the striped bass *Morone saxatilis* which occurs on the Atlantic coast of North America from the Gulf of St Lawrence south to Florida and the Gulf of Mexico. Typically, adult striped bass overwinter in rivers and estuaries, spawn in the river in spring and emigrate to sea for summer and autumn. The young drift downriver and grow in the lower river and estuary before moving to coastal areas. Striped bass are typically regarded as anadromous, but over the latitudinal range of their distribution they exhibit a range of lifecycle strategies. Between New England and Cape Hatteras adult striped bass primarily inhabit the marine environment and only move into inland waters to spawn (Merriman 1941). At the extreme portions of their range (south of Cape Hatteras, in the Gulf of Mexico; and in the north, the St Lawrence River area) striped bass are rarely found in the ocean (Coutant 1985). In some rivers of the southeastern United States and Gulf of Mexico, striped bass remain in rivers because, it appears, coastal waters may be too warm (Dudley *et al.* 1977; Wooley & Crateau 1983). In the extreme north, ocean waters may be too cold (Rulifson & Dadswell 1995). Striped bass are repeat spawners, mature at about 50 cm (usually age 2 for males and age 3–4 for females) and normally reproduce in late spring. Analysis of strontium to calcium ratios of striped bass is providing an informative method for examining individual variation in diadromy within and between striped bass populations. For example, in the Hudson River area, New York, both sexes emigrate into brackish and marine environments to grow and return to the river to spawn, but females tend to reside at higher salinities, typical of coastal waters, than males throughout their lifespan (Secor & Piccoli 1996).

Spawning normally occurs in swift-flowing rivers, not far above the tidal limit, but in some cases this may involve migrations of several hundred kilometres from the saline waters. The semi-buoyant eggs drift downstream and require suspension in the water column before hatching after 2–3 days. The young are also buoyant and drift into sheltered brackish and estuarine nursery areas, but may quickly enter coastal marine environments throughout much of the geographical range (Setzler *et al*. 1980). In the Bay of Fundy, the only self-sustaining population of striped bass spawns in the tidal-bore dominated Shubenacadie watershed, which is a particularly challenging environment, as abrupt changes in river elevation (several metres) and salinity (20‰) occur during the tidal bore on spring flood tides. In this environment, spawning of striped bass commences after water temperature reaches 18°C, but is cued to neap tides, when temperature and salinity show little variation and water clarity is greatest (Rulifson & Tull 1999).

Long-term tagging studies show that for most river populations a group of local fish migrate up and down estuaries and into coastal waters in summer (Setzler *et al*. 1980; Rulifson & Dadswell 1995). There is, however, wider movement of striped bass, with fish tagged in Nova Scotia and New Brunswick waters recaptured as far south as Virginia and North Carolina, and fish tagged in the Potomac River (Maryland) and Hudson River (New York) captured in the Bay of Fundy less than 1 year later (Nichols & Miller 1967; Rulifson & Dadswell 1995). Historically (prior to the 1960s) striped bass overwintered in the St Lawrence River and Lac St-Pierre, migrating about 150 km from the estuary to do so (Rulifson & Dadswell 1995). Spawning occurred downstream and upstream of this area, perhaps as far as Montreal, over 200 km from the estuary, in May and June. Adults moved downriver and out into the estuary in summer. Young striped bass migrated downstream to the estuary and overwintered there. Rivers such as the Miramichi, which drains into the Gulf of St Lawrence, are regular overwintering sites for adults averaging 1.4–1.8 kg, which move upriver in autumn and downriver in January and February. Striped bass with much lighter coloration (presumed to be different to the overwintering fish) migrate 40 km upstream in spring to spawn in May and June as ice break-up is finishing, at water temperatures of 12–14°C. According to Rulifson and Dadswell there have been reports of large adults (>15 kg) moving upstream in the Tabusintac in autumn, running milt and eggs. If correct, this would represent the only autumn spawning population of striped bass known.

The Roanoke River and Albermarle Sound lie near the southern limit for which adult striped bass exhibit extensive movements in marine waters and this population spends much of the year within the adjoining Albermarle Sound, migrating up the Roanoke River (Carmichael *et al*. 1998) in spring to spawn. The lower Roanoke River flows freely for 221 km between Roanoke Rapids Lake dam and the river mouth at the western end of Albermarle Sound. Carmichael *et al*. tagged mature striped bass overwintering in Albermarle Sound with surgically implanted acoustic transmitters. A total of 27 tagged striped bass entered the Roanoke River from Albermarle Sound during the 1994 spawning season and 23 fish (14 females, 9 males) reached the spawning grounds. In 1995, 14 of the same fish entered the Roanoke River, with 6 females and 5 males reaching the spawning grounds. Tagged bass typically began upriver migration in mid-late April but with substantial individual variation. In both years fish completed the 165-km upriver migration in about 1 week. Mean water temperature was 17.5°C when fish started migration and was similar when they arrived at the spawning grounds. There was a significant positive correlation between date of arrival and duration of residence on the spawning grounds for males. Males spent longer on the spawning grounds than females, 22 days compared with 8 days in 1994 and 21 days compared with 11 days in 1995 for males and females respectively (Fig. 5.10). This was principally a result of males arriving earlier than females, a similar pattern occurring in the Hudson River (McLaren *et al.* 1981).

A large number of freshwater populations of striped bass are known, most of which appear to have resulted from introduction into reservoirs or blockage of outward migration in impounded systems. In warm climates the temperatures in shallow-water reservoirs often approach the upper critical thermal limit for striped bass. A large number of studies have demonstrated clear seasonal behavioural thermoregulation through migration by striped bass in these environments. The most common pattern is for fish to move upstream in spring to spawn in reservoir tributaries and disperse throughout the main water body of the reservoir following spawning (Combs & Pelz 1982; Farquhar & Gutreuter 1989). They remain in shallower (upstream) water until water temperature rises above 25–27°C in summer and forces them to seek spatially restricted thermal refuge habitats (Coutant 1985; Zale *et al.* 1990). In other situations striped bass remain in or move into lotic thermal refuge habitats following spawning (Cheek *et al.* 1985; Wilkerson & Fisher 1997). Farquhar and Gutreuter (1989) tracked 30 adult (3.2–8.6 kg) striped bass, tagged with ultrasonic transmitters, some of which were temperature sensing, for up to 475 days in Lake Whitney, a 9510 ha Texas reservoir. In winter they occupied the warmest water available (7.4–8.8°C). In summer they migrated to an area around the dam which had the coolest water

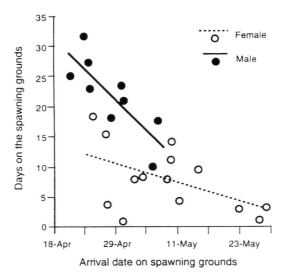

Fig. 5.10 Timing of arrival and duration of residence on the spawning grounds by radio-tracked striped bass *Morone saxatilis*, in the Roanoke River, North Carolina. Reproduced from Carmichael *et al.* (1998).

temperatures (27–29°C) with adequate oxygen (>4 mg O_2 per litre), providing a thermal refuge. Spring and late autumn fish were widely distributed throughout available temperatures. In September the water became isothermal but warm, the thermal refuge near the dam disappeared and fish were forced to tolerate temperatures as high as 29°C, above the typical critical thermal limit for striped bass. Wilkerson and Fisher (1997) determined the seasonal and diel summer movements of striped bass in Kerr reservoir, Oklahoma (18 000 ha) which has three tributary rivers. The Illinois River arm consists of a 15 km stretch of an Ozarkian river with gravel and mud substrate, low conductivity, clear and relatively cool water (annual mean temperature about 4°C lower than the main lake and other inflows). Striped bass showed strong fidelity to the Illinois River throughout this study, with 65% of all fish locations occurring there. Observed behaviour seems to have been a result of the highly stable habitat, and the presence of abundant prey fish. The river provided a thermal refuge in summer, but fish stayed, even when temperatures were similar to other lake and river areas in autumn. The differences between these two behaviour patterns may be due to variations in the spatial configuration of reservoir habitats and variations in thermal refuges available.

It has been postulated that in the Roanoke River–Albermarle Sound system appropriate summer habitat may be limiting for striped bass. Striped bass populations in this system have declined and are susceptible to disease. Deeper areas of the western sound might act as thermal and dissolved oxygen refuge, as might the Atlantic Ocean. However acoustic tracking and tagging have provided no evidence for a migration into the Atlantic (Haeseker et al. 1996). During 1993 and 1994 water temperature in Albermarle Sound rose well above suitable levels for striped bass but dissolved oxygen remained within tolerance limits (>3mg O_2 per litre). No stratification of temperature and dissolved oxygen occurred and no thermal refuges were located. Despite poor condition and lesions associated with *Aeromonas hydrophila*, as well as ectoparasites, fish remained in the sound and selected deep water and underwater structures.

The white perch *Morone americana* also displays significant migratory behaviour, with at least some populations migrating upriver from estuarine or marine environments to spawn in fresh and slightly brackish water on the Atlantic coast of North America (Mansueti 1964; Scott & Crossman 1973). Like striped bass, this anadromous behaviour seems most strongly developed in the middle latitudinal area of the white perch's range of distribution, while in the cool Canadian climate it is mostly a freshwater species. Spawning runs associated with wholly freshwater populations have also frequently been described (though not in detail), normally occurring during April and May. This species has recently become established in Lake Ontario, perhaps due to its mobility, but also due to its high fecundity. In lakes, white perch exhibit diel migratory behaviour, moving onshore at night and offshore during day, and also show a vertical migration, moving towards the surface at night and into deeper water during the day, often in large shoals (Sheri & Power 1969). This behaviour is thought to be associated with feeding. The white bass *Morone chrysops* is abundant throughout much of eastern North America and is less of a diadromous species than the striped bass or white perch, although it occurs in coastal waters in the Gulf of Mexico. Sexually mature fish often form unisexual shoals and move into estuaries to spawn, or in fresh water accumulate over shoals or reefs in spring (Scott & Crossman 1973). Normally males arrive on the spawning grounds before females (Hasler et al. 1958). After spawning, adults move out over deeper water but usually remain in the epilimnion, tending to remain in shoals.

The European seabass *Dicentrarchus labrax* and the Mediterranean seabass *D. punctatus* are both principally coastal marine species, spawning at sea. However, the juveniles of both species frequently enter estuaries and other brackish areas and sometimes into freshwater reaches of lowland rivers that provide key nursery areas for the larvae and juveniles of these species. In England 25–30 mm postlarval *D. labrax* arrive in estuaries between July and August, and frequently remain in sheltered, brackish water areas until their second summer (Pickett & Pawson 1994). A similar life history pattern occurs for *D. punctatus* in the Mediterranean and Black Sea areas. The Japanese seabass *Lateolabrax japonicus*, occurring around Japan and Korea, is somewhat catadromous in that the adults spawn at sea. However, juveniles, subadults and adults migrate into freshwater rivers in spring and overwinter at sea (McDowall 1988; Secor *et al.* 1999).

5.35 Temperate perches (Percichthyidae)

The temperate perches comprise about 20 species that occur mostly in freshwater habitats of temperate Australia and South America (Nelson 1994) and most exhibit potamodromous or catadromous migratory behaviour between habitat types, varying only in the scale of movement. The Australian bass *Macquaria novemaculeata* occurs in rivers of eastern Australia where it provides an important sport fishery (Harris 1984; Mallen-Cooper 1992). Adults migrate downstream to estuaries to spawn and the young progressively move back upstream, growing until maturation. This appears to be a strictly catadromous lifecycle since there are no wholly marine populations. The eggs require saline water for development and the occurrence of impoundments in river systems has caused dramatic declines in a number of populations (Harris 1984). Breeding landlocked populations of Australian bass are unknown. The related estuarine perch *M. colonorum* occurs over a similar geographical range but is principally an estuarine species and rarely moves into fresh water.

Several Australian temperate perches exhibit potamodromous migrations and are principally associated with the larger river systems such as the Murray-Darling river basin. Golden perch *M. ambigua* show characteristic upstream spawning migrations as adults and these are particularly apparent at times of high water, which is known to stimulate spawning in this species (Reynolds 1983; Mallen-Cooper 1994; McDowall 1996). Golden perch produce salt-intolerant, buoyant eggs, which take about 7 days to hatch and for the postlarvae to be capable of swimming and maintaining position in low flow areas (Reynolds 1983). Reynolds calculated that during this period, assuming a rate of travel of 3 km h^{-1} during high flows, eggs and larvae would need to be deposited at least 500 km upstream of the limit of saltwater intrusion to survive. In a mark recapture experiment, 3267 subadult and adult golden perch were tagged in the lower part of the Murray-Darling River by Reynolds (1983) and 704 were recaptured. Most fish that were reproductively mature (>35 cm long) were captured more than 10 km upstream of the tagging site and movement occurred during low and high water, but was especially pronounced during floods. Of 294 fish tagged during high flow periods 41.8% were recaptured. Rates of upstream migration during the first 40 days at liberty were mostly 2–4 km d^{-1}, but up to 15 km d^{-1}. Forty fish moved more than 60 km, 23 moved more than 200 km and one fish migrated 2300 km to the upper tributaries of the Darling River (Fig. 5.11). The Murray cod *Maccullochella peelii*, also from the Murray-Darling system, is usually re-

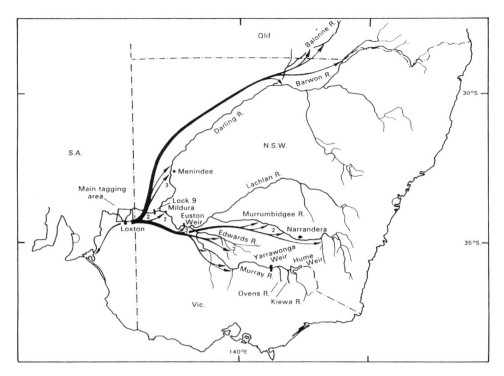

Fig. 5.11 Schematic plan of the spawning migration by adult golden perch *Macquaria ambigua* in the Murray-Darling River, southwest Australia, based on mark recapture studies. SA, South Australia; NSW, New South Wales; Vic, Victoria; Qld, Queensland. Reproduced from Reynolds (1983).

garded as potamodromous, but Reynolds (1983) found less evidence from mark recapture studies to demonstrate distinct migratory patterns, although some individuals were recovered up to 200 km from the site of tagging. In a survey of migratory fish species from coastal drainages of Southeast Australia, Harris (1984) listed Macquarie perch *M. australasica* as another potamodromous species.

5.36 Sunfishes (Centrarchidae)

Occurring naturally in the fresh waters of North America, the sunfishes comprise about 30 species, several of which are known as basses of different types. Some centrarchids (mostly basses) have been introduced to other parts of the world including Europe and southern Africa for their sporting qualities. They are predatory, perch-like fishes with moderately laterally compressed bodies, which build nests and exhibit parental care. Most species are less than 50 cm when mature, although the largemouth bass *Micropterus salmoides* may reach 80 cm in length.

Centrarchids are typically regarded as fishes which establish and occupy clear home ranges, outside of which they rarely move (Gerking 1950). To a degree, these views reflect that in some freshwater systems centrarchids exhibit little movement between habitats; but partly it is also because some of the earlier studies occurred during a single season or examined stream and pond environments only. However, the fact that these centrarchids exhibit clear homing characteristics when displaced also demonstrates the potential for migratory capacity (Gerking 1950, 1953; Funk 1957). Patterns of movement appear to be determined by the spatial distribution of resources. For smallmouth bass *Micropterus dolomieui* summer river habitat is typically large pools between riffles, with rock and gravel substrates, and cover in the form of boulders and logs. Lake-dwelling smallmouth bass also tend to use structured habitats in moderately shallow water (Scott & Crossman 1973). Winter habitat is deeper water with slow current and with hiding areas such as boulders, crevices and caves. Radio-tagged stream-dwelling smallmouth bass (24–45 cm in length) in Jacks Fork River, Missouri, exhibited extremely limited median movements of 0.2 km downstream and 1.3 km upstream, with maxima of 5.7 km and 7.5 km respectively. The fish remained in restricted home ranges for most of the year but tended to disperse in spring, when all tagged fish left their home pool to spawn (Todd & Rabeni 1989). Only short movements were needed between summer habitat, consisting of log and root cover during the day and also boulders at night, and the winter habitat, in which boulders were predominantly used. Movement patterns during floods were no different to those at normal discharges. By contrast, smallmouth bass in the Embarrass River, Wisconsin, exhibited much more extensive migrations between summer and winter habitats as determined by mark recapture, using angling returns and electric fishing (Langhurst & Schoenike 1990). Smallmouth bass older than 2 years of age (>20 cm) migrated downstream 69–87 km from the Embarrass River to the Wolf River when water temperatures fell below 16°C in autumn. A similar pattern was found with radio-tracked fish. They returned to the Embarrass River in April and May, and most individuals homed to the same 5 km stretch as where they were tagged. Adults, older than 4 years and larger than 28 cm, returned sooner than subadult fish. The upper Embarrass River has few deep pools and this appears to be the reason for the relatively long downstream overwintering movements. In lakes largemouth bass use distinct home ranges in summer, which they show high fidelity to and will return when translocated, but often display seasonal movements to overwintering sites (e.g. Ridgway & Shuter 1996).

Largemouth bass *Micropterus salmoides* also exhibit characteristic use of home ranges, typically varying between 0.1 and 5 ha in summer for adult fish with established home ranges in lakes, usually in areas of thick aquatic vegetation (Winter 1977; Mesing & Wicker 1986). Movements tend not to be extensive (most movements are less than 8 km), but homing to spawning areas and summer home ranges does appear to occur and may involve movements of several kilometres (Scott & Crossman 1973; Mesing & Wicker 1986). Largemouth bass from river systems tend to favour overwintering areas in backwaters and off-river areas, as long as oxygen and water levels are adequate (Gent *et al.* 1995; Raibley *et al.* 1997a). In at least some large river–floodplain ecosystems, such as the Illinois River, largemouth bass exhibit distinct spatial behaviour in response to flood pulses (Raibley *et al.* 1997b). Flocculent, silty substrates are unfavourable for nest-building centrarchids that prefer firmer substrate when spawning (Scott & Crossman 1973). Most Illinois River backwater lakes have soft, silty substrates, so spawning centrarchids may use annual spring floods to gain access to inundated terrestrial vegetation and previously dry, compact substrates on the floodplain. Raibley *et al.* (1997b) found that following spawning in May–June, high abundance of 0+ largemouth bass was associated with extended inundation of the floodplain. Regression of the number of days in flood between May and July explained 53% of variance in percentage catch of 0+ largemouth bass. However, long steady periods of inundation were needed for strong recruitment and when river levels fluctuate or rapidly decline during or soon after spawning, weak cohorts may result.

Largemouth bass utilise oligohaline brackish areas as juveniles and adults (Scott & Crossman 1973; Meador & Kelso 1989; Carlson 1992; Richardson-Heft et al. 2000) and in these cases may be more mobile, entering rivers to spawn and moving to distinct winter refuge areas (Carlson 1992). Even in large areas of brackish water such as Chesapeake Bay, largemouth bass exhibit home ranging behaviour and tend to exhibit homing if translocated (Richardson-Heft et al. 2000). This is significant because, as Richardson-Heft et al. explain, large sportfishing tournaments in the US tend to release fish at a few sites, which could result in 'stock-piling' of largemouth bass. The homing tendency of largemouth bass tends to prevent any such effect.

Rock bass *Ambloplites rupestris* are regarded as being relatively cryptic sedentary species but Kennen et al. (1994) reported catching them in salmon smolt traps between April and August and interpreted this as a 'downstream migration to Lake Ontario from ponds and beaver impoundments in the drainage'. Sunfish and bluegills (*Lepomis* spp.) are characteristic of lake and backwater habitats. They do not tend to move long distances, although they may exhibit clear seasonal movements between habitat types. Juvenile and adult redbreast sunfish *Lepomis auritus* and bluegill *L. macrochirus* aggregate to overwinter in deep water and move into shallow, nearshore areas to make nests and spawn (Scott & Crossman 1973). Acoustic telemetry of adult black crappie *Pomoxis nigromaculatus* (Guy et al. 1992) and white crappie *P. annularis* (Guy et al. 1994) in glacial South Dakota lakes has shown these species not to move very long distances, although both displayed distinct diel and seasonal variation in depth and distance from shore, utilising shallow, littoral habitat for spawning and then moving offshore in summer and back in autumn. Black crappie displayed diel onshore-offshore migrations, moving into shallow water at night. The young of sunfishes (*Lepomis*) and crappies (*Pomoxis*) become pelagic soon after hatching and drift in surface waters for several weeks before settling to the bottom and then moving inshore in shoals to shallow, highly vegetated bays. In large floodplain systems, these species seek very slow-flowing but adequately oxygenated water in winter in order to avoid displacement by high flows (Knights et al. 1995; see also section 3.2.2).

5.37 Perches (Percidae)

The percids comprise a range of typically perch-like fish with spiny dorsal fin and ctenoid scales, ranging from deep-bodied, laterally compressed percids such as *Perca* to elongated forms such as *Stizostedion*. The larger species, especially *Stizostedion* and to a lesser extent *Perca* are piscivores. They tend to be shoaling species, although large adults may become solitary. Yellow perch *Perca flavescens*, occurring in temperate North America, and Eurasian perch *Perca fluviatilis* are closely related species of percid found in clear lakes and backwaters, together occurring in an almost circumpolar distribution. In most cases they do not exhibit substantial movements, but like various centrarchid species, where habitats for specific conditions are widely separated perch may exhibit significant migrations. Eurasian perch in homogeneous environments exhibited much greater mobility than those in heterogeneous, more highly structured environments (Bruylants et al. 1986). Perch are principally lake fishes and although they do occur in lowland rivers there is little information available on the existence or nature of migratory behaviour. However, Fig. 4.2 shows that Eurasian perch occur in fish pass catches with a distinct peak in the spring which may coincide with the spawning period, although a second peak occurs in autumn in the River Dordogne, France (Travade *et al.* 1998). The fact that they occur in fish passes at a specific time suggests a degree of migratory behaviour.

Juvenile Eurasian perch *P. fluviatilis*, migrating down the River Angeran towards the Baltic Sea were active by night, whereas adults were active by day (Johnson & Müller 1978a). They suggest that the high nocturnal activity shown by juvenile Eurasian perch during the migratory period was a response to avoid predation by diurnally active Eurasian perch and northern pike *Esox lucius*. In lake environments both yellow perch and Eurasian perch overwinter in deep water and exhibit shoreward migrations in spring into the shallows or tributaries to spawn (Allen 1935; Thorpe 1974; Craig 1977, 1987). The eggs of perch are arranged in ribbons and attached to vegetation or debris, but the larvae are pelagic and drift in wind-generated currents. As the young develop they actively move inshore to the littoral zone in shallow bays, usually less than 3 m deep, and form shoals. As winter approaches they move into deeper water.

Generally perch *Perca* spp. are more active during daytime than walleye and pikeperch (*Stizostedion* spp.) which are principally active at night. The eyes of walleye (and pikeperch) have tapeta lucida and are more effective at low light levels than those of perch (Ali *et al.* 1977). Both groups also commonly exhibit high activity during twilight at dawn and dusk, and forage more during daytime in conditions of reduced light penetration due to high turbidity (Craig 1987). Acoustically tracked walleye *S. vitreum* in the Mississippi showed distinct twilight increases in irregular movement, interpreted as foraging (McConville & Fossum 1981). Hasler and Bardach (1949) noted that, in the summer, lake-dwelling yellow perch moved inshore about an hour before sunset and foraged along the 6-m contour until dark. This crepuscular activity occurs at higher temperatures and is principally a summer phenomenon (Craig 1987). YOY yellow perch often cease activity earlier in the evening than older conspecifics and Helfman (1979) hypothesised that YOY perch avoid predation through inactivity. A variety of other studies have shown that lake-dwelling perch often exhibit clear diel migrations, and that these are associated with foraging and risk from predation. Jansen and Mackay (1992) examined diel chronology of food consumption and catch composition of yellow perch in the littoral zone of Baptiste Lake, Alberta. Feeding intensity increased throughout day and peaked in evening, and almost ceased after sunset. Perch densities netted in the littoral zone followed this trend, showing that inshore movement was related to feeding. Despite the high abundance of YOY perch in the littoral zone in the evening, they were preyed upon much more heavily during the day by larger conspecifics, suggesting that the diel migration of YOY perch to the littoral zone provides a temporal refuge from predation while allowing feeding. Jansen and Mackay argue that the relative reactive distance of predators to prey may reduce more quickly for larger predators than smaller ones when light is reduced below the threshold. Patterns of diel migration and activity of YOY and older perch vary between studies (Helfman 1979; Jansen & Mackay 1992); in Helfman's study there was a wider range of predators and it is likely that diel migration and behaviour strategies vary between environments.

Several Eurasian perch *P. fluviatilis* and pikeperch *S. lucioperca* populations in the Baltic Sea (and formerly in the Aral Sea) exhibit slight anadromous behaviour, feeding and growing in coastal areas, for at least part of the year, but migrating substantial distances up rivers to spawn (Berzins 1949; Müller & Berg 1982; Müller 1986; Craig 1987; Koed et al. 2000; K. Aarestrup [pers. comm.]). Böhling and Lehtonen (1984) also reported migrations of up to 170 km by Eurasian perch in the Baltic Sea, and indicate that tagged perch were frequently recovered at their tagging places during the spawning periods over the following years, suggesting strong homing behaviour. Böhling and Lehtonen suggested that perch migrated over long distances in this environment because their populations were abundant and probably exceeded the food resources locally available in the vicinity of the spawning grounds. On the island of Lolland, southeastern Denmark, adult Eurasian perch spawn in the Flintinge stream and associated ponds, and soon after emigrate to the Baltic Sea to feed in coastal areas. This is followed by a downstream emigration of juveniles. In the autumn there is a return migration with perch overwintering in the stream and ponds, as well as another influx in spring, before spawning (K. Aarestrup [pers. comm.]). Broadly similar behaviour patterns are exhibited by yellow perch *P. flavescens* in brackish water bays and estuaries of North America (Craig 1987).

Off the southern Finnish coast pikeperch exhibit a distinct pattern of migration between spawning, feeding and wintering areas (synthesis in Lehtonen et al. 1996). Pikeperch aggregate in freshwater inlets to spawn in spring. Females leave the spawning ground while males guard the eggs and move to nearby summer feeding grounds, later followed by males. In early autumn, they move to deeper water where they overwinter and show little mobility (Lehtonen & Toivonen 1987). Migrations between winter and spawning habitats are usually less than 30 km, but they may extend over more than 200 km (Lehtonen 1983). Homing behaviour is frequent and causes the spawning populations to remain isolated from each other (Lehtonen 1983), but straying from the Finnish coast to Polish rivers, across the Baltic Sea may occur. As in many species, pikeperch populations of the Baltic Sea seemingly comprise 'migratory' and 'resident' fishes, the latter showing most restricted movements, but this may partly be accounted for by habitat structure and competition. Typically the salinities which can be tolerated by feeding and growing *Perca* spp. and *Stizostedion* spp. are up to about 10‰ and survival of young is often better in slightly brackish water than in fresh water (Craig 1987). Migrations of Eurasian perch between coastal waters and freshwater around Denmark occur on a significant scale and are exploited by commercial fishermen (K. Aarestrup [pers. comm.]). They may be particularly significant because these waters are more saline than those found in the upper Baltic Sea and are about 20‰. Presumably movements of 'anadromous' Danish perch are restricted to areas of distinct freshwater influence, but as yet this is unproven.

The genus *Stizostedion* occurs in North America and Europe and provides important recreational fisheries. Two species, walleye *Stizostedion vitreum* and sauger *Stizostedion canadense*, occur in North America and three species occur in Europe, of which the only well-studied European species is the widely distributed pikeperch or zander *Stizostedion lucioperca*. The pikeperch is widely distributed throughout Europe but has been introduced to areas of southern and western Europe, including eastern England. It occurs in lowland rivers and lakes and tolerates moderately turbid environments. This species generally exhibits significant migratory behaviour although this is not supported by some studies. The movements often appear to follow those of prey species, resulting in seasonal longitudinal migrations within river systems, moving to slower, deeper downstream areas in winter and moving upstream in spring. Overwintering areas may include estuarine and brackish water and this is well established for the Danish population in the River Gudenå (Koed *et al.* 2000). Fickling & Lee (1985) showed that introduced pikeperch in the Great Ouse Relief Channel, eastern England exhibited movements of up to 38 km which could possibly have been spawning migrations although it was also possible that these movements were due to dispersal of the introduced population or to prey searching behaviour.

Craig (1987) reports that pikeperch undertake major migrations for spawning, and there is evidence that populations may 'home' to the same spawning area year after year (Puke 1952), a behaviour pattern that is also common in walleye. Schmutz *et al.* (1998) report movements of significant numbers of adult pikeperch into the man-made Marchfeldkanal from the Danube via the Russbach, a distance of 37 km, although it is not clear whether this represents a spawning migration or a dispersal movement. Upstream movements of pikeperch in the Dordogne River fish elevator at Tuilières, France, peak in autumn (Travade *et al.* 1998). In the River Gudenå, Denmark, radio-tracked pikeperch exhibited a conspicuous downstream migration to the estuary and lower reaches of the river in autumn and moved upstream in spring (Koed *et al.* 2000). The spring migration was related to spawning behaviour, but the annual cycle of movement of pikeperch in the Gudenå also appears to be related to the availability of prey, since Koed *et al.* presented evidence of higher CPUE in the lower river in autumn and the upper study reach in spring (Fig. 3.2). In addition to main river channel movements pikeperch move into backwaters or lentic areas of rivers to spawn (Lelek 1987). In lakes there is a distinct movement into shallow water, often associated with areas of emergent vegetation. The eggs are laid in April–May in a nest made by the male, which then guards it until the young hatch. The young pikeperch have a tendency to passively drift from a fluvial to a lentic environment; they aggregate in shallow, slow-flowing areas, moving downstream at night in clear water or over a wide diel scale in turbid water (Craig 1987).

The walleye *S. vitreum* and the closely related sauger *S. canadense* occur widely throughout central and eastern North America, typically in lakes, but also in large, slow-flowing rivers. Both species are negatively phototaxic and occur in dimly lit or turbid water, for which their eyes, with light-gathering tapeta lucida, make them well-adapted. As adults, neither species usually moves much more than 5 km during summer, although there are reports of sauger moving up to 150 km in the Mississippi (Scott & Crossman 1973). However, in subarctic Canada, Dietz (1973) reported movements of marked walleye between the Peace-Athabasca delta and Lake Athabasca, an annual circuit of as much as 600 km. Pegg *et al.* (1997) demonstrated large-scale downstream prespawning movements of up to 200 km for some sauger in the Tennessee River (Kentucky-Alabama), while others moved upstream through locks. Both species spawn on coarse gravel and boulder areas in rivers or lakes, but walleye frequently exhibit distinct upstream spawning migrations into lake tributaries. Daily movements of walleye in Lake Superior at spawning time were up to 3 km d^{-1}, but more usually about 0.8 km d^{-1} (Ryder 1968), and most walleye spawning migrations are less than 20 km in distance. Some walleye populations spawn exclusively on rocky reefs in lakes while others migrate up tributaries; in both cases there is evidence of homing to historic sites (Scott & Crossman 1973; Olson *et al.* 1978). Forney (1963) provided evidence that, in Oneida Lake, three distinct walleye populations home to their own specific spawning areas year after year. It has been suggested that the existence of population-specific patterns of traditional spawning site use by walleye may result from learnt behaviour by subadults following experienced spawners

(Olson *et al.* 1978), but there is now also evidence from the recapture of biochemically marked larvae that these differences may have a substantial heritable component (Jennings *et al.* 1996).

In lakes, walleye *S. vitreum* larvae older than about 5 days become pelagic and drift on wind-generated currents. Like pikeperch, hatching walleye larvae drift downstream into receiving lakes and backwater areas where they begin to feed. Larval migration typically occurs over a 12–18 day period at an age of 2–6 days posthatch (Corbett & Powles 1986; Mitro & Parrish 1997). Larval migration from tributaries is passive, influenced by current velocity and is primarily nocturnal (Corbett & Powles 1986; Mitro & Parrish 1997). Newly hatched walleye larvae have swimming speeds of 0.03–0.04 m s^{-1} and drift uncontrolled at water velocities in excess of 0.07 m s^{-1} (Houde 1969). Adult walleye from Lake Champlain, Vermont migrate into the Poultney and Missiquoi Rivers to spawn. Using drift-sampling in Poultney River, Mitro and Parrish (1997) found that peak larval densities of 2 fish m^{-3} occurred immediately downstream of spawning areas at the onset of darkness, and occurred at a downstream station 3.5 h later. An estimated 528 000 walleye larvae migrated from Poultney River and 1 306 000 larvae migrated from Missiquoi River over a 12-day period. Migration of larvae takes several days, during which larvae settle at dawn and move up into the water column at the onset of darkness. Drifting at night may decrease predation risk from sight-feeding predators and increase survival rates.

The darters of North America are a diverse group (about 150 species) of small, elongated percids and are typically benthic in habit. They are common in small streams and lakes and in many species small migratory movements are a common feature of the lifecycle, although these are usually a matter of perhaps no more than a few hundred metres. Many of the riffle-dwelling species exhibit prespawning upstream movements to swift runs or riffles, spend the summer in riffle areas and move downstream to overwinter in deep pools (Craig 1987). On hatching, the young drift to slower pools where they remain until they are large and powerful enough to remain in riffle habitats where they are less susceptible to predation. Localised upstream spawning movements to gravel-bottomed pools by the blackside darter *Percina maculata* occur in spring and fantail darters *Etheostoma flabellare* move from the typical riffle habitat of adults to slow pools (Scott & Crossman 1973). While riffle-dwelling adult amber darter *Percina antesella* in the Conasauger River, Georgia, US exhibit seasonal variation in mesohabitat use they move negligible distances to achieve this (Freeman & Freeman 1994). However, larvae of closely related species, and probably also *P. antesella*, typically drift and find slower, appropriate nursery areas. Contiguous, unrestricted areas of these habitats are therefore required to enable lifecycle completion. Recapture of marked adult blackbanded darter *Percina nigrofasciata* over an 18-month period in a large coastal plain stream in the southeastern US suggested that they exhibit long-term residence of small areas, since most recaptures were less than 30 m from the site of release (Freeman 1995). However on occasions individuals moved substantial relative distances (200–420 m), and shifted between mesohabitats (boulder riffle, sand pool, gravel riffle). Localised migrations appear to occur in many lake-dwelling darter species, in particular associated with spawning, with shoreward migrations from deeper water noted for lake-dwelling Iowa darter *Etheostoma exile*, least darter *Etheostoma microperca* and logperch *Percina caprodes* (Scott & Crossman 1973). This behaviour is especially apparent in the latter species; *Percina* are larger and perhaps more mobile than other darter genera.

The various *Gymnocephalus* and *Zingel* species are small Eurasian percids occupying broadly similar habitats to the North American darters. The most widely distributed species is the ruffe *Gymnocephalus cernuus* which occurs throughout much of Europe, except the southern and northwestern areas, and has been introduced quite widely (including the Great Lakes of North America). It lives in shoals in low velocity areas of rivers and easily colonises perturbed environments. It is not reported as being migratory but shows some tendency for upriver movement in spring, entering fish passes at obstructions in small numbers (M. Lucas [unpubl. data]). Several other species of *Gymnocephalus*, *G. acerinus*, *G. baloni* and *G. schraetser* are rheophilic species, the former occurring in rivers to the north of the Black Sea and the remaining two species occurring in the River Danube. Balon's ruffe *G. baloni* exhibits a migration into the side branch of the Danube at Schonbuhel, Austria, between the end of April and the middle of June (Siligato 1999). This coincides with the spawning time for this species and so seems to be a spawning migration. Migratory behaviour has not been reported for *G. acerinus* or *G. schraetser*, which have been little studied. The various species of *Zingel*, also European, are also characteristic of swift-flowing water, but seem to be relatively sedentary, benthic species. Lelek (1987) reports that streber *Zingel streber* are more active at night when they move into shallow water.

5.38 Snappers (Lutjanidae)

Snappers comprise 103 species, of essentially marine fishes, spread all over the warm oceans of the world, where they seemingly exhibit limited home range. A few species enter estuaries of warm tropical rivers, or inhabit mangroves, such as the mangrove snapper *Lutjanus argentimaculatus* and *L. russellii*. Neither of these two species becomes mature in estuaries, and they are regarded as euryhaline wanderers (McDowall 1988). Larvae are more tolerant to intermediate than to high salinities, and estuaries or mangroves may thus serve as important nurseries. The Papuan black snapper *L. goldiei*, a large (1 m TL) popular sportfish, is endemic to Papua New Guinea, where it inhabits rivers and estuaries between the Fly River and Port Moresby District. It spends most of its life in fresh water, but McDowall suspected it could be a catadromous species, although evidence of a marine stage is lacking.

5.39 Drums (Sciaenidae)

The drums are a predominantly marine group of perciforms, but about 28 species are restricted to fresh water, mostly in South America, but with one species, the freshwater drum *Aplodinotus grunniens*, occurring throughout much of North America south of Québec. *Aplodinotus grunniens* is mainly a lacustrine species that is not reported to migrate long distances (Scott & Crossman 1973; Hesse *et al*. 1982). It produces buoyant eggs and pelagic larvae which are often the dominant group of ichthyoplankton in systems where they occur (see also section 7.3.1). In the middle Missouri River, freshwater drum comprised 70–90% of all fish larvae in drift samples, although many of these came from impoundments upriver (Hergenrader *et al*. 1982). *Plagioscion* spp. are South American batch spawners (Worthman 1982), with omnivorous-piscivorous habits, which inhabit lentic areas and show very limited or no migration (Agostinho *et al.* 1994).

5.40 Tigerperches (Terapontidae)

There are about 35 freshwater species of tigerperches, most of which occur in Australia and New Guinea. Potamodromous migratory behaviour is relatively common in this group and tends to be most commonly associated with a well-defined upstream migration of mature adults during periods of elevated water level, exemplified in the silver perch *Bidyanus bidyanus* and spangled perch *Leiopotherapon unicolor* of south eastern Australia (Llewellyn 1973; Lake 1978; Reynolds 1983; Harris 1984; Mallen-Cooper 1994). In a mark recapture programme on silver perch and several other species in the Murray-Darling River, Reynolds (1983) found that adults moved long distances, up to a maximum of 570 km for a fish tagged some 19 months earlier. Of the 660 silver perch tagged, 32 were recaptured and of these 73% were recaptured more than 10 km away from the tagging site, although similar numbers had moved upstream and downstream, but during a period without large floods. Observations at fish passes have recorded substantial upstream migratory behaviour by mature adults at times of elevated flow and since this species has semi-buoyant eggs, this migratory pattern seems appropriate. Some species of tigerperch, such as *Mesopristes kneri*, endemic to Fiji, are catadromous, the adults migrating downstream to estuarine and coastal habitats to spawn and the juveniles moving back into fresh water to grow to maturity (McDowall 1988).

5.41 Aholeholes (Kuhliidae)

The aholeholes (one genus: *Kuhlia*) are a small group (eight species) of perciforms which are common in tropical marine and brackish habitats throughout the Indo-Pacific region, but one species, the jungle perch *K. rupestris*, occurs primarily in fresh water (Nelson 1994) and several other species such as *K. sandvicensis* move long distances up rivers. Jungle perch are widespread from the eastern coast of Africa as far east as Tahiti in the Pacific. Available evidence indicates that the jungle perch is an obligate catadromous species. Juveniles grow in fresh water, feeding mainly on surface drift, and mature adults migrate downstream into estuaries and coastal waters to spawn (McDowall 1988). Dams interfere considerably with their capacity to maintain populations upstream of impoundments, as in the Mariana islands (Concepcion & Nelson 1999).

5.42 Cichlids (Cichlidae)

Cichlids are a large family of over 1300 fresh and brackish water species, spread over Central and Latin America, Africa and the southern part of India. Some species, chiefly tilapiines, have been introduced in a considerable number of water bodies for fisheries or aquaculture purposes. Worldwide, tilapias represent the third largest group of fish for consumption by humans, with about 1 million tonnes eaten per year. Considerable knowledge has been gained through study of the haplochromine species flocks from the Great Lakes of Eastern Africa, which host several hundred species, most of them being endemic to each lake, and sometimes found nowhere else other than on a single shore of a small island (review in Ribbink 1991). In addition to the African Great Lakes, cichlids are encountered in almost all freshwater and brackish habitats within their geographical range, including coastal lakes and estuaries, floodplain, inundated forests, rivers, including rapids (e.g. genera *Steatocranus* and *Teleogramma* in Africa, genus *Teleocichla* in the Brazilian highlands; review in Lowe-McConnell 1991). Most cichlids are equilibrium strategists with fast growth and small size at sexual maturity. They have evolved through a series of sophisticated courtship, mating and parental guard systems, which have made most of them less dependent of the need for finding appropriate conditions for their eggs and young, with an almost total independence for mouth-brooding cichlid species. This has resulted in a highly resident lifestyle for most species, reinforced by batch spawning and territoriality.

Nevertheless, cichlids inhabiting rivers of the Soudanian savannah region undertake lateral (and possibly longitudinal) seasonal migrations onto the inundated floodplain where their young find favourable environments for fast growth, then return to the river under receding waters (Bénech *et al.* 1983; Bénech & Penáz 1995). A recent biotelemetry study on the habitat use and movements of several cichlid species in the Namibian stretch of the Zambezi River near Kalimbeza Island has provided some intriguing results concerning spatial behaviour of cichlids (Økland *et al.* 2000). Over the period October 1999 to March 2000, which corresponds to the periods of rising and high waters in Namibia, threespot tilapia *Oreochromis andersonii* showed home range behaviour, consistently remaining within an area extending to no more than 200 m, on average. Another cichlid species, the pink happy *Sargochromis giardi*, tracked at the same time, occupied slightly larger home ranges (about 400 m), with a single fish moving 2 km to an upstream backwater. Økland *et al.* pointed out that in spite of the seemingly residential habits of these cichlids, only one of the eleven fish displaced by 350–1,000 m at the start of the study, had been found to home. Many factors such as post-tagging stress or inability to migrate or orientate might be invoked, but it is also likely that displaced fish might have been unable to regain their original territory, as suggested by Hert (1992) for another cichlid, *Pseutropheus aurora* (see details in section 2.2.2). The strong reputation of sedentarism in cichlids does not preclude their capacity of dispersal and colonisation of new habitats. Examples of fast and extensive dispersal have been provided by the colonisation of Gatun lake (Panama Canal) by *Cichla ocellaris*, a large predatory neotropical cichlid (Zaret 1980), or during mark recapture experiments in African lakes with some species moving an average 3 km d^{-1} over several weeks (e.g. Fryer 1961 in Fryer & Iles 1972).

In lacustrine environments, open waters are generally avoided by cichlids, this contributing to the development of geographical isolates as a first step to speciation. However, this does not apply to some pelagic cichlids, like *Cyrtocara* or *Cyprichromis* spp., for which genetic analyses revealed no distinct populations (Eccles 1986), suggesting these species were more mobile than other cichlids. In Lake Mweru, shoals of *Oreochromis macrochir* have been shown to undertake spawning migrations over several kilometres (Carey 1965 in Fryer & Iles 1972). Notwithstanding the aforementioned examples, cichlids can be regarded as having very limited migratory habits, although detailed information is lacking for most, especially riverine, species.

5.43 Southern rock cods (Bovichthyidae) and sandperches (Pinguipedidae)

The southern rock cods are primarily benthic, marine fishes occurring in cool southern temperate and subantarctic areas. One species, the tupong *Pseudaphritis urvillii*, occurs in fresh water in southeastern Australia and spawning occurs in estuaries (Allen 1989). The young move into fresh water, exhibiting a sequential movement upstream with age until they mature when they return to estuaries to spawn.

The torrentfish *Cheimarrichthys fosteri* (Cheimarrichthyinae, Pinguipedidae) is the only sandperch to occur in fresh water. It is endemic to New Zealand rivers and is reported to be amphidromous (McDowall 1988, 1990). Males and females are spatially segregated with females moving up to 150 km upriver to headwaters while males mostly occur at low elevations. It is thought that females migrate downstream to spawn in the lower reaches of the river. Larvae are found in coastal marine environments and it is thought that they drift downstream soon after hatching. After several months, the small juveniles (about 25 mm long) move into rivers, growing as they move upstream over a period of several years, resulting in increasing size of this species with distance from sea. Despite its name, the torrentfish tends not to be able to traverse areas of rapids, and long inland penetration (up to 289 km) is achieved in low gradient rivers only (McDowall 2000). Migration to elevated reaches (up to 700 m) occurs, but over relatively short distances inland. Inland penetration is further impeded by natural falls or man-made weirs and dams.

5.44 Fresh and brackish water dwelling gobioid fishes (Eleotridae, Rhyacichthyidae, Odontobutidae and Gobiidae)

The gobioids are among the most speciose taxonomic groups of fish (and vertebrates), with well over 2000 species, of which about 200 species occur in fresh water (Nelson 1994). As small fishes, rarely exceeding 10 cm in length, significant migratory movements might not be expected but diadromous behaviour occurs widely in this group which is most diverse in the tropical and subtropical Indo-Pacific region. Most of these species spawn in fresh water in typical gobioid nests and the larvae are swept to sea. The pelagic larvae may spend 1–2 months at sea before invading streams as tiny juveniles, often in large numbers. Despite their small size they are able to negotiate fast flows and obstacles using their characteristic pelvic fins which are modified to form a sucking disk. McDowall (1988) provides extensive consideration of diadromy in this group, and a limited review is given here.

The sleepers (Eleotridae) are a group of mostly tropical and subtropical gobioid species which extend as far north as the Atlantic coast of the United States and as far south as New Zealand. Most of the 150 or so species occur in fresh and brackish water environments. Six species occur in swift-flowing streams in New Zealand and are regarded as amphidromous (McDowall 1990), although the Cran's bully *Gobiomorphus basalis* seemingly is a truly freshwater species, with little migratory activity. The redfinned bully *Gobiomorphus huttoni* spawns in fresh water where the eggs hatch. The tiny larvae are swept downstream into the sea where they grow and re-enter fresh water as juveniles (about 20 mm long) in late spring and early summer. At this time large aggregations of juveniles at the mouths of rivers may occur. They move upstream into adult habitat, growing to maturity over another 1–2 years and reproduce annually. Upstream movement tends to occur throughout adult life, resulting in a positive correlation between fish size and distance upriver. Similar life histories occur for at least four other *Gobiomorphus* species in New Zealand and probably other species in Australia (McDowall 1988). Eleotrids of the genus *Kribia* are typically freshwater species that are spread over the Nile River basin, Central and Western Africa. The gobiid *Nematogobius maindroni* and the eleotrids *Dormitator lebretonis*, *Kribia* spp. have been recorded as far inland as in the headwaters of River Niger tributaries, several thousand kilometres from the sea (Harrison & Miller 1992a, b). Welcomme (1979, 1985) described *Dormitator lebretonis* as an anadromous species, entering western African tropical rivers to spawn on flooded vegetation in the lower reaches. Its occurrence in the headwaters of the River Niger, at considerable distances from the sea, suggests it may have adapted to a freshwater life history, but it is not known whether it exhibits potamodromous migrations. Welcomme suspected *D. latifrons* exhibited a similar lifestyle but McDowall (1988) suggested there was some contradictory evidence that it was a catadromous species. A few species of gobioids may exhibit catadromous migratory behaviour. Perhaps the best example is *Gobiomorus dormitor*, an eleotrid, which has been studied in Central America. In the Tortugueros estuary, Costa Rica, juveniles occur in the estuary and adults migrate downriver to spawn in the estuary (Nordlie 1981; McDowall 1988). The Mexican fat sleeper *Dormitator maculatus*, the largest eleotrid, reaching 60 cm in length, may be anadromous (Nordlie 1981).

The family Gobiidae includes several subfamilies that are frequently encountered in fresh water. Mudskippers (subfamily Oxudercinae, genera *Boleophthalmus*, *Periophthalmodon* and *Periophthalmus*) are found in coastal waters, mangrove habitats and estuaries, where they show amphibious habits (notably as oxygen debts may be paid more easily in air than in water). They are occasionally found in fresh water, but always at short distances from the coast, and they probably exhibit most restricted migrations in fresh water. The subfamily Sicydiinae occurs primarily in fresh water (Nelson 1994) and comprises many species with an amphidromous lifestyle (e.g. some species of the genera *Lentipes*, *Sicyopterus*, *Sicydium* and *Stenogobius*). The Hawaiian goby *Lentipes concolor* is known to ascend fast-flowing rivers and waterfalls and to spawn in fresh water. Eggs develop in fresh water and young are swept to the sea (Maciolek 1977). After about 1 month, 2-cm long juveniles return to rivers and ascend them as they grow. *Sicyopterus extraneus* occurs in tropical mid-Pacific areas and is also amphidromous (Manacop 1953). Spawning occurs in fresh water, the eggs hatch in 1–2 days and the tiny larvae, about 1 mm long, are swept into the sea. They return to freshwater after about 1 month when they are 2–3 cm in length, and move up rivers to grow and mature. This and other species provide a fishery for the returning juveniles known as 'ipon' or 'hipon' (review in Bell 1999). In Dominica, the so-called tritri fishery, which is mainly supported by the returning juveniles of *Sicydium punctatum*, has been running for centuries (Atwood 1791, in Bell 1999). However, it has received little attention until recent studies by K.N.I. Bell and his collaborators, who also were among the rare scientists to provide some information on newly hatched embryos of gobioids. The following description of its lifecycle is strongly inspired by the review of Bell (1999) (Fig. 5.12).

Adult *S. punctatum* build nests in coarse substratum, and small eggs hatch within 24 hours. During their riverine stage, embryos (about 2 mm in length) remain in the water column due to a pattern of vertical swimming and sinking, and they enter the sea where they undergo metamorphosis to the juvenile stage. Very little is known of the marine stage, except that it may last from 50 to 150 days, and that its duration varies on a seasonal basis, which is related to growth patterns and intensity of spawning. Transparent postlarvae of tritri (about 22 mm long) return to rivers all year round, but with a peak during August–November. Although river entry is pan-seasonal, it is lunar-phased as in many other gobioids (Manacop 1953), and schools start entering rivers on the fourth day after the last lunar quarter. This pulsed migration, which might be regarded as a strategy for taking advantage of large tide height or as an anti-predator tactic, makes the tritri fishery episodic, but on a regular basis. This periodicity is well-known and understood by the inhabitants who have made a part-time job of this regular supply (tritri being the most expensive fish retailed in Dominica). While in the river, schools disperse, postlarvae settle, become pigmented and they adopt benthic and solitary habits. Upstream migrations rarely exceed 14 km and 300 m in altitude, but these values are about the longest possible upstream migrations in the short rivers of Dominica (Bell 1999). A wide variety of other gobiid and eleotrid species with broadly similar migratory behaviour and lifecycles have been described with varying levels of certainty (McDowall 1988): *Sicyopterus* (6+ species), *Sicydium* (7+ species), *Sicyopus* (2+ species), *Lentipes* (2+ species), *Stiphodon* (2+ species), *Awaous* (2+ species), *Rhinogobius* (2+ species), *Acanthogobius* (1+ species), *Stenogobius* (1+ species) and *Eleotris* (3+ species), as well as a large number of other species that *may* be amphidromous.

Recently, additional knowledge has been gained on the lifestyles of gobioids in general, and stream gobies in particular, through the application of electron probe microanalysers to the study of Sr:Ca ratios in otoliths (see section 6.3.3.2 for further information on this method). Using this technology, Shen *et al.* (1998) provided evidence that among 11 species of gobies and sleepers collected in Longlong Brook, northeastern Taiwan, two (*Glossogobius*

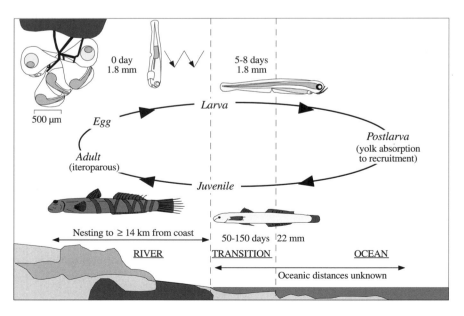

Fig. 5.12 Lifecycle of the amphidromous goby *Sicydium punctatum* in Dominica. Redrawn from Bell (1999).

biocellatus and *Papillogobius reichei*) were estuarine species and a single one, *Ctenogobius brunneus*, was a truly freshwater species. The other eight species (*Awaous melanocephalus, Eleotris acanthopoma, Glossogobius celebius, Oligolepis acutipennis, Redigobius bikolanus, Rhinogobius nagoyae formosanus, Sicyopterus japonicus* and *Stenogobius genivittatus*) migrated to a marine environment at an early age of between 1 month (*O. acutipennis*) and 6 months (*S. japonicus*), then returned to fresh water.

Anadromy has been confirmed in the tiny Japanese ice goby *Leucopsarion petersi* (McDowall 1988). The adults migrate into fresh water in spring, cease feeding and spawn, and when the larvae emerge they are carried into the sea, where all feeding and growth occurs. The loach gobies (Rhyacichthyidae) are represented by two species of *Rhyacichthys* occurring in freshwater streams of the western Pacific. *Rhyacichthys aspro* may be amphidromous, but neither species has been studied in detail. Relatively few gobies are known from freshwater environments in the temperate zones, but there is one striking example of deep inland penetration for the round goby *Neogobius melanostomus*, which now occurs as far upstream as rkm 861 of the Danube River, in Serbia (Simonovic *et al.* 1998). The same species has been accidentally introduced into the North American Great Lakes, and it has become one of the most abundant benthic fishes in the Lake Huron–Lake Erie system only 10 years after it was discovered first in the St Clair River. Round gobies mature at an early age, they exhibit batch spawning, and greater fecundity than native fishes inhabiting the same niche (the sculpins *Cottus bairdi* and *C. cognatus*), which promotes their rapid dispersal, and competitiveness with native species (review in MacInnis & Corkum 2000).

5.45 Fresh and brackish water dwelling flatfishes (Pleuronectidae, Soleidae, Achiridae, Paralichthyidae and Cynoglossidae)

There are approximately 570 species of flatfish (Pleuronectiformes), all exhibiting the characteristic larval metamorphosis to the bottom-living adult. Of these, about 20 species regularly enter fresh water and four species probably occur only in fresh water (Nelson 1994). In European waters the flounder *Platichthys flesus* is the predominant flatfish found in fresh water, spending much of its lifecycle in estuarine and brackish environments, but occurring well above the limit of salt intrusion in unobstructed lowland rivers. It occurs from the White Sea to North Africa, as well as in the Mediterranean and Black Seas. The adults spawn at sea in coastal areas in early spring, usually within 10–20 km of freshwater influence, and return to estuarine environments. They are therefore catadromous. The newly hatched young are intolerant to fresh water but rapidly develop a tolerance to low salt concentrations and may spend as little as 2 weeks at sea, actively moving into lower salinity areas (Dando 1984), or using tidal transport (e.g. Bos 1999). Otolith analyses of larvae from the Elbe River, Germany indicated that the travel time from the estuary to inland nurseries about 50 km upstream, could be as short as 6 to 9 days in April–May (Bos 1999). Further inland progression by 0+ flounder into fresh water tends to be a progressive behaviour and Beaumont and Mann (1984) found that in the lower River Frome, on the south coast of England, numbers increased through the autumn.

Young *P. flesus* stay in brackish or fresh water and then migrate with adults to the sea to spawn (Nikolsky 1961; Berg 1962). Tidal rivers are extremely important as nursery and feeding grounds (Summers 1980; Kerstan 1991) although off the west coast of Scotland the flounder is a predominantly marine species. Juveniles tend to occur further upstream in rivers than adults, the distance tending to depend on the degree of obstruction, and usually being greater in rivers with large tidal influence. Maitland and Campbell (1992) report young flounder as common in Loch Ness, 50 km from the sea. The young of plaice *Pleuronectes platessa*, also common on eastern Atlantic coasts frequently occur in European brackish water estuaries and bays, principally in their first year of life, but do not occur in fresh water. Another more northerly species *Pleuronectes glacialis*, occurring in the White and Barents Seas and northern Europe is reported to ascend rivers such as the Dvina (Lelek 1987).

In western Atlantic areas winter flounder *Pseudopleuronectes americanus* frequently occur in estuarine areas but do not appear to enter fresh water (van Guelpen & Davis 1979). In coastal areas of the northeastern and northwestern Pacific, the starry flounder *Platichthys stellatus* is tolerant to low salinities and exhibits seasonal movements in and out of fresh water (Hart 1973). The hogchoker *Trinectes maculatus*, a species of sole, occurs in eastern North America to Venezuela in South America and shows a clear annual migratory behaviour into fresh and brackish water, similar to that of European flounder. The adults migrate to coastal areas to spawn in summer and return to less saline areas for feeding and overwintering (Dovel *et al.* 1969; Peterson 1994). The newly metamorphosed young settle out in estuarine areas and move towards fresh water. There is a progressive upstream tendency for movement of young and adults through the autumn. The seaward migration distance increases with age, so that the largest fish are found furthest from fresh water in summer. This pattern occurs for the two subspecies characterising northern and southern latitudes, although the Gulf of Mexico population studied by Peterson (1994) spawned at salinities of 18‰, by contrast to those studied by Dovel *et al.* (1969) in the Patuxent River, Maryland, which were reported to have spawned in full-strength seawater. Rapid movement of young towards low-salinity water does not appear to be due to intolerance to high salinities, but may be explained by greater metabolic efficiency in this environment (Peterson-Curtis 1997).

In New Zealand, two species of *Rhombosolea* enter fresh water (McDowall 1990). Newly metamorphosed young of the black flounder *Rhombosolea retiara* enter rivers in spring and may move 50–100 km upstream (McDowall 1990). They live over sand and gravel in slow to swift-flowing water where they feed and grow to maturity before making a seaward migration to spawn. Unlike *P. flesus* and *T. maculatus*, it seems that adult black flounder do not make subsequent seasonal migrations into rivers.

In tropical and subtropical environments several flatfishes, especially soles (Soleidae), American soles (Achiridae) and tonguefishes (Cynoglossidae) enter or spend most of their lives in fresh water, although their lifecycles are poorly understood. Merrick and Schmida (1984) list three riverine species of sole *Synaptura salinarum*, *S. selheimi* and *Aseraggodes klunzingeri*, but migratory movements are not known. Five species of tonguefishes (Cynoglossidae) are known primarily from rivers (Nelson 1994). A species of *Citharichthys* and of *Pseudorhombus* (both Paralichthyidae) ascend rivers from the ocean in Africa (M. Desoutter in Daget *et al.* 1986). Some species of *Achirus* and *Trinectes*, as well as *Catathyridium jenyinsii*, may occur several hundred or even thousand kilometres up rivers such as the Amazon. The eggs of flatfish are initially buoyant and it is likely that in some of these freshwater species the eggs and larvae drift downstream into either fresh or estuarine water before movement of juveniles back upstream. These patterns of movement could be described as marine amphidromy or potamodromy, depending on the extent of egg and larval drift, but nevertheless represent large-scale migration.

Chapter 6
Methods for Studying the Spatial Behaviour of Fish in Fresh and Brackish Water

6.1 Introduction

Tagging of fish to study their movements has been carried out since at least the seventeenth century when, in *The Compleat Angler* (first published in 1653), Izaak Walton reported the attachment of ribbon tags to the tails of juvenile Atlantic salmon *Salmo salar* to determine their movements (Walton & Cotton 1921). The technological advances of the past few decades have resulted in dramatic improvements in the range of techniques available to study the spatial behaviour, including migration, of fishes. There is a broad range of methods which have been used to examine the movements of fish in fresh water and brackish environments (Fig. 6.1). Some others have been used in marine environments exclusively, but their application to freshwater environments can be foreseen in the very near future in view of recent technical developments (e.g. archival tags; Sturlaugsson 1995; Block *et al.* 1998).

All techniques do not perform equally well for addressing a particular question concerning fish migration in fresh- or brackish water environments. Nor is any one method applicable to all studies of fish spatial behaviour and migration, as there are species- or environment-specific limitations for all of them (e.g. see Table 6.1). Timescale also is a central problem in fish migration studies, as some methods are particularly suitable for short-term studies, whereas others are more adequate for long-term studies. In this context, there always is a trade-off between the accuracy of the information gathered, the duration of the study, the numbers of fish from which relevant information can be retrieved, disturbance by the method, and budget available for the study (e.g. see Table 6.1). However, almost all aspects of fish migration can be tackled by applying the right methods, alone or in combination, provided the biological or management issues have been defined properly, and the type of information needed has been identified. This chapter gives an overview of the methods for studying migration and other aspects of spatial behaviour of freshwater fishes, focusing on state-of-the-art techniques, which offer the most promising perspectives for comprehensive and quantitative approaches (hydroacoustics, telemetry, microchemistry). Rationale and case studies are also illustrated.

Although any classification of methods and techniques is always a questionable task, methods useful for studying the spatial behaviour of fishes broadly may be divided into two categories; methods that are capture-independent and those that are capture-dependent (Fig. 6.1). The former includes methods such as visual observation, the use of resistivity fish counters and hydroacoustics. The latter includes methods that rely on sampling marked or

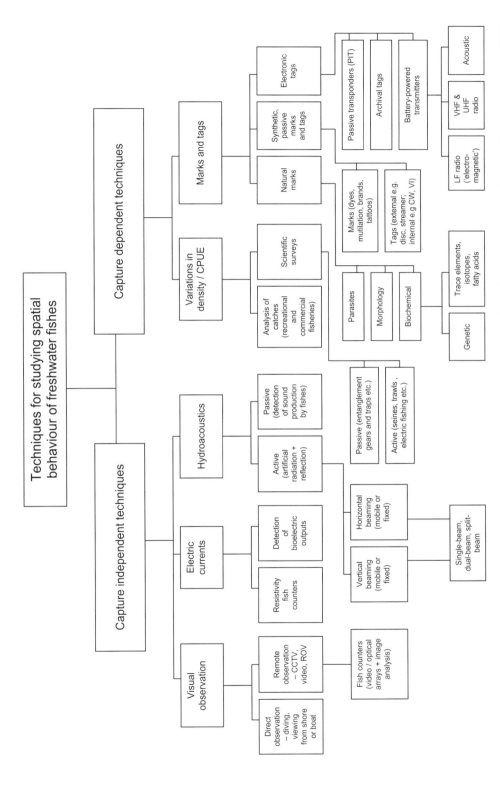

Fig. 6.1 Classification of the range of methods available for determining spatial behaviour of fishes in fresh and brackish water environments. Reproduced from Lucas and Baras (2000).

Table 6.1 Comparison of selected methods for use in studies of freshwater fish spatial behaviour. Modified from Lucas and Baras (2000).

	Reflected natural radiation (visual)	Reflected artificial radiation (hydroacoustics)	Active radiation from a transmitter		Active radiation from a transponder		Interference with electric field (resistivity fish-counters)
			VHF radio tagging	Acoustic tagging	Powered LF transponders	PIT tagging	
Situation	Clear water, restricted site. Shallow streams, rivers and lakes	Little noise or entrained air, few plants. Lowland rivers and lakes	Low conductivity (<500 µS cm^{-1}), shallow. Usually oligotrophic–mesotrophic streams and lakes	Low noise, low turbidity, little entrained air. Usually lakes and slow-moving rivers	Any aquatic environment. Usually deep, noisy and highly conductive, e.g. harbour, barrage	Any environment, so long as fish swims within range of antenna	Fresh water, must be set on weir or in pipe so fish swims within range
Location of sensor	Within sight	Fixed station or mobile on a boat	On land or boat or air	In water	Within range, usually on river bed	Within range	On river bed, on weir or in fish pass
Range (m)	1–10	20–200	20–5000	20–2000	2–30	0.1–1+	0.5
Typical lifespan (days)	For tags, limited by tag algal growth	No limits	20–600	10–300	30–300	>3000 (or life of fish if retained)	No limits
Water depth (m)	<30 for SCUBA, <2 for snorkelling	>1.5	Dependent on conductivity (normally <5)	Dependent on noise (usually 0.5–100)	Within range (usually 2–10)	Within range (generally <1)	Within range (<0.5)

Methods for Studying the Spatial Behaviour of Fish   233

Minimum fish size (cm)	Visible (c. 5)	1	12	12	30	5	20
*Technical demand	Low–moderate	High	Moderate	Moderate	High	Moderate-high	Moderate
*Sample size	10^2	No limits	10^2–10^3	10–10^2	10–10^2	10^2–10^5	No limits
Disadvantages	Poor range, relies on water clarity, poor for cryptic species, difficult to obtain long-term tracks	Poor species and individual identification. High data processing requirements	Lower directionality than acoustic systems. Poor range in deep, lowland waters. Tagging may influence behaviour (next 3 columns also) and lacks population scale measurement (next 2 columns also)	Shorter life than equivalent radio tags. Usually requires boat. Poor range in noisy environs. Sound reflections. Fewer tags can be operated cf. radio	Low range, narrow range of utility. Data collection limited to vicinity of antenna (usually non-mobile)	Very low range, data collection limited to antenna sites or with recaptured fish	Very low range, usually must be sited at structure. No individual identification, limited size sorting
Value for fish behaviour studies	Low	High	High	Medium-high (in freshwater)	Low	High (at by-passes/streams and for small fish)	Low
*Equipment costs ($US)	5×10^2–>10^3	10^4–10^5 (boat etc. extra)	>2×10^2 per tag 5×10^2–>10^4 for system	>3×10^2 per tag 10^3–4×10^4 for system	>4×10^2 per tag >10^4 for system	>4 per tag >2×10^3 for system	>10^4 (+ structure)

*Estimates for a field study 'typical' of those referred to in this chapter tagged fish (mark recapture) or unmarked fish (density estimates, catch per unit effort) over defined scales of space and time in order to obtain information on distribution and movement. Captured fish may also be tagged with electronic tags which usually radiate energy (e.g. radio or acoustic tags) enabling the fish to be tracked and/or environmental data to be gathered. Priede (1992) rightly defined 'wildlife telemetry' as 'all methods of obtaining information on living free-ranging animals by remote means' including, for example, direct observation of fishes through interception of reflected natural radiation within the visual spectrum of humans or by the use of cameras or image intensifiers. However, in this chapter we restrict the use of the term 'telemetry' to the familiar context of use of electronic tags for remote monitoring of wildlife. Biochemical analysis of samples from fish, requiring non-destructive sampling (genetic analysis, stable isotope/trace element analysis of non-vital tissue) or destructive sampling (otolith microchemistry) may also provide information on migration and ontogenetic processes.

6.2 Capture-independent methods

6.2.1 Visual observation

Helfman (1983) argued that direct observation is a valuable and frequently neglected tool in fisheries research. The most obvious form of direct observation is that employing naturally radiated energy within our visual spectrum (Table 6.1). Basic methods of visual observation involve snorkelling, SCUBA diving and observations from the shore or boats, as well as the deployment of remote still or video camera systems (Helfman 1983; Wardle & Hall 1994; Dolloff *et al.* 1996). The major advantage of observational methods is that as long as disturbance is minimised, fish behaviour will be as near normal as possible since fish are not manipulated in any way (Table 6.1). Direct visual observation (diving, boat- and bank-based viewing) has provided information on fish abundance, distribution, habitat preferences and behaviour in a variety of studies (Keenleyside 1962; Helfman 1981, 1983; Heggenes *et al.* 1993; Dolloff *et al.* 1996). However, observations are dependent on proximity to the fish, time of day (or availability of natural light), water depth, clarity, flow conditions, and small or bottom-dwelling fishes are often difficult to see. Site-specific information or evidence can be recorded on hand-held still or video cameras, although most underwater recording is usually carried out on plexiglass plates, or on a cylindrical 'diving cuff' fixed around the forearm (Dolloff *et al.* 1996). It is usually not possible to identify individual fish unless they are tagged with numbered or coloured external tags (Heard & Vogele 1968; Helfman 1981), and this requires fish capture or tagging *in situ*. In shallow water individual colour codes may be more easily distinguished than alphanumeric coding, but from a practical viewpoint the number of colour codes that can be distinguished is extremely limited (Table 6.1). These factors tend to make direct observations limited to special site monitoring (e.g. migration to shallow spawning grounds, Baras 1994; diel migration studies; Helfman 1981, 1983). These methods are of very limited value in some tropical rivers where turbidity may be high all year round (Fernandes & Mérona 1988), although visual observation has been used successfully in the Amazon on several species (Goulding 1980).

Closed-circuit television and video recording provides an effective approach for surveying fish and their behaviour under relatively clear water conditions (Collins *et al.* 1991; Wardle & Hall 1994). Modern charged-couple device (CCD) monochrome cameras are sensitive to low light levels and have been miniaturised to a high degree. Most applications of this approach in fresh water have been and continue to be located at specific sites past which fish are migrating, such as at fish passes, where detailed behavioural information can be obtained (Haro & Kynard 1997). These methods have been accompanied by the development of image analysis techniques for automatically sensing, counting and sizing fishes using video or light emitting diode (LED) arrays (Irvine *et al.* 1991; Fewings 1994; Larinier 1998). Recent attempts to develop optical fish counter systems using computer-driven real-time image capture and analysis have been successful for adult migratory salmonids and in some multispecies environments (Larinier 1998), but have been of limited success for smaller fish, where fish are in shoals, or where a variety of morphologically similar species occur (Irvine *et al.* 1991; Fewings 1994; Larinier 1998). Under very low light conditions, infra-red light sources, to which most freshwater fishes are insensitive, may be used in conjunction with infra-red sensitive cameras (including many monochrome CCD cameras) to monitor fish behaviour. Video cameras attached to remotely operated vehicles (ROVs) have proved useful for examining fish behaviour in deep lakes (Dolloff *et al.* 1996; Davis *et al.* 1997), although their high cost of operation has restricted their use. Three-dimensional stereographic and video tracking techniques (Boisclair 1992; Hughes & Kelly 1996) have proved useful in bioenergetic studies to examine local foraging behaviour and activity levels of fishes (Trudel & Boisclair 1996) and have been used to assess energy expenditure of upstream migrating sockeye salmon *Oncorhynchus nerka* through the measurement of tailbeat frequency under varying local water velocity conditions (Hinch & Rand 2000).

6.2.2 Resistivity fish counters and detection of bioelectric outputs

If an electric potential is set up between two electrodes in fresh water, a small current is passed, but this will be influenced by the presence of a large fish (or other large, dense organism) which has a lower electrical resistance than the water it displaces, in the vicinity of the electrodes. Resistivity fish counters, which detect characteristic changes in resistance in the overlying water column as a fish crosses one or more electrodes, can measure fish passage past specific points and have been used extensively in the monitoring of adult salmonid migrations through rivers and fish passes (Lethlean 1953; Bussell 1978; Dunkley & Shearer 1982; Reddin *et al.* 1992; Fewings 1994; Aprahamian *et al.* 1996; Smith *et al.* 1996). In open river channel environments, conventional resistivity fish counters are flatbed structures, usually placed on sloping structures such as weirs, the velocity profile of which encourages fish swimming upstream to pass close to the weir face on which the counter's electrodes are positioned (Table 6.1). Some success has been obtained with portable resistivity counters deployed on the stream bed (Smith *et al.* 1996), but sites have to be chosen carefully to maximise uninterrupted passage close to the electrodes. Resistivity fish counters are frequently coupled to video camera systems for validation, sizing and identification purposes (Fewings 1994; Aprahamian *et al.* 1996). These video systems may work in continuous time-lapse mode, or may be automatically triggered to switch from time-lapse to real-time mode by resistivity counts or by independent sensors (e.g. LED arrays). Resistivity counters tend to be much less efficient at recording downstream movement of fishes due to their wide variation in swimming depths and inconsistent path of travel over the counter electrodes (Dunkley & Shearer 1982; Smith *et al.* 1996).

Resistivity counter sites are normally inaccessible to most non-salmonid fish due to their lower swimming performance. This problem can partly be overcome by using resistivity counters of tubular construction, through which the fish swims (Lethlean 1953), but this limits their use to sites where a fish pass or culvert is present (Bussell 1978). Moreover, reducing the speed of upstream travel of fish past the electrodes encourages lingering or non-steady traversal of the electrodes, complicating reliable counting. Most resistivity counters are not capable of resolving small fish (less than about 25 cm in length) or fish migrating together in shoals (Aprahamian *et al.* 1996; Table 6.1). Variations in water depth and conductivity also complicate signal interpretation, although automatic calibration in response to environmental variables is now common. Resistivity counters become unusable in brackish or tidal conditions where the resistance signature of a passing fish becomes difficult to detect against the lower electrical resistance of saline water. Conversely they have a great potential in many tropical waters where conductivity is low. The suitability of using resistivity counters for studying fish migration is thus clearly dependent on fish size, behaviour and environment.

Several large taxa of freshwater fishes, principally the mormyroid (suborder Mormyroidea) and gymnotiform (order Gymnotiformes) electric fishes, produce distinct electric organ discharges (EODs) which can easily be recorded. Using specialised analysis techniques, the EODs can be discriminated and compared as these are often species- and sex-specific (Heiligenberg 1991), and in some cases are individually recognisable (Friedman & Hopkins 1996). The latter authors showed that small EOD differences between individuals of two species of *Brienomyrus* (Mormyridae) could be used to track their day-to-day movements in their natural stream habitat in West Africa, with fish normally returning to the same daytime refuge on successive days. Due to the limited range of signal detection (about 1 m), it is unlikely that this technique would enable 'tracking' of electric fishes over long distances and time periods, although Friedman & Hopkins (1996) did manage to track movements of *Brienomyrus* in excess of 100 m over several days using this technique.

6.2.3 *Hydroacoustics*

Sonar (the acronym of SOund NAvigation and Ranging) is a general term for any device, including echosounders and acoustic telemetry systems, that uses propagated acoustic energy to enable the remote detection and positioning of objects under water. Active sonar produces sound while passive sonar systems detect naturally occurring sound. A variety of teleost fishes produce sounds that are often distinct between species and sexes, and which can be located with a sensitive directional hydrophone (Hawkins 1993). Most of the better-known examples of sound production by fishes have been reported from marine environments and, although there is evidence that, in a few cases, individual fishes can recognise conspecifics of the same sex from their vocalisations (Hawkins 1993), this method does not yet appear to have been used to identify and locate individual fishes in the natural environment. Sonar methods that are most useful for examining distribution, behaviour and migration of fishes therefore use active sonar (Table 6.1).

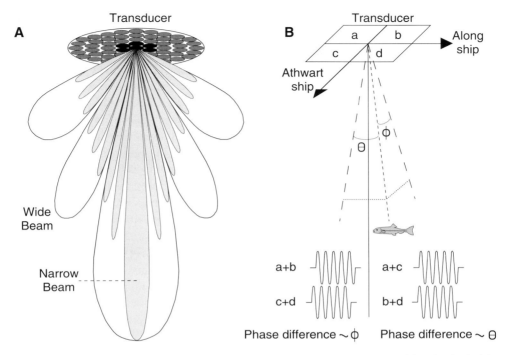

Fig. 6.2 Schematic illustration of the beam pattern in a dual-beam echosounder (left) and functional principle of a split-beam echosounder (right), with the target direction being determined from phase differences between quadrants a–d. Redrawn from MacLennan and Simmonds (1992).

The term 'echosounder' has been used to describe equipment which directs acoustic beams vertically downwards in deep waters (originally used for sounding water depth) while echorangers transmit beams near horizontally across shallow waters (Sissenwine *et al.* 1983), although the term 'echosounder' is used generically here. Acoustic energy is transmitted in pulses at a particular frequency, usually between 100 and 400 kHz, by means of a transducer producing a directional sound beam. Almost all echosounding transducers are composed of a series of elements which, through interference between the signals of the elements, usually results in the overall beam pattern incorporating a narrow sensitive cone on the acoustic axis, the main lobe, surrounded by several smaller side lobes, separated by signal null points. On encountering a fish target, in particular the swimbladder, the sound pulse is reflected in all directions and some is 'back-scattered' towards the transducer. The transducer detects the back-scattered sound (namely, the echo) and converts it to a quantified electrical signal. Where fishes are relatively well-spaced in the water column, echoes may be grouped as individual fish by software algorithms and the number of individual fish counted (MacLennan & Simmonds 1992). An alternative approach, particularly suitable where fish occur as shoals or cannot easily be identified as single targets, is the use of echo integration, whereby the total reflected energy received by the transducer is measured and reallocated into individual fish targets. The echosounding approach for measuring fish distribution and abundance is commonly referred to as 'hydroacoustics' and several detailed reference texts are available (MacLennan & Simmonds 1992; Brandt 1996). Hydroacoustic techniques strictly do not exclude other telemetric techniques which employ detection of acoustic transmitters attached to fish. Nevertheless, due to its widespread use in the literature 'hydroacoustics' is used here to refer to the reception of reflected artificial acoustic radiation and 'acoustic tracking/telemetry' is reserved for the detection of artificially generated sound from a tag or transponder.

Hydroacoustics is a well-established tool for fish studies in the marine environment where it has been used to assess fish stocks and examine fish spatial heterogeneity (Sissenwine *et al.* 1983; MacLennan & Simmonds 1992). Increasingly it has been used for similar purposes in fresh water (Brandt 1996). It has the advantage over many other techniques in not being intrusive, except for possible interference between the observing boat, where used, and the fish (Table 6.1). Noisy environments and entrained air or high suspended solid levels prohibit clear signal analysis and so, in fresh water, hydroacoustic techniques are generally restricted to lakes and non-turbulent areas of rivers. A key feature of echosounding is measurement of target strength, so that information on fish size and number can be obtained. Echosounders, particularly those of more advanced design, can measure individual target strengths over considerable distances (tens to hundreds of metres) and in fine detail.

Single-beam echosounders indicate the distance of a target from the transducer, but provide no information on the direction of movement or orientation of a fish. When using single-beam vertical transmission of sound, targets are in the same orientation (dorsal view) and sizing may be quite accurate, with correction applied for distance from the transducer. However, when single-beam transducers are used in horizontal orientation, reliable sizing is much more problematic due to the range of fish orientations (side-on, head-on, tail-on and variations between), which dramatically influence the strength of the echo, but with no immediate way of determining that orientation. Dual-beam echosounders, which utilise concentric rings of transducers that are alternately and separately activated and interrogated, can determine the distance of a target from the transducer and the distance off the axis of the sound beam, enabling correction for both distance and orientation, allowing improved sizing and fish discrimination characteristics (Fig. 6.2). Split-beam echosounders divide signal analysis of reflected sound to four quadrants of the transducer that are then compared through algorithms which examine the relative magnitude and timing of the signals (Fig. 6.2). Briefly, single targets can be discriminated from multiple targets by analysing the signal phase coherence, with each phase change corresponding to an additional target. Individual targets can be identified in true three-dimensional space within the sound beam, enabling accurate measurement of echo target strength and three-dimensional tracking through a limited volume of the water column insonified by the sound beam.

All types of echosounder rely on technical calibration of the relationships between echo strength for an object with known reflectance characteristics, usually a tungsten carbide sphere, in accordance with international practice (Foote *et al.* 1987; MacLennan & Simmonds 1992; Brandt 1996). The calibration process relies on on-axis sensitivity, time varied gain (TVG) function, equivalent beam angle (ψ) and standard target calibration. Details of such calibration procedures are given by MacLennan and Simmonds (1992). Beyond this, in order to determine the sizes of fishes which are being 'observed' acoustically, there is also a need to calibrate relationships between measured target strength for fishes (ideally live) of known size, species and orientation (Foote *et al.* 1987; MacLennan & Simmonds 1992; Brandt 1996). Echosounding does, however, have to be combined with live capture techniques to obtain information on species composition over the survey range (Table 6.1). Where the fish community is dominated by one or two species this is less problematic, as for hydroacoustic studies in Loch Ness, Scotland, where over 95% of pelagic fish are Arctic char *Salvelinus alpinus* (George & Winfield 2000) or in many rivers dominated by anadromous salmonids where most fish larger than about 30 cm are returning adults of one or a few species (e.g. Ransom *et al.* 1998).

6.2.3.1 Vertical beaming hydroacoustics

Until the early 1990s, freshwater hydroacoustics mostly used vertical beaming in relatively deep (>5 m), open water to determine abundance and distribution of pelagic fishes (Thorne 1979; Brandt 1996). Increasingly, in deep fresh water, hydroacoustics is being employed to answer questions concerning the factors influencing fish distribution and behaviour. A variety of studies have used hydroacoustics to examine vertical migratory behaviour of pelagic fishes in 'deep' lake environments, often in response to light variations and zooplankton distribution (Bohl 1980; Janssen & Brandt 1980; Levy 1991; Luecke & Wurtsbaugh 1993; Brandt 1996; George & Winfield 2000). Levy's (1991) study showed that diel vertical migrations of juvenile sockeye salmon *Oncorhynchus nerka* were intrinsically linked to the vertical movements of the major zooplanktonic prey *Mysis relicta*. In deep lake environments it has also been possible to record seasonal changes in vertical and horizontal distribution, reflecting seasonal migratory behaviour superimposed upon diel rhythms (e.g. Bohl 1980). With the advent of more sophisticated echosounders, capable of resolving near-bottom echoes, it is also possible to survey profundal fish species, using beams that are vertical or offset from vertical, although the ease of using such techniques is complicated by uneven lake bed characteristics or vegetation.

A detailed review of vertical beaming hydroacoustics with particular reference to freshwater habitats is given in Brandt (1996) and is not expanded upon here. Hydroacoustics using downward-oriented sound beams is inappropriate for many relatively shallow freshwater environments due to the small sampling volume and boat-avoidance behaviour by the fish. Use of horizontal beaming in fresh water with conventional transducers of older design is difficult due to the confounding effects of reverberations from surface and bottom boundaries, particularly from acoustic side lobes, resulting in low usable range (Kubecka 1996a). Following the commercial production of narrow-beamed transducers with negligible beam sidelobes, it has become possible to use echosounders horizontally in shallow waters with depths between 1.5 m and 5 m (Mesiar *et al.* 1990; Kubecka *et al.* 1992; Thorne & Johnson 1993; Kubecka 1996a). The development of echosounder transducers producing elliptical beams in cross-section has also been an important development for the use of hydroacoustics in horizontal beaming mode in shallow water, with the narrow axis of the ellipse orientated horizontally, giving a wider range over which to detect and, for split-beam systems, to track the fish's progress (Ransom *et al.* 1998). Horizontal sonar can be deployed in two sampling modes:

(1) by fixed location where the transducer is fixed at one location from which the sound beam is directed across a river or lake, and

(2) by mobile surveying where the transducer is attached to a rigid frame in front of a boat and the sound beam is directed across the river or lake while the boat is under way, usually close to the shore.

The majority of fish migration and behaviour studies using echo-monitoring devices in shallow fresh water have been confined to the former sampling mode but recent studies have shown the potential of the latter mode.

6.2.3.2 Fixed location studies for monitoring upstream migrations

Since the 1960s, fixed location acoustic techniques have been used to count, non-intrusively, upstream migrations of anadromous salmonids (mostly *Oncorhynchus* spp.) returning up the very large clearwater rivers on the west coast of North America and facing obstacles such as hydroelectric dams (Ransom *et al.* 1998; Thorne 1998). These systems are used to count and size fish passing a particular point and increasingly are routinely used to assist stock management of anadromous salmonids (Ransom *et al.* 1998). Fixed location hydroacoustic techniques have also been widely used at North American hydroelectric facilities to understand the behaviour of downstream-migrating juvenile salmonids in relation to turbine entrainment and bypassing facilities (Thorne & Johnson 1993).

Early (1970s) single-beam echosounders worked well for rivers with large migrations of big fish moving close to the river bank where the sound beam could be directed, but gave no information on direction of movement, fish speed or vertical distribution. Information on fish size was limited by the inability of single-beam systems to determine the orientation of a fish, which is required to enable conversion of target strength to fish size. The need to size the fish led to the use of dual-beam systems with narrow and wide beams from the same transducer and with signal processors which could detect peak echoes and discriminate single targets. Subsequently, the data could be processed to track individual fish and provide a mean target strength for each fish. With two dual-beam systems located side by side with slightly offset elliptical transducers and transmitting alternately, the direction of travel could be determined by observing which transducer beam the fish entered first. Most early studies on riverine dual-beam acoustic techniques were applied to adult salmonid migrations (Braithwaite 1971; Johnston 1985; Mesiar *et al.* 1990; Johnston & Hopelain 1990).

An important step forward for measuring fish sizes and echosounding of single targets was the development of the dual-beam digital signal processing system which detected, measured and saved for future analysis the echo signal peaks of single targets from both narrow and wide beams. A high proportion of fish targets in European lowland rivers at night were found to be single targets (Kubecka *et al.* 1992). The acoustic size or target strength of these single targets could be determined by methods described by Ehrenberg (1984). Thus, in rivers it became possible to distinguish whether the fish were moving as a dense shoal or loosely spaced individuals which could be sized. Furthermore, post-processing target tracking software was developed which grouped all the echoes from one individual target moving across the beam and gave a better feel for direction of movement (Johnston 1985). The appropriateness of the grouping of echoes by the tracking system could be checked on the echogram.

In the early 1990s, split-beam acoustic systems with the advantage of lower side-lobes and faster signal processors became commercially available for studies on salmonid migrations.

These are also capable of tracking fish targets in three dimensions in real time. In addition to the absolute direction of a fish's movement, the split-beam system gives its three-dimensional position within the sound beam, velocity of the fish target and less variable fish target strengths than other types of echosounder (Ehrenberg & Torkelson 1996; Ransom *et al.* 1998). Downstream movement of riverine debris on the surface of the water can be easily identified and their echo traces eliminated. There has been increased development of fully-automated fixed location hydroacoustic techniques for monitoring fish migrations, especially of salmonids, at dams (Steig & Iverson 1998) and at the cooling water intakes of some North American power plants which also incorporate high-frequency sound fish deterrents (Ross *et al.* 1993). Advanced systems may also incorporate modem options for controlling and transferring data from fixed hydroacoustic systems, particularly those acting as fish counters on rivers (Ransom *et al.* 1998). Advances in microcomputers, electronics and signal processing have made it possible to remotely operate and monitor hydroacoustic systems incorporating up to 16 transducers surveying discrete sites continuously or on programmable duty cycles, with substantial savings on human resources. Current developments include a system known as BATS (Behavioral Tracking System) in which robotically controlled split-beam transducers are mounted on high-speed dual-axis rotators and coupled to a computer (Hedgepeth *et al.* 1999). When a deviation of the target from a transducer's beam is detected, the computer uses a predictive tracking algorithm to realign the transducer. In the meantime, fish position, movement and target strength are recorded to hard disk. The main limitation of the system, which can also be used for tracking acoustic tags, is that narrow-beamed transducers require a relatively long time for scanning a radius of a certain length, making their use most suited to relatively shallow water.

Fixed location studies can be used to study diel behaviour and movements of fishes in lakes (Comeau & Boisclair 1998) and lowland rivers (Kubecka & Duncan 1998a) and, combined with field sampling, can provide information on feeding movements. Kubecka and Duncan (1998a) used this approach in a 24-hour study during June 1992 in the River Thames, England, where the fish community was dominated by three cyprinids, roach *Rutilus rutilus*, gudgeon *Gobio gobio* and dace *Leuciscus leuciscus*, and two percids, ruffe *Gymnocephalus cernuus* and Eurasian perch *Perca fluviatilis*. By siting one dual-beam horizontally directed transducer in the littoral zone (0.5 m deep, beaming to the river) and another in mid-river (*c.* 3 m deep, beaming across the river), the movements of fish were followed over 24 hours at hourly intervals and at three 1-m depth intervals in mid-river. The larger fish occurred in the littoral and the top depth stratum of the river during the night and early morning but moved to deeper layers during the day where they were not detectable by the horizontal beam as they were too close to the bottom. In the open river, all the fish oriented themselves to the river flow and swam upstream or downstream, as detected by the tracked angle of movement across the sound beam. In the littoral area, fish movement was more random in relation to river flow. As has been mentioned earlier (section 3.3), similar or diametrically opposed daily patterns of inshore–offshore migrations of fish have been recorded in several lakes and reservoirs by shore seining (Kubecka 1993) and in large rivers by electric fishing (Sanders 1992; Copp & Jurajda 1993; Baras & Nindaba 1999a, b).

6.2.3.3 The mobile horizontal hydroacoustic technique in shallow waters

The use of mobile horizontal-beaming hydroacoustics to survey the distribution and abundance of fishes in shallow freshwater habitats has rapidly developed over the last decade (Kubecka *et al.* 1992; Kubecka 1996a; Lyons 1998). At temperate latitudes the mobile horizontal-beaming hydroacoustic approach is normally carried out in summer when the fish are active, and at night when fish move up in the water column and shoals partially disaggregate (Kubecka *et al.* 1992; Gaudreau & Boisclair 1998; Kubecka & Duncan 1998a). Fish densities can be determined at short sampling intervals, enabling the characteristic patchiness of fish density distributions to be measured (Duncan & Kubecka 1996). The method has been used to examine abundance and distribution of fish communities over long stretches of various European lowland rivers such as the Thames (Duncan & Kubecka 1996; Hughes 1998), Trent (Lyons 1998), Yorkshire Ouse (Lucas *et al.* 1998), all in England, and the Elbe in the Czech Republic and Germany (Kubecka *et al.* 2000). Both dual-beam and split-beam echosounders have been employed, with the current trend towards the latter due to their better performance. Mobile split-beam hydroacoustic techniques have been used highly effectively to quantify daily inshore–offshore migrations of fishes in Canadian lakes (Gaudreau & Boisclair 1998). A common feature of all of these studies is that within apparently homogeneous lowland rivers or along the pelagic-littoral ecotone of lakes, fish are extremely clumped in distribution, with coefficients of variation for density estimates along a transect being high, often in excess of 100%.

Substantial seasonal changes in fish density distributions in a lowland river were demonstrated by Lucas *et al.* (1998) in a series of monthly mobile horizontal echosounding surveys over the same 27-km stretch of the Yorkshire Ouse, England. This study also demonstrated the potential impact of high river flow events on recorded fish densities in the river. During one night of September 1993 when river flow was five times greater ($63.5 \text{ m}^3 \text{ s}^{-1}$) than during the previous night ($12.7 \text{ m}^3 \text{ s}^{-1}$), mobile surveys showed that acoustically visible fish densities were three times lower. This was attributed to either downstream displacement of fish by high flows or, alternatively, that fish sought refuge where flows were reduced, on the bottom or in the margins, where they could not be detected by hydroacoustics. Hydroacoustic studies of fish aggregation during the main spring spawning period (April–June 1992) for cyprinids in the River Thames in 1992 revealed distinct, highly dense aggregations of fish of 3000–4000 fish ha^{-1} separated by areas of low densities (Duncan & Kubecka 1993). Duncan and Kubecka (1996) also used mobile surveys to provide evidence for fish aggregation at sewage outfalls in the Thames, and indicate their influence on fish diel migration patterns.

6.2.3.4 Fish sizes

Size discrimination of fish is relatively simple using vertically orientated echosounders, since the fish are all in the same approximate aspect (viewed from above) and appropriate calibrations are generally available (Love 1977; Foote *et al.* 1987; MacLennan & Simmonds 1992), although these can be influenced by tilt angle, body form and swimming behaviour of the fish (Brandt 1996). In order to produce a frequency distribution of the sizes of individual fish targets by horizontal beaming, fixed location is often necessary, since the orientation of the fish body or aspect being insonified cannot be tracked in a moving boat. Without tracking the fish across the sound beam, the acoustic sizes or target strengths cannot be converted to real sizes because the fish aspect (side, head or tail) is unknown. This is important, as the echo reflected from a side-aspect fish is much greater than the same fish in head or tail aspect. During mobile horizontal hydroacoustic surveys, regular fixed location studies with the boat anchored for a short period along the survey route enables the slope of the fish track across the horizontally-oriented beam to be estimated. In rivers fish tend to orientate to the flow (but not in lakes where there is little rheotropic stimulus), and mostly cross the horizontal acoustic beam perpendicularly to the acoustic axis as the boat moves upstream or downstream (Kubecka 1996a). Under these circumstances relatively accurate sizing can be achieved from a moving boat, particularly when the surveying modifications described by Kubecka *et al.* (2000) are employed. Most existing target strength to fish size relationships are for marine fishes, usually in dorsal aspect (Love 1977; Foote *et al.* 1987; MacLennan & Simmonds 1992). Increasingly conversions are available for freshwater fish species in side and other aspects (Kubecka & Duncan 1998b), although due to the range of echosounder types and frequencies used in freshwater research there is still a need for further calibration, especially on live fish. However, it should also be appreciated that where mixed fish communities predominate, detailed species-specific size calibration may be of limited value, and that fish aspect to the acoustic beam axis affects echo strength much more than differences between species for fish of similar size.

6.2.3.5 Further development of hydroacoustics

Hydroacoustic techniques provide an extremely useful approach for quantifying fish distribution, behaviour and migration in freshwater lakes and rivers as well as in estuaries and brackish water environments (Guillard 1998). Use of fixed-location split-beam hydroacoustics has become widely used for determining the magnitude, size, composition and timing of upstream salmonid migrations (Ransom *et al.* 1998). Although mobile hydroacoustic surveys have not yet been widely used for fish migration studies of lowland river fishes, the method's potential for surveys of whole stretches of rivers deeper than 1.5 m makes it an appropriate tool, particularly in combination with techniques such as radio tracking and direct sampling approaches, for studies of fish migration at the catchment scale.

Although maximum usable acoustic range may be short in many freshwater environments (usually 10–30 m for horizontal beaming), the total sampled volume during mobile sampling is very large, providing large data sets for statistical analysis and a continuous spatial record of absolute fish densities in the water column. It is important in shallow, freshwater environments that surveys include night work, since in these habitats this is when most fish species are active and somewhat disaggregated in the water column and so more easily insonified and recorded (Kubecka *et al.* 1992; Kubecka & Duncan 1998a). However, the influence of environmental factors on observed fish abundance is not yet fully understood. This is important because those fish which choose to remain close to the lake or river bed cannot always be discriminated from the bottom echo. There is now good evidence that lunar phase can have a major influence on the depth distribution of fish at night in fresh water and hence the number of echoes counted (Luecke & Wurtsbaugh 1993; Gaudreau & Boisclair 2000). As stated earlier, river flow and water temperature are also likely to be key factors influencing acoustic observation efficiency of freshwater fish in shallow habitats. Gas bubble production, especially by photosynthesising algae and submerged macrophytes, also causes difficulties in the use of the hydroacoustic technique in shallow water, as do the plants themselves.

There is substantial variation in the cost of hydroacoustic systems (Table 6.1). It is possible to obtain a high-quality scientific single-beam echosounder with signal processing software for as little as about $10 000, while single frequency dual-beam and split-beam systems cost about $30 000 and $40 000 respectively for those incorporating a single transducer, laptop computer and software. Increasingly, because of the refinement of narrow, small side-lobe, elliptical beams and advanced signal processing with split-beam hydroacoustic systems, these are the preferred choice for use in fresh water, especially when used in horizontal mode in shallow habitats. Most 'split-beam' systems can, in any case, normally be used in single-beam mode. Multi-transducer systems, multiplexed to a single computer station, cost in the order of $80 000 (for about five transducers, each split-beam transducer costs about $10 000), while similar systems incorporating robotically-driven transducers may be more expensive still.

6.3 Capture dependent methods

6.3.1 Variations in density and catch per unit effort

Variations of fish numbers in the same place over time usually result from birth, mortality, emigration and immigration. When mortality and birth (or recruitment) rates can be estimated, or when the time interval is short enough to neglect these parameters, movements of fish can be implied from measures of abundance of different lifecycle stages and species. These measures may be absolute estimates of density or may be related to fishing effort as catch per unit effort (CPUE) (Coles *et al.* 1985; Casselman *et al.* 1990; Cowx 1990; Hilborn & Walters 1992). These methods usually have low temporal resolution and they require large sample sizes and/or numerous sampling points. Additionally, capture efficiency may vary substantially between times of the day, seasons, sampling sites and environmental conditions, and to variable extents depending on the capture methods. There is a rarely a linear relationship between CPUE and absolute fish density. However, such methods may be useful for assessing movements where sampling is already being carried out for other purposes such as stock assessment, particularly where the sampling regime can be optimised to provide information at the appropriate spatial scale. Moreover, such methods may provide the only realistic option in many large tropical freshwater environments or for fish that are too small to tag or where hydroacoustic equipment cannot be employed.

6.3.1.1 Recreational and commercial catch data

Angling catch (creel) statistics are particularly influenced by environmental conditions (Hilborn & Walters 1992), as they depend on the appetite of fish, which is known to be extremely dependent on environmental conditions (Hickley 1996; Malvestuto 1996; Kestemont & Baras 2001). Commercial catch statistics are also influenced by environmental variations. Surveys of commercial catches (Fabrizio & Richards 1996) from fixed sites using specific fishing gear can also provide local indices of fish abundance, whereas it can be difficult to obtain reliable information of the location and effort of mobile fishing methods (e.g. gears set or fished from boats), able to fish wide areas. Recreational and commercial catch records can be useful indicators of local fish abundance and species composition (Hickley 1996), especially in situations where no other capture methods can be implemented efficiently, or for long-term ecological series. For example, Axford (1991) used angler catches to demonstrate that the percentages of migratory flounder *Platichthys flesus* in angling catches from the Yorkshire Derwent, England, dramatically decreased after the construction of a tidal barrage.

6.3.1.2 Trapping and netting

Passive capture techniques (entanglement and entrapment gears, see Hubert 1996 for review) can provide substantial information on patterns of movement by fishes, in that they rely on active movement by fishes for their capture. Nets or traps fished at particular sites can, therefore, be used to quantify migration or movement in a particular locality (Bénech & Peñaz 1995; Quiros & Vidal 2000). The use of ichthyoplankton nets in rivers provides the principal method of studying drift of eggs and larvae in rivers and is reviewed in detail by Kelso and Rutherford (1996). Another example of this approach is the use of fish-counting fences (also known as 'weirs') in shallow rivers, where migrating fish, usually salmonids, are channelled upstream or downstream into traps (Chadwick 1995; Hubert 1996). Catches made in this way enable assessment of migratory activity in relation to environmental factors and also provide fish for marking or tagging. Traps and fixed nets are also commonly used to assess patterns of activity on diel scales, often in relation to foraging behaviour (Rahel & Nutzman 1994; Hubert 1996). Rahel and Nutzman (1994) used bottle traps to demonstrate that central mudminnows *Umbra limi* regularly entered severely hypoxic bottom waters of a lake in northern Wisconsin, US, to forage on *Chaoborus* (Diptera) larvae during the day.

It must be recognised that, while increased CPUE in traps and static nets can be indicative of increased locomotor activity or movement into a particular area, susceptibility to capture is also strongly influenced by other features such as net avoidance (usually much higher at elevated light levels) and by the presence of conspecifics (Hubert 1996). Both static nets and traps often exhibit marked selectivity towards certain species or sizes of fish (Casselman *et al.* 1990; Hubert 1996). Passive netting (e.g. with gill nets) is more efficient in lakes and other slow-moving water bodies (Keast & Fox 1992) than in riverine environments, where floating or drifting vegetation or debris can damage the gear or interfere with sampling. However they also require fish being active, and most frequently show marked selectivity towards fish species or size. Traps can also be selective (e.g. Hubert 1996; Kubecka 1996b) but have, however, been used with some success in migration studies of non-salmonid fishes such as brook lamprey *Lampetra planeri* (Malmqvist 1980), three-spined stickleback *Gasterosteus aculeatus* (Harvey *et al.* 1997) and especially in freshwater eel (Anguillidae) migration studies (Jellyman 1977; Moriarty 1986; Baras *et al.* 1994b; White & Knights 1997). Traps are often operated at the upstream outlet of fish passes, where they can provide quantification of fish species which have successfully ascended, and information concerning the extent of and stimuli for upstream passage by several species (Larinier 1983, 1998; Baras *et al.* 1994a, b). However, such traps fail to quantify the proportion of fish unable to ascend the pass, or the behaviour of fish which may accumulate below the barrier but not ascend the pass. Their use may also stress or damage fishes, particularly more delicate species such as most clupeids and some cyprinids.

Active capture netting and trapping methods (e.g. seine nets, trawls, dredges, fish wheels) can give valuable information on local abundance of fishes, including sedentary species, but their efficiency is species- and size-dependent, and they do not perform equally well in all habitats (Hayes *et al.* 1996). For example, seine nets can hardly be operated in deep or fast-flowing water, while they show excellent performance in shallow, slow-flowing habitats such as lake shores or backwaters (Coles *et al.* 1985; Casselman *et al.* 1990). Kubecka *et al.* (1992) simultaneously used seine nets and acoustic assessment in shallow water and obtained excellent correlations between the two estimates. However, seine nets may have a lower efficiency for fish species with leaping habits, as may be the case in numerous assemblages (e.g. South American characiforms; Baras [unpubl.]). The use of active capture by netting, towed ichthyoplankton samplers and other approaches intended specifically for ichthyoplankton sampling is described by Kelso and Rutherford (1996).

6.3.1.3 Electric fishing

Because of its advantage over netting methods in terms of manpower and survey time reduction, electric fishing has become one of the most popular methods of catching fish in shallow freshwaters (streams, rivers, lake shores). However, it imposes additional risks to the fish as well as to the operators, and its efficiency is highly variable depending on water depth, conductivity, clarity, type of power generator, electrode size, operator experience, fish size and behaviour (Casselman *et al.* 1990; Cowx 1990; Harvey & Cowx 1996). Hence density or CPUE estimates derived from conventional electric fishing are definitely semi-quantitative and need correction factors, based on gear calibration, before any inference can be made on the actual population size (e.g. Bütticker 1992). Recent developments in electric fishing include boat-based multi-electrode arrays for sampling large rivers, use of ring electrodes and control boxes capable of sequentially energising each ring in the array (Cowx 1990). These improvements have reduced fish mortalities, increased capture efficiencies and enabled the capture of small fish (<20 mm) thus reducing the selectivity of electric fishing methods. Alternatively, point abundance sampling may provide a useful measure of relative abundance which can be quickly applied and enables changes in populations over short periods of time to be measured. This applies particularly well to larvae and 0+ fish (Copp 1989) and this technique has been used to demonstrate seasonal and diel shifts in distribution of lowland river fishes in natural and regulated systems (e.g. Copp & Jurajda 1993; Copp & Garner 1995; Garner 1995).

Most active capture methods, including electric fishing, may cause a fright bias (Cowx 1990; Hayes *et al.* 1996), of particular nuisance for point abundance sampling. Fixed electrodes or frames, energised by a.c. (Bain *et al.* 1985) or d.c. (Baras 1995a), have been used to reduce or suppress this bias, provided that a sufficient delay is allowed between the installation of the frame and its energisation. Additionally the high gradient voltage during operation of these frames results in minimal selectivity towards fish species or size. Using these methods, it is possible to obtain quantitative estimates at regular intervals within discrete microhabitats, based on *a priori* sampling design (Baras *et al.* 1995, 1996a). For example Baras and Nindaba (1999a, 1999b) demonstrated that the diel migrations of 0+ rheophilic cyprinids were size-structured. The main disadvantage of this electric fishing method is the time required to obtain sufficient sample sizes (frame installation plus recolonisation delay) and the small area sampled at each point, as well as the general limitations of electric fishing in deep waters.

6.3.2 Marks and tags

6.3.2.1 Mark recapture

Mark recapture is an important method in fisheries stock assessment because it allows the estimation of population size, mortality and independent assessments of growth rate, but it also provides information on patterns of movement between the sites at which the fish were marked and subsequently recaptured. Complementary to CPUE, this method has provided much important information on home range, migration and homing of diadromous and freshwater-resident fishes. Implicit in these studies are the assumptions that the marking procedure and mark or tag presence do not interfere with fish physiology or behaviour beyond a period of post-tagging stress, which should be as short as possible, and in every case determined as accurately as possible to validate such studies.

On some occasions, the outcomes of mark recapture studies have been validated by other techniques (e.g. parasite studies and radio-tracking of barbel *Barbus barbus*, Philippart & Baras 1996). However, often mark recapture approaches considerably underestimate the movements of fish, principally because of spatially limited recapture effort. Indeed, migratory patterns are inferred from recaptured fish only, and fail to take account of missing fish, which may have died but alternatively may have emigrated beyond the limits of sampling areas. Mark recapture also suffers from a rather poor temporal resolution of fish location (Gowan *et al.* 1994; Baras 1998), which generates further imprecision and underestimation of the propensity to migrate, especially for short-term migrations followed by homing behaviour. Additionally, the efficiency of recapture of tagged fish is likely to vary dramatically in time and space. In tropical floodplain systems fish may have a very high likelihood of recapture when crowded into small pools during the dry season(s) but are much less likely to be captured when they disperse during the rainy season(s). Other techniques, employing naturally occurring marks, may be used to reconstruct a fish's geographical origin or habitat use and may not necessitate recapture and are considered in section 6.3.3.2.

6.3.3 Types of marks and tags

Since the first tagging attempts, at least as early as the seventeenth century, there have been considerable developments in tag/mark design and analysis methods (Wydoski & Emery 1983; Parker *et al.* 1990; Nielsen 1992), along three main axes:

(1) increasingly sophisticated, longer-lasting and better-performing tags that enable the measurement of biological or environmental variables in relation to space use (e.g. telemetry tags);
(2) reduction in the size, bulk and weight of tags, enabling tagging of smaller fish and large numbers of fish in a short period;

(3) an increasing use of intrinsic or extrinsic marks naturally borne by the animal, and sophistication of identification and analysis techniques.

These improvements have helped to dramatically increase our knowledge of fish migration, and to reduce substantially the biases inherent to fish capture and/or recapture. Unfortunately, none of these attempts to achieve the perfect tag design proposed in Box 6.1 has succeeded completely. Some techniques are adapted to large samples but give low spatial and temporal resolution, others provide high resolution but for limited samples of large individuals only, whereas some others suffer from no environment or size restriction but overlook individual identification and resolution. Hence, studies of fish migration and spatial behaviour should focus their objectives, and select the most appropriate – or least maladapted – technique (Tables 6.1–6.4). This section briefly reviews the major types of marks and tags, advantages, limitations, and context of application for the study of fish migration, focusing on state-of-the-art techniques.

6.3.3.1 Synthetic extrinsic marks and tags

This grouping comprises those marks and tags which are applied to the fish, but which can only be identified by close inspection of the fish (Tables 6.2 and 6.3). External marks consist essentially of dyes or pigments (applied by balneation, aspersion, injection or tattooing),

---

Box 6.1 Characteristics of the ideal tag or mark (modified from Nielsen 1992)

(1) No risk of alteration during storage
(2) Easy and fast application, requiring no anaesthesia or specialised equipment
(3) High tagging/marking rate
(4) Minimum bulk and size, applicable to fish of all sizes
(5) Enabling individual identification
(6) Low cost
(7) 100% retention
(8) No alteration or fouling of tag material
(9) No effect on health, physiology, behaviour, performance and fitness of tagged fish
(10) No influence on the probability that the fish be preyed upon or captured by fishing or sampling gears
(11) No effect on fish appearance and saleability (e.g. for commercial fisheries)
(12) No need of specialised equipment or training for detection and identification
(13) No risk of confusion while identifying tag presence or code
(14) Requires no handling for post-tagging identification
(15) Can be detected and identified at any distance and at any time
(16) Should be relayed to the fish progeny

brands (generally freeze branding) or mutilation (McFarlane *et al.* 1990). Although they offer some possibilities for individual coding (e.g. freeze branding, combinations of fin clips or tattoo marks), they are better adapted to group and mass marking, essentially because of their low cost and fast application. However, external marks suffer from numerous specific drawbacks (Tables 6.2 and 6.3), including confusion with normal fish injury (fin clipping, freeze branding or dye) or melanophore development (black ink pigment) as well as risk for the operator (freeze branding) or the fish during the application (tattooing). Other chemical compounds (rare earth elements, calcein, tetracycline) can be used as internal markers of fish tissues, mainly in skeletal tissues (Hansen & Fattah 1986; Muncy *et al.* 1990), but their application may necessitate long-term stocking, and their identification frequently requires tissue biopsy or fish sacrifice (Table 6.2). Coloured latex or elastomer visible implant (EVI) tags (droplets of polymer compound, injected as a liquid into a transparent tissue, then polymerising as a solid compound within hours; anon. 1994) may be used to mark fishes by injecting the material into transparent tissues, so that the mark can be seen externally. Although different colours and sites may be used, these approaches are most-suited to batch marking since there is a limited number of recognition codes. Unfortunately tissue reactions may result in loss, break-up or overgrowing of latex or EVI tags. Although viewing with ultraviolet light aids detectability, EVI tags appear to be of limited value for mark recognition in excess of 1–2 years (Close 2000; Table 6.3). Additionally, both internal and external marks have a most limited individual resolution by comparison to most physical tags.

There are numerous types of external 'conventional' tags (reviews in Wydoski & Emery 1983; McFarlane et al. 1990; Nielsen 1992; Table 6.3), conspicuous in shape, size and colour, which can be fixed firmly to the fish's body (e.g. Petersen disk, Bachelor button, oval bars or straps) or dangle relatively freely at the end of a nylon loop or anchor attachment (anchor tags, Carlin tags). These are relatively cheap tags that can be individually coded, and have the major advantage of being easily detected by anyone. High tagging rates can be achieved for some of them, such as the Floy anchor tags (about 500–1000 fish per hour). Each tag type has specific limitations and drawbacks, consisting essentially of a reduction of growth for fish tagged with button or disk tags, and risks of entanglement, drag, and erosion of muscles for dangling tags. The retention rate and duration of external tag attachment varies greatly depending on tag type, operator experience, species and environment (structure, current). In general, the longer the migration or the faster the current, the higher the probability that external tags will be shed or affect fish performance, and this may cause substantial biases for studies of fish migration.

These limitations promoted the development of internal tags, some still being externally legible and requiring no specialised equipment for detection (Table 6.3), and others being truly internal and necessitating remote sensing (coded wire tags, electronic tags; Tables 6.1 and 6.3). Among the former category are visible implanted (VI) tags, which consist of small ($2.5 \times 0.9 \times 0.13$ mm) rectangular fluorescent polyester sheets coded with three alphanumeric digits and colour, and are usually inserted into the adipose eyelid of the fish (Haw *et al.* 1990; Table 6.3). Not all fish species have adipose eyelids as developed as those in salmonids, for which VI tags were originally designed, but other clear tissues including tissue overlying the operculum, jaws or dorsal neurocranium, or within fin membranes of large fish, can be used successfully. However, the body reactions to the implant, chiefly encapsulation or overlying by melanophores, may obscure the tag and render its detection or legibility more difficult,

Table 6.2 Types of chemical tags for use in studies of the spatial behaviour of freshwater fish.

	Dye and paint marks	Naturally occurring chemicals	Radio-isotope marks
Description	May include bathing fish in dye, use of dye-spot inoculators to batch mark fish, or utilising binary codes of marks to identify smaller numbers of individual fishes. Alcian Blue most appropriate dye in terms of recognition and longevity. Mercuric chloride introduced by hypodermic injection most effective for larval ammocoetes of lampreys. Subepidermal injections of acrylic paint may be used because they cause minimal disturbance and produce long-lasting marks and different colour combinations can be used to identify batches or individuals.	Determination of 'natural' chemical signatures in fish tissue, including radionuclides, stable isotopes and elements, or ratios between these. Depending on the chemical these may be sampled from tissue, scales or otoliths. Otoliths provide the most inert matrix, giving reliable temporal information.	Use of rare earth elements and their radioisotopes e.g. euridium (^{152}Eu and ^{155}Eu) to mark fishes by bathing them in solution. Screened by appropriate methods, ICPMS, atomic absorption spectrophotmetry (ASS), etc.
Advantages	Easy to apply, require a low handling time and can be used for small fish or early life stages. Do not normally affect fish behaviour.	Do not affect fish behaviour. Enable habitat reconstruction over fish's lifetime. Otolith sampling requires fish to be killed.	Easy to apply, require a low handling time and can be used for small fish or early life stages. May be 'diluted' with age. Do not affect fish behaviour.
Disadvantages	The main disadvantages are that individuals cannot be identified and, in the majority of cases, retention times are low. Small fish could be damaged by force of inoculators. HgCl$_2$ is too expensive and toxic for widespread use.	High cost of using many precise techniques such as laser ablation microprobe and inductively coupled plasma mass spectroscopy (ICPMS).	Cannot identify individuals. May be expensive to analyse large numbers.
References	Hart & Pitcher (1969); Schoonoord & Maitland (1983); Baras et al. (1996b); Gollmann et al. (1986); Knights et al. (1996)	Coutant (1990b); Nicolas et al. (1994); Campana (1999); Hobson (1999)	Hansen & Fattah (1986)

Table 6.3 Types of physical marks and passive tags for use in mark recapture studies of the spatial behaviour of freshwater fish.

	Conventional marks and tags	Coded tags	PIT tags
Description	Most widely used. Consist of fin clipping, branding and physical tags. Physical tags come in a plethora of shapes and sizes. Implanted tags (jaw tags) have been used in mark recapture studies with eels. Visible implanted (VI) tags, implanted underneath the epidermis of transparent tissues, popular for batch or individual identification of fish. Latex and elastomer, visible implanted tags are injected.	Coded wire tags of several forms are widely used. Consist of pieces of wire embedded subcutaneously. Can be detected at short range using a flux-gate magnetometer. Allow individual recognition by reading notches or alphanumeric code following dissection, or if present in translucent tissue. Notches can be read by X-ray but complicated by shadowing effects in head. In theory, retained for life, as the fish grows, but loss rates of up to 95% in some studies.	Passive integrated transponder (PIT) tag. Contains no power source and comprises a coil antenna and integrated circuit chip encapsulated in glass. Programmed with one of several billion possible codes. Usually implanted into body cavity. Tag transmits its code when energised by LF radio interrogation from an induction coil. In mark recapture studies, hand-held readers are used to scan and identify tagged fish.
Advantages	Can be used to individually mark large numbers of fish which can be identified and returned to the river for long-term studies. VI tags are 'internal' but readable and with combinations of colour and code, many fish may be individually identified.	Large numbers of fish can be quickly and easily tagged and tags have no effect on fish behaviour.	Collection of detailed information on large numbers of fish. Small, programmed with almost infinite number of individual codes; infinite life. Little effect on fish behaviour.
Disadvantages	Fin clipping – number of individual marking combinations is low. Branding – only used for scaled fish and deteriorates with time. Externally attached tags may cause disease and infection, attract predators or alter fishes' swimming ability and buoyancy control. Growth rates 50% lower in eels with jaw tags. EVI tags may have poor retention.	Cost of tagging and need to dissect fish to recover tag. Often needs independent marking, e.g. fin clip, to identify fish that are tagged, otherwise low tag recovery rates are likely.	Relatively expensive
References	Hunt & Jones (1974); Starkie (1975); Whelan (1983); Berg (1986a); Nielsen (1992)	Bergman et al. (1968, 1992); Jefferts et al. (1963); Buckley & Blankenship (1990); Haw et al. (1990); Crook & White (1995)	Prentice et al. (1990 a, b, c); Douglas & Marsh (1996)

especially by anglers or fishermen.

Coded wire tags (CWTs; Jefferts et al. 1963; Table 6.3) are the smallest tags available, among the cheapest ones ($US 0.1) and the most widely used, with more than 50 million fish tagged in this way each year. The tags are sections of stainless steel wire (0.25 mm in diameter), cut at a standard length (1.1 mm or 0.5 mm for half-length tags), magnetised, and marked with notches by a hand-held or automatic injector, enabling tagging rates of about 200 and 600 fish per hour, respectively. They can be injected into any part of the body, although the most frequent location is the ethmoid region of the cranium, with species-specific head moulds being used to standardise the tag positioning. Fully automatic injectors, requiring no fish handling at all, have recently been developed for salmonids and these enable tagging rates as high as 1500–3000 fish per hour for fish of similar body size. Tagged fish can be remotely detected as they pass through or near magnetic coils, but the identification of the tag code requires that the tag be extracted from the fish tissues and examined under a microscope ($20-30 \times$ magnification). This implies that the fish be sacrificed except for tags sited in translucid tissue (e.g. adipose eyelid) or non-vital parts of the fish body (e.g. adipose fin), or where X-radiography is used for identification. Most CWTs are used for group tagging, although individual tagging can be achieved by using sequential coding. Traditionally CWTs were notched to give binary coding, but they can now be obtained with etched alphanumeric codes which can be read in the field with a hand-held microscope.

Coded wire tags have no major drawbacks, except for the low temporal resolution that is characteristic of mark recapture techniques, the need for specialised equipment for tagging and detecting tagged fish, and the frequent need to sacrifice fish for batch or individual identification. Because of the tiny dimensions of CWTs, their impact on fish health and performance is almost restricted to capture and handling. However, there has been recent evidence that CWTs injected into the skulls of small juvenile pink salmon *Oncorhynchus gorbuscha* may damage their nervous system, and especially the olfactory rosettes, and later modify the orientation capacities of these fish during their homing migration (Habicht et al. 1998).

6.3.3.2 Natural intrinsic and extrinsic marks

Instead of tagging a fish to determine where it goes to, one can take advantage of natural marks carried by the fish to identify its origin and to reconstruct its migration path or spatial segregation (Table 6.4). These include truly intrinsic characteristics (genome, morphology), as well as extrinsic marks becoming incorporated in (chemical elements and isotopes, Table 6.2) or fixed to the animal during its migration or residency time (parasites). Key features of all these marks are the absence of marking/tagging bias, the need for specialised equipment for detection or identification, and their indirect nature. Indeed, migration can only be implied from a database on the spatial distribution of the measured features, the size and completeness of which frequently affects their accuracy.

Morphological characteristics (meristic variables, pigmentation patterns) are usually of extremely limited value to studies of spatial behaviour since they are frequently subject to ontogenetic changes and the influence of environmental conditions (Table 6.4). Persat (1982) demonstrated the identification of individual European grayling *Thymallus thymallus* from body markings and scale patterns. Symbionts or parasites that induce no major pathogenic sublethal effects can also be used, alone or in combination with other techniques to map

Table 6.4 Types of biological or natural tags for use in studies of the spatial behaviour of freshwater fish.

	Parasitic markers	Morphological markers	Genetic markers
Description	Parasites are specific to particular habitats and leave marks on host that can later be used for identifying groups or stocks of fish and for determining migration patterns. Mostly used in marine environments.	Meristic counts, pigmentation marks, differences in the shape and size of body parts or scales, etc. used to identify individuals or groups of individuals. Little used for studying migration in freshwater fish species. Species for which identification of individual fish has been demonstrated include *Esox lucius* and *Thymallus thymallus*.	In this method different populations can be distinguished by examining the loci of individual genes. Individuals can be recognised by examining hypervariable mini- or microsatellite DNA.
Minimum fish size	None	YOY fry (must be pigmented)	None
Advantages	Natural, low-cost, can be used on large water bodies.	Natural, low-cost, do not alter fish behaviour	Natural, do not alter behaviour of individuals.
Disadvantages	Time needed to research whether parasite can be used; cannot recognise individual fishes; requires well-trained personnel to identify parasites. Limited use where there is high stocking. Only suitable for the separation of relatively self-contained fish stocks.	Subject to ontogenetic and environmental influences (may change over time). Massive database of images required, together with objective discrimination system.	Expensive for studies involving large numbers of individuals, generally only suitable for determining population mixing or large-scale movements involving substantial components of the population.
References	Kabata (1963); MacKenzie (1983); Buckley & Blankenship (1990)	Fickling (1982); Persat (1982); Wydoski & Emery (1983); Buckley & Blankenship (1990)	Allendorf *et al.* (1975); Avise *et al.* (1986); Carvalho & Hauser (1994)

the habitats used by fish during their migration (e.g. Margolis 1982; Table 6.4). Analyses of genetic markers (Carvalho & Hauser 1994; Park & Moran 1994; Table 6.4), offer a much broader range of perspectives, either at the population (enzymatic polymorphism, mitochondrial DNA) or individual level (variable number of tandem repetitions [VNTRs]: i.e. hypervariable mini- or microsatellites). For example, VNTRs now make it possible to map the dispersal of the progeny of an individual fish. Coupling this analysis with tracking of the parent fish migration pattern would enable determination of the extent to which gene flow is dependent on migratory behaviour of the parents and/or progeny. A major advantage of the genetic approach is that material can be obtained from blood or fin tissue without killing the fish. Mitochondrial DNA analyses are in routine use for the determination of stock structure, and where populations are not completely isolated they provide valuable information on the level of dispersal and the integrity of migratory populations (Avise *et al*. 1986, 1990; Carvalho & Hauser 1994). Microsatellite genetic analysis in combination with mark recapture has been used to independently demonstrate limited dispersal of two species of barbel *Barbus barbus* and *B. meridionalis*, and its influence on the maintenance of a hybrid zone (Chenuil *et al*. 2000). Microsatellite analysis has also been used to investigate the role of dispersal in rapid speciation of haplochromine cichlids (Van Oppem *et al*. 1997) and, together with mitochondrial DNA analysis, to reconstruct recolonisation routes of Eurasian perch *Perca fluviatilis* into Norway following the last glaciation (Refseth *et al*. 1998).

Screening the hard parts of fish (cartilage, bone, scales, otoliths) for trace elements or isotopic ratios gives the opportunity to reconstruct habitat conditions during the fish's life (Coutant 1990b; Table 6.2). The chemical composition of water, sediments and food varies in space and time and, as a fish grows, some of these elements are incorporated into hard structures and may provide distinct chemical signatures associated with the fish's habitat (Coutant 1990b). The degree to which the composition of hard structures reflects past exposure to particular elements or isotopes depends on the incorporation of these and on the stability of the matrix into which they are incorporated. Scales can provide a valuable and convenient source of material without the need for killing the fish and have been used successfully in a number of microchemistry studies (Pender & Griffin 1996). However, scales can be replaced and material can be reabsorbed from original scales, as well as from bone and cartilage.

The crystalline structure of otoliths, largely formed from aragonitic calcium carbonate laid down on an organic matrix, appears to be relatively inert and retains a fine structure and chemistry that reflects deposition of material over the fish's lifetime (Campana & Neilson 1985). The majority of inorganic material laid down in otoliths and other hard parts of bony fishes is calcium, but other minor or trace elements are incorporated at levels which tend to reflect their availability and ultimately depend on the relationship between ambient water and otolith chemistries (Secor *et al*. 1995). Among the most significant of these minor constituents from the viewpoint of markers of spatial behaviour, is strontium, which is much more abundant in saline water than in fresh water and substitutes for calcium. Although diet and temperature may influence Sr/Ca ratios, it is thought that around 85% of variability in Sr/Ca ratio is due to changes in salinity (Secor *et al*. 1995). Ratios between strontium and calcium have, therefore, become important indicators of diadromous behaviour at the individual fish level and are providing exciting opportunities for studying lifetime spatial behaviour of a wide range of species. For European eel *Anguilla anguilla* and Japanese eel *Anguilla japonica*, historically regarded as obligately catadromous species, these techniques have demonstrated that substantial population components of these species may never enter fresh or brackish water (Tsukamoto *et al.* 1998). For freshwater fishes such as common bream *Abramis brama* and pikeperch *Stizostedion lucioperca*, fine-scale information is now emerging on the migratory behaviour of some population components between fresh water and brackish environments (Kafemann *et al.* 2000).

Barium, another divalent metal, with an ionic radius similar to that of calcium, also effectively replaces calcium. Barium tends to occur at higher relative concentrations in fresh water than seawater and can therefore often be used as an indicator of freshwater residence, although there may be substantial variations in barium levels between catchments, associated with differences in catchment geology. From analyses of barium and strontium levels in scales, Pender and Griffin (1996) demonstrated that barramundi *Lates calcarifer* from around the Mary River, northern Australia, display facultative catadromy and suggested that marine stocks remote from freshwater inflows probably have no freshwater phase. As well as being significant for our understanding of migratory behaviour, such findings have major implications for issues of resource allocation between freshwater sport fisheries and marine commercial exploitation. A variety of other trace elements, including heavy metals such as copper, lead and cadmium, often associated with industrial pollution, have also been suggested as markers in hard parts, although biologically unregulated elements such as lead and cadmium are likely to be most useful (Coutant 1990b; Campana 1999). Although whole-scale/otolith wet analysis by a variety of spectroscopic techniques can be used (Coutant, 1990b; Pender & Griffin 1996) fine-scale sampling by one of several methods, combined with incremental age analysis, is increasingly preferred to enable detailed habitat reconstruction to be achieved (Coutant 1990b; Tsukamoto *et al.* 1998; Campana 1999; Kafemann *et al.* 2000). Micro-drilling methods may be used to obtain small samples of material at discrete locations of scales/otoliths, but *in situ* use of a laser ablation microprobe and inductively coupled plasma mass spectrometry (LA-ICPMS) is increasingly the preferred (though expensive) method, enabling fine-scale, precise measurements of most elements and even isotopes, with minimal risk of sample contamination (Campana 1999).

Determination of variations in stable isotope ratios of elements such as carbon ($^{13}C/^{12}C$), nitrogen ($^{15}N/^{14}N$) and sulphur ($^{34}S/^{32}S$) in body tissues is increasingly well-known in ecology as a technique for examining trophic structure of communities, but is now also proving to be a valuable method for assessing the history of space use by individual animals including fishes (Hobson 1999). Dietary shifts and/or changes in trophic status are often intimately associated with habitat shifts and ontogeny, and may result in distinct changes in ratios of stable isotopes. This approach has been used to examine migratory fish in the Mackenzie River, Canada (Hesslein *et al.* 1991), and appears suitable for larval settlement studies (Herzka & Holt 2000). Similarly, the oxygen isotope fractionation ($^{18}O/^{16}O$) in tissues can be used as an indicator of growth in low and high salinities and at different temperatures (Coutant 1990b), while isotope ratios of elements such as strontium ($^{87}Sr/^{85}Sr$) in hard parts may also be useful habitat indicators (Campana 1999). Characteristic radionuclide signatures from nuclear power plants may also be used as markers of spatial behaviour of freshwater fishes (Nicolas *et al.* 1994).

6.3.4 Electronic tags – telemetry

The development of electronic tags has proved to be one of the most important advances for studying fish behaviour and migration (Trefethen 1956; Stasko & Pincock 1977; Baras 1991; Priede & Swift 1992; Winter 1996). They enable rapid, long-term and often long-range identification and positioning of fishes, with high temporal and spatial resolution, including in environments poorly accessible to human observers (Table 6.1). Lucas (in press) reviewed the development of telemetry studies of fish in fresh water on the basis of analysis of 498 references covering four decades between 1956 and 1996. The number of studies per year has grown steadily over this time, and although most studies have occurred in the northern hemisphere, a greater range of fish taxa have been studied in recent years (Fig. 6.3). While the purpose of most telemetry studies of fishes is to elucidate their movements, home range or habitat use, the technique has also increasingly been used in the assessment of a wide variety of specific problems associated with fish migration, behaviour and space use including evaluation of fish responses to obstructions (e.g. Lucas & Frear 1997; Hinch & Bratty 2000), establishing the efficacy of fish pass programmes (e.g. Travade et al. 1989; Bunt et al. 1999), quantifying survival during migration (Jepsen et al. 1998), measuring energy expenditure during foraging and migration (e.g. Lucas et al. 1991, 1993a; Hinch et al. 1996), identifying the responses of river fish to artificial freshets (e.g. Thorstad & Heggberget 1998) and acid episodes in rivers (e.g. Gagen et al. 1994), and obtaining information for specific conservation programmes (e.g. Moser & Ross 1995).

6.3.5 Passive electronic tags

Passive integrated transponders (PIT) tags, are small (currently commercially available as small as 10.3 mm long × 2.1 mm in diameter) glass cylinders comprising a coil and an integrated circuit, programmed to transmit one of some billions of codes (Prentice et al. 1990a, b, c; Box 6.1 and 6.3; Tables 6.1 and 6.3). Passive integrated transponders are interrogated with the field of an induction coil, commercially available at 125, 134 or 400 kHz, which energises and causes a tag to retransmit its code to the reader. Since PITs contain no power source, their life is theoretically infinite, and because their identity is electronically coded, they enable a fast and reliable identification of individual fish with minimum handling. Passive integrated transponders are normally used as internal tags that can be implanted into the peritoneal cavity of fish as small as 1–2 g, using syringe injectors (Prentice et al. 1990a) or surgery techniques (Baras et al. 1999), or into the musculature of larger fish (Bergersen et al. 1994). Automated tagging systems, consisting of an electronic balance, digitiser, tag detector and automatic tag injector activated by high pressure carbon dioxide have been developed for salmonids, and these enable tagging rates as high as 150 fish per working hour (Prentice et al. 1990c; Achord et al. 1996). Implanted PITs are unlikely to be shed once the incision has healed (3–15 days depending on fish species, age, size and temperature; Baras et al. 1999; Baras et al. 2000a), except for some rare cases of transintestinal expulsion, but these seem to be restricted to Siluriformes (Baras & Westerloppe 1999).

With respect to fish migration studies, PITs have often been used in mark recapture programmes (Douglas & Marsh 1996; Ombredane et al. 1998). However, because transponder interrogation and detection relies on inductive coupling, they can also be detected at some

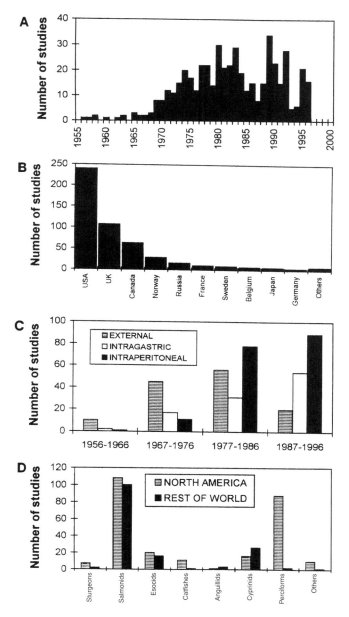

Fig. 6.3 Growth of telemetry studies on freshwater fishes, based on an analysis of four decades of publication, 1956–1996 (a), countries where work was carried out over this period (b), taxa studied over this period (c), and changes in telemetry tag attachment techniques (d). Further information is provided in the text. Modified from Lucas (in press).

distance (*c*. 20 cm for the smallest PITs; *c*. 50 cm or more for larger [32.5 × 3.8 mm] transponders; Castro-Santos *et al.* 1996), and their code automatically stored, together with passage time, by a PIT-tag data entry station (Prentice *et al.* 1990b; Brännas *et al.* 1994; Adam & Schwevers 1997). Most remote detection antennae are of tubular or square construction, and impose that fish swim through them, restricting their use to the monitoring of fish passage in special facilities like fish passes. More recently, flatbed antennae (1.0×0.3 m) have been developed (Armstrong *et al.* 1996) and these can be applied to the study of fish migration in more open environments (Armstrong *et al.* 1997) as well as fish passes (Lucas *et al.* 1999). A recent study using 23.1×3.9 mm PIT tags, together with loop antennae in the form of two adjacent 4×1.2 m rectangular frames at an 8-m wide stream site has demonstrated the ability to monitor movements of PIT-tagged fish across the width of whole streams with high efficiency (G. Zydlewski, K. Whalen, A. Haro & S. McCormick [pers. comm.]). Roussel *et al.* (2000) have described a new portable reading unit, incorporating a chest-mounted palmtop computer, a reader and 12 V battery enclosed in a backpack, connected to a 60-cm diameter coil mounted on a 4 m pole. The antenna is moved above or within the stream surface to search for tags and when used with 23.1×3.9 mm PIT tags the equipment has a range of 1 m.

It is worth noting that the reading efficiency of automatic stations may be less than 100%, for several reasons. Because inductive coupling depends on the respective orientations of the detection antenna and tag coil, PITs which are misaligned at the time of tagging or become misorientated during fish growth may compromise the reading efficiency (Pirhonen *et al.* 1998; Baras *et al.* 2000a), especially when the tag is close to the maximum detection range of the antenna. Irrespective of distance or tag orientation, the passage of several fish at a time may exceed the detection capacity of the data entry station (about 8 and 25 fish s^{-1} for 125 and 400 kHz, respectively), and not all individuals may be detected. Similarly, detection can be impaired when the residence time at the detector is less than the minimum time for interrogation and detection, and this may happen at fast swimming speeds (e.g. 5–7 ms^{-1} through a 1-m long antenna tunnel; Castro-Santos *et al.* 1996). Conversely, a fish sitting above or inside an antenna coil may block the detection system and prevent other passing fishes from being

---

Box 6.2   Propagation of actively radiated sound and radio

Acoustic signals propagate omnidirectionally in water with homogenous sound velocity (Voegeli & Pincock 1996). The spheric propagation mode implies that spreading losses (L_s) obey the inverse square law (i.e. $20 \log R$, where is R is range in metres). Absorption (a) losses are proportional to frequency (e.g. 0.3 dB km^{-1} at 30 kHz and 20 dB km^{-1} at 300 kHz in fresh water). This makes lower frequency of greater potential use for tracking. However, because the diameter of the ceramic transducer is inversely proportional to its resonant frequency, frequencies lower than 30 kHz are inpractical for most tracking purposes, except for very large fish. When an acoustic signal strikes a boundary between two media with different velocities (e.g. bubbles, salinity gradient), signal scattering occurs as a part of the energy is reflected. Similarly, acoustic signals are progressively bent in temperature gradients and reflected by thermoclines.

Because the reception range of an acoustic transmitter is a function of the signal to noise ratio, any factor increasing ambient noise reduces the operational detection range of an acoustic transmitter (e.g. increases of 10–15 dB noise in shallow or coastal water, 15–25 dB noise during rain; 18 dB or more for boat and water current). Additional noise can come from the equipment itself, but it can be reduced by proper hydrophone design and shielding of cables. Finally, demodulation from acoustic to audible signal requires a signal to noise ratio (S/N) of 6 dB for manual tracking and of about 12 dB for automatic positioning stations which cannot afford to miss any pulse. For example, if we operate a 300 kHz transmitter with an external radiated power of 161 dB re 1 µPa at a distance of 1 m (0.1 acoustic watt) in freshwater with moderate noise (N = 25 dB) and water velocity (V: loss = 18 dB), and try to detect it from 1 km of distance, the S/R ratio of the receiver will be:

$$S/R = 161 - 20 \log(1000) [L_s] - 0.02 \times 1000 [a] - (25 + 10 \log(1000)) [N] - 18 [V]$$
$$= 8 \text{ dB}$$

Radio signals also propagate omnidirectionally in the water but only the wave vectors almost ($< \pm 6°$) perpendicular to the air–water interface will cross it and travel through the air and be detected by an aerial antenna. Hence, the key variable is depth, and the received signal strength (RSS, dBm) from a radio transmitter in an open environment is given by (Velle et al. 1979):

$$RSS = ERP - Loss_{water} - Loss_{if} - Loss_{air} + Gain_{ra} - Loss_{tr}$$

where ERP is the external radiated power of the transmitter (dBm^{-1}); Loss$_{water}$ (dB) = $d_w \times (\partial L_w / \partial d_w)$, with d_w = depth of the transmitter (m) and $\partial L_w / \partial d_w$ = propagation loss depending on conductivity and frequency (e.g. c. 1.75 dB m^{-1} for each increment of 100 µS cm^{-1} at frequencies from 30 to 170 MHz); Loss$_{if}$ is the loss at the air–water interface (c. 30 dB); Loss$_{air}$ (dB) = 17.7 + 20 log (d_{air} / wavelength), with d_{air} being the linear distance between the transmitter and receiving antenna (m); Gain$_{ra}$ is the gain of the receiving antennna (dBd); and Loss$_{tr}$ is the loss in the transmission line from the antenna to the receiver (e.g. 0.0317 dB m^{-1} for a standard coaxial RG$_{58}$ cable). As for acoustic tags, the S/N ratio required for automatic pulse detection requires an additional 12 dB safety margin, and signal attenuation is greater for higher frequencies. It is worth pointing out that the use of internal coil antennae, increasingly preferred for animal health and welfare, cuts down the reception range by a 30–40% margin. For example, a transmitter with an ERP of 40 dBm^{-1} (i.e. 40 dB less than 1 mW) at 5 m depth can be detected by an automatic listening station with a 3-element Yagi antenna at about 400 m.

Because signal attenuation with increasing depth and conductivity is proportional to the carrying frequency of the radio signal, low frequencies (30–50 MHz) may be preferred in deep and/or highly conductive environments. However, the loss of range at high frequencies (>100 MHz) may be compensated for by the smaller size of directional antennae (e.g. 1 m for a Yagi antenna at 150 MHz, versus 3 m at 50 MHz) and their resultant ease of use for mobile tracking. Increased range can be obtained by using transmitters with external whip antennae instead of integral coiled antennae, but this may cause additional problems to fish health (see Box 6.3).

detected. This implies that antennae should ideally be installed so as to discourage fish from hiding or sitting close to them, and that the length of antenna tunnels should be adapted to the maximum expected swimming speed of fish, as well as to the numbers of tagged fish.

6.3.6 Signal propagation and detection of battery-powered transmitters

The detection of acoustic signals (usually at 30–300 kHz), from battery-powered transmitters or transponders requires a hydrophone to be immersed in the water, whereas VHF radio signals (usually at 30–170 MHz) can be detected by underwater or aerial antennae, making it possible to radio-track from boat, land or aircraft (McCleave *et al.* 1978; Table 6.1). Ultra-high-frequency (UHF) radio signals are rapidly attenuated in water but allow high rates of data transfer to orbiting satellites, including those of the ARGOS (Advanced Research and Global Observation by Satellite) system which allows wildlife telemetry applications. If a high-power UHF transmitter can gain at least three uplinks to an overpassing ARGOS satellite, its position can be located by Doppler shift, and stored data can be transferred (Priede 1992; Winter 1996). Individual acoustic or radio transmitters can be identified from their specific frequency, pulse rate or pulse coding sequence (Winter 1996). The smallest acoustic transmitters (also named pingers) usually are slightly larger than the smallest radio transmitters (0.5 g in air, $c.\ 12 \times 5 \times 5$ mm). They have higher power requirements and shorter life than equivalent-sized radio transmitters, but they can give greater accuracy while positioning the fish (in theory $c.\ 0.1$–0.2 m with advanced 3-D positioning systems versus $c.\ 0.5$–1.0 m for radio tags, though achieved precision in field conditions is often an order of magnitude less).

Acoustic and radio signals are variously affected by environmental features, which restrict their optimum application to particular sets of environmental conditions (see Box 6.2). Radio-tracking has become the preferred method for use in shallow (usually <5 m), low conductivity (usually <500 μS cm^{-1}) lakes, ponds, rivers and streams. In slow, deep rivers, lakes and reservoirs, and many lowland or brackish waters with high conductivity, acoustic tracking has continued to provide the most appropriate tracking technology. Combined acoustic and radio transmitters (CARTs), switching automatically between these modes using internal clock circuitry or a salinity sensor, have been developed for diadromous species (Solomon & Potter, 1988; Smith & Smith, 1997).

In some deep, highly conductive and noisy environments (e.g. deep and fast-flowing estuaries) or where complex physical structures occur (e.g. harbours), neither acoustic nor VHF radio signals can be detected confidently. In these circumstances, electromagnetic tags employing low-frequency radio waves (LF radio = 30–300 kHz) may be appropriate. Most passive integrated transponders work in this frequency range, but battery powered transponders or battery powered transmitters can also be useful (Table 6.1). While PIT tags have very low ranges, active transponders, interrogated by inductive coupling, in the same way as PIT tags but powered by a battery transmitting the signal to the receiving antenna, can achieve a range of a few tens of metres. Breukelaar *et al.* (1998) used this technology to identify the migration routes of sea trout *Salmo trutta* in the Rivers Rhine and Meuse delta in the Netherlands. Because of their extremely large size and weight (85 mm long $\times$ 15 mm in diameter, 25 g in water), these tags are restricted to large fish (>1.4 kg). However, battery powered coded LF radio transmitters such as those developed for tracking decapod crustaceans on artificial reefs have been made with dimensions of 40 mm diameter and 10 mm depth (Collins 1996; Smith *et al.* 1998) and, with modification, may be appropriate for localised tracking of moderate-sized fish under the adverse conditions described above.

6.3.7 Transmitter positioning

Following the line of the strongest signal is the most frequent way of tracking fish, information from which can later be plotted on maps using landscape marks or GPS technology. With both acoustic and radio signals, horizontal positioning can be obtained by triangulation, using directional hydrophones or antennae (e.g. loop, Yagi, Hadcock), with a minimum of two bearings, preferably taken at 50–120° from one another (Winter 1996).

While using acoustic signals, one may also achieve automatic positioning of a fish, and thus track its migration path more accurately, by measuring the relative arrival times of acoustic signals to a fixed or movable array of omnidirectional hydrophones (Hawkins *et al.* 1974; Lagardère *et al.* 1990). A minimum of three hydrophones give 2-D positioning, and four hydrophones enable 3-D positioning (inverse principle of hyperbolic navigation). The positioning of fish with omnidirectional radio antennae is unpractical, as radio signals travel much more rapidly (3×10^8 m s^{-1} versus *c*. 1.5×10^3 m s^{-1} for acoustic signals), and the measurements of signal time arrivals lack accuracy (e.g. a 1 ms difference in signal arrival time [current resolution of most systems] corresponds to *c*. 1.5 m for acoustic signals, and to 300 km for radio signals). Although multiple automated radio stations, each with a rotating directional antenna, may be used to fix radio transmitter positions by triangulation, precision is much poorer than with automated acoustic systems due to the limited directional sensitivity of radio antennae, including Yagi antennae. At restricted sites, such as dams, single radio receiver stations can switch between several antennae to localise transmitter position on the basis of relative signal strength (e.g. Gowans *et al.* 1999a). While using combined transmitters producing both radio and acoustic signals, fish can be located from a single bearing in polar coordinates, using a directional hydrophone or antenna to determine the bearing, and measuring the distance from the difference between the arrival times of the radio and acoustic signals (RAFIX system; Armstrong *et al.* 1988).

An alternative to active tracking and automated position fixing is to monitor the passage of fish at discrete sites with automatic listening stations (ALSs) coupled to fixed antennae or sonar buoys (Solomon & Potter 1988); this may be of great value in fish migration studies. This approach can be used to identify 'macro'-scale progress of fish along river systems, and especially at obstructions (Marmulla & Ingendahl 1996; Lucas & Frear 1997) or in remote environments (Eiler 1995). These ALSs may be remotely interrogated by modem, radio or satellite (Eiler 1995), saving time and effort for locating fish between sites covered by ALSs.

6.3.8 Telemetry of intrinsic and extrinsic parameters

Electronic transmitters can also be equipped with physiological or environmental sensors that change the pulse rate or pulse width of the transmitter proportionally to the measured values. A similar approach can be used for archival (data storage) tags that store information until the tag is recovered or the data is transmitted to a satellite (see section 6.3.11). Telemetry of environmental variables from fish can provide much information regarding responses to physical factors such as temperature (Coutant 1969; Snuccins & Gunn 1995), salinity (Priede 1982), depth (Williams & White 1990; Gowans et al. 1999b) and oxygen concentration (Priede et al. 1988), and is of great significance in seeking to understand the behaviour of fish in relation to natural variations in environmental conditions, as well as the influence of anthropogenic disturbances to the freshwater environment. Comparing internal and external temperatures obtained from a two-channel transmitter equipped with two temperature probes (one external and one internal), can also provide useful information on digestive processes, chiefly through the measurement of the heat increment resulting from specific dynamic action (SDA). Tunas are known to warm their viscera by several degrees after feeding (Gunn et al. 1994), but recent findings suggest that differences of 0.5–0.7°C can be measured in typical poikilothermic fish (Baras [unpubl.]).

Pressure sensors coupled to electronic tags provide information on ambient pressure, and thus on the swimming depth of the fish (up to 3000 m), but they require stability of calibration, which can be a problem in shallow aquatic environments (Williams & White 1990). Light sensors sufficiently sensitive to detect light down to several hundred metres in clear water can also be incorporated into electronic tags. While using archival tags for fish making long-range migrations, these sensors can provide key information on the longitude (time of sunrise and sunset) and latitude (daylength), provided water turbidity is homogenous and swimming depth is measured simultaneously. For these reasons, their use is best for pelagic fishes in oceans or large lakes rather than in rivers where turbidity changes, vegetation or physical shelters may influence ambient light intensity.

Simple tilt-switch transmitters which vary pulse rate with changes in the fish's body attitude provide an excellent means by which to quantify the activity of fish (e.g. barbel *Barbus barbus*, Baras 1995b; European eel *Anguilla anguilla*, Baras et al. 1998). Such transmitters have also proved effective for recording benthic feeding by fishes such as tench *Tinca tinca*, which tip up to feed on benthos (Perrow et al. 1996). Accelerometer transmitters, the output of which is directly proportional to fish movement, permit a finer discrimination between different behaviours such as redd cutting, quivering and aggressive charges in spawning Atlantic salmon *Salmo salar* (Johnstone et al. 1992; Økland et al. 1996). These transmitters can use either frequency modulation in response to changes in signal amplitude (Johnstone et al. 1992) or mercury droplet motion sensors which trigger additional pulses during acceleration events (Eiler 1995; Økland et al. 1996). For assessment of specific behaviours such as feeding, quivering and charging, these tags require a detailed series of calibrations of the correspondence between behaviours and transmitter output. Swimming direction can now be measured by sensors measuring compass heading with a 1° accuracy, provided the sensor is kept no more than a few degrees off horizontal, which is a major limitation for studying the vertical migrations of fish.

Further advances in telemetry of heart rate or electromyograms (EMGs) have enabled a much better appreciation of the internal status and physiology of free-swimming fishes. Physiological telemetry is increasingly being used as a method of estimating energy costs of fishes in the natural environment (Rogers et al. 1984; Lucas et al. 1991, 1993a; McKinley & Power 1992). Recent studies using EMG telemetry have identified the existence of costly localised activity (Demers et al. 1996) and evaluated the costs of migration through areas of river with different velocity regimes, including those for which passage is difficult (Hinch et *al.* 1996; Hinch & Bratty 2000). Further information on this subject is given in sections 2.3.3 and 2.3.8 and Box 2.2.

6.3.9 Attachment methods

Any method for studying fish spatial behaviour in the natural environment should not itself lead to changes in the behaviour or physiology of the individual being studied, and this applies particularly to telemetry tags, which exceed all other types of marks and tags in size and weight. With few exceptions in fresh water (e.g. some siluroids, cottids), fish maintain near neutral buoyancy by adjusting the volume of their swimbladder which, in fresh water, represents about 7% of the fish volume and has an adjustment capacity of about 25% (Alexander 1966). Consequently, it is usually recommended that the weight of the transmitter in water should be less than 1.75% of the fish body weight (Gallepp & Magnuson 1972; Stasko & Pincock 1977; Winter 1996; Baras *et al.* in press).

In early studies (Fig. 6.3), most transmitters were attached externally, as streamer tags, or using a pannier-type mount, usually adjacent to the dorsal fin. However, external transmitters can lead to a loss of postural equilibrium, increase drag and may be physically snagged, resulting in damage to the fish or premature loss of the transmitter (Ross & McCormick 1981; Mellas & Haynes 1985; Perrow *et al.* 1996). Because telemetry tags have become increasingly long-lived, surgical attachment has become the most popular method for tag attachment (Fig. 6.3). External attachment is now mostly restricted to applications where the sensor must remain in contact with the water (e.g. dissolved oxygen, light intensity, salinity), or where tag recovery is a high priority (e.g. archival tags).

Intragastric implantation has been widely used for transmitter attachment to adult anadromous salmonids, which do not feed during their return freshwater migration, and largely resulted in their increased use in the 1970s and 1980s (Fig. 6.3). In other species or ontogenetic stages, intragastrically inserted transmitters are likely to interfere with feeding, and in some species, are quickly regurgitated, leading to premature loss of the transmitter (Lucas & Johnstone 1990; Armstrong *et al.* 1992; Armstrong & Rawlings 1993). However, intragastric transmitters may still prove extremely valuable, when they can be voluntarily ingested by fish which are difficult to access or capture (e.g. deep lake fishes).

The implantation of telemetry tags into the peritoneal cavity, close to the centre of gravity of the fish, has the greatest potential for long-term studies (Lucas, in press). In fish species where the urogenital ducts lead to the body cavity (e.g. female salmonids, sturgeons, dipnoans, bowfins, male hagfish or agnathans), tags can be inserted into the body cavity through the gonoduct (Peake *et al.* 1997). In all other species, intraperitoneal implantation requires surgery (Box 6.3). It is clear that whenever surgery is involved fish will be subjected to disturbance, the duration and extent of which varies substantially depending on fish species, age and tag to body weight ratio. Surgically implanted tags can be shed through the incision (Marty & Summerfelt 1986), through an intact part of the body wall (Summerfelt & Mosier 1984; Lucas 1989) or through the intestine (Marty & Summerfelt 1986; Baras & Westerloppe 1999), though the latter process seems almost restricted to siluriform fishes. Hence it is important to critically evaluate implantation methods prior to their application in telemetry studies on a given fish species. Baras *et al.* (in press) provide a review of the most appropriate techniques and considerations to be met (summarised in Box 6.3). It is also important to bear in mind that in many countries surgical implantation of transmitters is a regulated procedure, often requiring a licence. Guidance on national regulations for surgical procedures is given at http://www.hafro.is/catag, the website of the recent EU-funded project on tagging of fishes for scientific research. Additionally transnational problems may occur

---

Box 6.3 Recommended practice for surgical implantation of electronic tags in fish (summarised from Baras *et al.* in press)

Anaesthesia. Except in extremely cold water, surgery requires that fish are chemically anaesthetised. Quinaldine (10–40 mg L^{-1}), tricaine (25–100 mg L^{-1}) and 2-phenoxyethanol (0.25–0.40 ml L^{-1}) are among the most popular anaesthetics (MacFarland & Klontz 1969, Bonath 1977; Summerfelt & Smith 1990). Fish are immersed in an anaesthetic bath until the tolerance stage, then placed in dorsal recumbency in a support adapted to their morphology, with head and gills immersed in the anaesthetic solution.

Incision site and length should be selected on the basis of several criteria, including innocuity, healing dynamics and minimum expulsion risks. Midventral incisions reduce the risks of damaging the viscera when the fish is upside down, and striated muscles, which require longer healing (Roberts *et al.* 1973; Knights & Lasee 1996). Lateral incisions are worth considering in fish with midventral ridges which prevent midventral tag insertion (e.g. serrasalmines; Baras 2000), but they are prone to puncture the gonads (Schramm & Black 1984), to damage striated muscles, and to cause bleeding, which is involved in adhesions and expulsion processes (Rosin 1985). Incision length should be as short as possible to minimise trauma, and to limit the risks of tag exit via the incision. Recommended incision length to tag diameter ratios depend on the flexibility of the fish body wall, and thus on species and incision site: e.g. 1.4–1.5 in catfishes, 1.6–1.8 in salmonids, cyprinids and cichlids, and up to 2.5 for ventrolateral incisions in serrasalmines.

Internal positioning of the implant should be done so as to minimise the risk of damage to internal organs arising from tag movement inside the body cavity (Chamberlain 1979; Bidgood 1980; Schramm & Black 1984), and minimise pressure over abdominal tissue to reduce expulsion risks. Tag placement over the pelvic girdle is the most frequent position. Suturing the transmitter to the body wall was effective in Atlantic cod *Gadus morhua* (Pedersen & Andersen 1985), but led to expulsion in channel catfish *Ictalurus punctatus* (Marty & Summerfelt 1986).

Incision closure is traditionally achieved with separate stitches (Hart & Summerfelt 1975; Summerfelt & Smith 1990). Choice of absorbable (e.g. catgut, Dexon) or non-absorbable (e.g. nylon, braided silk) suture material is often a trade-off between risk of expulsion through an unhealed incision when the filament dissolves, and the risk of infection due to the presence of transcutaneous foreign bodies (e.g. Thoreau & Baras 1997). Surgical staples permit quicker incision closure (Mulford 1984; Mortensen 1990), but they require removal of scales with resultant greater risk of infection. Medical or commercial grade tissue adhesives give fast closure and so help to suppress the inflammatory response (Nemetz & MacMillan 1988; Petering & Johnson 1991; Baras & Jeandrain 1998) but the adhesive can be shed within a few days only, and is difficult to apply innocuously in small fish with a narrow body wall (Baras *et al.* 1999).

Healing rate is proportional to the growth potential of the fish, and is thus more rapid in fast-growing species, proportionally faster in juveniles than in adults (Thoreau & Baras 1997; Baras & Westerloppe 1999), and quicker at warm than at cold temperatures (Knights & Lasee 1996). Juvenile tropical catfishes can heal abdominal incisions within 4 days (Baras & Westerloppe 1999), whereas adults of temperate (*Barbus barbus*; Baras 1992) or cold water species (*Gadus morhua*; Pedersen & Andersen 1985) require 4–6 weeks for complete healing. Permanent transcutaneous bodies, such as non-absorbable suture filaments or externally trailing antennae, frequently promote a chronic inflammatory response (Roberts *et al.* 1973).

Implant encapsulation and exit. Irrespectively of their coating (Helm & Tyus 1992), implanted tags frequently become encapsulated into host tissues in a classical reaction to foreign bodies. Tags free in the body cavity or encapsulated in connective tissue may be shed through an unhealed incision, or through the intact body wall, as a result of proliferating granulation tissue, and contraction of myofibroblasts in the capsule (Marty & Summerfelt 1986, 1990; Lucas 1989), that force the tag through the route of least resistance (fish tissues are only slightly denser than water, and their abdominal region is not developed to cope with gravity or unusual internal pressure). When the capsule adheres to the intestine, the intestinal wall may become disrupted, the tag enters the intestine, and is expelled by peristalsis. This process seems mainly restricted to siluroid fishes (Marty & Summerfelt 1986; Baras & Westerloppe 1999). Implant exit is favoured by all factors inducing internal pressure, such as large or heavy tags, enlarged gonads, and infection, making prophylactic measures highly recommended, and methods of positioning the tag far from the incision worthwhile (Ross & Kleiner 1982).

Postoperative recovery should be as short as possible to prevent any detrimental effect of confinement on fish health and behaviour (Otis & Weber 1982), but should incorporate appropriate release methods to minimise predation risk during the release phase, particularly for small fishes.

Postoperative perturbation may extend over variable periods of time, and may affect fish in different respects including posture (Chamberlain 1979; Thoreau & Baras 1997), activity (Diana 1980), predator avoidance (Adams *et al.* 1998), swimming capacity and migration (Haynes & Gray 1979; Mellas & Haynes 1985; Moore *et al.* 1990; Adams *et al.* 1998).

in terms of inadequate consolidation of use of frequencies for animal tracking, which has resulted in research groups tracking other groups' fish when they have migrated across borders, or of interference problems for other users of radio frequencies.

6.3.10 Limitations of telemetry systems

Major drawbacks of telemetry systems relate to the costs of individual tags and detection equipment, numbers of fish tracked at a time, tag endurance, and minimum fish size. A simple radio or acoustic tag costs $200 or $300 respectively (Table 6.1). The cost of simple receivers for manual tracking ranges from $500 to $2000, and ALSs including data logging computers range from $5000 to more than $10 000 depending on software and attachments included. Automatic positioning systems with sonarbuoys and remote links may exceed $40 000. Currently, the minimum size of VHF radio and acoustic tags limits the lower size of fish that can be tagged to about 12 cm (Table 6.1).

The battery typically constitutes more than 80% of the transmitter mass and more than 50% of transmitter volume (Winter 1996). For a given power output, adapted to the study environment, transmitter life is proportional to battery size of a given type (e.g. 3-V $LiMnO_2$), restricting long-term studies to large fish only. However, longer life can be obtained without compromising range by programming a delayed start or longer interpulse interval (Voegeli & McKinnon 1996) or to operate on long-term duty cycles (see section 6.4.3).

An additional problem is the rather limited number of transmitters that can be tracked at one time. For acoustic transmitters, receiver bandwidth limits the number of frequencies to about six to ten, which can adequately be spaced over a range of about 15–20 kHz around the receiver's nominal frequency, and multipath effects limit the number of pulse rates of simultaneously operating tags to no more than two to three (Stasko & Pincock 1977). Radio frequencies enable larger numbers of frequencies to be used, usually with a 5 kHz or greater spacing, although national controls on radio frequencies may restrict the range of frequencies available. Provided that migrating fish tend to remain solitary, standard transmitters can also be differentiated by pulse rates, with no more than three or four different pulse rates per frequency for easy identification by an operator. However, if advanced digital acoustic or radio receivers, with a capability of identifying the interpulse period to the nearest millisecond, are used, then tens of tags with different pulse periods at each frequency can be used, as long as tags are not close to one another. The use of coded radio or acoustic transmitters, each emitting an identifying code of one or more brief pulse(s) interrupting the normal longer pauses, provides a better method for identification of 10–20 transmitters at each frequency, increasing the numbers of tags which can be tracked by nearly an order of magnitude (Eiler 1995; Voegeli & McKinnon 1996). Because these tags have longer interpulse intervals, their life is also greater, but their identification requires a sophisticated receiver. However, these are of particular value for studies of fish migration at obstructions, where tagged fish may accumulate in high concentrations.

Care must also be taken in the sampling strategy of any tracking study. For example Baras (1998) argued that the timing of relocating fish in telemetry studies at intervals longer than a day generates a bias in results, particularly in studies of home range movements or homing after a short-term migration. In some species, the loss of accuracy can be predicted and corrected but only in the river or lake under study (Baras 1998). Hence, preliminary studies on a daily basis should be carried out to determine the effects of different time intervals between position fixes on the interpretability of results, and the optimum positioning intervals to be used later in long-term studies relying on the use of transmitters working on pre-programmed daily, weekly or monthly duty cycles, or archival tags. Similar issues must be considered in relation to the effect of time of day when fish are located; tracking at the same time of day may give a very biased impression of the fish's activity. Clough and Ladle (1997) showed that during the summer, dace *Leuciscus leuciscus* in a small chalk-stream exhibited localised use of different areas during day and night, with rapid movements between these at dawn and dusk. Daytime or night-time tracking of these fish would have given the false impression that they were very sedentary. These issues are less of a concern where long-distance migration lasting days or weeks occur, but factors influencing the diel pattern of that migratory behaviour still demand consideration. Description of data analysis and statistical methods appropriate for telemetry and other spatial data are beyond the scope of this chapter but a wide range of information and advice is now available elsewhere (Kenward 1987; White & Garrott 1990; Priede & Swift 1992; Larkin & Halkin 1994).

6.3.11 Archival tags

Archival (data storage) tags were recently developed to record large temporal series of environmental characteristics along the migration routes of fish travelling through contrasting and poorly accessible environments such as the open sea (Metcalfe *et al.* 1992; Gunn *et al.* 1994; Sturlaugsson 1995; Karlsson *et al.* 1996; Metcalfe & Arnold 1997; Sturlaugsson & Johansson 1996; Sturlaugsson *et al.* 1998). Pre-programmed duty cycles of operation result in the periodic measurement of environmental and/or internal variables which are stored in a high memory capacity RAM chip instead of being immediately transmitted.

In principle, any kind of sensor can be coupled to an archival tag, and combinations of sensors enable reconstruction of fish tracks with acceptable accuracy, provided there is sufficient knowledge of environmental variables in the fish's presumed home range. For example, coupling of pressure and temperature sensors to archival tags attached to plaice *Pleuronectes platessa*, a marine flatfish, enabled demonstration of selective tidal stream transport over an unprecedented timescale, together with accurate track reconstruction for this species in the North Sea (Metcalfe & Arnold 1997). Similarly, the coupling of light, pressure and temperature sensors permitted CSIRO scientists to highlight the long-range migrations of southern bluefin tuna *Thunnus maccoyii* between Australia and South Africa (Gunn *et al.* 1994). Sturlaugsson (1995) first demonstrated the potential of archival tags on adult Atlantic salmon *Salmo salar* during coastal migration, and the technique has now been used to examine river to sea, and return, movements of adult sea trout *Salmo trutta* (Sturlaugsson & Johansson 1996) and Arctic char *Salvelinus alpinus* (Sturlaugsson *et al.* 1998). To date, archival tags are still relatively large units, and their use is restricted to fish exceeding 300–500 g, but improvements in size reduction and increased storage capacity can reasonably be foreseen.

Archival tags are relatively expensive units (at least $300 each, and ten times this for data-transmitting tags), and the number recovered may be low. However, this is compensated for by the enormous amount of information that can be retrieved from a single tag. Recovery of tagged fish or tag information is thus a key factor in research programmes relying on archival tags. This may be achieved through an extensive recapture effort, including the assistance of fishermen for commercially important species. The probability of efficient tag recovery by fishermen requires easy identification of fish tagged with archival tags, and thus that these tags be attached externally (Metcalfe & Arnold 1997), or that the fish be double-tagged with an external passive tag or mark when internal archival tags are used (Thorsteinsson 1995).

Additionally, tag recovery is improved by adequate publicity (advertisement in local and international newspapers, public presentations and local contacts) and by incentives to declare tag recovery (monetary rewards, gifts, lottery, recognition). When the probability of recapture is exceedingly low, detachable tags are worth considering. Pop-up tags, that detach themselves from the fish after a preset interval (Baba & Ukai 1996), have been used successfully on bluefin tuna (*Thunnus thynnus*) in the Atlantic Ocean (Block *et al.* 1998; Lutcavage *et al.* 1999). These are low-drag, positively buoyant units with a float and 16 cm aerial antenna. When the tag pops up to the surface, the float maintains the antenna in an upright position, and a limited amount of archived data (temperature, depth, geolocation estimated from light level data, etc.) can be transmitted to an ARGOS satellite, which can also geolocate the transmitting tag. The tags are therefore commonly known as 'pop-up archival transmitting' (PAT) tags.

Similar applications can reasonably be foreseen in large lakes, with the absence of waves easing the recovery of archived data, but bulk and weight (about 60 g) of pop-up tags restrict their application to very large fish only (>5 kg). Fishes such as Nile perch *Lates niloticus* and lake sturgeon *Acipenser fulvescens* may be appropriate freshwater species for which the technique could be applicable. Prospects of development of much smaller 'pop-up' radio tags or 'communicating history acoustic transponder' (CHAT) tags that could be interrogated by automated underwater or land-based receivers are considered in a useful review by Moore *et al.* (2000). Nevertheless, the major problem with the use of archival tags is that migration paths can only be reconstructed from environmental data (light, pressure, temperature, salinity). Implicit in their application is that the environment is variable and documented sufficiently to permit this reconstruction. There also are additional limitations specific to each sensor type (see 6.3.8), including the risks of fouling of external sensors. Foreseeable applications to freshwater fish are thus mostly restricted to large lakes, and to diadromous species.

6.4 Choice of methods in fish migration studies

As mentioned throughout this chapter, all techniques and methods for investigating the migration of freshwater fish suffer from intrinsic, environmental and specific limitations (Box 6.1; Tables 6.1–6.4). Therefore, before starting any study on fish migration, one should define its objectives most clearly, and adapt existing techniques, alone or in combination, to the study environment and target species.

Generally, telemetry techniques can almost always prove valuable for species large enough to be tagged with transmitters. Beyond bringing direct knowledge on fish migration, activity patterns or energy expenditure, they can also help to delimit more efficiently the areas, seasons and times of the day to sample with capture or hydroacoustic techniques, or to look for tagged fish in the course of mark recapture studies, including those involving archival tags. They can also pinpoint which sites may represent obstacles to fish migration from the fish's point of view (Hinch *et al.* 1996; Lucas & Frear 1997; Hinch & Bratty 2000), and help further study to focus on these. However, telemetry may not always be the best way of investigating fish migration, especially when large samples are needed or when the access to the environment is so difficult that tracking would not bring more information than recreational or commercial fish catches.

Catch per unit effort and mark recapture techniques are most efficient where long-term fishery or monitoring studies are in place and data on crude spatial and temporal scales are adequate. They also have the advantage of low technical requirements and low equipment costs. Where targeted studies with specific management or ecological questions are pertinent, recapture-independent techniques may be more appropriate. Telemetric methods can provide high-resolution information at the individual level, while hydroacoustics is increasingly providing information at the population level in large lake and river environments. Several studies have incorporated telemetry and hydroacoustics to enable simultaneous interpretation of individual- and population-scale behaviour patterns (e.g. Malinin *et al.* 1992). Biochemical methods are becoming increasingly useful in determining the extent of segregation of migratory populations through DNA analysis, and for study of habitat history and ontogenetic changes through microchemistry of hard parts and stable isotope analysis.

The following paragraphs provide some examples of how techniques might be chosen to solve some problems of a fundamental or applied nature concerning migration and movements of fishes in fresh water.

6.4.1 Functional delimitation of fisheries districts

In most tropical rivers with intensive fisheries, like the Niger River Inner Delta, fish capture is highly local or regional, with few fish landing places. Neighbouring landing or market places are frequently grouped as fisheries districts, the management of which is more or less independent of their neighbours. However, there is little evidence that the current administrative delimitation matches the delimitation of fish populations, considering that fish, including tilapias and catfishes, can migrate between these. Shallow depth and abundant vegetation make hydroacoustics impractical, whereas these conditions are favourable for seine netting, trapping in side channels and electric fishing in rivers with sufficient conductivity. The mean size of captured fish (45–50 g) restricts the possibilities of tracking fish efficiently (small transmitters with a limited range and thus poorly detectable in a floodplain where terrestrial and boat tracking are difficult). In these circumstances, tagging with conventional passive tags and screening of discrete landing places may be the best solution to test for the functional delimitation of fisheries districts. Considering that the entire programme relies on tag recovery, tags should be external and each recovery rewarded to avoid biases.

6.4.2 Lateral and longitudinal migrations of large catfishes in a South American assemblage

Large South American pimelodid catfishes (e.g. genera *Brachyplatystoma* or *Pseudoyplatystoma*, locally called 'bagre' or 'sorubim') usually live in deep and/or sheltered bays, where they prey on other fish species, but they apparently can travel hundreds of kilometres during their spawning migrations (Goulding 1980; Welcomme 1985). Whether these take place in the main river exclusively, or also involve excursions in tributaries, remains to be determined. Because of their large size, these catfishes cannot easily be captured by electric fishing, but large seine or fyke nets can be used successfully. There are few constraints on tagging these large fish (0.5–1.0 m), and almost any kind of tag or mark could be used successfully. Considering that these species are of great value for inland fisheries, external tags would be preferred to encourage tag identification by fishermen, and a reward should ideally be given for each return.

However, conventional tagging would be of limited value for several reasons. Firstly, the temporal resolution would probably be too low for these fast-moving fish. Secondly, South American rivers and streams, especially those flowing from the Andes, have more contrasting flood regimes than most European rivers of similar width, and fishing effort is very restricted during high waters, when fish are believed to migrate. Thirdly, capture of these fish during the spawning season has been prohibited, and information from occasional captures by poachers would probably not be relayed. Fourthly, roads are scarce in tropical foothills, and are poorly accessible during the rainy season, preventing regular access to field stations for scientific samples. The same limitations apply to any remote sensing method, including fish tracking. Therefore, this situation could be viewed as a poorly accessible environment like those for which archival tags have been developed. Because foothill streams, the main river and the floodplain have contrasting thermal regimes, and most frequently different conductivity and clarity, archival tags recording these three variables together with depth could be of great value, provided that sufficient mapping of habitat characteristics is made. For the reasons described above, tags should be externally attached, ideally streamlined to minimise drag, and a substantial reward granted for each tag's return.

6.4.3 Fidelity of fish to spawning grounds

Whether iteroparous fish show repeated fidelity to the same spawning grounds is of fundamental importance to understanding those populations, and of practical importance when spawning grounds are destroyed by human activities. This applies to European rheophilic cyprinids such as barbel *Barbus barbus*, that dig spawning pits in gravel bars targeted by gravel abstraction industries (Baras 1994; Baras *et al.* 1996a). Assessment of spawning site fidelity can be achieved by an extensive marking effort using conventional tags (e.g. oval bar tags), and successive recaptures by electric fishing during subsequent spawning seasons.

However, mature females (about 700 g) can travel over substantial distances (up to 25 km), they stay on the spawning ground or in its immediate vicinity for a limited period of time (<48 h) only, and this period varies between years (from late April to mid-June), making it difficult to efficiently sample a long river stretch within a few days. Fish straying outside of the study area would also have little chance of being detected. Additionally, an extensive sampling effort during the spawning season is questionable from a conservation viewpoint, as it may jeopardise recruitment of barbel and other fish species in the study area. For these reasons, it is preferable to tag fish with radio tags (because the rivers are shallow and turbulent) operating on pre-programmed long-term duty cycles, with the transmitter automatically switching on during the prespawning and spawning period (*ca.* 2 months) and off for the rest of the year. With this duty cycle, a radio tag for a 1 kg fish has a theoretical minimum life of 3–4 years. Tag cost ($200 each) would limit sample size, but enough fish should be tagged to take into account natural mortality and angling-related mortality over this extended period of time.

Chapter 7
Applied Aspects of Freshwater Fish Migration

7.1 A broad view of the impact of man's activities on freshwater fish migration

Since the seventeenth century European rivers and streams have become increasingly modified for navigation, hydropower and water regulation purposes, and similar patterns have occurred worldwide, in some cases considerably pre-dating those in Europe (Baxter 1977; Dudgeon 1992; Northcote 1998). These have considerably impacted upon fish assemblages and continue to do so, as they interfere with one or several crucial steps in their lifecycles (Fig. 2.1), either directly, or indirectly through additional effects on other biota. Connectivity of habitats on a range of spatial and temporal scales is viewed as critical to the integrity of aquatic ecosystems and the communities of fishes and other biota (Jungwirth *et al*. 2000; Schiemer 2000), and has been expounded in the river continuum concept and the more recent extended serial discontinuity concept. Dams and weirs have drastically modified the landscape, the distribution of physical habitats and their physicochemical characteristics, including a greater propensity for warming and deoxygenation (Baxter 1977). Intriguingly, in the north temperate zone, damming by beavers *Castor* spp., and by natural processes such as woody debris accumulation, is increasingly regarded as a key component in maintaining stream habitat diversity on a range of spatial and temporal scales (Schlosser & Kallemeyn 2000). But of course these dams are short-lived on a biological timescale and they promote habitat heterogeneity rather than homogeneity.

Rectification of the longitudinal and vertical profiles in river channels has further contributed to the loss of habitat diversity and reduced lateral connectivity in floodplain habitats. Natural flow patterns, which represent major cues for fish undertaking migrations (see section 2.2.2.3), have been strongly modified in regulated rivers. Dams and weirs have also contributed to reduce the longitudinal connectivity in many rivers. This obviously is the case for dams with a difference of water levels of several metres, but even obstacles as small as 20 cm may represent a barrier to the upstream movement of small fishes (e.g. cottids, see section 5.32; Utzinger *et al*. 1998). Obstacles interfere not only with upstream spawning migrations but also with compensatory upstream movements after displacement by flood or emigration from habitats that become temporarily unsuitable (see section 3.2.4). Accumulation of migrants below dams makes fishes more susceptible to exploitation by inland fisheries, and in some cases has been widely utilised through the building of fishing weirs in seasons when fish ascend rivers. This approach has been used in the capture of beluga sturgeon *Huso*

huso at the Iron Gate corridor in the Danube River, and is still continued in many tropical countries, for example, the catfish fisheries in the lower Ouémé River, southern Benin. Building of barrages and tidal power schemes may also have a major influence on diadromous fishes and those that rely on brackish nursery areas (Dadswell & Rulifson 1994; Russell *et al.* 1998).

In rivers and streams open to navigation, or flowing through areas densely populated by humans, dredging or removal of any obstacle to flow such as gravel bars or fallen trees is a common practice, which interferes further with the reproductive cycle of lithophilous and phytolithophilous spawners, especially when banks have been modified in such a way that they no longer offer any alternative spawning sites. Modified flow patterns as a result of regulation further interfere with the natural reconstruction of gravel bars during winter spates, as low flows prevents coarse substratum from being moved downstream. Small substrate particles, such as silt, are far less affected, still drift and eventually cause the siltation of many gravel or cobble areas. A substratum comprising a high proportion of fine particulate material makes it more difficult for lithophilous spawners exhibiting brood hiding or pit digging (e.g. salmonids, diadromous clupeids, and some cyprinids such as *Barbus barbus*) to dig spawning pits, and it exposes their eggs to more hypoxic conditions which impact on their survival or development. Such negative effects can be compounded by the lower oxygen content in waters that warm as a result of reduced flows, and also by eutrophication, especially in densely populated habitats.

Finally, the larvae or juveniles of many fishes may find no suitable nursery habitat nearby in rivers or streams where bank rectification has removed most or all nearshore habitats, and reduced the lateral connectivity with the floodplain which, when intact, offers numerous refuges against displacement by flow, protection against predation, and abundant food. Drift of early life stages upon flood pulses in regulated rivers might thus be more frequent, or take place earlier in the season than in natural environments, with compensatory upstream movements being impaired by the reduction of the longitudinal connectivity in rivers with weirs and dams.

Other direct impacts of man's activities include the recreational use of rivers, logging operations, mineral extraction, water pumping for irrigation, industrial or urban use, fisheries and the introduction of non-native species, with Nile perch *Lates niloticus*, common carp *Cyprinus carpio*, brown trout *Salmo trutta* and rainbow trout *Oncorhynchus mykiss* being commonly cited as some of the best-known examples of 'eco-disasters' that may arise from such fish introductions (see section 4.4. for Nile perch). Indirect or side effects comprise, among others, the chemical or organic pollution arising from agriculture, urban or industrial development, and the changes brought about by deforestation and civil works.

The assumption that changes brought about by damming are just a matter of local management is wholly incorrect. The Aswan Dam, one of the world's most famous civil engineering works, has caused drastic modifications of salinity in the Nile River Delta, several hundreds of kilometres away, with considerable impact on the fish fauna and fisheries in the delta and Mediterranean Sea (Baxter 1977). Other examples include damming of the River Meuse which originates in France, crosses Belgium and the Netherlands, and flows into the North Sea. Damming in Belgium blocked the spawning runs of many anadromous species moving into the Belgian Ardenne, resulting in a marked decline of the populations exploited by Dutch fishermen, and eventually in the extinction of the original populations, of which some (salmonids) are now being reconstructed (Philippart *et al.* 1994). It had been considered that fish species which complete their lifecycles in fresh water are less severely affected by damming than diadromous species (Baxter 1977), although this view has proven to be oversimplistic (Northcote 1998). The extent of disruption on holobiotic freshwater fish species clearly depends mainly on their migratory patterns, reproductive habits, sensitivity to water quality, growth rate, and size attained at the onset of the harsh season (drought in tropical assemblages or winter in temperate/arctic assemblages).

A comprehensive synthesis of how these changes might impact on aquatic biota through variations of environmental factors has been provided by Cowx and Welcomme (1998), and many case studies can be found in Leclerc *et al.* (1996). These two books also provide criteria and techniques for assessing the effects of man-made changes, and restoring the diversity of aquatic habitats. Within the scope of this book, we focus here on how man-made changes interfere with fish migration.

7.2 Impact of man's activities on the diversity of fish assemblages in different geographic regions, focusing on damming

The general picture presented above has principally been drawn from European fish assemblages inhabiting rivers and streams, which have been modified for decades or tens of decades. However, it applies to most freshwater ecosystems and estuaries throughout the world, notably in the tropics or in temperate mountain regions where huge hydroelectric dams have been built (Baxter 1977; Dudgeon 1992; Stanford & Hauer 1992; Northcote 1998). This brief overview, based on selected examples in different geographic regions, illustrates the degree of severity of the changes brought about by obstacles to longitudinal and lateral connectivity, with respect to the migratory habits of fishes.

7.2.1 Eurasia

In Lelek's (1987) review, 133 of the 200 European freshwater fish species were regarded as 'stable' (i.e. widespread and abundant), and 67 others as endangered, threatened or vulnerable (N = 25, 10 and 32, respectively). The endangered status refers to species near extinction throughout all or a significant portion of this geographical area. The vulnerable status is granted to species highly sensitive to environmental disturbances, which deserve careful monitoring. The threatened status is an intermediate category. Northcote (1998) reports that the building of dams and weirs accounts for 55–60% of the known causes behind these degrees of endangerment. More detailed information, including country-by-country assessments, can be found in Kirchhofer and Hefti (1996). The general trend for European countries is that most endangered, threatened or vulnerable species are diadromous or potamodromous species, and that their status largely is a consequence of their migrations being impaired or blocked by civil works. A partial list of European migratory freshwater fishes is provided by Cowx and Welcomme (1998). Extreme examples where all diadromous species, except for the European eel *Anguilla anguilla,* have become extinct as a result of habitat fragmentation can be found in several European countries. For example, shads *Alosa* spp., European Atlantic sturgeon *Acipenser sturio*, Atlantic salmon *Salmo salar*, sea trout *Salmo trutta*, sea lamprey *Petromyzon marinus* and river lamprey *Lampetra fluviatilis* have disappeared from Swiss river systems (Peter 1998). In many cases dams, even where fish passage facilities are present, have resulted in near-elimination of upriver-migratory elements of diadromous populations, with the majority of remaining fishes migrating to spawn in tributaries downstream of the main barriers, as for sea trout *S. trutta* in the River Gudenå, Denmark (Aarestrup & Jepsen 1998).

Dams have also been held responsible for declines in the commercial river lamprey fisheries in Finland (Tuunainen *et al.* 1980). Lampreys do not swim rapidly and therefore are unable to utilise most fish passes designed for teleost fishes (Laine *et al.* 1998a). Prior to the construction of a barrage on the River Leven, Scotland, adult sea lampreys were occasionally reported in Loch Lomond (Lamond 1931). Reports of anadromous lampreys in the Yorkshire Ouse system, England, suggest that they tend to aggregate below river barriers (Lucas *et al.* 1998). Estuarine barrages may also inhibit movement, especially in an upstream direction, of diadromous fishes (Russell *et al.* 1998). River and sea lampreys and eels ascend the Denil fishway at the Tees Barrage, Northeast England, only during spring high tides when the estuarine water floods the pass (M. Lucas [unpubl.]).

Peter (1998) drew further attention to the idea that weirs or dams do not need to be impressive to constitute true obstacles to fish migration. He observed that above a 40-cm high log weir on Sagentobel Creek (Glatt River system, Switzerland), only the brown trout *Salmo trutta*, which exhibits leaping behaviour, was abundant, whereas nine other species (seven cyprinids plus the European eel and the three-spined stickleback *Gasterosteus aculeatus*) were present or abundant below the obstacle. Smaller weirs, as low as 18–20 cm in height, have also been shown to represent major obstacles to bullhead *Cottus gobio* and stone loach *Barbatula barbatula* (Bless 1981; Jungwirth 1996; Utzinger *et al.* 1998). Peter (1998) points out that because small artificial barriers are most frequent in Swiss streams (about one every 50 m), they may have contributed to the local decline or extinction of bullhead and small migratory cyprinid species, including the gudgeon *Gobio gobio*, the spirlin *Alburnoides bipunctatus* and the blageon *Leuciscus souffia*. In the Czech Republic smaller dams installed in rivers for bypass hydropower stations were also shown to have a significant impact on fish communities (Kubecka *et al.* 1997).

Comprehensive reviews of how obstacles have interfered with the distribution of anadromous and potamodromous species in the Meuse and Rhine Rivers have been given by Philippart and Vranken (1983), Philippart *et al.* (1988) and Admiraal *et al.* (1993). Baçalbasa-Dobrovici (1985) and Waidbacher and Haidvogl (1998) have produced similar reviews for the Danube, and Jungwirth (1998) provides a general picture of man-made changes in lateral, longitudinal and vertical connectivity. As a result of the construction of a large hydropower facility at the Iron Gate (rkm 931) in the early 1970s, the migration of most anadromous fish species in the Danube has been shortened substantially. This applies to the pontic shad *Alosa pontica*, which was historically encountered in Hungary as far as 1600 km from the estuary, and to three sturgeon species (the beluga *Huso huso*, the Russian sturgeon *Acipenser gueldenstaedtii*, and the stellate sturgeon *Acipenser stellatus*) which were reported to be frequent spawners in the Austrian and German tributaries of the Danube (up to rkm 2580). Damming is also cited as the major cause behind the decline of the Eurasian huchen *Hucho hucho*, a salmonid, in the upper reaches of the Danube (Holcik *et al.* 1988). Other historical accounts from the Danube River system illustrate the considerable influence of damming and hydropower development on potamodromous cyprinids such as the barbel *Barbus barbus* and the nase *Chondrostoma nasus*. Waidbacher and Haidvogl (1998) report that these two species almost disappeared from the upper reaches of the Inn River in Bavaria after the construction of the Jettenbach hydropower station in 1923. A fish pass was built later, but the populations never recovered. Penáz and Stouracova (1991) also showed reductions in the abundance, biomass and angling catches of barbel after construction of the Dalesice Hydropower and Dukovany nuclear power stations on the River Jihlava in what was former Czechoslovakia. Axford (1991) reported a similar decline in the CPUE of small cyprinids (mainly dace *Leuciscus leuciscus* and gudgeon *Gobio gobio*) after the installation, in 1978, of a flow-gauging weir on the River Nidd, England. Damming on Russian rivers such as the Don (draining to the Sea of Azov) and the Volga and Terek (draining to the Caspian Sea) has strongly affected migrations of cyprinids between these seas and the rivers (Welcomme 1985; see section 5.12). Fish catches in the River Volga have fallen by 90%.

Far less information is available for Asian fish assemblages, although impacts similar to those observed in Europe might be reasonably expected on large heavily dammed rivers. In China, habitat destruction and dam construction in the Yangtze River have blocked the upstream spawning migrations of the Chinese paddlefish *Psephurus gladius* which is now regarded as the most endangered fish in China (Wei *et al.* 1997). Damming of the Hanjiang River in China to form Danjiang Reservoir resulted in the local extinction of the cyprinids *Coreius heterodon* and *Rhinogobius typus* in the upper reaches of the river, and it also blocked the spawning runs of several species of indigenous carps (Liu & Yu 1992). River regulation also appears to be having a major effect on migration, dispersal and recruitment of Himalayan mahseer *Tor putitora* and related species in Himalayan rivers (Bhatt *et al.* 2000).

Damming, construction of diversion canals and large-scale abstraction of water from rivers draining to the Aral Sea (Kazakstan and Uzbekistan), to provide irrigation for the cotton-growing industry, has resulted in near-complete destruction of the Aral Sea ecosystem (Micklin 1988). From an area of 68 000 km^2 in 1960, the Aral Sea has been reduced in size by over 50%, salinity has increased from 10‰ to over 30‰ and there has been functional disconnection of rivers from the sea. All of the migratory populations of fishes which used to leave the Aral Sea to spawn or overwinter, including the ship sturgeon *Acipenser nudiventris*, Aral barbel *Barbus brachycephalus*, roach *Rutilus rutilus*, common bream *Abramis brama*, and pikeperch *Stizostedion lucioperca* (Nikolsky 1963) have disappeared. Migrations of *B. brachycephalus* are reported to have extended 1000 km up the Amudarya River, Uzbekistan and Turkmenistan (Nikolsky 1963). The number of fish species in the Aral Sea has fallen from 27, to just four euryhaline, eurytopic species and fisheries have completely collapsed. There are, of course, many examples of ecosystem damage impinging on migratory freshwater fishes, but few cases could instil such a sense of sheer incredulity, as that obtained by comparing the accounts of Nikolsky (1963) and Micklin (1988). We urge all those with an interest in aquatic ecology and management to read these accounts and see how easy it is to devastate an ecosystem in just a few decades.

7.2.2 North America

Cada (1998) reported there were as many as 2350 hydropower dams in the US, most of them run by private citizens or companies, and the total number of so-called 'minor' obstacles is unknown to us. As in Europe, the construction of dams has long been held responsible for the decline of sea lampreys *Petromyzon marinus* in parts of North America (Morman *et al.* 1980), but on several occasions this has been intentional. In the Laurentian Great Lakes drainage basin, where invasion by the sea lamprey *Petromyzon marinus* has had a major impact on the indigenous fish fauna, low-head dams have been built specifically to deny access by sea lampreys to spawning tributaries (Hunn & Youngs 1980). An expansion in construction of low-head dams for sea lamprey control is planned but there is evidence that these may limit the migration potential for various potamodromous species (Porto *et al.* 1999). Similarly, where fish passes in the lower parts of Great Lakes river systems are present, they are generally of truncated design with an overhanging lip, to prevent access by lampreys but allow salmonids to leap into the fish pass (Clay 1995). While helping to exclude sea lampreys, these fish passes may also deny access to non-leaping fishes. Further research may enable low-head barrier and fishway design which will pass a greater range of fishes, but not *P. marinus* (Porto *et al.* 1999).

Intense damming of the Fraser River, Canada, had caused the abundant upriver populations of pink salmon *Oncorhynchus gorbuscha* to become virtually extinct (Northcote 1998). Some recovery took place after efficient fish passes were provided although there has been recent evidence that upstream passage at Hell's Gate remains a problem. Sockeye salmon *O. nerka* expend much energy trying to migrate through Hell's Gate and those which fail to successfully ascend the pass on their first attempt appear unable to do so later (Hinch *et al.* 1996; Hinch & Bratty 2000). Northcote (1998) also reported the massive decline of the chinook salmon *O. tshawytscha* in the heavily dammed Columbia River system, in British Columbia. Before mainstem dam construction, spawning runs were estimated at 10 to 16 million salmon annually; they had declined to less than 3 million in the mid 1980s, and now are estimated at about half a million in spite of upstream and downstream bypass facilities, and stocking of hatchery-reared smolts.

In the Rocky Mountain river systems, the distribution of many species (e.g. flannelmouth sucker *Catostomus latipinnis*, Colorado pikeminnow *Ptychocheilus lucius*) has been modified by many dams that blocked historic migrations routes (e.g. flannelmouth sucker at Glen Canyon Dam, Arizona; McKinney *et al.* 1999; Colorado pikeminnow at Taylor Draw Dam on the White River, Irving & Modde 2000; see also sections 3.2.1 and 5.12). Stanford and Hauer (1992) provide an informative review of the historical development of river regulation in the Flathead River catchment, Montana, and of its effects in preventing migration of and isolating native fish populations comprising salmonines, coregonines, cyprinids and catostomids. Rivers of the Californian coast have also been heavily dammed, restricting the distributions of several species formerly migrating over hundreds of kilometres, such as the splittail *Pogonichthys macrolepidotus* in the San Joaquin River (Meng & Moyle 1995).

The most recent study on freshwater fishes in the Southern United States (Warren *et al.* 2000) indicated that 28% of the 662 species surveyed across 51 major drainage units of this region were vulnerable, threatened or endangered, this proportion representing a 125% increase in jeopardised species over the last 20 years. Generally speaking many species in jeopardy are those with a diadromous life history, as well as several others making long potamodromous migrations, and here too, damming is invoked as a major factor, acting through the blockage of migration routes and habitat modifications. Paradoxically, the other main group affected are non-migratory, small-bodied fish such as most cyprinidontiforms, species of which often occur as populations restricted to just one or a few small water bodies, and which are therefore at high risk of habitat damage (Minckley & Deacon 1991). Nearly all large rivers of the Southern United States are dammed, and small to medium-sized rivers in this region have been dammed, and affected by urban and industrial development.

7.2.3 Australasia and Oceania

All of the 27 native freshwater fish species in New Zealand are diadromous or of marine origin (McDowall 1990). Australia contains at least 183 native species that rely on fresh water for completion of their lifecycles and about 150 'marine' species that also utilise fresh and/or brackish water (Allen 1989). Over 95% of Australian freshwater fishes are diadromous or derived from predominantly marine families (Allen 1989). Almost all are now considered to be migratory and many rely on access between estuarine/coastal waters and fresh water for population maintenance (Harris *et al.* 1998). A similar situation occurs for New Guinea and other islands in the region. The ichthyofauna of the rivers in Australasia and Oceania is thus more susceptible to damage by damming than in many other parts of the world (McDowall 1995; Harris *et al.* 1998). Although many New Zealand fish species such as anguillid eels and galaxiids exhibit impressive climbing abilities, their migrations have been hindered by the succession of dams in river systems, and substantial proportions of river drainages have been made inaccessible to them (e.g. Simons 1992 and Jellyman 1993 in Northcote 1998). A similar statement has been provided by McDowall (2000) regarding the distribution of the torrentfish *Cheimarrichthys fosteri* in New Zealand.

In Australia, the situation may be worse. Urbanisation and rapid agricultural development, combined with an arid and rather unpredictable climate, have promoted the building of a multitude of small dams and weirs, and further barriers are currently being built to address the need to control streambed erosion. Harris *et al.* (1998) estimated that in New South Wales alone, there were at least 1500 man-made instream obstacles that could significantly affect the movements of fish, and another 1500 smaller obstacles, which were deemed to cause no major impact but might still interfere with the movements of small species or young stages. Many of Australia's migratory freshwater fishes are catadromous and for young life stages even small obstacles pose great problems for ascent. Such obstacles can only be ascended by small fishes when the barriers are submerged at high flows, but the fast currents often exceed their swimming capacities and preclude this (Harris & Mallen-Cooper 1994). Harris (1984) further reported that less than 10% of these dams were equipped with fish pass facilities, the functionality of which was highly questionable in many cases (see also Harris & Mallen-Cooper 1994). In the Murray-Darling system which flows over several thousand kilometres in Australia, the blockage of migration routes has been demonstrated in several species and suspected in many others (Reynolds 1983; McDowall 1996). The alteration of aquatic habitat as a result of damming is invoked as a major cause behind the marked declines in distribution or abundance of many endemic freshwater fishes (review in Humphries *et al.* 1999), and systematically for endangered species (Lake 1971 in Northcote 1998).

Where diadromy is represented by catadromy or amphidromy, as throughout much of Oceania (see section 4.4.1 and McDowall 1988), the impact of obstacles may be more severe than for anadromous fishes as upstream migrants are of a smaller size. For example, Concepcion and Nelson (1999) observed that above a small dam in Guam (Mariana Islands) the small jungle perch *Kuhlia rupestris* was absent, and populations of the mountain goby *Stiphodon elegans* were far less abundant than below the dam. Similar depauperate fish communities have been seen above natural waterfalls in Malaysia by Martin-Smith and Laird (1998). However these authors pointed out that some species (e.g. *Crossocheilus, Gastromyzon* and *Protomyzon* spp.) that managed to climb these waterfalls, could do so because the surface was graded and with irregularities, unlike the smooth, often near-vertical, man-made obstacles. McDowall (1988) provides a good synthesis of the amazing climbing abilities of several fish species which have been found above waterfalls as high as the Victoria Falls in Africa, or observed during their climbing process alongside the concrete walls of dams (most or all anguillid eels, the lamprey *Geotria australis*; galaxiids *Galaxias* spp.; and gobioids *Gobiomorphus, Lentipes, Sicydium* and *Sicyopterus* spp.). Most species possessing adhesive organs, such as the homalopterids or sisorids, may share this ability. The extent of predation by piscivorous birds during these ascents is unknown.

7.2.4 Tropical South America, Africa and Asia

South American rivers have become increasingly dammed as a consequence of hydropower development, and notably the Paraná River (see also section 7.3.2). In the early 1960s the total dammed area amounted to about $1000\,km^2$, and has increased by 20 times over the past 35–40 years. Prior to the closure of the Itaipú reservoir (about $1,460\,km^2$) in 1982, more than 110 fish species were found in the upper reaches, but within the 5 years following impoundment the assemblage had declined to 83 species (Agostinho *et al.* 1994). Large migratory characins (*Brycon orbignyanus* and *Piaractus mesopotamicus*) that were previously abundant in the reaches downstream of the impounded area disappeared, probably because their upstream movements were blocked and because they could no longer feed on drifting fruits. Other large (*Salminus maxillosus*) or medium-sized characins (*Leporinus elongatus, L. obtusidens, Prochilodus lineatus*) and the large pimelodid *Pseudoplatystoma coruscans*, all migratory species, declined but maintained populations partly because floodplain nurseries and lotic environments are found upstream of the reservoir. The fishery, originally based on characins, turned to a catfish orientated activity, essentially because these species had become the most abundant ones. Among these were several opportunistic small-sized species with short lifespan and batch spawning, such as the sciaenid *Plagioscion squamosissimus*, and the catfishes *Auchenipterus nuchalis* and *Hypophthalmus edentatus*, all of which have been reported as exhibiting very little migratory activity (see chapter 5). However, despite river regulation from the upper basin dams, resulting in a greatly attenuated flood pulse regime, most fish species in the middle Paraná have maintained seasonal cycles of abundance, reflecting potamodromous behaviour, similar to those evident in the pre-regulation period (Quiros & Vidal 2000).

Many tributaries of the Paraná River have also been affected by hydropower development, notably the Mogi Guaçu River where the lifecycle of *Prochilodus lineatus* was elucidated by Godoi and collaborators, and in which the area available to this species has now shrunk by about 20%. Similarly there have been changes in the migration patterns of *P. nigricans* in the Tocantins River, as described by Carvalho and Mérona (1986) after the building of a large dam at Tucurui, which most probably impacted on the fish assemblage in a way similar to that described for the Itaipú, although most of the evaluation remains to be done. For

South American species undertaking basin-wide migrations, such as the large pimelodid catfishes (*Brachyplatystoma* and *Pseudoplatystoma*), consequences of obstructions might be catastrophic (Bayley & Petrere 1989, Barthem *et al.* 1991), and somewhat similar to those for diadromous species. Also these species support both non-targeted and targeted fisheries (Ruffino & Barthem 1996; Barthem & Goulding 1997), and the socio-economic impact of their decline would be significant. Finally these large predatory catfishes consume enormous amounts of food, mainly fishes, and their decline would impact over the entire river ecosystem.

Africa also has many huge dams (Lemoalle 1999), the number of which continues to increase, mainly to address the growing needs for energy resulting from the steep increase of human populations in these regions. As depicted above for South American ecosystems, building of huge dams and creation of large impounded areas has caused profound changes in many fish assemblages and fisheries that were originally dominated by seasonal strategists undertaking long potamodromous migrations (the so-called white fishes in Africa; Welcomme 1979; Régier *et al.* 1989). After impoundment, assemblages and fisheries progressively became dominated by opportunistic and equilibrium strategists, notably cichlids which exhibit hypoxia resistance, parental care and restricted migrations. This general scheme has been observed in many impounded areas, such as in the Volta Lake, where characins declined a few years after the dam was built on the Volta River (1964) and were replaced by cichlids (mainly the tilapias *Oreochromis niloticus* and *Tilapia zillii*) and clupeids (*Pellonula leonensis*, *Odaxothrissa mento* and *Sierrathrissa leonensis*) (Petr 1986; see also Lowe-McConnell 1999).

In the Niger River, the building of several dams has strongly affected the fish fauna, with the Markala Dam, Mali, being well known for having interfered with the tinéni *Brycinus leuciscus* migration (Daget 1952; Laé 1992). Another well-documented example from the Niger River is the evolution of the fish assemblage after the building, in 1968, of the hydropower Kainji Dam in Nigeria (Ita 1984). Mormyrids and the mochokid catfishes which exhibit riverine habits and potamodromous migrations (sections 4.4 and 5.16) underwent a strong decrease in local diversity (35–40%) and a marked decrease in biomass. Prior to the impoundment they formed about 40% of the catches, by comparison to less than 4% after the impoundment. Characins and schilbeid catfishes, which also are migratory species (sections 5.15 and 5.16) maintained a similar diversity, but their relative abundance declined by about 50%. Conversely, cichlids, which originally represented negligible proportions of the catches prior to the dam being built, amounted to over 40% in the impounded area, occurring mainly in inshore habitats.

A different story comes from Lake Kariba, built in 1958 on the Zambezi River, essentially as this man-made lake is oligotrophic, and has the lowest morphoedaphic index of all African man-made lakes, due to its depth, reduced shoreline and rapid replacement time (Lowe-McConnell 1999). As in other man-made lakes, cichlid populations (mostly *Oreochromis mossambicus* and *Serranochromis condringtonii*) soared, but mormyrids and mochokids maintained their abundance. Characins, which were initially abundant, declined then rebounded (mostly the predatory tigerfish *Hydrocynus vittatus*) after the clupeid *Limnothrissa miodon* was introduced from Lake Tanganyika in the late 1960s to support inland fisheries. *Limnothrissa miodon* does just that as this species has represented about 85–90% of the landings in Lake Kariba since 1978, catches having soared from 5000 to about 30 000 tonnes per year since then, with no obvious sign of overexploitation of this clupeid (Lévêque & Paugy 1999; Lowe-McConnell 1999).

The more recent building of large dams on the Senegal River in Western Africa has also brought further knowledge on how civil works can interfere with the life history of migratory tropical freshwater fishes (syntheses in Albaret & Diouf 1994; Lévêque & Paugy 1999). The Diama Dam, completed in 1986, has been built at rkm 50 to minimise the entry of salt waters further inland and provide a large reserve of fresh water for irrigation purposes. The diversity of the estuarine fish assemblage, which comprises over 100 species, has been preserved as salinities did not change below the dam, except upon water release. However, the estuarine area, which originally extended over 200 km, has been constricted by about 75%, and most euryhaline species that spawned in areas where salinity ranged from 5‰ to 15‰ now suffer from a major decrease in habitat availability. Also, the blockage of salt water by the Diama Dam, together with the increasing use of fertilisers for agriculture, has resulted in a major growth of the water hyacinth *Eichornia crassipes* population which, in turn, has affected the oxygen content in the lower Senegal River. Two years after the Diama Dam was operational, another dam, the Manantali Dam, was completed at rkm 1250, on a major tributary, the Bafing River, in Mali, also for irrigation, water regulation and hydroelectric purposes. Water regulation schemes permitted by the Manantali Dam resulted in a major reduction in the duration of the high flow periods, which were further shortened due to water evaporation from the impounded area. Except for the blockage by the dam, the reproduction of most potamodromous migrants is deemed to be little affected, but recruitment seems to be impaired as fish have less access to the floodplain, which also becomes poorer as less nutrients are flooded from upstream reaches.

River impoundment and development in tropical Asia has had marked effects on migratory fish populations with the loss of many populations of anadromous tropical shads *Hilsa kelee* and *Tenualosa* spp., as well as the decline of large catfishes such as the giant Mekong catfish *Pangasianodon gigas* (Dudgeon 1992). As in Europe or North America, where most large rivers have become channelised, levees, embankments, polders and impoundments have been widely constructed throughout the floodplains of tropical rivers, to control or prevent flooding in agricultural and urban areas. The immense floodplain of the Ganges River encompasses most of Bangladesh and protection from flooding is a priority in this country, which is one of the world's most densely populated regions. In some regions, pumping stations and sluices have been installed to control water levels, and these operate along well-defined schemes, with sluices gates being closed during most of the high water season to prevent overspill (so-called flood control drainage and irrigation [FCDI] schemes). Halls *et al.* (1998) have highlighted how these civil works, although highly respectable, might impact heavily on fish assemblages by restricting the movements of fishes onto and off the floodplain.

To test this hypothesis, they combined tagging studies with surveys of fish diversity in areas outside and inside FCDI schemes in the Pabna district (northwest Bangladesh, rivers Padma and Jamuna), and found that the diversity inside FCDIs was lowered by 25 species. Most species absent inside FCDI schemes, or present in much lower numbers than outside FCDI schemes (Fig. 7.1) were migratory whitefishes, among which were several large catfishes (*Aila, Heteropneustes, Mystus, Silonia* and *Wallago* spp.), cyprinids (mainly *Catla, Labeo* and *Cirrhinus* spp.), clupeids (*Coroca, Gudusia* and *Hilsa*), the mugilid *Rhinomugil corsula* and some gobioids (*Brachygobius* and *Glossogobius* spp.). Species more abundant inside the FCDI schemes essentially comprised small, blackwater or greywater fishes such as the glassperch *Chanda nama* (Ambassidae), the minnow *Salmostoma phulo* (Cyprinidae) and the needlefish *Xenentodon cancila* (Belonidae). These were less abundant outside FCDI schemes, probably as they are preyed upon by several whitefish species. As these small spe-

Fig. 7.1 Average abundance (seine net catch per unit effort) of fish species sampled inside (solid bars) and outside (open bars) flood control drainage and irrigation (FCDI) schemes in northwestern Bangladesh (see text for details). Only species contributing to 75% of the cumulative average dissimilarities are shown, with decreasing differences from the top to the bottom of the figure. Modified from Halls *et al.* (1998).

cies have a low market price, Halls *et al.* estimated, that in addition to losses in biodiversity, FCDI schemes in northwest Bangladesh resulted in net monetary losses of about 25% to fishermen. However, they pointed out that irrespective of ecological and economical perspectives, these anthropogenic changes might contribute to improve the health of local populations, since small fish are eaten whole, thus providing higher levels of fat, calcium and vitamins.

7.3 Other impacts on fish migration resulting from man's activities

Obstruction of upstream migration by damming, loss of suitable habitat resulting from impoundment, and chemical pollution are the most frequently implicated causes of changes in fish assemblages, diversity and abundance. Here, we review some other types of impacts and mechanisms less obvious than those described so far. These include effects of water abstraction mainly as regards entrainment and impingement, changes in flow regime, temperature and oxygen, increased water velocity in culverts and road crossings, and modifications to genetic diversity or life history.

7.3.1 Entrainment and impingement

Turbine operation is the most frequently cited anthropogenic activity where fish are entrained and damaged. Whitney *et al.* (1997 in Coutant & Whitney 2000) estimated that mortality of fish passing at standard spill bays ranges from 0 to 2%, whereas it varies from 5 to 15% for turbine passage. Damage is due to several effects including mechanical strike, grinding, shear, variations in pressure which affect the swimbladder, and cavitation. Most turbine manufacturers claim their machines are designed not to cavitate. However, Turnpenny (1998) argues that this merely depends on how close to the design point they operate, and that levels of cavitation which might be negligible from the engineer's viewpoint might turn out to be harmful to the fish. The relative impact of these factors varies with fish behaviour and size (Fig. 7.2), and it also varies with flow regulation at the dam. Although most diadromous species and especially salmonids are frequently reported to be damaged by turbines, the same applies to many so-called 'coarse' fish species of lesser economic importance that are undertaking potamodromous migrations. Barus *et al.* (1984, 1985, 1986) investigated fish drift through hydroelectric turbines from Czechoslovakian reservoirs. Fish that passed through turbines were exposed to considerable physical trauma, and mortality of some species, particularly European eel *Anguilla anguilla*, was high. Their studies showed that migration from reservoirs consisted of both passive drift of juvenile stages and active migration of adults. Fish that survived passage through the turbines were shown to make a substantial contribution to the biomass in the river downstream of the dam and many were engaged in spawning activity. Berg (1986b) found similar effects on fish passage through Kaplan turbines at a power plant on the River Neckar, Germany. The most affected species there was the European eel, with the rate of lethal injuries reaching 50% even at relatively low flows (40 m^3 s^{-1}). Rates of injuries were in fact higher at low flows during the day when an adjacent sluice was opened. During the night, when the sluice was closed, water flow through the turbine was higher, leading to a higher relative opening of the runner blades and a reduced number of injuries.

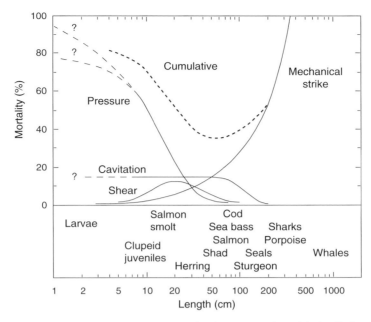

Fig. 7.2 Variations of mortality rate and its causes, resulting from passage through hydraulic, low-head turbines depending on the animal body length. Redrawn from Dadswell and Rulifson (1994).

Coutant and Whitney (2000) have produced a brilliant review of the impact of turbines and of the biological attributes affecting turbine passage and mortality during turbine passage, and Turnpenny (1998) has provided the mathematical bases and experimental support for appraising damage to fishes in low-head turbines. Both are principally salmonid-based articles, but this merely reflects the fact that the greatest amount of knowledge regarding turbine passage has been gathered on these species. They provide the functional bases for understanding how and to what extent fish are attracted or entrained into turbines, and the probability that they become damaged (Fig. 7.2). Ways of improving turbine design for reducing fish mortality and of diverting fish from turbines will be considered in section 7.4.

In addition to turbines, there has been growth in the number of installations removing substantial amounts of water for domestic and industrial supply, and many eggs, larvae, juveniles and larger fish are entrained at these. In most cases screens are used to retain debris, causing most medium-sized or large-sized fish to become impinged and die. Smaller fish and debris not retained by screens are trapped further away on finer grids, which are passed on rotating drums, and removed by high pressure jets that cause immediate death or lethal or sublethal injuries. Entrainment/impingement at water abstractions is a fate shared by many fishes throughout the world, especially at the cooling systems of power plants. The number of fish being impinged or entrained is proportional to the pumping rate supplying the condensers, and inversely proportional to fish size, by virtue of the relationship between sustained swimming speed and body size in fish (section 2.3). Fish entrained during floods, and eggs, larvae or juveniles making downstream migrations are the most frequent victims of these abstractions, as exemplified by several surveys originating from Europe and especially North

America, where this issue has been considered for decades. Forthcoming surveys from other regions of the world will probably contribute to make the picture worse than it already is.

Solomon (1992) reported that over 1000 1+ juveniles, 87 000 larvae 25–35 mm in length and an unknown quantity of fish larvae less than 18 mm in length (mostly roach *Rutilus rutilus*, dace *Leuciscus leuciscus* and chub *L. cephalus* for all groups) were captured in a louvre screen trap installed in the Walton Waterworks intake on the River Thames, England, between 25 April and 9 September 1989. Over 77% of entrainment occurred in 3 weeks between 15 June and 6 July, during normal, steady flows and since efficiency of capture was estimated to be 3–10%, total estimates of entrained 0+ fish over this 3-week period are 0.67–2.23 million 0+ fish. At Fort Calhoun power station on the Missouri River, Hergenrader *et al.* (1982) estimated that 1.5 million larval fishes, mostly freshwater drum *Aplodinotus grunniens* and several catostomid species, were killed passing through the intake each day in summer 1977, representing a mortality rate of 3.22% per day due to entrainment. Investigations in 1976–1977 in two power plants (Chalk Point and Morgantown) sited in Chesapeake Bay, Maryland indicated a combined rate of impingement of about 3.8 million fish per year, of which 56% consisted of Atlantic menhaden *Brevoortia tyrannus* (Clupeidae) which spawns in estuaries (Richkus & McLean 2000). At the Calvert Cliffs nuclear power plant, Maryland, a 21-year survey indicated that an average of 1.3 million fish were impinged each year, with annual peaks as high as 9.6 million fish (Ringger 2000). Michaud (2000) estimated that annual fish impingement at the three major Wisconsin electric power plants on Lake Michigan (Point Beach, Port Washington and Oak Creek) amounted to 5.85 million fish, among which were about 5 million alewife *Alosa pseudoharengus* (Clupeidae) and 0.8 million rainbow smelt *Osmerus mordax* (Osmeridae), which are known to undertake long migrations in rivers. The corresponding estimates for entrained larvae and fertile eggs was 9.2 million larvae (of which 68.5% were rainbow smelt larvae) and 17.6 million eggs (of which 98% were eggs of alewife). In other circumstances, the ratio between impinged and entrained fish can be much higher, and surveys of impinged fish might give a more biased picture. For example, at the intake of Presque Ile power station, impingement in the late 1970s amounted to less than 5000 fish per year, most of them also being alewives. However, as many as 8.8 to 11.0 million larvae were entrained there, among which 65% were burbot *Lota lota* (Gadidae) (thus giving a ratio of 1 : 2000 between fish being impinged and those being entrained; Michaud 2000).

Impingement rates show strong seasonal and daily variations. Richkus and McLean (2000) found that high impingement episodes in Calvert Cliffs and Morgantown corresponded to low dissolved oxygen, which corresponded to lesser swimming capacities. Ringger (2000) also reported that sudden changes in weather conditions, resulting in a sudden drop of water temperature of 3–4°C, also made fish more susceptible to impingement at Calvert Cliffs. At Chalk Point power plant, high impingement episodes typically corresponded to seasonal migrations (Richkus & McLean 2000), and so were the peaks of captures peaking in late June–early July observed on the River Thames (Solomon 1992). Similarly an ichthyoplankton net suspended in the water intake of the De Gijster Reservoir on the River Meuse, the Netherlands, captured 12 468 0+ fish, between 8 May and 18 July 1996 (Ketelaars *et al.* 1998). From this they estimated that 45.3 million 0+ fish were pumped into the reservoir over this period. Catches comprised mostly pikeperch *Stizostedion lucioperca* (36.5% by number), common bream *Abramis brama* (27.7%), roach *Rutilus rutilus* (23.7%) and Eurasian perch *Perca fluviatilis* (11.9%), with percids predominating initially and cyprinids later in the study period. Drift studies also provide evidence that entrainment mainly takes place at night. In Ketelaars *et al.*'s study, night catches were higher than in the day for all species except for pikeperch and constituted approximately 80% of the mean 24-h catch, a similar pattern to that found in other entrainment studies (e.g. Hergenrader *et al.* 1982). There is much evidence from different riverine environments that young fish mainly drift at night (e.g. Pavlov 1994) although this can extend over the entire 24-h cycle in turbid waters (e.g. Pavlov *et al.* 1995).

Irrespective of how impressive these values are, one should pay attention to how adverse or normative they are with respect to fish abundance and community structure. Coutant (2000) considers that 'if an intake structure does not move the aquatic ecosystem outside the normative range [based on expressions of normality discussed in his article], then no adverse impact has occurred'. Basically this concern relates to whether there is a continuation of a balanced community or not, this being intimately dependent on how far the community has already been modified by man's other activities, and on the life-history strategies of the species vulnerable to entrainment and impingement. An illustration of this statement is provided by Van Winkle (2000), who compared the potential impacts of power plants on different fish species with contrasting ages at sexual maturity, fecundities and mortality rates. He came to the conclusion that sturgeons would be affected to a greater extent than paddlefish *Polyodon spathula* and striped bass *Morone saxatilis*, and to a much greater extent than anchovies (Engraulidae), flounders (Pleuronectidae) or other species with high fecundity and small size at maturity, including many sciaenids and clupeids. For example, Spicer *et al.* (2000) estimated that about 10 million eggs and 5 million larvae of freshwater drum *Aplodinotus grunniens*, and about 9 million larvae and 0.6 million juveniles of threadfin shad *Dorosoma petenense* were entrained each year at the intake of Comanche Peak Steam Electric Station, Texas. They equated that the losses of these species due to entrainment represented less than the removal of 100 and 1000 adults, respectively.

7.3.2 Hydropeaking, changes in temperature and oxygen

In almost all countries, urbanisation and industrialisation have resulted in the multiplication of eutrophic or chemically polluted effluents, the proportion of which are processed in wastewater treatment stations varying substantially between countries. Effluents generally cause an increase in river temperature and increased biochemical oxygen demand, the amplitude of which is dependent on the density of human population, degree of industrialisation, and flows from tributaries. Dams reduce water velocity and often cause a further increase of temperature in non-thermally stratified rivers. Hydropower plants often pump the water from the bottom layer, which contains less oxygen, and may render downstream reaches more hypoxic.

The River Meuse, which flows from France, across Belgium and then the Netherlands, is a typical case of most of the problems that can be encountered in large obstructed rivers, including some unexpected issues (Descy & Empain 1984; Philippart *et al.* 1988). The Belgian River Meuse valley has become heavily industrialised over the past 150 years; it supports a high population density, and three nuclear power plants are sited in Tihange (rkm 230). Along the course of the river, a natural thermal regime is maintained from France until Namur, then the river warms as the River Sambre, a major tributary, brings the outflows of Charleroi's industrial basin. About 35 km downstream, when the Meuse reaches Tihange, the river has recovered a near-normal regime. The effluents of the Tihange nuclear power plants cause the water temperature to increase by 3–5°C, depending on the season, and it is not unusual for the mean water temperature to reach 20°C as early as mid-April, with peaks of over 27–28°C during summer. As the Meuse enters Liège, about 30 km further downstream, the cool water inflow from a major tributary, the River Ourthe, restores a near-normal temperature in the Meuse. While crossing Liège and the industrialised area downstream of it, the temperature rises again by several degrees and no major tributary flows into the river until the Dutch border, less than 20 km from Liège. The hydroelectric power plant sited at Lixhe Dam, on the Belgian border, is equipped with four fixed turbines each having a capacity of 80 m^3 s^{-1}. When the flow of the River Meuse is less than 320 m^3 s^{-1} one or several turbines are switched on and off at regular intervals, causing pulses of hypoxic water to be released at regular intervals. The oxygen concentration below the dam at Lixhe can be as low as 2 mg per litre, and temperatures are 2–3°C warmer than in natural situations. These conditions are deemed to impact heavily on migratory species, primarily on the coldwater oxyphilic salmonids, but also on many rheophilic oxyphilic cyprinids such as the barbel *Barbus barbus*. Fish approaching the dam may show avoidance behaviour, or incur an oxygen debt which could compromise their chances of ascending one of the two bypass facilities. These factors account partly for the relatively low numbers of rheophilic fishes ascending these bypasses compared to facilities further upstream in the Meuse (e.g. Prignon *et al.* 1998). The fluctuating discharge of the River Meuse downstream of Lixhe has also raised many issues in the Netherlands. Recent studies found that the large amplitude of discharge fluctuations (80 m^3 s^{-1} and sometimes 160 m^3 s^{-1}) was at least five times greater than the acceptable value of 15 m^{3}s^{-1} for maintenance of the biota (Salverda *et al.* 1996).

Different alterations of the natural thermal regime have been observed in other dammed river systems across the world, such as in the Czech Republic (e.g. Svratka River, Penáz *et al.* 1968; Malse River, Krivanec & Kubecka 1990; Vltava River, Kubecka & Vostradovsky 1995). In these situations release of hypolimnetic water from deep valley reservoirs results in water cooler than normal downstream of the reservoir in spring and summer and warmer water in autumn and winter. The lower water temperature and decreased trophic potential below reservoirs in the summer growth season has resulted in the substitution of the original fish community with coolwater fishes. The cascade of five reservoirs on the Vltava has resulted in greater abundance of large fish with increasing distance (and water temperature) downstream of the reservoirs (e.g. Kubecka & Vostradovsky 1995; Slavík & Bartos 1997). Slavík (1996b) also showed a significant relationship between species diversity and temperature in the Vltava River near Prague, but that immigration to downstream reaches via the Podbaba navigation channel substantially increased the abundance of fish there.

The benefits or damage to inland fisheries, of such habitat modifications, also depend on the possibility that fish living in impounded areas or below dams can access spawning grounds and that their offspring can find appropriate nurseries. Generally speaking, this is more problematic for lithophilous oxyphilic species than for others, as reduced water velocities in regulated rivers promote the siltation of gravel beds, which are also targeted by mineral extraction industries, and by civil works in rivers open to navigation. The scarceness of spawning habitat imposes additional constraints on lithophilous species, especially on brood hiders, including longer spawning migrations, the necessity of clearing several obstacles before reaching a spawning ground in upstream tributaries and migration delay (see case studies for barbel *Barbus barbus* in Penáz & Stouracova 1991; Baras *et al.* 1994a, 1996a; Lucas & Batley 1996; Lucas & Frear 1997). In large channelised rivers, many fish species have been found to spawn in the downstream pool's bypass facilities which were the only places where coarse substratum could be found (E. Baras & J.-C. Philippart [pers. comm.]).

The issue of larvae finding suitable nursery habitats, as described above, remains. Provided that food is abundant, warm temperatures normally favour fast growth, even in cool or temperate species, as the thermal growth optimum is generally higher in larvae and small juveniles than in adults (Jobling 1994). Nevertheless, early life stages have low absolute swimming capacities and are likely to be swept downstream upon water releases or hydropeaking, especially in channelised rivers with continuous embankment, thus resulting in longer upstream migration at an older age in species exhibiting homing behaviour. In this respect, Copp (1997) emphasised the role of marinas and off-channel refuges for young fish in regulated rivers. The problems associated with hydropeaking in a broader context, including habitat modifications and variations in species diversity, have been outlined in several publications (e.g. Petts 1984; Cushman 1985; Bain *et al.* 1988). Scheidegger and Bain (1995) provide a good example of how the distribution of fish larvae varies between free-flowing and regulated rivers. In mountain streams and tributaries, large hydropower dams have also been built, and a vast number of them operate a discontinuous discharge mode (e.g. as many as 144 dams of this kind in France; Liebig *et al.* 1996). The reservoirs are generally deeper than those of large regulated rivers, and flood pulses thus essentially consist of cold (4°C) water releases. Beyond the expected effects on habitat availability and use by fish (see above), low temperatures resulting from hydropeaking can impact on fish growth, as demonstrated for *Salmo trutta* in the Oriège stream (French Pyrenees; Liebig *et al.* 1996). They may also impair swimming performance of fishes (see section 2.3), thus restricting their capacity for migration.

Although hypoxia is the most frequently cited consequence of dams as regards the concentration of dissolved gases in the water, supersaturation may have effects that are just as severe, especially for characins and catfishes which are notoriously sensitive to supersaturation. Supersaturation below large dams may occur during large floods, when the incoming flood exceeds the capacity of turbines, and flows through overflow spillways. This has been found to happen at the two large dams built on the Paraná River during the 1980s, at Itaipú (frontier of Paraguay and Brazil) and Yacyretá (frontier of Paraguay and Argentina). The Yacyretá Reservoir is a shallow (averaging 7–8 m in depth) large (21 000 km^2) reservoir, limited downstream by a 64-km long, 20-m high dam, equipped with 20 Kaplan turbines, each of which can discharge 850 m^3 s^{-1} and produce 160 MW. They pass almost all incoming water during periods of normal flows (mean annual discharge of about 12 000 m^3 s^{-1}), and water pumped through the turbines is near-saturated with gases. It is equipped with two gated spillways (totalling 34 gates) with mobile valves to control flow and air ducts to reduce cavitation effects. This causes a substantial increase in the amount of air incorporated as the water plunges 20 m to the stilling basins below the spillways (Bechara *et al.* 1996). This applies to periods of high flows, and applied to a greater extent before all turbines were operational (1998). On each side of the powerhouse, which is sited next to the main spillway, are two large elevators that continuously transport fish above the dam. As described in section 4.4 most seasonal strategists undertaking long potamodromous migrations do so under increas-

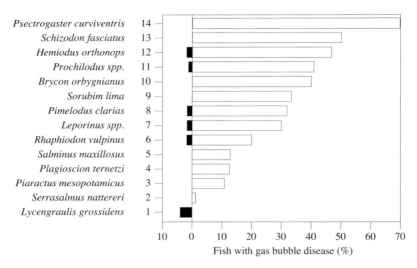

Fig. 7.3 Proportion of fish with gas bubble disease below the Yacyretá Dam, Paraná River (open bars) compared with a control site, 70 km downriver (solid bars). All fish species are potamodromous migrants (except for 2, and possibly 1), and support inland fisheries (except for 2). Characiforms: anostomids (7, 13), characids (2, 3, 5, 10), curimatids (11, 14), cynodontids (6) and hemiodontids (12); non-characiforms: engraulids (1), pimelodid catfishes (8, 9), sciaenids (4) (drawn based on data presented by Bechara *et al.* 1996).

ing water levels and high flows, i.e. at times when the risks of supersaturation below the dam are high. Bechara *et al.* (1996) compared the frequency of gas bubble disease between fish captured within 2 km below Yacyretá dam and fish of the same species collected 70 km further downstream (Fig. 7.3). They found that almost all large seasonal strategists, which also are key species in inland fisheries, were frequently affected, both macroscopically (emphysema, exophthalmia and/or haemorrhages) and microscopically (essentially lesions to the gills), with the herbivorous/iliophagous characins (anostomids and curimatids) being proportionally more affected than others. Bechara *et al.* also indicate from unpublished results that external emphysema might appear after 16 h exposure at 115% saturation. The duration of sojourn by migrants below the dam is unknown, but periods of more than one day might be reasonably expected, and in view of the lesions depicted above, there is little doubt that the migratory capacities of these fish are seriously impaired. Massive fish kills as a result of supersaturation were exceptional (Domitrovic *et al.* 1994 in Bechara *et al.* 1996), but dead fish were frequently found following peaks of supersaturation.

7.3.3 Culverts and road crossings

In some areas, watercourses are often directed through culverts under roads or navigation channels and these may act as a barrier to fish migration (review in Powers & Osborne 1984)

Road crossings, varying in design from simple fords to large concrete culverts, are potential barriers to the movements of small-stream fishes. Some may preclude movement by all fishes and have effects that are similar to those of impassable dams, differing only in the scale of the river environment and differences in fish communities (Winston *et al.* 1991). Other crossing types may act as seasonal barriers to fish movement, in a similar manner to riffles (Matthews *et al.* 1994). Due to the high frequency with which road crossings over streams occur in developed countries, the cumulative effect on migration and dispersal movements in such streams could be considerable. However, there has been little research to identify the extent of such damage or as to which structures are least inhibiting to movement. Warren and Pardew (1998) used mark recapture to examine the effects of four types of road crossings on fish movement during spring base flows in small streams of the Ouachita Mountains, Arkansas. They marked 6 113 individuals of 26 species and eight families of fishes and assessed movements for 21 fish species in seven families through culvert, slab, open-box and ford crossings and through natural reaches. No recaptured fishes (of three species) were found to have moved upstream or downstream past the slab weir which consisted of a low dam across the stream, with a 25 cm vertical drop-off at the downstream edge into the receiving pool. Of recaptured species 44% passed through culverts, 58% of species moved through open-box crossings, 77% through fords and 83% through natural reaches. For tests of association with passage at weir types the authors considered four families: Centrarchidae, Cyprinidae, Fundulidae and Percidae. Fish movement through crossings was bi-directional; no differences were detected between upstream and downstream movement across crossing types and natural reaches. Total fish movement was an order of magnitude lower through culverts than through open-box and ford crossings or natural reaches. Water velocity at crossings was inversely related to fish movement; culvert crossings consistently had the highest velocities and open-box crossings the lowest. Similarly, Toepfer *et al.* (1999) found that water currents in most corrugated-pipe culverts through road crossings in Oklahoma were much faster than those sustainable by the threatened leopard darter *Percina pantherina*. They concluded that culverts might represent true obstacles to the migrations of this species, depending on flow regimes and between-year variability in flows, thereby affecting small, localised populations.

A way to reduce water velocity in culverts is to obstruct the flow inside the culvert, taking advantage of shear stress that reduces water velocity in the boundary layer. This can be achieved through fitting culverts with baffles or using corrugated (roughened) pipes, with annular or spiral corrugations. Bates and Powers (1998) demonstrated that corrugated pipes enabled juvenile coho salmon *Oncorhynchus kisutch* to ascend higher slopes than smooth pipes, and that spiral corrugations were more effective than annular corrugations. However, they pointed out that salmon did so only when turbulence was low, and stressed the need for defining threshold levels of turbulence, and how these affect fish of different sizes. Warren and Pardew (1998) also provided evidence that for some fishes water velocity was not the sole determinant of passage. Topminnows (Fundulidae) exhibited low levels of movement through culvert and open-box crossings; the former with high water velocities, but the latter with low velocities. Similarly, some culvert crossings inhibit passage of alosine shads, not because of excessive water velocity but due to low ambient light levels (M. Moser [pers. comm.]; see also section 7.4.7.3). Thus behavioural responses to physical structures are of great importance, a fact which has to date received inadequate consideration.

7.3.4 *Changes in genetic diversity and life history*

Even when biodiversity and biomass are maintained in man-modified environments, one may not assume that environmental changes have had no impact at all. There is increasing evidence that a large number of fish species undertaking diadromous or potamodromous migrations exhibit homing behaviour, of which the functional mechanisms and ecological advantages have been discussed in detail in chapter 2. This picture reflects situations where fish have evolved mechanisms in response to more or less predictable environmental conditions, or rhythmic (daily and/or seasonal) patterns of environmental changes. As long as the environmental characteristics that have elicited the selection of particular traits prevail, homing will remain the most adaptive strategy, and will contribute to further genetic differentiation, as the species increasingly becomes divided into a series of reproductively isolated populations. A consequence of such strong site fidelity is that gene flow is largely restricted to within the population of fish that home to that location, and this too will favour the selection of traits that enhance homing by the offspring. In these circumstances, only individuals which 'accidentally' find their way to a different spawning site (namely the strayers) will maintain any gene flow between populations (Wootton 1990). Straying during diadromous or potamodromous migrations might serve to maintain heterozygosity within a population, and serve to maintain a large gene pool. This has been considered to be beneficial in terms of improved fitness for survival in the event of small or larger scale environmental changes (synthesis in Carvalho & Pitcher 1994). At the same time, genetic differentiation in closed populations is thought to reflect local adaptations (Wootton 1990). Straying may also be an appropriate strategy in harsh or unstable environments, where the probability of conditions having become less suitable during absence from the home site, is higher (Quinn 1984; see also sections 4.2 and 5.23). In the 'chess game' that fish play against Nature, both homing and straying are conservative and predictive strategies that may serve to anticipate the moves of the adversary, whether these consist of stasis or change.

Implicit in this view is that fish have the opportunity of migrating freely in between places, thus reinforcing the role of longitudinal and lateral connectivity in river systems. Within this context, Gollmann *et al.* (1998) investigated the effects of damming on the genetic variability of several European cyprinids between and within reaches downstream and upstream of man-made obstacles (the Iberian nase *Chondrostoma polylepis* in tributaries of the Tejo River, Portugal; the chub *Leuciscus cephalus* in the Danube, the Rhone and two Greek rivers, the Ardas and the Argitis, roach *Rutilus rutilus* in the Danube and the Rhone, and nase *Chondrostoma nasus* in the Danube). The working hypothesis was that variations between reaches, combined with low variation within reaches, reflects fragmentation and inbreeding. In the Rhone and Danube, no loss of genetic variability in nase, chub or roach was found as a result of damming, even for areas having been physically separated from others for about 100 years. Regarding the Rhone, it has been suggested that this maintenance of heterozygosity reflected fish being flushed from upstream reaches during large floods (Pattée 1988). Genetic variability of chub in the small Greek rivers was low, and the heterozygosity of the Iberian nase above the Castelo de Bode Dam, the largest dam in Portuguese inland water, was about half that of other places. Gollmann *et al.* suggested that the impact of damming on the River Tejo had interfered with the genetic flow there to a greater extent than at other sites, precisely because of minimal influences on genetic structure from further upstream, and because the populations were smaller. The latter factor has also been invoked to account for the low degree of heterozygosity among headwater populations of the European grayling *Thymallus thymallus* (Bouvet *et al.* 1992).

Obstruction by dams may result in changes of migratory history among populations, the extent of the change being dependent on how restrictive obstacles on the migration route are to upstream and downstream movements. Morita *et al.* (2000) compared the migratory behaviour of the anadromous white-spotted char *Salvelinus leucomaenis* in accessible below-dam river sections and in inaccessible above-dam river sections in Hokkaido, Japan. The proportion of char descending the river as smolts was greater in the below-dam than above-dam section, and vice versa for the proportions of 'residents'. Char translocated above the dam prior to possible smoltification also became resident, and showed faster growth than those left below the dam, but not faster than those originally found above the dam. Morita *et al.* concluded that residency in young white-spotted char was promoted by fast growth and low density, thereby reflecting phenotypic traits not genotypic differences. Nevertheless, it is acknowledged that genotype plays a part in determining variations in migratory behaviour between populations of many species (see section 2.2.1.1). In rivers where obstacles are few, migrants can achieve access past potential obstacles (i.e. through bypass facilities), but where the occurrence of numerous obstacles is likely to jeopardise the anadromous strategy, a similar shift to residency or potamodromous strategy may take place (e.g. *Salmo trutta*, Box 7.1). Finally, as mentioned in the preceding sections, damming may contribute to potamodromous migrations becoming longer than in natural environments, as a consequence of river profiles and habitats being more homogeneous, and spawning places scarcer, notably for lithophilous species.

7.4 Mitigation of hazards and obstacles to fish migration

The growing list of impacts of man's activities on fish species, and especially on migratory species, has led to a recent shift in emphasis of seeking to understand the effects of barriers to migration from just a few species to the wider migratory fish community (Cowx & Welcomme 1998; Jungwirth *et al.* 1998). In so doing, questions concerning the impact of obstacles on freshwater fishes are increasingly being framed in terms of fish ecology and behaviour rather than predetermined notions or legislative prescriptions of 'migratory' and 'non-migratory' species. The Biodiversity Convention may have played a valuable part in this context, since it has encouraged governments to consider environmental impacts on species and communities which traditionally may not have been economically important. It has stimulated an ethos, and funding, for determining and tackling environmental effects at community and ecosystem levels. This inclusive approach is appropriate to maintaining the ecological integrity of freshwater communities by considering, as applied to freshwater fishes, the range of access to habitats required by all fishes rather than just for a few favoured species.

In the present section, we focus on water abstraction sites and physical barriers to migration. However, while not acting as physical obstructions, chemical or thermal pollution effects may also act as barriers to fish migration (e.g. Priede *et al.* 1988). In many cases, these problems are exacerbated during warm, dry periods when flows are low, exacerbating problems of oxygen depletion. Such conditions may delay river entry by anadromous fishes or, in freshwater conditions, may result in trapping of fishes in hypoxic regions. Water release from dams or use of aeration equipment may improve water quality in these conditions and enable migration to more suitable habitat.

Box 7.1 Facultative diadromy of *Salmo trutta* in man-modified environments

In the trout *Salmo trutta*, three major forms or ecotypes have been identified: the sea trout *S. trutta m. trutta*, which is anadromous, the lake trout *S. trutta m. lacustris* which makes spawning runs into lake tributaries, and the riverine brown trout *S. trutta m. fario*, which can spend its entire life in a small stream or brook. Their morphology and genetic traits are similar (Guyomard 1991), and different forms may coexist within the same river (e.g. Baglinière *et al.* 1989). When reviewing the literature on migratory trout, Northcote (1978) reported that a part of the progeny of sea trout could become freshwater-resident. The opposite phenomenon was reported by Guyomard *et al.* (1984), who found that the three forms of trout were encountered in the rivers of the Kerguelen Islands, in the southern hemisphere, where eggs from a single progeny of brown trout had been introduced. These observations support the idea that these are indeed forms adopting different responses to environmental conditions, and not different subspecies of trout (review in Baglinière & Maisse 1991).

It is generally assumed that facultative diadromy in a species indicates that none of the space utilisation patterns provides a definitive strategic advantage over the other(s) (McDowall 1988). The sea trout is viewed as gaining a positive trade-off of enhanced growth in the nutrient-rich marine environment set against the hazards of long migration, whereas the opposite trade-off applies to the stream-'resident' brown trout (Whelan 1993). In river systems where freshwater habitat becomes less suitable for maximising lifetime reproductive output and survival of the trout's progeny (e.g. northernmost latitudes, where growth potential in fresh water is very low), anadromy might prevail. By contrast, in man-modified environments where numerous obstacles interfere with the return of migrants, stream residency becomes the most adaptive strategy. Recent telemetry studies have further documented that individual resident trout could become anadromous or vice versa (Arnekleiv & Krabol 1996), indicating intra-individual variability in behaviour and life history.

Increasingly, there have been reports of sea trout-like fish (large, silvery trout) being found in the lower reaches of many large regulated rivers, and which are regarded as returning adult sea trout. Radio-tracking of some of these trout over several months in the heavily dammed River Meuse Basin (there are at least 25 dams on the Dutch and Belgian stretches of Meuse) revealed that individual fish displayed restricted home ranging behaviour within 100 m downstream of a given dam (M. Ovidio, J.-C. Philippart & E. Baras [unpubl.]). These fish frequently chased through schools of small cyprinids (mainly bleak *Alburnus alburnus* and roach *Rutilus rutilus*) that accumulated below the dam. Trout exhibiting this behaviour were found to grow as fast as 17.5% BW per month over 5–6 months, which is much higher than for any typical 'resident' brown trout, and as fast as the growth of sea trout during their growth period at sea, where their diet is dominated by fish (Pemberton 1976; Berg & Jonsson 1990). It is therefore assumed that the restricted migration pattern exhibited by these trout (dubbed 'dam-sitters') is an adaptation to environments with limited longitudinal connectivity, enabling dam-sitters to gain equivalent nutritional advantages to the anadromous migrants, but with lower risks of predation or migration failure.

7.4.1 Fish passes

The most appropriate conditions for diverse and balanced freshwater fish communities include good longitudinal and lateral connectivity of river systems. In ameliorating the effects of physical barriers to freshwater fish (including diadromous species) migration, Cowx & Welcomme (1998) consider that:

> 'generally the soundest solution ecologically is to remove structures as this not only restores longitudinal connectivity but can also lead to the more general restoration of the habitat'.

The return to pristine habitat conditions is a challenge which might be desirable from an ecological viewpoint, but which can rarely be achieved for many reasons owing to the development of human populations and their activities (Cowx & Welcomme 1998; Lucas & Marmulla 2000). Where this is not feasible, trapping and hauling of fish past barriers might be considered, and this is the case for about 16% of the hydroelectric power plant dams in the United States (Cada 1998). Fish passes, including fish locks and elevators may be used to mitigate difficulties of passage past physical barriers, and so can the proper management of navigation locks in regulated rivers open to boat traffic. In many countries, legislation is compulsory in restoring the longitudinal connectivity at man-made obstacles, and approval is only granted if design specifications are appropriate, and increasingly, only if the pass is shown to operate properly (e.g. United Kingdom: Cowx 1998; France: Larinier 1998; US: Cada 1998). However, there are still limitations here since, for example, in the UK such approval schemes are rigidly applied only with diadromous and holobiotic salmonid fishes in mind, rather than the broader migratory fish communities. Moreover, post-appraisal studies are frequently lacking in many countries, and many projects have no detailed performance criteria and no performance monitoring requirements. Also, it has become obvious that most fish passes existing for decades in Europe and North America had been conceived to facilitate the passage of salmonids and some other anadromous species, with no or little consideration for those freshwater species deemed to be resident. In many cases, monitoring schemes have been exclusively focused on the diadromous species, until very recently. Even though managers or authorities have become increasingly concerned with the idea that fish passes should also serve to allow the migration of potamodromous species, they have identified the dearth of knowledge on the migratory capacities of these species, size of migrants, seasonality of their movements, and the need for tools for monitoring their movements, as factors limiting provision of fish passes for these species.

Whichever design is used, fish passes should meet several major criteria (see Box 7.2, modified from Cowx & Welcomme 1998).

Fish passes have long been used in attempts to aid the movement of fish across obstructions in riverine systems but there have been considerable improvements in fish pass technology, design and evaluation over the past 20 years, for which an impressive number of good quality reviews and comprehensive syntheses have been produced (e.g. Mills 1989; Larinier 1992; Clay 1995; Cowx & Welcomme 1998; Jungwirth *et al.* 1998; Odeh 1999). Hence, only a brief overview of these fish pass systems is given here, and we encourage the reader more specifically interested in fish pass technology and design to refer to these indispensable references.

> **Box 7.2 Recommended criteria for design of fish passes (modified from Cowx & Welcomme 1998)**
>
> (1) Provision of comfortable passage for all migratory species, including the poorer swimmers, over the entire length of the fish pass, therefore requiring provision of refuges from fast currents at regular intervals, the spacing of which should be intimately conditioned by fish swimming endurance (see section 2.3)
> (2) Year-round functionality, under different flow regimes, temperatures and oxygen levels, notably to enable fish displaced by floods to return to their initial position
> (3) Sufficient carrying capacity to meet requirements for massive upstream ascents during reproductive or trophic migrations, thereby emphasising the need for a deeper knowledge of the fish community, fish stocks and stimuli that elicit migratory behaviour (see sections 2.2 and 4.4)
> (4) Positioning of the entrance of the fish pass so that it is readily accessible to migrants
> (5) Attraction of fish to the fish pass entrance and deterring them from dead-ends or dangerous places
> (6) Positioning of the upstream outlet of fish passes far enough from spillways and turbines to minimise the risks that fish having cleared the pass be swept downstream or damaged

7.4.2 Fish ladders

Fish ladders are the most common measures implemented to mitigate the effect of physical obstacles on upstream passage. As well as requiring a good knowledge of the behavioural attraction of fishes to the entrance and into the pass, sadly lacking in many cases (see section 7.5), they demand knowledge of the swimming performance of the fish. Such tests may be carried out in flumes (e.g. Beamish 1978) or from muscle physiology measurements (e.g. Wardle 1977) as described in section 2.3, but these tests usually apply only to laminar flow or still-water conditions. Because fish ladders operate by utilising the shear stress imparted by obstacles to the water flow to reduce water velocity, they are inherently turbulent environments, though to varying degrees. These environments increase drag forces on fish swimming and may also cause difficulties in orientation for some fish species, inhibiting upstream progress through the fish ladder. Extrapolation of swimming performance measurements from water flume or physiology tests to working fish ladders is, at best, fraught with error, and will at worst severely overestimate the performance of fish under natural conditions. For these reasons testing of actual fish pass designs under controlled laboratory or field conditions is important in allowing the development of suitable fish pass designs. Importantly, it can allow observation of the behaviour of different species and sizes of fish within the pass. Such an approach has been adopted at several research laboratories concerned with studying migration of freshwater and diadromous fishes and providing mitigating measures at barriers (e.g. Monk *et al.* 1989; Mallen-Cooper 1992, 1994; Castro-Santos *et al.* 1996; Haro *et al.* 1999). In these cases fish ascend the experimental pass at their own volition, rather than being forced to swim in a flume or tank. Water discharge and other conditions can normally be manipulated to enable fish ladder performance to be measured under varying conditions (Fig. 7.4). This fundamental difference can still cause problems in assessing fish pass designs, since motivation to swim through a fish ladder varies between and within individuals, depending on physiological state and environmental conditions, and requires careful thought to experimental design. For these reasons, field monitoring under natural conditions remains a vital aspect of fish ladder development, as well as for other fish passage facilities such as fish locks and surface bypass systems (see section 7.5).

The type of fish ladder used most frequently is the 'pool fish pass', which consists of a series of pools in steps leading from the foot of the obstruction to the top. Walls separating the pools have weirs, notches, vertical slots or submerged orifices which control the water level in each pool and the water discharge in the pass. Pools serve to provide resting areas for fish and ensure proper dissipation of the energy of water flowing through the pass. The slope of the fishway usually varies between 1% and 15%, depending on budget and space available at the time of construction, although increasingly gradients of less than 5% are preferred for passing the broader migratory fish community rather than targeted species. The drop per pool varies from a few centimetres to several tens of centimetres. Larinier (1998), by reference to fish assemblages in French rivers and streams, considers that drops of 30–60 cm are particularly suitable for salmonids, whereas drops less than 30 cm are better adapted to the migrations of shads (*Alosa* spp.) and 'resident' (potamodromous) fish species, including cyprinids, esocids and percids. Vertical slot designs are particularly suitable in rivers or streams undergoing strong variations of discharge, as they remain operational at a greater range of water column heights than most other designs (Clay 1995).

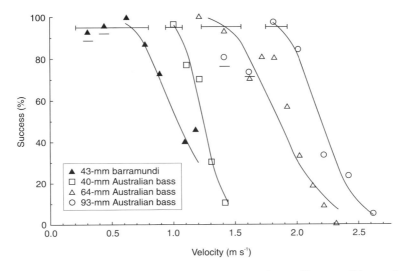

Fig. 7.4 Percentages of juvenile Australian bass *Macquaria novemaculeata* and barramundi *Lates calcarifer* successfully ascending a three-chamber section of an experimental vertical-slot fishway at different water velocities. Underlined data points not included in best-fit lines. Redrawn from Mallen-Cooper (1992).

Fishways using baffle designs were pioneered in Belgium in 1909 by Denil, and increasingly have been used in a large number of countries. Baffle technology has been progressively refined, from the plane design originally proposed by Denil, to multi-plane (herringbone patterned) designs using thin baffles made of prefabricated steel or chevron designs using square-sectioned timber baffles (Denil 1938; Larinier 1983, 1992). Fishways using plain baffles may operate at slopes between 15 and 22%, and the width of plane baffles ranges from 0.6 to 1.2 m. Fishways with herringbone or chevron designs may have baffles placed on the bottom only, and not exceeding 0.2 m in height, with the sides of the fishway remaining smooth. Their slope does not exceed 16% (after Larinier 1998). The 'Alaska steep pass' design of Zeimer, developed in 1962, has chevron baffles placed on the bottom and side baffles, and is designed to be prefabricated in bolt-together sections for use in remote areas. Fishways with baffles have originally been designed to facilitate the upstream passage of anadromous salmonids, and their efficiency is generally greater for large fast-swimming fish than for smaller fish or species with lesser swimming capacities, essentially as water velocities in this type of fishway under the normal operating discharge range (0.25–1.40 $m^3 s^{-1}$) are close to, or exceed, the burst speeds of those fishes with low absolute swimming performance. Nevertheless, fishways with baffle designs have been shown to permit the ascent of many non-salmonid species (e.g. Belgian River Meuse, Baras *et al.* 1994a), especially when they are operated at water flows much lower than normal, and resemble small pool-type fishways. Baras *et al.* (1994b, 1996b) so demonstrated that about 10 000 European eel *Anguilla anguilla*, sometimes as small as 10–12 cm, ascended yearly a Denil fishway operating at about 0.05 $m^3 s^{-1}$ on the Ampsin-Neuville Dam on the River Meuse, whereas a few dozens or hundreds were found ascending another fishway of the same design and sited on the same dam, but operating under a normal flow range.

7.4.3 Fish locks and elevators

Fish locks (Borland's design or modified) and elevators are not the most commonly implemented mitigation measures to physical obstructions, but Cada (1998) reported that they represented as much as 12% of the facilities provided at hydropower dams in the US, and a greater proportion at tall dams. Fish locks and gallery collections have been constructed in several countries, including Scotland, Ireland, the US, France and Norway. In France, the installation of future fish locks is no longer considered, essentially as holding pools are restricted in size, excessively turbulent and their efficiency varies considerably between species, with some species preferring to remain in the lock instead of passing into the forebay (Travade & Larinier 1992; Larinier 1998). Grande and Matzow (1998) describe a pressure chamber fishway enabling salmonids to pass through the Rygene Dam (Nidelva River, Norway) during their upstream movement. Fish enter an antechamber, then a pressure chamber and finally a rear-chamber with an exit into the upstream reservoir. This design, which had been tested previously in the US (McNary Dam, Columbia River) is simpler and less expensive than a standard Borland fish lock, but information on its efficiency is generally scarce. Adam and Schwevers (1998) provide an example of a collection gallery on the Lahn River, Germany, with the gallery leading fish to large ($3 \times 1.5 \times 1.5$ m) fine mesh weir baskets which are then transported upstream by remotely operated cranes, this being the basis for simple fish elevators.

Fish elevators are deemed to represent the most cost-effective mitigation measures for tall dams, both for biological and economic reasons. As the fish are moved upstream with exogenous energy, the chances of clearing the physical obstacle are theoretically independent of its height. Similarly, their construction cost is (almost) independent of the height of the dam, and they can operate at variable water levels. However, where large numbers of fish are expected to attempt to pass the obstacle, large holding pools are preferred to baskets or hoppers. Fish trapped in the holding pools are pushed by remotely operated vertical screens into a tank, which is raised and emptied upstream of the dam, the minimum operational cycle being about 10 minutes. Such elevators have been operating since the early 1980s in many locations, including the Holyoke Dam on the Connecticut River, US, and the Golfech and Tuillières dams, both in France, on the Garonne and Dordogne rivers, respectively. Travade *et al.* (1998) have compared the efficiency of pool-type fishways and elevators on these two French rivers, and concluded that elevators permitted the passage of all fish species occurring in the downstream reaches, although their efficiency was lower for small species, including the European eel *Anguilla anguilla* than for larger species. Bellariva and Belaud (1998) provide a more detailed case study of the passage of allis shad *Alosa alosa* in the Golfech elevator, emphasising the role of water temperature and discharge on the between-year variability of shad passages, although they indicate that these factors might impact on fish motivation and not on the efficiency of the bypass structure.

At the Holyoke Dam, two elevators have been constructed, one in the powerhouse tailrace in the mid-1950s, and another one at the base of the spillway in 1976. Kynard (1998) reports that these elevators represent one of the most efficient bypass structures on the Atlantic coast of North America, as they annually transport 0.5–1.0 million fish belonging to seven diadromous species, and several thousands of potamodromous fish. The major drawback of elevators is their functioning costs, including maintenance. Therefore not all elevators are operated continuously, nor is their lifting frequency maximal at any time of the year. As such, their efficiency may vary considerably between species, with the emphasis being laid again on species of commercial importance, to the detriment of other migratory fishes. The way the Holyoke elevators have been operated from 1976 to 1996 is a good example of this (Kynard 1998). The elevators have been operated over two main periods each year, from late April to early July for passing American shad *Alosa sapidissima* and Atlantic salmon *Salmo salar*, and during September and October for the autumn run of Atlantic salmon. During springtime, the two elevators were operated 7 days per week with over 30 lifts per day, whereas during autumn, only the tailrace elevator was functional over 5 days per week, and was lifted no more than 3–4 times daily. Kynard found that shortnose sturgeon *Acipenser brevirostrum* essentially occurred at the dam from late May to early October, and clearly suggested that changes in the normal operation schedule would increase the number of sturgeon ascents, and permit fish to pass over the obstacle in better condition. This probably applies to many other fish species not sharing the same seasonal migration patterns as shads and salmons.

7.4.4 Shipping locks and elevators

The use of navigation locks by fish is probably more widespread than suspected in regulated rivers open to boat traffic, although there has been little quantification of the effectiveness of these structures for fish passage. Klinge (1994) showed that roach *Rutilus rutilus* and common bream *Abramis brama* were capable of passing through a shipping lock on the Rhine in the Netherlands but that migration was considerably impeded with only 7 out of 100 marked fish passing through. On the basis of acoustic telemetry investigations of adult American shad *Alosa sapidissima* at a low-elevation dam on the Cape Fear River, Maryland, navigation locks provided a much more efficient method of passing shad than a steep pass fishway, especially when lock operation was modified to suit shad behaviour (Moser *et al.* 2000). Jolimaître (1992, in Larinier 1998) demonstrated that more than 10 000 shad *Alosa* spp. (mainly the Rhone shad *Alosa ficta rhodanensis*) passed through the Beaucaire navigation lock on the Rhone River in 1992, and Zylberblat and Menella (1996) have reviewed the conditions promoting their upstream passage through this structure. While investigating the ascents of eel *Anguilla anguilla* at the Ampsin-Neuville Dam on the River Meuse in the course of a mark-recapture study, Baras *et al.* (1996b) concluded that the vast majority (>95%) of eels passing this physical obstacle were probably doing so through the shipping locks. This was rendered possible as those locks were operated before sunrise and after sunset, thus matching the two major daily activity patterns of eels, what might not be the case in other places.

Ship elevators are not as frequent as navigation locks, but they too should be considered as possible ways of enabling fishes to bypass physical obstacles. The largest ship elevator in Europe has been recently constructed on the Canal du Centre at Strépy-Thieu, Western Belgium, to transport ships of up to 1350 t over a difference of altitude of 73 m. It will begin operating early in the twentyfirst century, and could be operated in a manner which would assist fish migration.

7.4.5 Nature-like fish passes

Over the past decades, there has been an increasing trend towards the design of nature-like fish pass systems, which include inclined planes or rocky ramps at weirs (see descriptions in Cowx & Welcomme 1998). Rock-ramps mimic the flow conditions in rapids, cascades or riffles, although the slope (about 1 : 20) is generally steeper than in natural riffles. As they are built with locally available material, they have a low construction cost and may be stable, even at high flows, when properly designed and constructed. The main advantages of these structures is their low cost and positioning inside the river channel, which makes their entrance more easily located by migrating fish. Coarse substratum might enhance the ascent of fish species unable to migrate against the current over smooth structures, such as bullhead *Cottus gobio* or stone loach *Barbatula barbatula*. Also, they offer a wide range of water velocities, thereby permitting the passage of fish of different sizes. The same approach has also been used to aid passage of fish through, and habitat availability in, long box culverts (Slawski & Ehlinger 1998). Harris *et al.* (1998) evaluated four rock-ramp fishways in two rivers of New South Wales, Australia and concluded that they were efficient for most fish species, including small-sized fish or species. They found no difference between the body length distributions below and above the weirs for the striped gudgeon *Gobiomorphus australis*, the empire gudgeon *Hypseleotris compressa* and the flathead gudgeon *Philypnodon grandiceps*, all of them about 8 cm in fork length). At the Goondiwindi weir on the McIntyre River, Australia, the entrance of the rock-ramp fishway was further downstream of the weir crest, and it was found to be less effective at high flow, except when a fence was used to guide the fish at the entrance of the pass.

For obvious reasons, the use of slopes or rock-ramps is restricted to weirs with relatively little difference in height. At taller weirs or dams, nature-like bypass channels surrounding the obstacle can be built, provided that room is available, which is not always the case in densely populated valleys or industrialised regions. These structures have received growing attention over the past 10–15 years (Jungwirth & Pelikan 1989; Schmutz *et al.* 1995; Jungwirth 1996), and they have been implemented in several European countries, notably in Austria, Germany and the Netherlands. In the Netherlands, seven high dams that had prevented the upstream migrations of anadromous fish species for decades, have been or are being bypassed in this way.

Among the reasons for the success of these passes is their capacity to offer permanently a wide range of water velocities that enable fish of different sizes to make spawning, trophic or compensatory upstream movements, as well as downstream migrations. Additionally, because of the naturalistic design of the channel these bypasses provide habitat for resident fish (Jungwirth 1996), and may provide valuable spawning places for lithophilic or phytophilic species in channelled and impounded rivers. Their cost is essentially dependent on manpower requirements, and is thus highly variable between countries and climates, but they are generally regarded as cheaper than formally engineered fish passes, both as regards construction and maintenance (Gebler 1998). These arguments suggest that such structures should receive increasing attention in tropical regions, although maintenance costs may be more variable depending on climate. Parasiewicz *et al.* (1998) provide the conceptual guidelines for defining and planning such bypass channels, and detailed examples from the Austrian and German experiences can be found in Schmutz and Unfer (1996), Eberstaller *et al.* (1998), Gebler (1998), Mader *et al.* (1998) and Steiner (1998).

7.4.6 *Typical downstream passes: surface bypass systems*

Basically each and every fish pass system depicted above can be used for upstream or downstream passage, although these systems are generally aimed to facilitate upstream movements, and their innocuousness for downstream passage might be questionable (e.g. baffled pass designs). As concerns downstream movements, there have been numerous reports that passage of fish over dams of moderate heights caused little damage, and in any case little damage compared to the risks incurred during turbine passage (see section 7.3.1). Surface bypass systems are defined as 'surface-oriented forebay outlets that provide sufficient depth, velocity and volume to attract and pass fish (downstream)' (Ferguson *et al.* 1998). Their efficiency relies on the basic principle that fish migrating downstream are generally oriented to the upper portion of the water column, which is generally the case for salmonid smolts but not for many non-salmonid species, which show different patterns of water column utilisation between species, life stages and between seasons (review in Coutant & Whitney 2000).

Biological criteria for designing surface bypass systems have been reviewed by Ferguson *et al.* (1998) and Larinier and Travade (1999), including passage efficiency, fish injury during passage, increased susceptibility to predation and migration delays, and their variations depending on flow regimes at different seasons. Of utmost importance in the design of surface bypass systems is the hydraulic profile in the so-called zone of influence, upstream of the system, notably as regards turbulence. For example, Ingendahl *et al.* (1996) attributed many aborted passages of salmonids (Atlantic salmon *Salmo salar* and sea trout *S. trutta*) in the bypass of Soeix Dam, in the French Adour River system, to turbulence and current accelerations. They increased flow discharge in the bypass and suppressed upwelling from the headrace bottom, resulting in more regular hydraulic conditions, and the efficiency of the bypass increased from 32% to 58% for salmon and from 35% to 70% for sea trout. Haro *et al.* (1998), Chanseau *et al.* (1999) and Croze *et al.* (1999) have provided further examples of how important hydraulic conditions are in operating surface bypass systems. Not surprisingly, most of the knowledge on how these systems operate has been collected over the past 10–15 years, as a result of the considerable technological advances in the field of fine monitoring of fish behaviour through remote sensing techniques, and especially hydroacoustics and biotelemetry (e.g. Johnson *et al.* 2000; see also chapter 6). At present, most available information refers to salmonids, although there is an increasing interest in potamodromous species (Larinier & Travade 1999; Coutant & Whitney 2000).

7.4.7 Attracting and deterring fishes

In addition to the success or otherwise of a fish pass in enabling the comfortable upstream or downstream passage of fish, its efficiency also depends on how attractive the structure is to fishes, and how long they take to find the entrance and to enter the pass. Cowx and Welcomme (1998) provide a comprehensive assessment of the appropriate siting of fish pass entrances on oblique or perpendicular obstacles, and more detailed considerations can be found elsewhere (Larinier 1992; Clay 1995; Larinier & Travade 1999). The Ampsin-Neuville Dam on the River Meuse is a good example of an initially acceptable design (two Denil fishways sited on the left and right sides of the spillway) which subsequently became poorly adequate after a small hydropower plant was built between the spillway and the left bank. Rheophilic species are attracted by the main outflows of turbines, and only find the entrance of a fish pass during high flows, when water is also flowing from the spillway. The spawning migration of several native rheophilic cyprinids takes place during an increase in water temperature (see section 2.2) and coincidence between high flows and temperature increase is exceptional, thereby reducing considerably the probability they find the entrance of the pass (Baras *et al.* 1994a). A similar conclusion has been drawn for upstream passage of migratory European 'warm-water' fish species at low-head dams and weirs, where physical access is adequate during the elevated flows of early spring, but is limited by poor swimming performance at the low temperatures which are characteristic of this period. By contrast, despite enhanced swimming performance at higher temperatures later in spring, upstream passage may be physically inhibited by declining discharge at the structure (Lucas *et al.* 2000).

Many telemetry studies have demonstrated that the passage at physical barriers was substantially delayed as conditions were inappropriate for traversing the obstacle (when a fish pass was absent) or fish did not find the entrance of the fish pass immediately or did so infrequently (Haynes & Gray 1980; Webb 1990; Lucas & Frear 1997; Bunt *et al.* 1999; Gowans *et al.* 1999a). Schmutz *et al.* (1998) radio-tracked 15 adult pikeperch *Stizostedion lucioperca* below the weir and bypass channel of the Marchfeldkanal, Austria and found a maximum range of movement of about 1.5 km over a period of up to 38 days during the key spring migration period for other fishes on this system. Six fish passed the fish pass entrance 28 times but none entered, so it would appear that further movement might have been limited by a behavioural aversion to moving through the fish pass rather than a lack of migratory behaviour. Continued reproductive maturation of delayed female Arctic grayling *Thymallus arcticus* influenced subsequent migratory behaviour with fish migrating shorter distances to spawn than controls (Fleming & Reynolds 1991). Fishes engaged in spawning migrations may exhibit widespread resorption of gametes when migration is delayed (Shikhshabekov 1971) and eggs retained after ovulation may be less viable than under normal circumstances (Sakai *et al*. 1975). Fleming and Reynolds (1991) recommended a delay of the spawning migration of Arctic grayling of no more than 3 days. Even though there are guidelines and recommendations for the correct positioning of fish entrances, local constraints (e.g. roads, railways on the banks) might force engineers to adopt a less appropriate design. Also, where hydropower dams are concerned, the attractiveness of the bypass may change depending on how many turbines are operating, and on the position of the operating turbines when a limited number operates. Finally, it is worth reiterating the view of a renowned fish pass expert that one may never be certain that the entrance of a fish pass is positioned optimally until it has been demonstrated that fish effectively find and enter it. This is because so much remains to be discovered on how fish orientate and make progress in structurally complex environments, so emphasising the need to look at fish passes from the fish's perspective.

The increasing application of hydroacoustics and biotelemetry at dams has partly helped to bridge this gap (see chapter 6), and so have the many studies where attractants or deterrents, to modify fishes' behaviour towards fishways for upstream passage, have been evaluated. The same applies to downstream migration, where it is important to drive fish away from dead-ends or hazardous water intakes, and/or to attract them to site(s) where they can pass the obstacle with maximum comfort. Therefore, behavioural methods of directing fish to the desired place will be dealt with together here, although it should be remembered that the equipment producing such behavioural signals might be positioned further from a hazardous area when downstream migration and entrainment is of concern. As for most other aspects dealt with in this section, the bulk of the literature available concerns salmonids or other diadromous species.

We initially intended to produce a much longer review as to how fish may be deterred or attracted by using physical barriers or behavioural stimuli. However, several excellent reviews of this topic have been produced over the past 2 years (Popper & Carlson 1998; Larinier & Travade 1999; Coutant & Whitney 2000; Taft 2000), and we will rely heavily on these and some selected examples in this overview.

7.4.7.1 Screens and physical barriers

Physical barriers have received considerable attention with respect to entrainment and impingement in cooling water systems or turbine intakes. Available technologies include fish collection systems, fish diversion systems and physical barriers (reviews in Odey & Orvis 1998; Taft 2000, and detailed evaluations in reports of the Electric Power Research Institute, EPRI, Palo Alto, California).

Physical barriers have variable efficiency depending on fish size, water flow, amount of debris and subsequent clogging. Radial wells, artificial filter beds, porous dikes and Gunderboom (full-water-depth filter curtains) have not been designed for screening large flow volumes, and their efficiency may be limited in most places, notably at cooling water intake structures (CWIS). Barrier nets using fine mesh size, and thus minimising the chance of fish becoming gilled in the net, are deemed to represent a viable technology at CWIS, but only under relatively low velocities (about 0.3 m s^{-1}), and provided debris loading is relatively low. They require permanent survey and immediate repair. Travelling water screens or rotating drum screens may reduce the proportion of fish being impinged, although they are generally regarded as having a limited efficiency for early life stages with a low swimming capacity. Among the types of screens and physical barriers, cylindrical wedgewire screens may offer the best technology available for the broadest range of fish sizes. They operate passively in ambient currents, with a low approach velocity and smooth surface, minimising risks of abrasion for fish (mesh of 0.5–2.0 mm). The finer the mesh, the greater the efficiency for avoiding impingement of early life stages, but also the greater the risks of clogging with fine sediments or biofouling and the need for water jets, thereby increasing the risks of damaging small fish. Ehrler and Raifsnider (2000) reported that the use of wedgewire screens at the Logan generating plant on the Delaware River, New Jersey, reduced impingement of striped bass *Morone saxatilis* larvae by an order of magnitude.

Fish diversion systems rely on the basic principle that under suitable hydraulic conditions fish can be guided by immersed physical structures, and thus diverted away from CWIS or turbine intakes. These systems have been primarily designed for diverting fish from turbines, and their efficiency for CWIS has rarely been evaluated. Louvres and bar racks consist of arrays of bars set diagonally to the main current, and are generally used to divert fish to spillways or bypass systems. Although they do not meet the maximum efficiency criterion, they have been implemented at many hydropower dams, and exhibit a relatively stable and high efficiency (80–95%) over a wide range of species and hydraulic conditions. Angled screens generally use large meshes and their efficiency in diverting fish varies between species, but is generally high. Eicher screens, which are passive pressure screens, have a very high efficiency (96–100%) in diverting steelhead trout *Oncorhynchus mykiss* and Pacific salmons *Oncorhynchus* spp. smolts at hydroelectric projects in the Northwest Pacific coast of the US (EPRI 1992 in Taft 2000). Modular inclined screens are set at an angle of 10–20° to the flow, and divert fish to a transport pipe. Laboratory studies using these systems were found to divert almost 100% of the fish (juvenile centrarchids, clupeids, cyprinids, ictalurids, percids and salmonids) at currents of up to 1.8 m s^{-1}, and a prototype field study on the Hudson River, New York, gave a similar diversion efficiency (EPRI 1992 in Taft 2000). Submersible travelling screens have been used to guide smolts of Pacific salmons *Oncorhynchus* spp. during their seaward migration at Bonneville Dam, Columbia River (Gessel *et al.* 1991). Guidance efficacy was as high as 75% for yearling coho salmon *O. kisutch* and chinook salmon *O. tshawytscha*, whereas it turned out to be as low as 60% or 30% for subyearling chinook salmon during spring and summer, respectively. Whether or not these structures have potential in diverting downstream migrating fish from hazardous places, their efficiency is extremely low, or zero, for early life stages (eggs and larvae).

Several systems have been designed to collect and transport fish during their downstream movements. Among these are travelling water screens equipped with low-pressure water jets that gently rinse the fish into a collecting basket which is transported into a safe release area. As for the other screen systems described above, their efficiency is proportional to mesh size and shape (e.g. see Ronafalvy *et al.* 2000), and inversely proportional to operation time, water velocity and fish size, as water jets acceptable to juveniles or adults cause more serious damage to early life stages. Taft (2000) points out that 'depending on species and life stage, mortality from impingement (by these fine-mesh screens) can exceed entrainment mortality', thereby limiting the overall adequacy of these systems to situations where they can be operated continuously, with a most limited time interval (a matter of a few minutes) between successive operations.

7.4.7.2 Air bubble curtains and electric screens

Bubble screens produced from submerged pipes have been used for decades to divert fish from water intakes, but there is an increasing bulk of evidence from field studies that bubbles do not cause a consistent avoidance behaviour *per se*. The relative success of bubble screens with some fish species might be attributed to the sound they produce and not to the presence of bubbles (Kuznetsov 1971 in Popper & Carlsson 1998), or to bubbles being identified as visible obstacles by fish when illuminated, especially with strobe lights (Patrick *et al.* 1985). Taft (2000) reported that, as a result of their low efficiency, air bubble curtains had been removed from service at sites where they had been evaluated on the North American Great Lakes.

Electric screens have also been used for decades, but they have been abandoned in many places, due to their relatively low efficiency, and hazards inherent to the application of electrical fields in water, both for fish and humans. Whereas some success has been achieved in directing upstream migrants (Ruggles 1991 in Popper & Carlson 1998; Gosset *et al.* 1992), most studies where electric screens were evaluated as ways of deterring downstream migrants have given equivocal results. One reason for this is that the electrical field is limited to regions between the electrodes, thereby restricting the repulsion efficiency of this system to the immediate vicinity. This might produce too short an effective range for efficiently diverting fish moving downstream in fast currents, as is generally the case at water intakes. For example, the a.c. electric screens deployed by Kynard and O'Leary (1993) failed to modify the downstream runs of the American shad *Alosa sapidissima* at the Holyoke Dam (Connecticut River, US).

Equivalent or greater success has been achieved in deploying behavioural attraction or diversion systems, based on the use of simple light and sound systems, which might be perceived at greater distances by fish. For example, Gosset and Travade (1999) observed that 130 V electric screens had a maximum efficiency ranging from 5 to 28% in successfully guiding Atlantic salmon *Salmo salar* smolts at the Halsou hydropower dam on the French Nive River, whereas the efficiency was as high when the bypass entrance was illuminated only with a mercury vapour light.

7.4.7.3 Light

Lighting at night has been used for centuries for attracting and catching fish, and it might serve the same purpose for attracting migrating fish of a variety of species to safe places. Conversely, it has been proposed that many fish enter water intakes at night as light intensity is so low that these structures are not seen by the fish. However, at least for some species and sites, migration at fish pass structures and obstacles is principally a nocturnal behaviour (Lucas & Frear 1997; Lucas 2000). Therefore, the reader is reminded that currently we know very little about the environmental cues determining fish behaviour at obstructions and that the following discussion may be species- and site-specific. Two main types of lighting are used for fish control: mercury vapour lights, which produce continuous lighting, and strobe (pulsing) lights. Whichever lighting system is used, lighting is generally regarded as more efficient at night than during the day, as the ratio with background illumination differs. Also light transmission is a function of water turbidity, and is proportional to wavelength.

Phototropism and photokinesis are the most frequently cited responses to continuous lighting. Haymes *et al.* (1984) demonstrated that juvenile alewives *Alosa pseudoharengus*, rainbow smelt *Osmerus mordax* and gizzard shad *Dorosoma cepedianum* were attracted by mercury vapour light. Hadderingh (1982) tested whether increased light levels around the water intake of the Bergum power plant, the Netherlands, might attract fish away from the intake and thereby reduce their rate of impingement. He found that impingement was reduced for all species, but that the improvement was more marked for ruffe *Gymnocephalus cernuus* and Eurasian perch *Perca fluviatilis* than for other species, including roach *Rutilus rutilus*, common bream *Abramis brama* and European eel *Anguilla anguilla*. There have been many other studies that have demonstrated between-species differences in the reaction to light, as well as within-species differences, sometimes owing to differences in light intensity, spectrum or frequency. For example, hatchery-reared juveniles of chinook salmon *Oncorhynchus tshawytscha* and coho salmon *O. kisutch* were found to avoid strobe light and mercury vapour lights at full intensity, but chinook salmon were attracted to dim mercury light (Nemeth & Anderson 1992).

Gehrke (1994) also reported that the reactions of larvae of silver perch *Bidyanus bidyanus* and golden perch *Macquaria ambigua* were dependent on light intensity, and also on wavelength. Differences in sensitivity to the visible light spectrum may be subtle, and determining the relative importance of different wavelengths requires a deeper knowledge of the spectral sensitivities of retinal and extraretinal photoreceptors of fishes (reviews in Daemers-Lambert 2000; Kusmic & Gualtieri 2000). Mercury vapour lamps produce light from the violet to the orange regions, but with peaks in the green (540 nm) and yellow (580–600 nm) regions. It is generally agreed that the light spectrum perceived by freshwater fish ranges from 400 to 700 nm, but absorption maxima of visual pigments vary substantially between freshwater fishes (Fig. 7.5). For example, the retinal rods of the Eurasian perch *Perca fluviatilis* have a maximum absorption at 540 nm, which is precisely one of the peaks of the mercury vapour light, but other species and notably cyprinids have absorption peaks at shorter or longer wavelengths. This might partly account for the variable success of mercury vapour lights in Hadderingh's (1982) study. Spectral sensitivity also varies seasonally, as demonstrated for rainbow trout *Oncorhynchus mykiss* and some cyprinids (e.g. *Notemigonus* spp.), as the proportion of pigments sensitive to different wavelengths varies. Similarly diadromous species undergo ontogenetic and seasonal changes during their migration, which 'prepare' them for living in a different environment, and this also involves changes in the proportion of pigments. For example, *Oncorhynchus* salmons caught at sea predominantly possess rhodopsin (peak of absorbance at *c*. 503 nm), and there is a gradual change to porphyropsin (peak of absorbance at *c*. 527 nm) as they penetrate fresh water. Conversely, both the American and European eels, *Anguilla rostrata* and *A. anguilla* respectively, contain visual pigments absorbing at 501 and 523 nm during their freshwater stage. As they metamorphose into silver eels, the porphyropsin pigment vanishes, and a new opsin pigment is formed, which confers them a sensitivity to shorter wavelengths (*c*. 482 nm) (reviews in Daemers-Lambert 2000; Kusmic & Gualtieri 2000). As a corollary, neither salmon nor eel are equally sensitive during their ascent or descent movements, and such knowledge should be incorporated into the design of appropriate lighting systems aimed at controlling their behaviour at physical obstructions and water intakes.

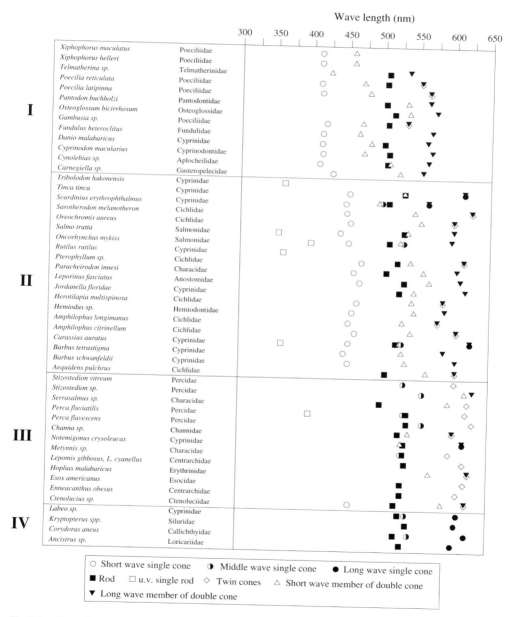

Fig. 7.5 Absorption peaks of visual pigment in different species or genera of freshwater fishes. Groups I, II, III and IV refer to ecological habits. I. Strictly diurnal species inhabiting shallow environments or using surface layers. II. Diurnal midwater species. III. Crepuscular and nocturnal species, mainly predators. IV. Crepuscular and nocturnal species, mainly with benthic habits. Modified from Kusmic and Gualtieri (2000).

Strobe lights and mercury vapour lights have been used with variable success (see recent case studies in Croze & Larinier 1999; Gosset & Travade 1999; Michaud & Taft 2000; Sager *et al.* 2000). Strobe lights, by virtue of their pulsing nature, are considered as having a greater potential in orientating the movements of migratory fish (e.g. Nemeth & Anderson 1992), especially now that underwater strobe lighting systems are replacing the standard above-water systems. Evaluations of optimal light intensity and pulse frequency are under study for many fish species, as not all fish may show avoidance to strobe light. Brown (2000) has produced a detailed review of the recent successes of strobe light systems with salmonid and non-salmonid fishes. As for mercury vapour lamps, their efficiency is limited in turbid waters, and greater at night than during the day, and as for all diversion systems, it is further limited by hydraulic conditions. At present, there is little information on the short- and long-term impact of bright light or light pulses at high frequency, but this too should receive attention, notably as the fish retina requires a delay for habituating to different light levels, and this may impact indirectly on orientation mechanisms or capacities for escaping predation.

7.4.7.4 Sound (including infrasound and ultrasound)

As for light, results with behavioural control systems based on sound have produced contrasting results with different species, depending on sound frequency (review in Popper & Carlson 1998). The perception of particle motion in the fish's near-field (within a few body lengths of the fish) is known to be a common phenomenon, as it involves the lateral line. In this respect, the studies of Knudsen *et al.* (1992, 1994) demonstrating that parr of Atlantic salmon *Salmo salar* exhibited an awareness reaction and avoidance of low frequency sounds (5–10 Hz) have led to growing interest in the use of infrasound barriers at water intakes. Popper and Carlson (1998) report that avoidance to infrasound (<50 Hz) has been found in several North American cyprinids, gadids and moronids. More recently, infrasound (here 11.8 Hz) has been shown significantly to deter silver eels *Anguilla anguilla* during their downstream movements (Sand *et al.* 2000). However, Sand *et al.* also found that intense infrasound was highly stressful, as experimental eels were found to show startle responses and prolonged tachycardia after exposure to infrasound.

Responses to sound in the audible range are more variable, notably because the hearing capacities of fish vary substantially between species (Hawkins 1993). Hearing specialists (Ostariophysi) have an extended range of hearing, from 50 Hz to about 3000 Hz, compared to non-specialists which have a narrower range of sensitivity of 50 Hz to about 500 Hz (sometimes less). The sensitivity of ostariophysian specialists is greater than that of non-specialists, as they possess an intimate connection – the Weberian apparatus – between the otoliths and the swimbladder, which acts as an efficient aid to hearing. For example, the Atlantic salmon *Salmo salar* hears sounds between 30 and 400 Hz, with the maximum sensitivity at *c.* 150 Hz, whereas the goldfish *Carassius auratus* can hear sounds from 50 Hz to about 3000 Hz, with a maximum sensitivity at *c.* 500 Hz. The lowest sound levels that can be heard at the frequency of maximal sensitivity by the two species are about 100 and 50 dB re 1 µPa respectively. Popper and Carlson (1998) provide a list of examples of the relative successes of sound barriers, essentially in fishes with non-specialist hearing capabilities (salmonids and clupeids). Irrespective of whether fish behaviour can be controlled by sound production, one should pay attention to the possibility that the sound levels (180–200 dB re 1 µPa) utilised to make sound barriers effective to hearing non-specialists are likely to cause deleterious effects to the lateral line, inner ear or generate variable stress responses, especially in the specialist fishes. These fishes represent the vast majority of the potamodromous fishes in tropical regions (Characiformes, Cypriniformes, Siluriformes) and any transposition of sound barrier technology to these environments should be made with extreme caution.

To date, the only sound barriers that are proven to produce consistent and repeatable results in a wide range of locations are ultrasound barriers, used to deter alosine shads. High-frequency (110–140 kHz) sounds emitted at a high sound pressure level (180 dB at 1 m of the acoustic transducer) have been found to elicit significant avoidance responses by blueback herring *Alosa aestivalis*, at distances of about 60 m, with avoidance persisting up to 1 h (Nestler *et al.* 1992). In laboratory studies, alewives *Alosa pseudoharengus* were shown to avoid pulsed tones and continuous tones within the 110–125 kHz band, as well as pulsed broadband sounds between 117 and 133 kHz. Whereas they habituated to tones, they avoided pulsed broadbands more consistently (Dunning *et al.* 1992). Reactions were stronger during the day, when alewives schooled, than at night when they did not school or swim actively. Reactions to ultrasound have been demonstrated in several alosine shads, but at present it is still unknown how they sense ultrasound and whether other fishes could be controlled by ultrasound barriers (Popper & Carlson 1998), and this should be a research priority in the near future.

7.5 Installation, monitoring and efficiency of fish passes

Historically, many dams and weirs have been built without adequate assessment of their environmental impact and river managers have not been as proactive as they might have been in questioning the necessity for such structures in the first place (Lucas & Marmulla 2000). Today, cost–benefit analysis should provide a first approach for identifying the likely sources and extent of damage by such structures and, where appropriate, for strongly resisting their construction. One of the problems with this approach is that economic evaluation of fish communities is usually applied only to the relatively few commercially important fish species (typically less than 5% of fish species in a given catchment) and there is still argument as to the true value of biodiversity. Also, our ability to construct fish passes to allow migration past an obstacle should not be interpreted as a view that multiplication of dams and weirs can be achieved without significant harm (Larinier *et al.* 1994). This is a false premise that has, and continues to be responsible for substantial damage to migratory freshwater fish populations.

Nevertheless, cost–benefit appraisal may be used to compare the sensitivity of sites, to identify less damaging alternatives to a particular scheme or, where necessary, can be used to argue the case for adequate funds to provide mitigation where river obstruction is unavoidable (Lucas & Marmulla 2000). In the latter case, 'adequate' funding does not just mean construction of a fish pass. It should, we strongly recommend, include costs of pre-obstruction monitoring, fish pass construction, post-obstruction monitoring, fish pass maintenance and, where necessary, the costs of modification to fish passage facilities and/or the working practices at the dam or weir. Where fish passes are added as 'retro-fits' to existing weirs and dams, the same comments apply. In too many cases fish passes are added to existing obstructions, usually when other maintenance is being carried out, simply on the basis that 'some fish passage must be better than none' even though the suitability of the fish pass for the fish community has not been established. Pre- and post-installation monitoring tends to be the exception rather than the rule. In the medium and long term the installation (often publicly funded) of fish passes, without monitoring, is of doubtful benefit. Equally, there are examples of extremely good pre-project monitoring, planning and post-project appraisal in the northern and southern hemispheres, some of which are provided in Jungwirth *et al.* (1998).

Methods appropriate to monitoring of fish migration prior to construction of an obstruction, or prior to fitting of a fish pass at an existing obstruction, vary with site, environmental conditions and fish species (see chapter 6). In pre-impoundment conditions, telemetry and hydroacoustic approaches are likely to provide the most useful information on the behaviour of fish passing through the area of interest, although mark-recapture methods may also be useful. Hydroacoustic and direct captures approaches can provide quantitative information on the numbers of fish present and their seasonal fluctuations, while only direct capture or visual observation can enable a detailed assessment of fish community composition. Where obstructions already exist, telemetry and hydroacoustic approaches can provide information on fish behaviour and aid the planning and siting of fish passage facilities. Monitoring of the effectiveness of fish passes can be carried out using a similar array of techniques. Traps are often employed routinely at fish passes so that fishes that have successfully passed through can be identified and enumerated. Resistivity fish counters and/or video systems may also be installed at fish ladder exits and give information on stock escapement. However, on their own these methods do not provide crucial information on the efficiency of the fish pass at the individual or population level. From the information given earlier in this section it is apparent that predictions of the effectiveness of a fish pass may not be confirmed under natural conditions.

Two principal components of the effectiveness of a fish pass are its ability to attract fish to the entrance, and the ability of fishes which have located the entrance to successfully pass though the fishway, termed attraction and passage efficiency respectively. Fish pass efficiency has not generally been formally defined in terms of minimum standards (Larinier 1998), but it is generally considered that attraction and passage efficiencies should be 90–100% for diadromous fishes and we recommend this as a target for fishes displaying marked potamodromy. For eurytopic species without specialised habitat requirements, substantially lower passage efficiencies are likely to be acceptable. Given the long history of fish pass design and use, it is surprising how few studies of fish pass efficiency have been carried out or reported (Clay 1995).

Details of appropriate methods for examining spatial behaviour of fishes in the natural environment are presented in chapter 6, and they are outlined here only in relation to assessing fish pass effectiveness. Traps may be employed above and below fish passes and this approach has been used in Australia to assess the effectiveness of rock-ramp fishways for passing a wide range of native species (Harris *et al.* 1998). The disadvantages of this approach are that such traps exhibit capture selectivity, record only a proportion of migrating fishes, and may operate effectively in a narrow range of conditions. Another method is to design the fish pass so that chambers at the upper and lower ends of the fishway can be blocked off simultaneously, enabling paired samples of the fishes to be taken by netting or electric fishing (Stuart & Mallen-Cooper 1999). With adequate replication, this can enable a sound statistical assessment of effectiveness for a wide range of taxa and sizes, and so is particularly suited to situations where assessment at the community level is important. Although helpful in providing information at the community level, this and similar approaches cannot provide true efficiency estimates for passage and attraction. Conventional marking or tagging of fish downstream of the fish pass, together with recapture at or above the fishpass, has been used in several studies (e.g. Schwalme *et al.* 1985; Linlokken 1993). This method can provide information on attraction efficiency, but passage efficiency is difficult to measure independently of attraction efficiency. Video monitoring at several locations in a fishway enables monitoring of passage efficiency at the population scale and provides the advantage of being able to directly observe fish behaviour (Haro & Kynard 1997; Larinier 1998), but the method is limited by ambient light level, and individual fish identification is rarely possible.

Telemetry provides the most informative approach to assessing attraction and passage efficiency at the individual and population scales for fishes that are large enough to tag (e.g. Webb 1990; Bunt *et al.* 1999; Gowans *et al.* 1999b; Moser *et al.* 2000). Radio antennae or acoustic transducers (depending on noise levels and conductivity) can be placed at the entrance and exit of the fish pass and at intermediate sites along its length, while fish can also be tracked downstream and upstream of the pass. However, for large fish samples this approach can be expensive (see Table 6.1). The energy cost of attempted passage can be estimated using physiological telemetry techniques (Lucas *et al.* 1993a; Hinch *et al.* 1996). Passive integrated transponder telemetry shows great promise for assessing passage efficiency, due to the large number and small size of fish which can be tagged (Castro-Santos *et al.* 1996). Again this relies on positioning at least one detector at the entrance and one at the exit of the fish pass. This method has been used to examine passage efficiency and behaviour of lowland-river fishes at a Denil pass on the Yorkshire Derwent, England (Lucas *et al.* 1999; Lucas 2000), where measurements of passage efficiency of 17%, during summer, were made. However, some fishes made repeated attempts at ascent, usually at night (Fig. 7.6) so that the total number of successful passages as a fraction of all entries was 4%. These technological developments should enable improvements in fishway design and operation to be made, and provide a formal method for assessing their effectiveness.

7.6 Conclusions

Although considerable knowledge has been gained over the past decades regarding the impact of man on aquatic ecosystems in general, and fish migration in particular, many issues of regional or international importance remain to be addressed, notably as regards the long-term monitoring of anthropogenic modifications, as these might extend over decades. There is also an urgent need for all obstacles to fish migration to receive greater attention, including the so-called 'minor obstacles', which are most frequently regarded as having no impact at all. As stated on many occasions in this book, the viewpoint of what constitutes 'an impact' is biased as it is usually considered in terms of how it affects large fish species of commercial importance. Similarly, fish pass design should be considered so as to provide multi-species fishways, appropriate to the local conditions and fish community. For example, historical attempts at direct transposition of technologies designed for northern hemisphere salmonids have proved to be of little value in application to Australian river systems (Harris & Mallen-Cooper 1994). Natural bypass channels, when feasible, represent one of the best mitigation

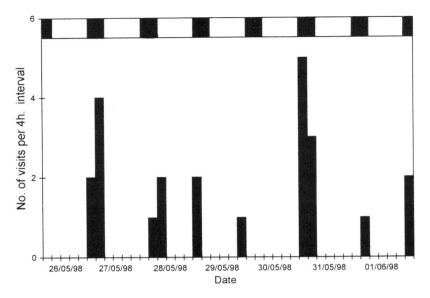

Fig. 7.6. An illustration of the frequency of entry into a Denil fish pass by a PIT tagged 45-cm adult male chub *Leuciscus cephalus* in reproductive condition when tagged on 22 May 1998. Shaded portions of the horizontal bar reflect night-time (civil twilight–civil twilight). This fish did not successfully ascend the fish pass. Modified from Lucas *et al.* (1999).

strategies (except when the obstacle can be removed), as they enable migration in a more natural way, under a wide range of flow conditions, and also offer alternative habitats in impounded areas. Their use may be just as valuable in tropical environments, as their construction and maintenance costs essentially are a matter of manpower.

The various additional mitigation measures, including ways of attracting or deterring fishes, should definitely receive more attention, notably the behavioural systems, which may turn out to be highly cost-effective in the long run, provided they are operated properly, alone or in combination. At present, most of these systems operate on a fixed and almost permanent design, but it is not inconceivable that they might operate in different ways at different seasons corresponding to the seasonal migrations of other fish species or life stages, with different responses to light or sound. In this respect, it is crucial that fundamental research on how fish sense their environment and react to it be continued and targeted to the broadest possible range of species, also focusing on habituation to stimuli and learning. In many situations, especially tropical systems, such research has yet to be initiated. Also there is an increasing amount of information being collected on the daily periodicity of fish migration, which generally is insufficiently used towards the friendly management of hydropower plants and similar facilities. Many fish are active at twilight periods, either when moving upstream or downstream. Hence, and although no generalisation is being made here, stopping or reducing hydropower operations during these two short periods of the day might represent a substantial contribution to improving the efficiency of fish pass systems. This surely has a cost, as do all of the mitigation systems evoked in this chapter, but the question remains 'How valuable is biodiversity?'

Chapter 8
Conclusion

Migration of fishes in freshwater environments is undoubtedly a widespread and important phenomenon for enhancing lifetime reproductive success. Until the last few decades the great majority of information concerning the migratory behaviour of fishes in fresh water was restricted to relatively few species, mostly diadromous ones, in North America and Europe. The recent shift in emphasis towards understanding ecology of fishes at the community scale, together with the need to understand causes of declines in fish populations and of how to rehabilitate damaged freshwater ecosystems in developed countries, has changed this. There has been a major expansion in the recorded number of freshwater-resident species for which migration plays an important role. There has also been a recognition that, for many fish species, migratory behaviour is a 'plastic', adaptive strategy that is evident in some catchments, but not in others, because it is strongly dependent on the biotic and abiotic environmental conditions which determine the fitness value of migratory behaviour. The finding that the migratory history of fish species generally varies with environment and climate should definitely be considered within the context of global warming, as populations nowadays adopting restricted home ranging behaviour might increasingly turn to migratory behaviour.

The advent of advanced techniques, especially telemetry, hydroacoustics and microchemistry, has revolutionised our ability to study the spatial behaviour of fishes. These advances have also greatly contributed to the increasing awareness of individual and population differences in migratory behaviour and the factors that influence it. We strongly expect that these techniques and specific advances, such as the increased use of archival tags, will continue to produce new information on the behaviour of diadromous and holobiotic freshwater fishes. Additionally these approaches are extremely useful in the development and testing of new ideas concerning behaviour and physiology of migratory fishes. For example, iteroparous freshwater fishes provide a good model for examining the significance of the degree and precision of homing within and between individuals of a population. Linking telemetry and molecular genetics provides an approach for linking individual-based models of behaviour to reproductive success and the partitioning of genetic material within a population. Equally, the use of archival tags is already providing new evidence concerning the cues used for orientation during migration by adult fishes, notably in marine environments or at the marine–freshwater interface. However, the use of such tags also offers great promise within freshwater systems, notably in large lake environments.

Nevertheless, although there have been progressive advances in our understanding of the migratory behaviour of freshwater fishes from the temperate northern hemisphere, development of knowledge in other geographical regions has tended to be more recent. Great strides have been made in understanding the migratory behaviour of fishes occurring in fresh water in southern hemisphere countries such as Australia and New Zealand, where the pattern of migratory behaviour is rather different to that exhibited throughout most of the northern hemisphere. Indeed, it has taken several decades for the fundamental differences in life-history patterns exhibited by these fish faunas compared to north temperate latitudes, and the implications of these for sensitive river management, to be fully acknowledged by the global scientific community, because of the historic legacy of experience from north temperate latitudes. There have also been major developments in the study of migratory behaviour of fishes in arctic and subarctic environments, especially in North America and Scandinavia, although there have been fewer reported studies (that we are aware of) from equivalent environments in Russia and countries formerly belonging to the USSR. Recent work on the genetic differentiation of coregonines and chars (*Salvelinus* spp.) from the Russian Arctic and subarctic (Politov *et al*. 2000; Salmenkova *et al* 2000), as well as reports of expeditions to rivers draining to the Russian Arctic (Novikov *et al*. 2000) may stimulate further study of migration and dispersal processes in this region.

However, it remains that relatively little is known about the migratory behaviour of fishes in tropical rivers and lakes, particularly since the diversity of fishes in these systems far outstrips that in other freshwater environments worldwide. Even for fishes such as *Prochilodus*, regarded as one of the most characteristic of migratory tropical freshwater fishes, information is mostly based on seasonal variations in CPUE data and, in a few cases, conventional tagging. These are methods which now tend to be somewhat frowned upon in field studies of fish behaviour in North America and Europe. Work on freshwater fish migration in the tropics is of key significance, because of the importance of fisheries in these regions, and their close association with seasonal inundation of floodplain habitats. Increased river regulation in tropical systems is likely to damage the integrity of these habitats through reduced longitudinal and lateral connectivity. Although effects on fisheries may, in many cases, be more greatly associated with changes in catch composition rather than quantity, highly valued migratory species may be lost. Moreover, there is increasing evidence that aquatic biodiversity is strongly damaged by river regulation in tropical systems. The extent of fish migration and its significance for lifecycle completion therefore needs to be addressed on two levels. Much more information is needed on the extent of local and large-scale migration for fish species occurring in communities, much of which may be gathered using simple techniques. To gain a better understanding of the individual and population-level strategies of migration used by key fish species, application of advanced techniques such as telemetry to these systems is appropriate. Otolith microchemistry may also provide the opportunity for detailed lifetime habitat reconstruction for those species which move between blackwater and whitewater environments. Since fieldwork may be the most difficult aspect of research in many tropical freshwater habitats, habitat reconstruction by the use of microchemistry or archival tags offers an attractive approach.

Worldwide, there is a growing awareness that river management practices need to become more sensitive to the needs for migration, dispersal and recolonisation by freshwater fishes over seasonal (or shorter) and generation timescales. These requirements are evident both for rehabilitation programmes on rivers that have already been damaged, and especially for rivers where development is beginning. An understanding of the spatial behaviour of fishes is also important to the sensitive management of fisheries and for carrying out surveys of fish biomass and diversity. Much ongoing development, including river regulation, is occurring in developing countries. With few exceptions, these are in regions for which the least is known about the extent and significance of migration to the lifecycles of indigenous fishes. There is a need to reassess the suitability of traditional fish passage facilities in these regions. Use of natural bypass facilities may be more economic and effective in developing countries than the more formally engineered structures that have been prevalent to date. However, these facilities, while mitigating deficiencies in longitudinal connectivity, do little to solve problems of reduced lateral connectivity. Moreover, at present many projects pay little attention to appropriate fish passage facilities at all, and in others, lack of regulation of fishing activities close to or at fishways impedes their functioning. It is an unfortunate fact that these types of problems continue to receive inadequate attention in many 'developed' countries, although a more enlightened approach by some authorities, as well as international co-operation, are beginning to show what can be achieved.

Finally, there is a need for dissemination of information on the significance of migration within freshwater ecosystems, improvements in methods for studying fish migration and in methods of assisting connectivity within river catchments. In the latter two subjects there have been several recent conferences and workshops which have served this purpose very effectively. However, there is still an urgent need to develop training and information exchange concerning these issues, especially in developing countries. One such approach currently being developed by the FAO, the French IRD and the University of Liège, is to provide hands-on training in telemetry techniques, targeted for use in developing countries, with the first workshop having been held in Séligué, Mali, in early 2001. Similar opportunities exist for sharing skills and experience with fundamental and applied issues concerning fish migration across the full range of biotopes. The constitution of an Internet database on fish migration might also serve this purpose. We hope that in the next decade, the advances to be made in understanding the significance and nature of fish migration in fresh and brackish water environments will reflect greater experience and collaboration on a global scale.

References

Aarestrup, K. & Jepsen, N. (1998) Spawning migration of sea trout (*Salmo trutta* (L)) in a Danish river. *Hydrobiologia*, **371/372**, 275–281.

Abakumov, V. A. (1956) The mode of life of the Baltic migratory lamprey. *Journal of Ichthyology*, **1956** (G), 122–128.

Achord, S., Matthews, G. M., Johnson, G. M. & Marsh, D. M. (1996) Use of Passive Integrated Transponder (PIT) tags to monitor migration timing of Snake River chinook salmon smolts. *North American Journal of Fisheries Management*, **16**, 302–313.

Adam, B. & Schwevers, U. (1997) Funktionsüberprüfung von fishwegen: einsatz automatischer kontrollstationen unter anwendung der transponder-technologie. *Schriftenreihe des Deutsche Verbandes für Wasserwirtschaft und Kulturbau e.V.*, **119**, Bonn, 79 pp.

Adam, B. & Schwevers, U. (1998) Monitoring of a prototype collection gallery on the Lahn River. In: *Fish Migration and Fish Bypasses* (eds M. Jungwirth, S. Schmutz & S. Weiss), pp. 246–254. Fishing News Books, Blackwell Science Ltd, Oxford.

Adams, N. S., Rondorf, D. W., Evans, S. D., Kelly, J. E. & Perry, R. W. (1998) Effects of surgically and gastrically implanted radio transmitters on swimming performance and predator avoidance of juvenile chinook salmon. *Canadian Journal of Fisheries and Aquatic Sciences*, **55**, 781–787.

Adams, S. R., Keevin, T. M., Killgore, K. J. & Hoover, J. J. (1999) Stranding potential of young fishes subjected to simulated vessel-induced drawdown. *Transactions of the American Fisheries Society*, **128**, 1230–1234.

Admiraal, W., van der Velde, G., Smit, H. & Cazemier, W. G. (1993) The rivers Rhine and Meuse in the Netherlands: present state and signs of ecological recovery. *Hydrobiologia*, **265**, 97–128.

Agostinho, A. A., Borghetti, J. R., Vazzoler, A. E. A. de M. & Gomes, L. C. (1994) Itaipu reservoir: impacts on the ichthyofauna and biological bases for its management. In: *Environmental and Social Dimensions of Reservoir Development and Management in the La Plata River Basin. United Nations Centre for Regional Development Research Report Series*, **4**, 135–148.

Agostinho, A. A., Vazzoler, A. E. A. de M., Gomes, L. C. & Okada, E. K. (1993) Estratificación espacial de *Prochilodus scrofa* en distintas fases del ciclo de vida, en la planicie de inundación del alto rio Paraná, y el embalse de Itaipu, Paraná, Brasil. *Revue d'Hydrobiologie Tropicale*, **26**, 79–90.

Aker, E. & Koops, H. (1973) Untersuchungen über Aalbestände in der Duetschen Bucht. *Archiv für Fischwisserei*, **24**, 19–39.

Albaret, J. J. (1982) Reproduction et fécondité des poissons d'eau douce de Côte d'Ivoire. *Revue d'Hydrobiologie Tropicale,* **15**, 347–381.

Albaret, J.-J. & Diouf, P. S. (1994) Diversité des poissons des lagunes et estuaires ouest-africains. In:

Diversité Biologique des Poissons des Eaux Douces et Saumâtres d'Afrique. Synthèses Géographiques (eds G. G. Teugels, J.-F. Guégan & J.-J. Albaret). *Annales du Musée Royal d'Afrique Centrale de Tervuren, Zoologie,* **275**, 165–177.

Alexander, R. M. (1966) Physical aspects of swim bladder function. *Biological Reviews,* **41**, 141–176.

Ali, M. A., Ryder, R. A. & Antcil, M. (1977) Photoreceptors and visual pigments as related to behavioural responses and preferred habitats of perches (*Perca* spp.) and pikeperches (*Stizostedion* spp.). *Journal of the Fisheries Research Board of Canada,* **34**, 1475–1480.

Allan, J. D. (1978) Trout predation and the size composition of stream drift. *Limnology and Oceanography,* **23**, 1231–1237.

Allen, G. R. (1989) *Freshwater Fishes of Australia.* T.F.H. Publications, Neptune City, New Jersey.

Allen, G. R. & Cross, N. J. (1982) *Rainbowfishes of Australia and Papua New Guinea.* Angus & Robertson, Sidney, 141 pp.

Allen, K. R. (1935) The food and migration of the perch (*Perca fluviatilis*) in Windermere. *Journal of Animal Ecology,* **4**, 264–273.

Allendorf, F. W., Utter, F. M. & May, B. P. (1975) Gene duplication within the family Salmonidae: detection of the genetic control of duplicate loci through inheritance studies and the examination of populations. In: *Isozymes IV. Genetics and Evolution* (ed. C. L. Markert). Academic Press, New York.

Allouche, S., Thévenet, A. & Gaudin, P. (1999) Habitat use by chub (*Leuciscus cephalus* L. 1766) in a large river, the French Upper Rhone, as determined by radiotelemetry. *Archiv für Hydrobiologie,* **145**, 219–236.

Almeida, P. R. (1996) Estuarine movement pattern of adult thin-lipped grey mullet, *Liza ramada* (Risso) (Pisces: Mugilidae), observed by ultrasonic tracking. *Journal of Experimental Marine Biology and Ecology,* **202**, 137–150.

Alt, K. T. (1977) Inconnu, *Stenodus leucichthys*, migration studies in Alaska. *Journal of the Fisheries Research Board of Canada,* **34**, 129–133.

Altringham, J. D. & Johnston, I. A. (1986) Energy cost of contraction in fast and slow muscle fibres isolated from an elasmobranch and an Antarctic teleost fish. *Journal of Experimental Biology,* **121**, 239–250.

anon. (1994) *Elastomer Visible Implant Product Literature.* Northwest Marine Technology International Ltd, Shaw Island, Washington.

Aoyama, J., Hissmann, K., Yoshinaga, T., Sasai, S., Uto, T. & Ueda, H. (1999) Swimming depth of migrating silver eels *Anguilla japonica* released at seamounts of the West Mariana Ridge, their estimated spawning sites. *Marine Ecology Progress Series,* **186**, 265–269.

Aparicio, E. & De Sostoa, A. (1999) Pattern of movements of adult *Barbus haasi* in a small Mediterranean stream. *Journal of Fish Biology,* **55**, 1086–1095.

Applegate, V. C. & Brynildson, C. L. (1952) Downstream movement of recently transformed sea lampreys, *Petromyzon marinus*, in the Carp Lake River, Michigan. *Transactions of the American Fisheries Society,* **81**, 275–290.

Aprahamian, M. W. (1988) Age structure of eel *Anguilla anguilla* L., populations in the River Severn, England, and the River Dee, Wales. *Aquaculture and Fisheries Management,* **19**, 365–376.

Aprahamian, M. W., Nicholson, S. N. & McCubbing, D. (1996) The use of resistivity fish counters in fish stock assessment. In: *Stock Assessment in Inland Fisheries* (ed. I. G. Cowx), pp. 27–43. Fishing News Books, Blackwell Science Ltd, Oxford.

Arai, T., Jellyman, D. J., Otake, T. & Tsukamoto, K. (1999a) Differences in the early life history of the Australasian shortfinned eel *Anguilla australis* from Australia and New Zealand, as revealed by otolith microstructure and microchemistry. *Marine Biology,* **135**, 381–389.

Arai, T., Limbong, D., Otake, T. & Tsukamoto, K. (1999b) Metamorphosis and inshore migration of tropical eels *Anguilla* spp. in the Indo-Pacific. *Marine Ecology Progress Series,* **182**, 283–293.

Araujo-Lima, C. & Goulding, M. (1997) *So Fruitful a Fish: Ecology, Conservation and Aquaculture of the Amazon's Tambaqui*. Columbia University Press, 191 pp.

Araujo-Lima, C. A. R. M. (1994) Egg size and larval development in central Amazonian fish. *Journal of Fish Biology,* **44**, 371–389.

Araujo-Lima, C. A. R. M. & Oliveira, E. C. (1998) Transport of larval fish in the Amazon. *Journal of Fish Biology,* **53** (Suppl. A), 297–306.

Araujo-Lima, C. A. R. M., Savastano, D. & Jordao, L. C. (1994) Drift of *Colomesus asellus* (Teleostei: Tetraodontidae) larvae in the Amazon River. *Revue d'Hydrobiologie Tropicale,* **27**, 33–38.

Armstrong, J. D. (1998) Relationships between heart rate and metabolic rate of pike: integration of existing data. *Journal of Fish Biology,* **52**, 223–442.

Armstrong, J. D., Braithwaite, V. A. & Huntingford, F. A. (1997) Spatial strategies of wild Atlantic salmon parr: exploration and settlement in unfamiliar areas. *Journal of Animal Ecology,* **66**, 203–211.

Armstrong, J. D., Braithwaite, V. A. & Rycroft, P. (1996) A flat-bed passive integrated transponder antenna array for monitoring behaviour of Atlantic salmon parr and other fish. *Journal of Fish Biology,* **48**, 539–541.

Armstrong, J. D., Johnstone, A. D. F. & Lucas, M. C. (1992) Retention of intragastric transmitters after voluntary ingestion by captive cod, *Gadus morhua* L. *Journal of Fish Biology,* **40**, 135–137.

Armstrong, J. D., Lucas, M., French, J., Vera, L. & Priede, I. G. (1988) A combined radio and acoustic transmitter for fixing direction and range of freshwater fish (RAFIX). *Journal of Fish Biology,* **33**, 879–884.

Armstrong, J. D. & Rawlings, C. E. (1993) The effect of intragastric transmitters on feeding behaviour of Atlantic salmon, *Salmo salar*, parr during autumn. *Journal of Fish Biology,* **43**, 646–648.

Armstrong, M. L. & Brown, A. V. (1983) Diel drift and feeding of channel catfish alevins in the Illinois River, Arkansas. *Transactions of the American Fisheries Society,* **112**, 302–307.

Armstrong, R. H. & Morrow, J. E. (1980) The Dolly Varden. In: *Charrs – Salmonid Fishes of the Genus Salvelinus* (ed. E. K. Balon), pp. 99–140. Dr W. Junk, The Hague.

Arnekleiv, J.V. & Krabol, M. (1996) The effects of induced floods on the upstream migration of adult brown trout, and the effects of water release on the postspawning downstream migration in a regulated Norwegian River. In: *Underwater Biotelemetry* (eds E. Baras & J. C. Philippart), pp. 172. University of Liège, Belgium.

Aronson, L. R. (1951) Orientation and jumping behaviour in the gobiid fish *Bathygobius soporator*. *American Museum Novitates,* **1486**, 22 pp.

Assunçao, M. I. da S. & Schuwassmann, H. O. (1992) Modos de reproduçao do poraque *Electrophorus electricus* (L.) (Gymnotiformes, Electrophoridae). XIX Congresso Brasiliero de Zoologia, Resumos, pp. 88. UFPA/MPEG.

Auer, N. A. (1996a) Importance of habitat and migration to sturgeons with emphasis on lake sturgeon. *Canadian Journal of Fisheries and Aquatic Sciences,* **53** (Suppl. 1), 152–160.

Auer, N. (1996b) Response of spawning lake sturgeons to change in hydroelectric facility operation. *Transactions of the American Fisheries Society,* **125**, 66–77.

Avise, J. C., Helfman, G. S., Saunders, N. C. & Hales, L. S. (1986) Mitochondrial DNA differentiation in North American eels: population genetic consequences of an unusual life history pattern. *Proceedings of the National Academy of Sciences of the USA,* **83**, 4350–4353.

Avise, J. C., Nelson, W. S., Arnold, J., Koehn, R. K., Williams, C. & Thorsteinsson, V. (1990) The evolutionary genetic status of Icelandic eels. *Evolution,* **44**, 1254–1262.

Axford, S. N. (1991) Some factors affecting angler catches in Yorkshire rivers. In: *Catch Effort Sampling Strategies* (ed. I. G. Cowx), pp. 143–153. Fishing News Books, Blackwell Science Ltd, Oxford.

Azuma, M. (1973) Studies on the variability of the land-locked ayu-fish, *Plecoglossus altivelis* T. and S., in Lake Biwa. II. On the segregation of populations and the variations in each population. *Japanese Journal of Ecology,* **23**, 126–139.

Baade, U. & Fredrich, F. (1998) Movement and the pattern of activity of the roach in the River Spree, Germany. *Journal of Fish Biology,* **52**, 1165–1174.

Baba, N. & Ukai, T. (1996) Intelligent tag and its recovery system for studying the behavior of free-ranging salmon in the ocean. *Bulletin of the National Research Institute of Aquaculture,* Suppl. **2**, 29–32.

Baçalbasa-Dobrovici, N. (1985) The effects on fisheries of non-biotic modifications of the environment in the East-Danube river area. In: *Habitat Modification and Freshwater Fisheries* (ed. J. S. Alabaster), pp. 13–27. Butterworths, London.

Backiel, T. (1985) Fall of migratory fish populations and changes in commercial fisheries in impounded rivers in Poland. In: *Habitat Modification and Freshwater Fisheries* (ed. J. S. Alabaster), pp. 28–41. Butterworths, London.

Baerends, G. P. (1986) On causation and function of pre-spawning behaviour of cichlid fish. *Journal of Fish Biology,* **28** (Suppl. A), 107–121.

Baggerman, B. (1957) An experimental study on the timing of breeding and migration in the three-spined stickleback (*Gasterosteus aculeatus* Linnaeus). *Archives Néerlandaises de Zoologie,* **12**, 105–317.

Baggerman, B. (1960) Salinity preference, thyroid activity and the seaward migration of four species of Pacific salmon (*Oncorhynchus*). *Journal of the Fisheries Research Board of Canada,* **17**, 295–322.

Baglinière, J. L. & Maisse, G. (eds) (1991) *La Truite: Biologie et Ecologie*. INRA Editions, Paris, 303 pp.

Baglinière, J. L., Maisse, G., Lebail, P.Y. & Nihouarn, A. (1989). Population dynamics of brown trout (*Salmo trutta* L.) in a tributary in Brittany (France): spawning and juveniles. *Journal of Fish Biology*, **34**, 97–110.

Baglinière, J. L., Maisse, G., Le Bail, P. Y. & Prévost, E. (1987) Dynamique de la population de truite commune (*Salmo trutta* L.) d'un ruisseau breton (France): les géniteurs migrants. *Acta Oecologica, Oecologia Applicata,* **8**, 201–215.

Bain, M. B. (1997) Atlantic and shortnose sturgeons of the Hudson River: common and divergent life history attributes. *Environmental Biology of Fishes,* **48**, 347–358.

Bain, M. B., Finn, J. T. & Booke, H. E. (1985) A quantitative method for sampling riverine microhabitats by electrofishing. *North American Journal of Fisheries Management,* **5**, 489–493.

Bain, M. B., Finn, J. T. & Booke, H. E. (1988) Streamflow regulation and fish community structure. *Ecology,* **69**, 382–392.

Bainbridge, R. (1958) The speed of swimming fish as related to size and to the frequency and amplitude of the tail beat. *Journal of Experimental Biology,* **35**, 109–133.

Baker, R. R. (1978) *The Evolutionary Ecology of Animal Migration.* Hodder & Stoughton, London.

Bakshtansky, A. I., Pupyshev, V. A. & Nesterov, V. (1977) Behaviour of pike and pike's influence on the downstream migration of young salmon in the period of light ripples. *International Council for the Exploration of the Sea, C.M.* 1977/M:**6**, 1–13.

Balirwa, J. S. & Bugenyi, F. W. B. (1980) Notes on the fisheries of the River Nzoia, Kenya. *Biological Conservation,* **18**, 53–58.

Balon, E. K. (1975) Reproductive guilds of fishes: a proposal and definition. *Journal of the Fisheries Research Board of Canada,* **32**, 821–864.

Balon, E. K. (ed.) (1980) *Charrs: Salmonid Fishes of the Genus Salvelinus.* Dr. W. Junk Publishers, The Hague.

Balon, E. K. (1981) Additions and amendments to the classification of reproductive styles in fishes. *Environmental Biology of Fishes,* **6**, 377–389.

Balon, E. K. (1984) Life histories of Arctic charrs: an epigenetic explanation of their invading ability and evolution. In: *Biology of the Arctic Charr* (eds L. Johnson & B. L. Burns), pp. 109–141. University of Manitoba Press, Winnipeg.

Balon, E. K. & Bruton, M. K. (1994) Fishes of the Tatinga River, Comoros, with comments on freshwater amphidromy in the goby *Sicyopterus lagocephalus. Ichthyology and Exploration of Freshwaters,* **5**, 25–40.

Bams, R. A. (1976) Survival and propensity for homing as affected by presence or absence of locally adapted paternal genes in two transplanted populations of pink salmon (*Oncorhynchus gorbuscha*). *Journal of the Fisheries Research Board of Canada,* **33**, 2716–2725.

Banister, K.E. (ed.) (1986) *Les Poissons.* Equinox Ltd, Oxford.

Banks, J. W. (1969) A review of the literature on the upstream migration of adult salmonids. *Journal of Fish Biology,* **1**, 85–136.

Baras, E. (1991) A bibliography on underwater telemetry, 1956–1990. *Canadian Technical Report of Fisheries and Aquatic Sciences,* **1819**, 55 pp.

Baras, E. (1992) Étude des stratégies d'occupation du temps et de l'espace chez le barbeau fluviatile, *Barbus barbus* (L). *Cahiers d'Ethologie,* **12**, 125–442.

Baras, E. (1993a) Étude par biotélémétrie de l'utilisation de l'espace, chez le barbeau fluviatile, *Barbus barbus* (L). Caractérisation et implications des patrons saisonniers de mobilité. *Cahiers d'Ethologie,* **13**, 135–138.

Baras, E. (1993b) A biotelemetry study of activity centres exploitation by *Barbus barbus* in the River Ourthe. *Cahiers d'Ethologie,* **13**, 173–174.

Baras, E. (1994) Constraints imposed by high densities on behavioural spawning strategies in the barbel, *Barbus barbus. Folia Zoologica,* **43**, 255–266.

Baras, E. (1995a) An improved electrofishing methodology for the assessment of habitat use by young-of-the-year fishes. *Archiv für Hydrobiologie,* **134**, 403–415.

Baras, E. (1995b) Seasonal activities of *Barbus barbus*: effect of temperature on time-budgeting. *Journal of Fish Biology,* **46**, 806–818.

Baras, E. (1996) Commentaire à l'hypothèse de l'éternel retour de Cury (1994): proposition d'un mécanisme fonctionnel dynamique. *Canadian Journal of Fisheries and Aquatic Sciences,* **53**, 681–684.

Baras, E. (1997) Environmental determinants of residence area selection and long term utilisation in a shoaling teleost, the common barbel (*Barbus barbus* L.). *Aquatic Living Resources,* **10**,

195–206.

Baras, E. (1998) Selection of optimal positioning intervals in fish tracking: an experimental study on *Barbus barbus*. *Hydrobiologia,* **371/372**, 19–28.

Baras, E. (2000) Day-night alternation prevails over food availability in synchronising the activity of *Piaractus brachypomus* (Characidae). *Aquatic Living Resources,* **13**, 115–120.

Baras, E., Birtles, C., Westerloppe, L., *et al.* (in press) A critical review of surgery techniques for implanting telemetry devices into the body cavity of fish. In: *Proceedings of the Fifth European Conference on Wildlife Telemetry* (eds Y. Le Maho & T. Zorn), CEPE/CNRS, Strasbourg, France.

Baras, E. & Cherry, B. (1990) Seasonal activities of female barbel *Barbus barbus* (L.) in the River Ourthe (Southern Belgium) as revealed by radio-tracking. *Aquatic Living Resources,* **3**, 283–294.

Baras, E., Debra, L., Grajales A. Q., Hahn von H., C., *et al.* (1997) Definition of the biological bases for the conservation, restoration and culture of highly valuable characids of the Magdalena River Basin, Colombia. European Union Research Report, Contract CI1-CT94–0032 (DG XII HSMU), second annual report, May 1997. 87 pp.

Baras, E. & Jeandrain, D. (1998) Evaluation of surgery procedures for tagging eel *Anguilla anguilla* (L.) with biotelemetry transmitters. *Hydrobiologia,* **371/372**, 107–111.

Baras, E., Jeandrain, D., Serouge, B. & Philippart, J. C. (1998) Seasonal variations of time and space utilisation by radio-tagged yellow eels *Anguilla anguilla* (L.) in a small stream. *Hydrobiologia,* **371/372**, 187–198.

Baras, E., Lambert, H. & Philippart, J. C. (1994a) A comprehensive assessment of the failure of *Barbus barbus* (L.) migrations through a fish pass in the canalised River Meuse (Belgium). *Aquatic Living Resources,* **7**, 181–189.

Baras, E., Malbrouck, C., Houbart, M., Kestemont, P. & Mélard, C. (2000a) The effect of PIT tags on growth and some physiological factors of age-0 cultured Eurasian perch *Perca fluviatilis* of variable size. *Aquaculture,* **185**, 159–173.

Baras, E., Ndao, M., Maxi, M. Y. J., *et al.* (2000b) Sibling cannibalism in dorada under experimental conditions. I. Ontogeny, dynamics, bioenergetics of cannibalism and prey size selectivity. *Journal of Fish Biology,* **57**, 1001–1020.

Baras, E. & Nindaba, J. (1999a) Seasonal and diel utilisation of inshore microhabitats by larvae and juveniles of *Leuciscus cephalus* and *Leuciscus leuciscus*. *Environmental Biology of Fishes,* **56**, 183–197.

Baras, E. & Nindaba, J. (1999b) Diel dynamics of habitat use by riverine young-of-the-year *Barbus barbus* and *Chondrostoma nasus* (Cyprinidae). *Archiv für Hydrobiologie,* **146**, 431–448.

Baras, E., Nindaba, J. & Philippart, J. C. (1995) Microhabitat use in a 0+ rheophilous cyprinid assemblage: quantitative assessment of community structure and fish density. *Bulletin Français de la Pêche et de la Pisciculture,* **337/338/339**, 241–247.

Baras, E. & Philippart, J. C. (1999) Adaptive and evolutionary significance of a reproductive thermal threshold in *Barbus barbus*. *Journal of Fish Biology,* **55**, 354–375.

Baras, E., Philippart, J. C. & Nindaba, J. (1996a) Importance of gravel bars as spawning grounds and nurseries for European running water cyprinids. In: *Proceedings of Ecohydraulics 2000* (eds M. Leclerc *et al.*), Vol. A, pp. 367–378. INRS-Eau, Québec.

Baras, E., Philippart, J. C. & Salmon, B. (1994b) Évaluation de l'efficacité d'une méthode d'échantillonnage par nasses des anguilles jaunes (*Anguilla anguilla*) en migration dans la Meuse. *Bulletin Français de la Pêche et de la Pisciculture,* **335**, 7–16.

Baras, E., Philippart, J. C. & Salmon, B. (1996b) Estimation of migrant yellow eel stock in large rivers through the survey of fish passes: a preliminary investigation in the River Meuse (Belgium). In: *Stock Assessment in Inland Fisheries* (ed. I. G. Cowx), pp. 82–92. Fishing News Books, Blackwell Science Ltd, Oxford.

Baras, E. & Westerloppe, L. (1999) Transintestinal expulsion of surgically implanted tags in African catfishes *Heterobranchus longifilis* of different size and age. *Transactions of the American Fisheries Society*, **128**, 737–746.

Baras, E., Westerloppe, L., Mélard, C., Philippart, J. C. & Bénech, V. (1999) Evaluation of implantation procedures for PIT tagging juvenile Nile tilapia. *North American Journal of Aquaculture*, **61**, 246–251.

Bardonnet, A. & Gaudin, P. (1990) Diel pattern of emergence in grayling (*Thymallus thymallus*, Linnaeus, 1758). *Canadian Journal of Zoology*, **68**, 465–469.

Bardonnet, A., Gaudin, P. & Persat, H. (1991) Microhabitats and diel downstream migration of young grayling (*Thymallus thymallus* L.). *Freshwater Biology*, **26**, 365–376.

Barthem, R. B. (1981) Experimenting fishing with gillnets in Central Amazonean lakes. M.Sc. thesis, Instituto Nacional de Pesquisas de Amazonia, Manaus, 84 pp. [in Portuguese].

Barthem, R. B. & Goulding, M. (1997) *The Catfish Connection – Ecology, Migration and Conservation of Amazon Predators*. Columbia University Press, New York, 144 pp.

Barthem, R. B., Ribeiro, M. C. L. de B. & Petrere, M. Jr. (1991) Life strategies of some long-distance migratory catfish in relation to hydroelectric dams in the Amazon Basin. *Biological Conservation*, **55**, 339–345.

Barton, B. A. (1980) Spawning migrations, age and growth, and summer feeding of white and longnose suckers in an irrigation reservoir. *Canadian Field-Naturalist*, **94**, 300–304.

Barus, V., Gajdusek, J., Pavlov, D. S. & Nezdolij, V. K. (1984) Downstream fish migration from two Czechoslovakian reservoirs in winter conditions. *Folia Zoologica*, **33**, 167–181.

Barus, V., Gajdusek, J., Pavlov, D. S. & Nezdolij, V. K. (1986) Downstream fish migration from the Mostiste and Vestonice reservoirs (CSSR) in spring-summer. *Folia Zoologica*, **35**, 79–93.

Barus, V., Pavlov, D. S., Nezdolij, V. K. & Gajdusek, J. (1985) Downstream fish migration from the Mostiste and Vestonice reservoirs (CSSR) in spring. *Folia Zoologica*, **34**, 75–87.

Bates, K. & Powers, P. (1998) Upstream passage of juvenile coho salmon through roughened culverts. In: *Fish Migration and Fish Bypasses* (eds M. Jungwirth, S. Schmutz & S. Weiss), pp. 192–202. Fishing News Books, Blackwell Science Ltd, Oxford.

Baxter, R. M. (1977) Environmental effects of dams and impoundments. *Annual Reviews in Ecology and Systematics*, **8**, 255–283.

Bayley, P. B. (1973) Studies on the migratory characin, *Prochilodus platensis* Holmberg 1889 (Pisces, Characoidei) in the River Pilcomayo, South America. *Journal of Fish Biology*, **5**, 25–40.

Bayley, P. B. (1983) Central Amazon fish populations: biomass, production and some dynamic characteristics. DPhil thesis, Dalhousie University, Halifax, 330 pp.

Bayley, P. B. (1988) Factors affecting growth rates of young tropical floodplain fishes: seasonality and density-dependence. *Environmental Biology of Fishes*, **21**, 127–142.

Bayley, P. B. (1995) Understanding large river-floodplain ecosystems. *BioScience*, **45**, 153–158.

Bayley, P. B. & Petrere, M. (1989) Amazon fisheries: assessment methods, current status and management options. *Canadian Special Publications of Fisheries and Aquatic Sciences*, **106**, 385–398.

Beadle, L. C. (1943) An ecological survey of some inland saline waters of Algeria. *Journal of the Lin-*

nean Society (Zoology), **41**, 218–242.

Beadle, L. C. (1981) *The Inland Waters of Tropical Africa. An Introduction to Tropical Limnology.* Longman, London.

Beamish, F. W. H. (1978) Swimming capacity. In: *Fish Physiology,* Vol. VII (eds W. S. Hoar & D. J. Randall), pp. 101–187. Academic Press, New York.

Beamish, F. W. H. (1980) Swimming performance and oxygen consumption of the charrs. In: *Charrs – Salmonid Fishes of the Genus Salvelinus* (ed. E. K. Balon), pp. 739–748. Dr. W. Junk Publishers, The Hague.

Beaumont, W. R. C. & Mann, R. H. K. (1984) The age, growth and diet of a freshwater population of the flounder *Platichthys flesus* (L.) in southern England. *Journal of Fish Biology,* **25**, 607–616.

Bechara, J. A., Domitrovic, H. A., Quintana, C. A., Roux, J. P., Jacobo, W. R. & Gavilán, G. (1996) The effects of gas supersaturation on fish health below Yaciretá Dam (Paraná River, Argentina). In: *Ecohydraulics 2000* (eds M. Leclerc *et al.*), Vol. A, pp. 3–12. INRS-Eau, Québec.

Beddow, T. A., Deary, C. A. & McKinley, R. S. (1998) Migratory and reproductive activity of radio-tagged Arctic char (*Salvelinus alpinus* L.) in northern Labrador. *Hydrobiologia,* **371/372**, 249–262.

Bell, K. N. I. (1999) An overview of goby-fry fisheries. *Naga,* **22** (4), 30–36.

Bellariva, J. L. & Belaud, A. (1998) Environmental factors influencing the passage of allice shad *Alosa alosa* at the Golfech fish lift on the Garonne River, France. In: *Fish Migration and Fish Bypasses* (eds M. Jungwirth, S. Schmutz & S. Weiss), pp. 171–179. Fishing News Books, Blackwell Science Ltd, Oxford.

Bemis, W. E. & Kynard, B. (1997) Sturgeon rivers: An introduction to acipenseriform biogeography and life history. *Environmental Biology of Fishes,* **48**, 167–184.

Bénech, V., Durand, J. R. & Quensière, J. (1983) Fish Communities of Lake Chad and Associated Rivers and Floodplains. In: *Lake Chad – Ecology and Productivity of a shallow Tropical Ecosystem* (eds J. P. Carmouze, J. R. Durand & C. Lévêque). *Monographiae Biologicae,* **53**, 293–356.

Bénech, V. & Ouattara, S. (1990) Rôle des variations de conductivité de l'eau et d'autres facteurs externes dans la croissance ovarienne d'un poisson tropical, *Brycinus leuciscus* (Characidae). *Aquatic Living Resources,* **3**, 153–162.

Bénech, V. & Peñaz, M. (1995) An outline of lateral migrations within the Central Delta of the River Niger, Mali. *Hydrobiologia,* **303**, 149–157.

Bénech, V. & Quensière, J. (1983) Migrations de poissons vers le lac Tchad à la décrue de la plaine inondée du Nord-Cameroun. 2. Comportement et rythme d'activité des principales espèces. *Revue d'Hydrobiologie Tropicale,* **16**, 79–101.

Bénech, V. & Quensière, J. (1989) Dynamique des Peuplements Ichtyologiques de la Région du Lac Tchad. Influence de la Sécheresse. ORSTOM, Paris, Vol. I, 428 pp. ; Vol. II, 195 pp.

Berg, L. S. (1962) *Freshwater Fishes of the USSR and Adjacent Countries.* Israel Program for Scientific Translations, Jerusalem, 3 volumes.

Berg, O. K. & Berg, M. (1993) Duration of sea and freshwater residence of Arctic char (*Salvelinus alpinus*) from the Vardnes River in northern Norway. *Aquaculture,* **110**, 129–140.

Berg, O. K. & Jonsson, B. (1990) Growth and survival rates of the anadromous trout, *Salmo trutta,* from the Vardnes River, northern Norway. *Environmental Biology of Fishes,* **29**, 145–154.

Berg, R. (1986a) Field studies on eel *Anguilla anguilla* L. In Lake Constance: tagging effects causing retardation of growth. *Vie et Milieu,* **36**, 285–286.

Berg, R. (1986b) Fish passage through Kaplan turbines at a power plant on the River Neckar and subsequent eel injuries. *Vie et Milieu,* **36**, 307–310.

Bergeron, N. E., Roy, A. G. & Chaumont, D. (1998) Winter geomorphological processes in the Sainte-Anne River (Québec) and their impact on the migratory behaviour of Atlantic tomcod (*Microgadus tomcod*). *Regulated Rivers: Research and Management,* **14**, 95–105.

Bergersen, E. P., Rogers, K. B. & Conger, L. V. (1994) A livestock hormone pellet injector for implanting PIT tags. *North American Journal of Fisheries Management,* **14**, 224–225.

Bergersen, E. P., Cook, M. F. & Baldes, R. J. (1993) Winter movements of burbot (*Lota lota*) during an extreme drawdown in Bull Lake, Wyoming, USA. *Ecology of Freshwater Fish,* **2**, 141–145.

Bergman, P., Jefferts, K., Fiscus, H. & Hager, R. (1968) A preliminary evaluation of an implanted coded wire fish tag. *Washington Department of Fisheries Resource Paper,* **3**, 63–84.

Bergman, P. K., Haw, F., Blankenship, H. L. & Buckley, R. M. (1992) Perspectives on design, use, and misuse of fish tags. *Fisheries,* **17**, 20–25.

Bergstedt, R. A. & Seelye, J. G. (1995) Evidence for lack of homing by sea lampreys. *Transactions of the American Fisheries Society,* **124**, 235–239.

Berman, C. H. & Quinn, T. P. (1991) Behavioural thermoregulation and homing by spring chinook salmon, *Oncorhynchus tshawytscha* (Walbaum), in the Yakima River. *Journal of Fish Biology,* **39**, 301–312.

Bernatchez, L. & Dodson, J. J. (1987) Relationship between bioenergetics and behavior in anadromous fish migrations. *Canadian Journal of Fisheries and Aquatic Sciences,* **44**, 399–407.

Berra, T. M. (1982) Life history of the Australian grayling *Prototroctes maraena* (Salmoniformes: Prototroctidae) in the Tambo River, Victoria. *Copeia,* **1982**, 795–805.

Berra, T. M. (1987) Speculations on the evolution of life history tactics of the Australian grayling. *American Fisheries Society Symposium,* **1**, 519–530.

Berzins, B. (1949) On the biology of Latvian perch (*Perca fluviatilis* L.). *Hydrobiologia,* **2**, 64–71.

Bettaso, R. H. & Young, J. N. (1999) Evidence for freshwater spawning by striped mullet and return of the Pacific tenpounder in the lower Colorado River. *California Fish and Game,* **85**, 75–76.

Beukema, J. J. (1968) Predation by the three-spine stickleback (*Gasterosteus aculeatus*): the influence of hunger and experience. *Behaviour,* **31**, 1–126.

Bhatt, J. P., Nautiyal, P. & Singh, H. R. (2000) Population structure of Himalayan mahseer, a large cyprinid fish in the regulated foothill section of the river Ganga. *Fisheries Research,* **44**, 267–271.

Bidgood, B. F. (1980) Field surgical procedures for implantation of radio tags in fish. *Fisheries Research Report of the Fish and Wildlife Division of Alberta,* **20**, 10 pp.

Biette, R. M. & Geen, G. H. (1980) Growth of underyearling sockeye salmon (*Oncorhynchus nerka*) under constant and cyclic temperatures in relation to live zooplankton ration size. *Canadian Journal of Fisheries and Aquatic Sciences,* **37**, 203–210.

Bigelow, H. B. & Schroeder, W. C. (1953) Fishes of the Gulf of Maine. *Fisheries Bulletin of the United States Fish and Wildlife Service,* **53**, 1–577.

Birstein, V., Waldman, J. R. & Bemis, W. E. (eds) (1997) Sturgeon biodiversity and conservation. *Environmental Biology of Fishes,* **48**, 440 pp.

Bishop, K. A. & Forbes, M. A. (1991) The freshwater fishes of northern Australia. In: *Moonsonal Australia: Landscape, Ecology and Man in the Northern Lowlands* (eds C. D. Haynes, M. G. Ridpath & M. A. J. Williams), pp. 79–107. A. A. Balkema, Rotterdam.

Bishop, K. A., Pidgeon, R. W. J. & Walden, D. J. (1995) Studies on fish movement dynamics in a tropical floodplain river: prerequisites for a procedure to monitor the impacts of mining. *Australian*

Journal of Ecology, **20**, 81–107.

Blaber, S. J. M. (1988) Fish communities of South-East African coastal lakes. In: *Biology and Ecology of African Freshwater Fishes* (eds C. Lévêque, M. N. Bruton & G. W. Ssentongo), pp. 351–362. Editions de l'ORSTOM, Travaux et Documents n° 216, Paris.

Blache, J. & Milton, F. (1962) Première contribution à la connaissance de la pêche dans le bassin hydrographique Logone-Chari Lac Tchad. *Mémoires ORSTOM,* **4** (1), 142 pp.

Blackburn, M. (1950) The Tasmanian whitebait, *Lovettia seali* (Johnston), and the whitebait fishery. *Australian Journal of Marine and Freshwater Research,* **1**, 155–198.

Bless, R. (1981) Unterschungen zum Einflass von gewasserbaulichen nassnehmen auf die Fischfauna in Mittelbirgsbachen. *Natur und Landschaft,* **56**, 243–252.

Bless, R. (1990) Die Bedtung von gewässerbaulichen Hinderinessen im Raum-Zrit-System der Groppe (*Cottus gobio* L.). *Natur und Landschaft,* **65**, 581–585.

Block, B. A., Dewar, H., Farwell, C. A. & Prince, E. D. (1998) A new satellite technology for tracking the movements of Atlantic bluefin tuna (*Thunnus thynnus thynnus*). *Marine Technological Society Journal,* **32**, 37–46.

Bodaly, R. A. (1980) Pre- and post-spawning movements of walleye, *Stizostedion vitreum*, in Southern India Lake, Manitoba. *Canadian Technical Report of Fisheries and Aquatic Sciences,* **931**, 30 pp.

Bodaly, R. A., Reist, J. D., Rosenberg, D. M., McCart, P. J. & Hecky, R. E. (1989) Fish and Fisheries of the Mackenzie and Churchill River Basins, Northern Canada. *Canadian Special Publication of Fisheries and Aquatic Sciences,* **106**, 128–144.

Bohl, E. (1980) Diel pattern of pelagic distribution and feeding in planktivorous fish. *Oecologia,* **44**, 368–375.

Böhling, P. & Lehtonen, H. (1984). Effect of environmental factors on migrations of perch (*Perca fluviatilis* L.) tagged in coastal waters of Finland. *Finnish Fisheries Research*, **5**, 31–40.

Boisclair, D. (1992) An evaluation of the stereocinematographic method to estimate fish swimming speed. *Canadian Journal of Fisheries and Aquatic Sciences,* **49**, 523–551.

Boisneau, P., Mennesson, C. & Baglinière, J. L. (1985) Observations sur l'activité de migration de la grande alose *Alosa alosa* L. en Loire (France). *Hydrobiologia,* **128**, 277–284.

Bok, A. H. (1979) The distribution and ecology of two mullet species in some freshwater rivers in the Eastern Cape, South Africa. *Journal of the Limnological Society of South Africa,* **5**, 97–102.

Bonath, K. (1977) *Narkose der Reptilien, Amphibien und Fische*. Paul Parey, Berlin. 148 pp.

Bonetto, A. A., Canon Cerón, M. & Roldán, D. (1981) Nuevos aportes al conocimiento de las migraciones de peces en el rio Paraná. *ECOSUR,* **8**, 29–40.

Bonetto, A. A., Dioni, W. & Pignalberi, C. (1969) Limnological investigations on biotic communities in the Middle Parana River Valley. *Verhanlungen Internationale Vereinigung für Theoretische und Angewandte Limnologie,* **17**, 1035–1050.

Bonetto, A. A., Pignalberi, C., Cordiviola, de Yuan, E. & Oliveros, O. (1971) Informaciones complementarias sobre migraciones de peces de la cuenca del Plata. *Physis,* **30** (81), 505–520.

Bos, A. R. (1999) Tidal transport of flounder larvae (*Pleuronectes flesus*) in the Elbe River, Germany. *Archive of Fishery and Marine Research,* **47**, 47–60.

Bouvet, Y., Pattée, E. & Maslin, J. L. (1992) Comparaison de la variabilité génétique de deux espèces de poissons, l'ombre commun et le gardon, dans un fleuve aménagé. *Bulletin Français de la Pêche et de la Pisciculture,* **323**, 50–76.

Bowmaker, A. P. (1973) Potamodromesis in the Mwenda River, Lake Kariba. In: *Man-made Lakes*

— *Their Problems and Environmental Effects* (eds W. C. Ackermann, G. F. White & E. B. Worthington). *Geophysical Monographs Series,* **17**, 159–164.

Braithwaite, H. (1971) A sonar fish counter. *Journal of Fish Biology,* **3**, 73–82.

Brandt, S. B. (1996) Acoustic assessment of fish abundance and distribution. In: *Fisheries Techniques,* 2nd edn. (eds B. R. Murphy & D. W. Willis), pp. 385–432. American Fisheries Society, Bethesda, Maryland.

Brännäs, E., Lundqvist, H., Prentice, E., Schmitz, M., Brännäs, K. & Wiklund, B. S. (1994) Use of the Passive Integrated Transponder (PIT) in a fish identification and monitoring system for fish behavioural studies. *Transactions of the American Fisheries Society,* **123**, 395–401.

Brannon, E. L. (1984) Influence of stock origin on salmon migratory behaviour. In: *Mechanisms of Migration in Fishes* (eds J. D. McCleave, G. P. Arnold, J. J. Dodson & W. H. Neill), pp. 103–111. Plenum, New York.

Braum, E. & Junk, W. J. (1982) Morphological adaptation of two Amazonian characoids (Pisces) for surviving in oxygen deficient waters. *Internationale Revue der Gesamten Hydrobiologie,* **67**, 869–886.

Breder, C. M. Jr. & Rosen, D. E. (1966) *Modes of Reproduction in Fishes.* Natural History Press, New York.

Breeser, S. W., Stearns, F. D., Smith, M. W., West, R. L. & Reynolds, J. B. (1988) Observations of movements and habitat preferences of burbot in an Alaskan glacial river system. *Transactions of the American Fisheries Society,* **117**, 506–509.

Bregazzi, P. R. & Kennedy, C. R. (1980) The biology of the pike, *Esox lucius* L., in a southern eutrophic lake. *Journal of Fish Biology,* **17**, 91–112.

Brett, J. R. (1964) The respiratory metabolism and swimming performance of young sockeye salmon. *Journal of the Fisheries Research Board of Canada,* **21**, 1183–1226.

Brett, J. R. (1971) Energetic responses of salmon to temperature. A study of some thermal relations in the physiology and freshwater ecology of the sockeye salmon (*Oncorhynchus nerka*). *Journal of the Fisheries Research Board of Canada,* **28**, 409–415.

Brett, J. R. (1972) The metabolic demand for oxygen in fish, particularly salmonids and a comparison with other vertebrates. *Respiratory Physiology,* **14**, 151–170.

Brett, J. R. (1983) Life energetics of sockeye salmon, *Oncorhynchus nerka*. In: *The Cost of Survival in Vertebrates* (ed. W. P. Aspey & S. I. Lustick), pp. 29–63. Ohio State University, Columbus.

Brett, J. R. (1986) Production energetics of a population of sockeye salmon, *Oncorhynchus nerka*. *Canadian Journal of Zoology,* **64**, 555–564.

Brett, J. R. & Groves, T. D. D. (1979) Physiological energetics. In: *Fish Physiology,* Vol. VIII (eds W. S. Hoar, D. J. Randall & J. R. Brett), pp. 279–352. Academic Press, New York.

Breukelaar, A. W., bij de Vaate, A. & Fockens, K. T. W. (1998) Inland migration of sea trout (*Salmo trutta*) into the rivers Rhine and Meuse (the Netherlands), based on inductive coupling radio telemetry. *Hydrobiologia,* **371/372**, 29–33.

Britz, R., Kirschbaum, F. & Heyd, A. (2000) Observations on the structure of larval attachment organs in three species of gymnotiforms (Teleostei: Ostariophysi). *Acta Zoologica,* **81**, 57–67.

Brown, A. V. & Armstrong, M. L. (1985) Propensity to drift downstream among various species of fish. *Journal of Freshwater Ecology,* **3**, 3–17.

Brown, D. J. A. (1979) The distribution and growth of juvenile cyprinid fishes in rivers receiving power station cooling water discharges. *Proceedings of the First British Freshwater Fisheries Conference,* University of Liverpool, pp. 217–229.

Brown, R. (2000) The potential of strobe lighting as a cost-effective means for reducing impingement and entrainment. *Environmental Science & Policy,* **3** (Suppl.), 405–416.

Brown, R. S. & Mackay, C. (1995) Spawning ecology of cutthroat trout (*Oncorhynchus clarki*) in the Ram River, Alberta. *Canadian Journal of Fisheries and Aquatic Sciences,* **52**, 983–992.

Brown, R. S., Power, G., Beltaos, S. & Beddow, T. A. (2000) Effects of hanging ice dams on winter movements and swimming activity of fish. *Journal of Fish Biology,* **57**, 1150–1159.

Bruton, M. N. (1979a) The survival of habitat dessication by air-breathing clariid catfishes. *Environmental Biology of Fishes,* **3**, 273–280.

Bruton, M. N. (1979b) The breeding biology and early development of *Clarias gariepinus* (Pisces, Clariidae) in Lake Sibaya, South Africa, with a review of breeding in species of the subgenus *Clarias* (*Clarias*). *Transactions of the Zoological Society of London,* **35**, 1–45.

Bruton, M. N. (1996) Alternative life-history strategies of catfishes. *Aquatic Living Resources,* **9**, 35–41.

Bruton, M. N., Bok, A. H. & Davies, M. T. T. (1987) Life history styles of diadromous fishes in inland waters of Southern Africa. *American Fisheries Society Symposium,* **1**, 104–121.

Bruton, M. N. & Boltt, R. E. (1975) Aspects of the biology of *Tilapia mossambica* (Pisces, Cichlidae) in a natural freshwater lake (Lake Sibaya, South Africa). *Journal of Fish Biology,* **7**, 423–445.

Bruylants, B., Vandelannoote, A. & Verheyen, R. (1986) The movement pattern and density distribution of perch, *Perca fluviatilis* L., in a channelled lowland river. *Aquaculture Fisheries Management,* **17**, 49–57.

Buckland, F. (1880) *Natural History of British Fishes*. Unwin, London.

Buckley, J. & Kynard, B. (1985) Yearly movements of shortnose sturgeons in the Connecticut River. *Transactions of the American Fisheries Society,* **114**, 813–820.

Buckley, R. M. & Blankenship, H. L. (1990) Internal extrinsic identification systems: overview of implanted wire tags, otolith marks and parasites. *American Fisheries Society Symposium,* **7**, 173–182.

Bulow, F. J., Webb, M. A., Crumby, W. D. & Quisenberry, S. S. (1988) Effectiveness of a barrier dam in limiting movement of rough fishes from a reservoir into a tributary stream. *North American Journal of Fisheries Management,* **8**, 273–275.

Bunt, C. M., Katapodis, C. & McKinley, R. S. (1999) Attraction and passage efficiency of white suckers and smallmouth bass by two Denil fishways. *North American Journal of Fisheries Management,* **19**, 793–803.

Bussell, R. B. (1978) *Notes for guidance in the design and use of fish counting stations*. Department of the Environment, London, 70 pp.

Bütticker, B. (1992) Electrofishing results corrected by selectivity functions in stock size estimates of brown trout (*Salmo trutta* L.) in brooks. *Journal of Fish Biology,* **41**, 673–684.

Buzby, K. M. & Deegan, L. A. (2000) Inter-annual fidelity to summer feeding sites in Arctic grayling. *Environmental Biology of Fishes,* **59**, 319–327.

Cada, G. F. (1998) Fish passage mitigation at hydroelectric power projects in the United States. In: *Fish Migration and Fish Bypasses* (eds M. Jungwirth, S. Schmutz & S. Weiss), pp. 208–219. Fishing News Books, Blackwell Science Ltd, Oxford.

Cadwalladr, D. A. (1965) Notes on the breeding biology and ecology of *Labeo victorianus* Blgr (Cyprinidae) of L. Victoria. *Revue de Zoologie et de Botanique Africaine,* **72**, 109–134.

Caffrey, J. M., Conneely, J. J. & Connolly, B. (1996) Radio telemetric determination of bream (*Abramis brama* L.) movements in Irish canals. In: *Underwater Biotelemetry* (eds E. Baras & J. C. Philippart), pp. 59–65. University of Liège, Belgium.

Cala, P. (1970) On the ecology of the ide *Idus idus* (L.) in the River Kävlingeån, South Sweden. *Reports of the Institute of Freshwater Research of Drottningholm,* **50**, 45–99.

Camargo, A. F. M. & Esteves, F. A. (1995) Influence of water level variation on fertilization of an oxbow lake of Rio Mogi-Guaçu, state of Sao Paulo, Brazil. *Hydrobiologia,* **299**, 185–193.

Cambray, J. A. (1990) Adaptive significance of a longitudinal migration by juvenile freshwater fish in the Gantoos River System, South Africa. *South African Journal of Wildlife Research,* **20**, 148–156.

Campana, S. E. (1999) Chemistry and composition of fish otoliths: pathways, mechanisms and applications. *Marine Ecology Progress Series,* **188**, 263–297.

Campana, S. E. & Neilson, J. D. (1985) Microstructure of fish otoliths. *Canadian Journal of Fisheries and Aquatic Sciences,* **42**, 1014–1032.

Carbine, W. F. (1942) Observations on the life history of the northern pike, *Esox lucius* L., in Houghton Lake, Michigan. *Transactions of the American Fisheries Society,* **71**, 149–164.

Carl, L. M. (1995) Sonic tracking of burbot in Lake Opeongo, Ontario. *Transactions of the American Fisheries Society,* **124**, 77–83.

Carline, R. F., DeWalle, D. R., Sharpe, W. E., Dempsey, B. A., Gagen, C. J. & Swistock, B. (1992) Water chemistry and fish community responses to episodic stream acidification in Pennsylvania, USA. *Environmental Pollution,* **78**, 45–48.

Carlson, D. M. (1992) Importance of wintering refugia to the largemouth bass fishery in the Hudson River estuary. *Journal of Freshwater Ecology,* **7**, 173–180.

Carmichael, J. T., Haeseker, S. L. & Hightower, J. E. (1998) Spawning migration of telemetered striped bass in the Roanoke River, North Carolina. *Transactions of the American Fisheries Society,* **127**, 286–297.

Carmouze, J. P., Durand, J. F. & Lévêque, C. (1983) *Lake Chad – Ecology and Productivity of a Shallow Tropical Ecosystem. Monographiae Biologicae,* **53**, Junk, The Hague, 575 pp.

Carvalho, G. R. & Hauser, L. (1994) Molecular genetics and the stock concept in fisheries. *Reviews in Fish Biology and Fisheries,* **4**, 326–350.

Carvalho, G. R. & Pitcher, T. J. (eds) (1994) Molecular genetics in fisheries. *Reviews in Fish Biology and Fisheries,* **4**, 269–399.

Carvalho, J. L. de & Mérona, B. de (1986) Estudos sobre dois pexes migratorios do baixo Tocantins, antes do fechamento da barragem de Tucurui. *Amazoniana,* **9**, 595–607.

Casselman, J. M., Penczak, T., Carl, L., Mann, R. H. K., Holcik, J. & Woitowich, W. A. (1990) An evaluation of fish sampling methodologies for large river systems. *Polish Archives of Hydrobiology,* **37**, 521–551.

Castro-Santos, T., Haro, A. & Walk, S. (1996) A passive integrated transponder (PIT) tag system for monitoring fishways. *Fisheries Research,* **28**, 253–261.

Cerri, D. R. (1983) The effect of light intensity on predator and prey behaviour in cyprinid fish: factors that influence prey risk. *Animal Behaviour,* **31**, 736–742.

Chadwick, M. (1995) Index rivers: a key to managing anadromous fish. *Reviews in Fish Biology and Fisheries,* **5**, 38–51.

Chamberlain, A. (1979) Effects of tagging on equilibrium and feeding. *Underwater Telemetry Newsletter,* **9**, 2–3.

Champion, A. S. & Swain, S. (1974) A note on the movements of coarse fish passing through the Ministry's trapping installation on the River Axe, Devon. *Journal of the Institute of Fisheries Manage-*

ment, **5**, 89–92.

Chang, C. W., Tzeng, W. N. & Lee, Y. C. (2000) Recruitment and hatching dates of grey mullet (*Mugil cephalus* L.) juveniles in the Tanshui estuary of northwest Taiwan. *Zoological Studies,* **39**, 99–106.

Chanseau, M., Larinier, M. & Travade, F. (1999) Efficiency of a downstream bypass as estimated by the mark-recapture technique and behaviour of Atlantic salmon (*Salmo salar* L.) smolts at the Bedous water intake on the Aspe River (France) monitored by radiotelemetry. *Bulletin Français de la Pêche et de la Pisciculture,* **353/354**, 99–120.

Chapman, C. A. & Mackay, W. C. (1984) Versatility in habitat use by a top aquatic predator, *Esox lucius* L. *Journal of Fish Biology,* **25**, 109–115.

Chapman, L. J., Chapman, C. A., Ogutu-Ohwayo, R., Chandler, M., Kaufman, L. & Keiter, E. A. (1996) Refugia for endangered fishes from an introduced predator in Lake Nabugabo, Uganda. *Conservation Biology,* **10**, 554–561.

Chapman, L. J., Kaufman, L. S. & Chapman, C. A. (1994) Why swim upside down? A comparative study of two mochokid catfishes. *Copeia,* **1994**, 130–135.

Chapman, L. J., Kaufman, L. S., Chapman, C. A. & McKenzie, F. E. (1995) Hypoxia tolerance in twelve species of East African cichlids: potential for low oxygen refugia in Lake Victoria. *Conservation Biology,* **9**, 1274–1288.

Chapman, L. J. & Liem, K. F. (1995) Papyrus swamps and the respiratory ecology of *Barbus neumayeri*. *Environmental Biology of Fishes,* **44**, 183–197.

Charnov, E. L. (1976) Optimal foraging: the marginal value theorem. *Theoretical Population Biology,* **9**, 129–136.

Cheek, T. E., Van Den Avyle, M. J. & Coutant, C. C. (1985) Influences of water quality on distribution of striped bass in a Tennessee River impoundment. *Transactions of the American Fisheries Society,* **114**, 67–76.

Chenuil, A., Crespin, L., Pouyard, L. & Berrébi, P. (2000) Movements of adult fish in a hybrid zone revealed by microsatellite genetic analysis and capture-recapture data. *Freshwater Biology,* **43**, 121–131.

Clapp, D. F., Clark, R. D. & Diana, J. S. (1990) Range activity and habitat of large, free-ranging brown trout in a Michigan stream. *Transactions of the American Fisheries Society,* **119**, 1022–1034.

Claridge, P. N., Potter, I. C. & Hughes, G. M. (1973) Circadian rhythms of activity, ventilatory frequency and heart rate in the adult river lamprey, *Lampetra fluviatilis*. *Journal of Zoology,* **171**, 239–250.

Clark, C. F. (1950) Observations on the spawning habits of the northern pike, *Esox lucius*, in northwestern Ohio. *Copeia,* **4**, 285–288.

Clay, C. H. (1995) *Design of Fishways and other Fish Facilities*, 2nd edn. Lewis Publishers, Boca Raton.

Clay, D. (1977) Preliminary observations on salinity tolerance of *Clarias lazera* from Israel. *Bamidgeh,* **29** (3), 102–109.

Close, T. L. (2000) Detection and retention of postocular visible implant elastomer in fingerling rainbow trout. *North American Journal of Fisheries Management,* **20**, 542–545.

Clough, S. & Beaumont, W. R. C. (1998) Use of miniature radio-transmitters to track the movements of dace, *Leuciscus leuciscus* (L.) in the River Frome, Dorset. *Hydrobiologia,* **371/372**, 89–97.

Clough, S. & Ladle, M. (1997) Diel migration and site fidelity in a stream-dwelling cyprinid, *Leuciscus leuciscus*. *Journal of Fish Biology,* **50**, 1117–1119.

Clough, S., Garner, P., Deans, D. & Ladle, M. (1998) Postspawning movements and habitat selection of dace in the River Frome, Dorset, southern England. *Journal of Fish Biology*, **53**, 1060–1070.

Clugston, J. P., Foster, A. M & Carr, S. H. (1995) Gulf sturgeon *Acipenser oxyrinchus desotoi* in the Suwannee River, Florida, USA. In: *Proceedings of the Second International Symposium on the Sturgeon* (eds A. D. Gershanovich & T. I. J. Smith), pp. 215–224. VNIRO Publishing, Moscow.

Coates, D. (1990) Biology of the rainbowfish, *Glossolepis multisquamatus* (Melatoniidae) from the Sepik River floodplains, Papua New Guinea. *Environmental Biology of fishes*, **29**, 119–126.

Coe, M. J. (1966) The biology of *Tilapia grahami* in lake Magadi, Kenya. *Acta Tropica,* **23**, 146–177.

Coe, M. J. (1969) Observations on *Tilapia alcalina* in lake Natron on the Kenya-Tanzania border. *Revue de Zoologie et de Botanique Africaine*, **80**, 1–14.

Coles, T. F., Wortley, J. S. & Noble, P. (1985) Survey methodology for fish population assessment within Anglian Water. *Journal of Fish Biology,* **27** (Suppl. A), 175–186.

Colgan, P. (1993) The motivational basis of fish behaviour. In: *Behaviour of Teleost Fishes,* 2nd edn. (ed. T. J. Pitcher), pp. 31–56. Chapman and Hall, London.

Collin, S. P. & Potter, I. C. (2000) The ocular morphology of the southern hemisphere lamprey *Mordacia mordax* Richardson with special reference to a single class of photoreceptor and a retinal tapetum. *Brain Behavior and Evolution,* **55**, 120–138.

Collins, K. (1996) Development of an electromagnetic telemetry system for tracking lobsters on an artificial reef. In: *Underwater Biotelemetry* (eds E. Baras & J. C. Philippart), pp. 225–234. University of Liège, Belgium.

Collins, N. C., Hinch, S. G. & Baia, K. A. (1991) Non-intrusive time-lapse video monitoring of shallow aquatic environments. *Canadian Technical Report of Fisheries and Aquatic Sciences,* **1821**, 35 pp.

Combs, D. L. & Peltz, R. L. (1982) Seasonal distribution of striped bass in Keystone Reservoir, Oklahoma. *North American Journal of Fisheries Management,* **2**, 66–73.

Comeau, S. & Boisclair, D. (1998) Day-to-day variation in fish horizontal migration and its potential consequence on estimates of trophic interactions in lakes. *Fisheries Research,* **35**, 75–81.

Concepcion, G. B. & Nelson, S. G (1999) Effects of a dam and reservoir on the distributions and densities of macrofauna in tropical streams of Guam (Mariana islands). *Journal of Freshwater Ecology,* **14**, 447–454.

Cook, M. F. & Bergersen, E. P. (1988) Movements, habitat selection and activity periods of northern pike in Eleven Mile Reservoir, Colorado. *Transactions of the American Fisheries Society,* **117**, 496–502.

Cooke, S. J. & Bunt, C. M. (1999) Spawning and reproductive biology of the Greater Redhorse, *Moxostoma valenciennesi*, in the Grand River, Ontario. *Canadian Field-Naturalist,* **113**, 497–502.

Cooper, J. C. & Hasler, A. D. (1974) Electroencephalographic evidence for retention of olfactory cues in homing coho salmon. *Science,* **183**, 336–338.

Copp, G. H. (1989) Electrofishing for fish larvae and 0+ juveniles: equipment modifications for increased efficiency with short fishes. *Aquaculture and Fisheries Management,* **20**, 453–462.

Copp, G. H. (1990) Shifts in the microhabitat of larval and juvenile roach *Rutilus rutilus* (L.) in a floodplain channel. *Journal of Fish Biology*, **36**, 683–692.

Copp, G. H. (1997) Importance of marinas and off-channel waters bodies as refuges for young fishes in a regulated river. *Regulated Rivers: Research and Management,* **13**, 303–307.

Copp, G. H. & Cellot, B. (1988) Drift of embryonic and larval fishes, especially *Lepomis gibbosus* (L.), in the upper Rhone River. *Freshwater Ecology,* **4**, 419–424.

Copp, G. H. & Garner, P. (1995) Evaluating the microhabitat use of freshwater fish larvae and juveniles with point abundance sampling by electrofishing. *Folia Zoologica,* **44**, 145–158.

Copp, G. H. & Jurajda, P. (1993) Do small riverine fish move inshore at night? *Journal of Fish Biology,* **43**, 229–241.

Copp, G. H. & Jurajda, P. (1999) Size-structured diel use of river banks by fish. *Aquatic Sciences,* **61**, 75–91.

Corbett, B. W. & Powles, P. M. (1986) Spawning and larval drift of sympatric walleyes and white suckers in an Ontario stream. *Transactions of the American Fisheries Society,* **115**, 41–46.

Courtenay, S.C., Quinn, T. P., Dupuis, H. M. C., Groot, C. & Larkin, P. A. (1997) Factors affecting the recognition of population-specific odours by juvenile coho salmon. *Journal of Fish Biology,* **50**, 1042–1060.

Coutant, C. (1969) Temperature, reproduction and behaviour. *Chesapeake Science,* **10**, 261–274.

Coutant, C. C. (1985) Striped bass, temperature, and dissolved oxygen: a speculative hypothesis for environmental risk. *Transactions of the American Fisheries Society,* **114**, 31–61.

Coutant, C. C. (1990a) Temperature-oxygen habitat for freshwater and coastal striped bass in a changing climate. *Transactions of the American Fisheries Society,* **119**, 240–253.

Coutant, C. C. (1990b) Microchemical analysis of fish hard parts for reconstructing habitat use: practice and promise. *American Fisheries Society Symposium,* **7**, 574–580.

Coutant, C. C. (2000) What is 'normative' at cooling water intakes? Defining normalcy before judging adverse. *Environmental Science & Policy,* **3** (Suppl.), 37–42.

Coutant, C. C. & Whitney, R. R. (2000) Fish behavior in relation to passage through hydropower turbines: a review. *Transactions of the American Fisheries Society,* **129**, 351–380.

Cowx, I. G. (ed.) (1990) *Developments in Electric Fishing.* Fishing News Books, Blackwell Science Ltd, Oxford.

Cowx, I. G. (1991) The use of angler catch data to examine potential fishery management problems in the lower reaches of the River Trent, England. In: *Catch Effort Sampling Strategies* (ed. I. G. Cowx), pp. 154–165. Fishing News Books, Blackwell Science Ltd, Oxford.

Cowx, I. G. (1998) Fish passage facilities in the UK: issues and options for future development. In: *Fish Migration and Fish Bypasses* (eds M. Jungwirth, S. Schmutz & S. Weiss), pp. 220–235. Fishing News Books, Blackwell Science Ltd, Oxford.

Cowx, I. G. & Welcomme, R. L. (eds) (1998) *Rehabilitation of Rivers for Fish.* FAO and Fishing News Books, Blackwell Science Ltd, Oxford.

Craig, J. F. (1977) Seasonal changes in the day and night activity of adult perch, *Perca fluviatilis* L. *Journal of Fish Biology,* **11**, 161–166.

Craig, J. F. (1987) *The Biology of Perch and Related Fish.* Croom Helm, London.

Craig, P. C. & Poulin, V. A. (1975) Movements and growth of arctic grayling (*Thymallus arcticus*) and juvenile arctic charr (*Salvelinus alpinus*) in a small arctic stream, Alaska. *Journal of the Fisheries Research Board of Canada,* **32**, 689–697.

Craig-Bennett, A. (1931) The reproductive cycle of the three-spined stickleback *Gasterosteus aculeatus* L. *Philosophical Transactions of the Royal Society, London B,* **219**, 197–279.

Crampton, W. G. R. (1998) Effect of anoxia on the distribution, respiratory strategies and electrical signal diversity of gymnotiform fishes. *Journal of Fish Biology,* **53** (Suppl. A), 307–330.

Creutzberg, F. (1961) On the orientation of migrating elvers (*Anguilla vulgaris,* Turt.) in a tidal area. *Netherlands Journal of Sea Research,* **1**, 257–338.

Crisp, D. T. & Mann, R. H. K. (1991) Effects of impoundment on populations of bullhead *Cottus gobio* L. and minnow *Phoxinus phoxinus* L. in the basin of Cow Green Reservoir. *Journal of Fish Biology,* **38,** 731–740.

Crisp, D. T., Mann, R. H. K. & Cubby, P. R. (1984) Effects of impoundment upon fish populations in afferent streams at Cow Green Reservoir. *Journal of Applied Ecology,* **21,** 739–756.

Crook, D. A. & White, R. W. G. (1995) Evaluation of subcutaneously implanted Visual Implant Tags and Coded Wire Tags for marking and benign recovery in a small scaleless fish, *Galaxias truttaceus* (Pisces, Galaxiidae). *Marine and Freshwater Research,* **46,** 943–946.

Crossman, C. J. (1990) Reproductive homing in muskellunge, *Esox masquinongy. Canadian Journal of Fisheries and Aquatic Sciences,* **47,** 1803–1812.

Croze, O., Chanseau, M. & Larinier, M. (1999) Efficiency of a downstream bypass for Atlantic salmon (*Salmo salar* L.) smolts and fish behaviour. *Bulletin Français de la Pêche et de la Pisciculture,* **353/354,** 121–140.

Croze, O. & Larinier, M. (1999) A study of Atlantic salmon (*Salmo salar* L.) smolt behaviour at the Pointis hydroelectric powerhouse water intake on the Garonne River and an estimate of downstream migration over the Rodere dam. *Bulletin Français de la Pêche et de la Pisciculture,* **353/354,** 141–156.

Cunjak, R. A. & Power, G. (1987) Cover use by stream resident trout in winter: a field experiment. *North American Journal of Fisheries Management,* **7,** 539–544.

Cury, P. (1994) Obstinate nature: an ecology of individuals. Thoughts on reproductive behavior and diversity. *Canadian Journal of Fisheries and Aquatic Sciences,* **51,** 1664–1673.

Cushman, R. M. (1985) Review of ecological effects of rapidly varying flows downstream of hydroelectric facilities. *North American Journal of Fisheries Management,* **5,** 330–339.

Dadswell, M. J. (1979) Biology and population characteristics of the shortnose sturgeon, *Acipenser brevirostrum* LeSueur 1818 (Osteichthyes: Acipenseridae) in the Saint John River estuary, New Brunswick, Canada. *Canadian Journal of Zoology,* **57,** 2186–2210.

Dadswell, M. J., Klauda, R. J., Moffitt, C. M., Saunders, R. L., Rulifson, R. A. & Cooper, J. E. (eds) (1987) Common strategies of anadromous and catadromous fishes. *American Fisheries Society Symposium,* **1,** American Fisheries Society, Bethesda, Maryland.

Dadswell, M. J. & Rulifson, R. A. (1994) Macrotidal estuaries: a region of collision between migratory marine animals and tidal power development. *Biological Journal of the Linnean Society,* **51,** 93–113.

Daemers-Lambert, C. (2000) Perception et communication en milieu marin: quatre leçons de physiologie animale. *Cahiers d'Ethologie,* **19,** 265–492.

Daget, J. (1952) Mémoires sur la biologie des poissons du Niger. 1. Biologie et croissance des espèces du genre *Alestes. Bulletin de l'Institut Français d'Afrique Noire,* **14,** 191–225.

Daget, J. (1954) Les Poissons du Niger Supérieur. *Mémoires de l'Institut Français d'Afrique Noire,* **36,** Dakar, 391 pp.

Daget, J., Gosse, J. P. & Thys van den Audenaerde, D. F. E. (eds) (1986) *Check-list of the Freshwater Fishes of Africa.* Cloffa II. ORSTOM, Paris; MRAC, Tervuren.

Dames, H. R., Coon, T. G. & Robinson, J. W. (1989) Movements of channel catfish between the Missouri River and a tributary, Perche Creek. *Transactions of the American Fisheries Society,* **118,** 670–679.

Dando, P. R. (1984) Reproduction in estuarine fish. In: *Fish reproduction – Strategies and Tactics* (eds G. W. Potts & R. J. Wootton), pp. 155–170. Academic Press, London.

Daniels, R. A. & Moyle, P. B. (1983) Life history of splittail (Cyprinidae: *Pogonichthys macrolepidotus* [Ayres]) in the Sacramento-San Joaquin estuary. *U. S. National Marine Fisheries Service Fishery Bulletin*, **81**, 647–654.

Dat, C. G., Leblond, P. H., Thomson, K. A. & Ingraham, W. J. Jr. (1995) Computer simulations of homeward-migrating Fraser River sockeye salmon: Is compass orientation a sufficient direction-finding mechanism in the north-east Pacific Ocean? *Fisheries Oceanography*, **4**, 209–216.

Davidson, F. A. (1937) Migration and homing in Pacific salmon. *Science*, **86**, 1–4.

Davis, C. L., Carl, L. M. & Evans, D. O. (1997) Use of a remotely operated vehicle to study habitat and population density of juvenile lake trout. *Transactions of the American Fisheries Society*, **126**, 871–875.

Davis, R. M. (1967) Parasitism by newly transformed anadromous sea lampreys on landlocked salmon and other fishes in a coastal Maine lake. *Transactions of the American Fisheries Society*, **96**, 11–16.

Davis, T. L. O. (1985) Seasonal changes in gonad maturity and abundance of larvae and early juveniles of barramundi, *Lates calcarifer* (Bloch), in Van Diemen Gulf and the Gulf of Carpentaria. *Australian Journal of Marine and Freshwater Research*, **36**, 177–190.

Davis, T. L. O. (1986) Migration patterns in barramundi, *Lates calcarifer* (Bloch), in northern Australia. *Australian Journal of Marine and Freshwater Research*, **37**, 673–689.

DeCicco, A. L. (1989) Movements and spawning of adult Dolly Varden charr (*S. malma*) in Chukchi Sea drainages of northwestern Alaska: evidence for summer and fall spawning populations. *Physiology and Ecology Japan*, Special Vol. **1**, 229–238.

DeCicco, A. L. (1992) Long-distance movements of anadromous Dolly Varden between Alaska and the U.S.S.R. *Arctic*, **45**, 120–123.

Dedual, M. & Jowett, I. G. (1999) Movement of rainbow trout (*Oncorhynchus mykiss*) during the spawning migration in the Tongariro River, New Zealand. *New Zealand Journal of Marine and Freshwater Research*, **33**, 107–117.

Deelder, C. L. (1952) On the migration of the elver (*Anguilla vulgaris*, Turt.) at sea. *Journal du Conseil Permanent International pour l'Exploration de la Mer*, **18**, 187–218.

Deelder, C. L. (1954) Factors affecting the migration of silver eel in Dutch inland waters. *Journal du Conseil Permanent International pour l'Exploration de la Mer*, **20**, 177–185.

Deelder, C. L. (1958) On the behaviour of elvers (*Anguilla vulgaris* Turt.) migrating from sea in to freshwater. *Journal du Conseil Permanent International pour l'Exploration de la Mer*, **24**, 136–146.

Deelder, C. L. & Tesch, F. W. (1970) Heimfindevermögen von Aalen (*Anguilla anguilla*) die über große Entlerungen verpflante unurden. *Ästuarische und Marine Biologie Berlin*, **6**, 81–92.

Demers, E., McKinley, R. S., Weatherley, A. H. & McQueen, D. (1996) Activity patterns of largemouth and smallmouth bass determined by electromyogram biotelemetry. *Transactions of the American Fisheries Society*, **125**, 434–439.

Denil, G. (1938) *La Mécanique du Poisson de Rivière*. Annales des Travaux Publics de Belgique, Bruxelles, Belgium.

Descy, J.P. & Empain, A. (1984) Meuse. In: *Ecology of European Rivers* (ed. W.A. Whitton), pp. 1–23. Blackwell, Oxford.

Detenbeck, N. E., DeVore, P. W., Niemi, G. J. & Lima, A. (1992) Recovery of temperate-stream fish communities from disturbance: a review of case studies and synthesis of theory. *Environmental Management*, **16**, 33–53.

Diamond, M. (1985) Some observations of spawning by roach *Rutilus rutilus* L., and bream *Abramis brama* L., and their implications for management. *Aquaculture and Fisheries Management,* **16**, 359–367.

Diana, J. S. (1980) Diel activity pattern and swimming speeds of northern pike (*Esox lucius*) in Lac Ste Anne, Alberta. *Canadian Journal of Fisheries and Aquatic Sciences,* **37**, 1454–1458.

Diana, J. S. (1984) The growth of largemouth bass, *Micropterus salmoides* (Lacepede), under constant and fluctuating temperatures. *Journal of Fish Biology,* **24**, 165–172.

Diana, J. S., Mackay, W. C. & Ehrman, M. (1977) Movements and habitat preference of northern pike (*Esox lucius*) in Lac Ste. Anne, Alberta. *Transactions of the American Fisheries Society,* **106**, 560–565.

Dickhoff, W. W. & Darling, D. S. (1983) Evolution of thyroid function and its control in lower vertebrates. *American Zoologist,* **23**, 697–707.

Diebel, C. E., Proksch, R., Green, C. R., Neilson, P. & Walker, M. M. (2000) Magnetite defines a vertebrate magnetoreceptor. *Nature,* **406**, 299–302.

Dietz, K. G. (1973) The life history of walleye (*Stizostedion vitreum vitreum*) in the Peace-Athabasca Delta. Alberta Department of Lands and Forests, Fish and Wildlife Division, 52 pp.

Dill, L. M., Ydenberg, R. C. & Fraser, A. H. G. (1981) Food abundance and territory size in juvenile coho salmon (*Oncorhynchus kisutch*). *Canadian Journal of Zoology,* **59**, 1801–1809.

Dillinger, R. E. Jr., Birt, T. P. & Green, M. (1992) Arctic cisco, *Coregonus autumnalis*, distribution, migration and spawning in the Mackenzie River. *Canadian Field-Naturalist,* **106**, 175–180.

Dingle, H. (1980) Ecology and evolution of migration. In: *Animal Migration, Orientation and Navigation* (ed. S. A. Gauthreaux), pp. 1–101. Academic Press, New York.

Dingle, H. (1996) *Migration: The Biology of Life on the Move.* Oxford University Press, Oxford.

Dittman, A. H. & Quinn, T. P. (1996) Homing in Pacific salmon: mechanisms and ecological basis. *Journal of Experimental Biology,* **199**, 83–91.

Dodson, J. J. (1988) The nature and role of learning in the orientation and migratory behavior of fishes. *Environmental Biology of Fishes,* **23**, 161–182.

Dodson, J. J., Lambert, Y. & Bernatchez, L. (1985) Comparative migratory and reproductive strategies of the sympatric anadromous coregonine species of James Bay. *Contributions in Marine Science,* **27** (Suppl.), 296–315.

Dodson, J. J. & Leggett, W. C. (1973) Behaviour of adult American shad (*Alosa sapidissima*) homing to the Connecticut River from Long Island Sound. *Journal of the Fisheries Research Board of Canada,* **30**, 1847–1860.

Dodson, J. J. & Young, J. C. (1977) Temperature and photoperiod regulation of rheotropic behavior in prespawning common shiners, *Notropis cornutus. Journal of the Fisheries Research Board of Canada,* **34**, 341–346.

Dolloff, A., Kershner, J. & Thurow, R. (1996) Underwater observation. In: *Fisheries Techniques,* 2nd edn. (eds B. R. Murphy & D. W. Willis), pp. 533–554. American Fisheries Society, Bethesda, Maryland.

Dolloff, C. A., Flebbe, P. A. & Owen, M. D. (1994) Fish habitat and fish populations in a Southern Appalachian watershed before and after Hurricane Hugo. *Transactions of the American Fisheries Society,* **123**, 668–778.

Dombeck, M. P. (1979) Movement and behavior of the muskellunge determined by radio-telemetry. *Technical Bulletin,* **113**. Department of Natural Resources, Madison, Wisconsin.

Dominey, W. J. (1984) Effects of sexual selection and life history on speciation. Species flocks in African cichlids and Hawaiian *Drosophila*. In: *Evolution of Fish Species Flocks* (eds A. A. Echelle

& I. Kornfield), pp. 231–249. University of Maine Press, Orono.

Donald, D. B. (1997) Relationship between year-class strength for goldeyes and selected environmental variables during the first year of life. *Transactions of the American Fisheries Society,* **126**, 361–368.

Donald, D. B. & Kooyman, A. H. (1977a) Migration and population dynamics of the Peace-Athabasca delta goldeye population. *Canadian Wildlife Service Occasional Paper,* **31**, 21 pp.

Donald, D. B. & Kooyman, A. H. (1977b) Food, feeding habits and growth of goldeye, *Hiodon alosoides* (Rafinesque), in waters of the Peace-Athabasca delta. *Canadian Journal of Zoology,* **55**, 1038–1047.

Doucett, R. R., Power, M., Power, G., Caron, F. & Reist, J. D. (1999) Evidence for anadromy in a southern relict population of Arctic charr from North America. *Journal of Fish Biology,* **55**, 84–93.

Douglas, M. E. & Marsh, P. C. (1996) Population estimates / population movements of *Gila cypha*, an endangered cyprinid fish in the Grand Canyon region of Arizona. *Copeia,* **1996**, 15–28.

Douglas, M. R. & Douglas, M. E. (2000) Late season reproduction by big-river Catostomidae in Grand Canyon (Arizona). *Copeia,* **2000**, 238–244.

Dovel, W. L. & Berggren, T. J. (1983) Atlantic sturgeon of the Hudson estuary, New York. *New York Fish and Game Journal,* **30**, 140–172.

Dovel, W. L., Mihursky, J. A. & McErlean, A. J. (1969) Life history aspects of the hogchoker, *Trinectes maculatus*, in the Patuxent River estuary, Maryland. *Chesapeake Science,* **10**, 104–119.

Døving, K. B., Nordeng, H. & Oakley, B. (1974) Single unit discrimination of fish odours released by char (*Salmo alpinus* L.) populations. *Comparative Biochemistry and Physiology,* **47** (A), 1051–1063.

Døving, K. B., Westerberg, H. & Johnsen, P. B. (1985) Role of olfaction in the behavioural and neuronal responses of Atlantic salmon *Salmo salar* to hydrographic stratification. *Canadian Journal of Fisheries and Aquatic Sciences,* **42**, 1658–1667.

Downhower, J. F., Lejeune, P., Gaudin, P. & Brown, L. (1990) Movements of the chabot (*Cottus gobio*) in a small stream. *Polskie Archivum Hydrobiologii,* **37**, 119–126.

Dudgeon, D. (1992) Endangered ecosystems: a review of the conservation status of tropical Asian rivers. *Hydrobiologia,* **248**, 167–191.

Dudley, R. G., Mullis, A. W. & Terrell, J. W. (1977) Movements of adult striped bass (*Morone saxatilis*) in the Savannah River, Georgia. *Transactions of the American Fisheries Society,* **106**, 314–322.

Duncan, A & Kubecka, J. (1993) Hydroacoustic methods of fish surveys. *Research and Development Note,* **196**. National Rivers Authority, Bristol, UK.

Duncan, A. & Kubecka, J. (1996) Patchiness of longitudinal distributions in a river as revealed by a continuous hydroacoustic survey. *ICES Journal of Marine Sciences,* **53**, 161–165.

Dunkley, D. A. & Shearer, W. M. (1982) An assessment of the performance of a resistivity fish counter. *Journal of Fish Biology,* **20**, 717–737.

Dunning, D. J., Ross, Q. E., Geoghegan, P., Reichle, J. J., Menezes, J. K. & Watson, J. K. (1992) Alewives avoid high-frequency sound. *North American Journal of Fisheries Management,* **12**, 407–416.

Durand, J. R. & Loubens, G. (1970) Observations sur la sexualité et la reproduction des *Alestes baremoze* du bas Chari et du lac Tchad. *Cahiers ORSTOM, Série Hydrobiologie,* **4**, 61–81.

Eberstaller, J., Hinterhofer, M. & Parasiewicz, P. (1998) The effectiveness of two nature-like bypass channels in an upland Austrian River. In: *Fish Migration and Fish Bypasses* (eds M. Jungwirth, S. Schmutz & S. Weiss), pp. 363–383. Fishing News Books, Blackwell Science Ltd, Oxford.

Eccles, D. H. (1986) Is speciation of demersal fishes in Lake Tanganyika restrained by physical limnological conditions? *Biological Journal of the Linnean Society,* **29**, 115–122.

Edwards, R. J. (1977) Seasonal migrations of *Astyanax mexicanus* as an adaptation to novel environments. *Copeia,* **1977**, 770–771.

Egglishaw, H. J. & Shackley, P. E. (1985) Factors governing the production of juvenile Atlantic salmon in Scottish streams. *Journal of Fish Biology,* **27**, 27–33.

Ehrler, C. & Raifsnider, C. (2000) Evaluation of the effectiveness of intake wedgewire screens. *Environmental Science & Policy,* **3** (Suppl.), 361–368.

Ehrenberg, J. E. (1984) A review of in situ target strength estimation techniques. *FAO Fisheries Report,* **300**, 85–90.

Ehrenberg, J. E. & Torkelson, T. C. (1996) Application of dual-beam and split-beam target tracking in fisheries acoustics. *ICES Journal of Marine Science,* **53**, 329–334.

Eiler, J. H. (1995) A remote satellite-linked tracking system for studying Pacific salmon with radio telemetry. *Transactions of the American Fisheries Society,* **124**, 184–193.

Eldredge, N. & Gould, S. J. (1972) Punctuated equilibria: an alternative to phyletic gradualism. In: *Models in Paleobiology* (ed. T. J. M. Schopf), pp. 82–115. Freeman, San Francisco.

Elliott, J. M. (1986) Spatial distribution and behavioural movements of migratory trout *Salmo trutta* in a Lake District stream. *Journal of Animal Ecology,* **55**, 907–922.

Enderlein, O. (1989) Migratory behaviour of adult cisco, *Coregonus albula* L., in the Bothnian Bay. *Journal of Fish Biology,* **34**, 11–18.

Ensign, W. E., Leftwich, K. N., Angermeier, P. L. & Dolloff, C. A. (1997) Factors influencing stream fish recovery following a large-scale disturbance. *Transactions of the American Fisheries Society,* **126**, 895–907.

Fabricius, E. & Gustafson, K. J. (1955) Observation on the spawning behaviour of the grayling, *Thymallus thymallus* (L.). *Reports of the Institute of Freshwater Research of Drottningholm,* **36**, 75–103.

Fabrizio, M. C. & Richards, R. A. A. (1996) Commercial fisheries surveys. In: *Fisheries Techniques,* 2nd edn. (eds B. R. Murphy & D. W. Willis), pp. 625–650. American Fisheries Society, Bethesda, Maryland.

Farmer, G. J. & Beamish, F. W. H. (1969) Oxygen consumption of *Tilapia nilotica* in relation to swimming speed and salinity. *Journal of the Fisheries Research Board of Canada,* **26**, 2807–2821.

Farquhar, R. B. & Gutreuter, S. (1989) Distribution and migration of adult striped bass in Lake Whitney, Texas. *Transactions of the American Fisheries Society,* **118**, 523–532.

Federov, A. V., Afonyushkina, E. V. & Alfeev, K. M. (1966) Contributions to the study of fish migrations in the Upper Don. *Rab Nauch-Issled Rybokhoz Laboratory Voronezh University,* **3**, 34–64.

Feldmeth, C. R. (1983) Costs of aggression in trout and pupfish. In: *Behavioural Energetics – the Cost of Survival in Vertebrates* (eds W. P. Asprey & S. I. Lustick), pp. 117–138. Ohio State University Press, Columbus.

Ferguson, J. W., Poe, T. P. & Carlson, T. J. (1998) Surface-oriented bypass systems for juvenile salmonids on the Columbia River, USA. In: *Fish Migration and Fish Bypasses* (eds M. Jungwirth, S. Schmutz & S. Weiss), pp. 281–299. Fishing News Books, Blackwell Science Ltd, Oxford.

Fernandes, C. C. (1997) Lateral migration of fishes in Amazon floodplains. *Ecology of Freshwater Fishes,* **6**, 36–44.

Fernandes, C. C. & Mérona, B. de (1988) Lateral migrations of fishes in a floodplain system in the central Amazon (Careiro Island, Lake of Rei) Am. Br. Prelimininary analyses. *Memoria de la*

Sociedad de Ciencias Naturales La Salle, **48** (Suppl.), 409–431 [in Spanish].

Fernandes, M. N., Rantin, F. T., Kalinin, A. L. & Moron, S. E. (1994) Comparative study of gill dimensions of three erythrinid species in relation to their respiratory function. *Canadian Journal of Zoology*, **72**, 160–165.

Fernandez-Pasquier, V. (1999) *Acipenser sturio* L. in the Guadalquivir river, Spain. Water regulation and fishery as factors in stock decline from 1932 to 1967. *Journal of applied Ichthyology*, **15** (4–5), 133–135.

Fernando, C. H. (1994) Zooplankton, fish and fisheries in tropical freshwaters. *Hydrobiologia*, **272**, 105–123.

Fernando, C. H. & Holcik, J. (1982) The nature of fish communities: a factor influencing the fishery potential and yields of tropical lakes and reservoirs. *Hydrobiologia*, **97**, 127–140.

Fernando, C. H. & Holcik, J. (1989) Origin, composition, and yield of fish in reservoirs. *Archiv fur Hydrobiologie*, **33**, 637–641.

Feunteun, E., Acou, A., Laffaille, P. & Legault, A. (2000) European eel (*Anguilla anguilla*): prediction of spawner escapement from continental population parameters. *Canadian Journal of Fisheries and Aquatic Sciences*, **57**, 1627–1635.

Fewings, A. (1994) *Automatic salmon counting technologies: a contemporary review*. Atlantic Salmon Trust, Pitlochry, Scotland, 66 pp.

Fickling, N. J. (1982) The identification of pike by means of characteristic marks. *Fisheries Management*, **13**, 79–82.

Fickling, N. J. & Lee, R. L. (1985) A study of the movements of the zander, *Lucioperca lucioperca* L., populations of two lowland fisheries. *Aquaculture and Fisheries Management*, **16**, 377–393.

Fischer, P. (1999) Otolith microstructure during the pelagic, settlement and benthic phases in burbot. *Journal of Fish Biology*, **54**, 1231–1243.

Fischer, S. & Kummer, H. (2000) Effects of residual flow and habitat fragmentation on distribution and movement of bullhead (*Cottus gobio* L.) in an alpine stream. *Hydrobiologia*, **422**, 305–317.

Fish, G. R. (1956) Some aspects of the respiration of six species of fish from Uganda. *Journal of Experimental Biology*, **33**, 186–195.

Fishelson, L., Goren, M., van Vuren, J. & Manelis, R. (1996) Some aspects of the reproductive biology of *Barbus* spp., *Capoeta damascina* and their hybrids (Cyprinidae, Teleostei) in Israel. *Hydrobiologia*, **317**, 79–88.

Fleming, D.F. & Reynolds, J.B. (1991) Effects of spawning-run delay on spawning migration of Arctic grayling. *American Fisheries Society Symposium*, **10**, 299–305.

Flore, L. & Keckeis, H. (1998) The effect of water current on foraging behaviour of a rheophilic cyprinid, *Chondrostoma nasus* (L.), during ontogeny: evidence of a trade-off between energetic gain and swimming costs. *Regulated Rivers: Research and Management*, **14**, 141–154.

Foerster, R. E. (1968) The sockeye salmon. *Bulletin of the Fisheries Research Board of Canada*, **162**, 422 pp.

Fontaine, M. (1975) Physiological mechanisms in the migrations of marine and amphihaline fish. *Advances in Marine Biology*, **13**, 241–335.

Fontenele, O. (1948) Contribuiçao para o conhecimento da biologia do pirarucu "*Arapaima gigas*" (Cuvier) em cativeiro (Actinopterigii, Osteoglossidae). *Revista Brasiliera Biologica*, **8**, 445–459.

Fontenele, O. (1951) Contribuiçao para o conhecimento da biologia do Apaiari "*Astronotus oscellatus*" (Spix) (Pisces, Cichlidae) em cativeiro: aparelho de reproduçao, hábitos de desova e incubaçao. *Revista Brasiliera Biologica*, **11**, 467–484.

Fontenele, O. (1952) Hábitos de desova do pirarucu *Arapaima gigas* (Cuvier) (Pisces: Isospondyli, Arapaimidae), e evoluçao de sua larvae. *Boletin Tecnico DNOCS Fortaleza,* **153**, 1–22.

Fontenot, Q. C. & Rutherford, D. A. (1999) Observations on the reproductive ecology of pirate perch *Aphredoderus sayanus. Journal of Freshwater Ecology,* **14**, 545–549.

Foote, K. G., Knudsen, H. P., Vestnes, G., MacLennan, D. N. & Simmonds, E. J. (1987) Calibration of acoustic instruments for fish density estimation: a practical guide. *ICES Cooperative Research Report,* **144**, 57 pp.

Forney, J. L. (1963) Distribution and movement of marked walleyes in Oneida Lake, New York. *New York Fish and Game Journal,* **13**, 146–167.

Forseth, T., Naesje, T. F., Jonsson, B. & Harsaker, K. (1999) Juvenile migration in brown trout: a consequence of energetic state. *Journal of Animal Ecology,* **68**, 783–793.

Fortin, R., Léveillé, M., Laramée, P. & Mailhot, Y. (1990). Reproduction and year-class strength of the Atlantic tomcod (*Microgadus tomcod*) in the Sainte-Anne River, at La Pérade, Québec. *Canadian Journal of Zoology,* **68**, 1350–1359.

Foster, A. M. & Clugston, J. P. (1997) Seasonal migration of Gulf sturgeon in the Suwannee River, Florida. *Transactions of the American Fisheries Society,* **126**, 302–308.

Fox, D. A., Hightower, J. E. & Paruka, F. M. (2000) Gulf sturgeon spawning migration and habitat in the Choctawhatchee River system, Alabama-Florida. *Transactions of the American Fisheries Society,* **129**, 811–826.

Franklin, D. R. & Smith, L. L. Jr. (1963) Early life history of the northern pike, *Esox lucius* L., with special reference to the factors influencing the numerical strength of year classes. *Transactions of the American Fisheries Society,* **92**, 91–110.

Fraser, F. D. & Emmons, E. E. (1984) Behavioural responses of blacknose dace (*Rhinichthys atratulus*) to varying densities of predatory creek chub (*Semotilus atromaculatus*). *Canadian Journal of Fisheries and Aquatic Sciences,* **41**, 364–370.

Fraser, N. H. C., Metcalfe, N. B. & Thorpe, J. E. (1993) Temperature-dependent switch between diurnal and nocturnal foraging in salmon. *Proceedings of the Royal Society of London, Series B – Biological Sciences,* **252**, 135–139.

Freadman, M. A. (1981) Swimming energetics of striped bass (*Morone saxatilis*) and bluefish (*Pomatomus saltatrix*): hydrodynamic correlates of locomotion and gill ventilation. *Journal of Experimental Biology,* **90**, 253–265.

Fredrich, F. (1996) Preliminary studies on the daily migration of chub (*Leuciscus cephalus*) in the Spree River. In: *Underwater Biotelemetry* (eds E. Baras & J. C. Philippart), *Proceedings of the First Conference and Workshop on Fish Telemetry in Europe,* p. 66. University of Liège, Belgium.

Freeman, B. J. & Freeman, M. C. (1994) Habitat use by an endangered riverine fish and implications for species protection. *Ecology of Freshwater Fish,* **3**, 49–58.

Freeman, M. C. (1995) Movements by two small fishes in a large stream. *Copeia,* **1995**, 361–367.

Fried, S. M., McCleave, J. D. & LaBar, G. W. (1978) Seaward migration of hatchery-reared Atlantic salmon, *Salmo salar*, smolts in the Penobscot River estuary, Maine: riverine movements. *Journal of the Fisheries Research Board of Canada,* **35**, 76–87.

Friedman, M. A. & Hopkins, C. D. (1996) Tracking individual mormyrid electric fish in the field using electric organ discharge waveforms. *Animal Behaviour,* **51**, 391–407.

Froese, R. & Pauly, D. (2000) *Fishbase*. World Wide Web publication. www.fishbase.org

Frost, W. E. (1950) The eel fisheries of the River Bann, Northern Ireland and observations on the age of the silver eels. *Journal du Conseil Permanent International pour l'Exploration de la Mer,* **16**,

358–383.

Fryer, G. & Iles, T. D. (1972) *The Cichlid Fishes of the Great Lakes of Africa*. Oliver & Boyd, Edinburgh.

Fukushima, M. (1994) Spawning migration and redd construction of Sakhalin taimen, *Hucho perryi* (Salmonidae) on northern Hokkaido Island, Japan. *Journal of Fish Biology,* **44**, 877–888.

Fuller, W. A. (1955) The inconnu (*Stenodus leucichthys mackienziei*) in Great Slave Lake and adjoining waters. *Journal of the Fisheries Research Board of Canada,* **12**, 768–780.

Funk, J. L. (1957) Movement of stream fishes in Missouri. *Transactions of the American Fisheries Society,* **85**, 39–57.

Furse, M. T., Kirk, R. C., Morgan, P. R. & Tweddle, D. (1979) Fishes: distribution and biology in relation to changes. In: *Lake Chilwa – Studies of Change in a Tropical Ecosystem* (eds M. Kalk, A. J. McLachlan & C. Howard-Williams), pp. 209–229. Monographiae Biologicae, Junk, The Hague.

Gagen, C. J., Sharpe, W. E. & Carline, R. F. (1994) Downstream movement and mortality of brook trout (*Salvelinus fontinalis*) exposed to acidic episodes in streams. *Canadian Journal of Fisheries and Aquatic Sciences,* **51**, 1620–1628.

Gagen, C. J., Standage, R. W. & Stoeckel, J. N. (1998) Ouachita madtom (*Noturus lachneri*) metapopulation dynamics in intermittent Ouachita mountain streams. *Copeia,* **1998**, 874–882.

Gallepp, G. W. & Magnuson, J. J. (1972) Effects of negative buoyancy on the behavior of the bluegill, *Lepomis macrochirus* Rafinesque. *Transactions of the American Fisheries Society,* **101**, 507–512.

Ganapati, S. V. (1973) Ecological problems of man-made lakes of south India. *Archiv für Hydrobiologie,* **71**, 363–380.

Garner, P. (1995) Suitability indices for juvenile 0+ roach (*Rutilus rutilus* [L.]) using point abundance sampling by electrofishing data. *Regulated Rivers: Research and Management,* **10**, 99–104.

Garner, P. (1996) Diel behaviour of juvenile 0-group fishes in a regulated river: the Great Ouse, England. *Ecology of Freshwater Fish,* **5**, 175–182.

Garner, P. (1997) Effects of variable discharge on the velocity use and shoaling behaviour of *Phoxinus phoxinus*. *Journal of Fish Biology,* **50**, 1214–1220.

Garner, P., Bass, J. A. B. & Collett, G. D. (1995) The effects of weed cutting upon the biota of a large regulated river. *Aquatic Conservation: Marine and Freshwater Ecosystems,* **171**, 1–9.

Gaudreau, N. & Boisclair, D. (1998) The influence of spatial heterogeneity on the study of fish horizontal daily migration. *Fisheries Research,* **35**, 65–73.

Gaudreau, N. & Boisclair, D. (2000) Influence of moon phase on acoustic estimates of the abundance of fish performing daily horizontal migration in a small oligotrophic lake. *Canadian Journal of Fisheries and Aquatic Sciences,* **57**, 581–590.

Gebler, R.-J. (1998) Examples of near-natural fish passes in Germany: drop structure conversions, fish ramps and bypass channels. In: *Fish Migration and Fish Bypasses* (eds M. Jungwirth, S. Schmutz & S. Weiss), pp. 403–419. Fishing News Books, Blackwell Science Ltd, Oxford.

Gee, J. H., Tallman, R. F. & Smart, H. J. (1978) Reactions of some great plains fishes to progressive hypoxia. *Canadian Journal of Zoology,* **56**, 1962–1966.

Geen, G. H., Northcote, T. G., Hartman, G. F. & Lindsey, C. C. (1966) Life histories of two species of catostomid fishes in Sixteenmile Lake, British Columbia, with particular reference to inlet spawning. *Journal of the Fisheries Research Board of Canada,* **23**, 1761–1788.

Gehrke, P. C. (1994) Influence of light intensity and wavelength on photoactive behavior of larval silver perch *Bidyanus bidyanus* and golden perch *Macquaria ambigua* and the effectiveness of light traps. *Journal of Fish Biology,* **44**, 741–751.

Gent, R., Pitlo, J. Jr. & Boland, T. (1995) Largemouth bass response to habitat and water quality rehabilitation in a backwater of the Upper Mississippi River. *North American Journal of Fisheries Management,* **15**, 784–793.

George, D. & Winfield, I. J. (2000) Factors influencing the spatial distribution of zooplankton and fish in Loch Ness, UK. *Freshwater Biology,* **43**, 557–570.

Gerking, S. D. (1950) Stability of a stream fish population. *Journal of Wildlife Management,* **14**, 193–202.

Gerking, S. D. (1953) Evidence for the concepts of home range and territory in stream fishes. *Ecology,* **34**, 347–365.

Gerking, S. D. (1959) The restricted movement of fish populations. *Biological Reviews,* **34**, 221–242.

Gessel, M. H., Williams, J. G., Brege, D. A., Krcma, R. F. & Chambers, D. R. (1991) Juvenile salmonid guidance at the Bonneville Dam second powerhouse, Columbia River, 1983–1989. *North American Journal of Fisheries Management,* **11**, 400–412.

Giamas, M. T. D., Santos, L. E. & Vermulem, H. (1983) Influencia de fatores climaticos sobre reproducao de manjuba, *Anchoviella lepidentostole* (Fowler, 1911) (Teleostei: Engraulidae). *Boletin de Instituto Pesca,* **10**, 95–100.

Gibson, R. N. (1986) Intertidal teleosts: life in a fluctuating environment. In: *The Behaviour of Teleost Fishes* (ed. T. J. Pitcher), pp. 388–408. Croom Helm, London.

Giroux, F., Ovidio, M., Philippart, J. C. & Baras, E. (2000) Relationship between the drift of macroinvertebrates and the activity of brown trout *Salmo trutta* (L.) in a small stream. *Journal of Fish Biology,* **56**, 1248–1257.

Glebe, B. D. & Leggett, W. C. (1981a) Temporal, intrapopulation differences in energy allocation and use by American shad (*Alosa sapidissima*) during the spawning migration. *Canadian Journal of Fisheries and Aquatic Sciences,* **38**, 795–805.

Glebe, B. D. & Leggett, W. C. (1981b) Latitudinal differences in energy allocation and use during the freshwater migrations of American shad (*Alosa sapidissima*) and their life history consequences. *Canadian Journal of Fisheries and Aquatic Sciences,* **38**, 806–820.

Gliwicz, Z. M. & Jachner, A. (1992) Diel migrations of juvenile fish: a ghost of predation past or present. *Archiv für Hydrobiologie,* **124**, 385–410.

Godin, J.-G. (1982) Migrations of salmonid fishes during early life history phases: daily and annual timing. In: *Salmon and Trout Migratory Behavior* (eds E. L. Brannon & E. O. Salo), pp. 22–50. University of Washington College of Fisheries, Seattle.

Godoy, M. P. de (1959) Age, growth, sexual maturity, behaviour, migration, tagging and transplantation of the curimbatá (*Prochilodus scrofa* Steindachner, 1881) of the Mogi Guassu River Sao Paulo State, Brazil. *Annales de la Academia Brasiliera de Ciencias,* **31**, 447–477.

Godoy, M. P. de (1967) Dez anos de observaçoes sobre periodicidade migratória de peixes do Rio Moxi-Guaçu. *Revista Brasiliera Biologica,* **27**, 1–12.

Godoy, M. P. de (1972) Brazilian tagging experiments, fishes migration, and upper Paraná River basin ecosystem. *Revista Brasiliera Biologica,* **32**, 473–484.

Godoy, M. P. de (1975) *Peixes do Brasil, sub-ordem Characoidei, bacia do Rio Mogi-Guaçu.* Editora Franciscana, Piracicaba, 4 volumes, 846 pp.

Goldspink, C. R. (1977) The return of marked roach (*Rutilus rutilus* [L.]) to spawning grounds in Tjeukemeer, the Netherlands. *Journal of Fish Biology,* **11**, 599–604.

Goldspink, C. R. (1978) A note on the dispersion pattern of marked bream *Abramis brama* released into Tjeukemeer, the Netherlands. *Journal of Fish Biology,* **13**, 493–497.

Gollmann, G., Bouvet, Y., Brito, R. M., *et al.* (1998). Effects of river engineering on genetic structure of European fish populations. In: *Fish Migration and Fish Bypasses* (eds M. Jungwirth, S. Schmutz & S. Weiss), pp. 113–123. Fishing News Books, Blackwell Science Ltd, Oxford.

Gollmann, H. P., Kainz, E. & Fuchs, O. (1986) Marking and tagging of fish with particular regard to the application of dyes, especially of Alcian Blue 8 GS. *Österreichische Fisherei,* **39**, 340–345.

Gosset, C. & Travade, F. (1999). Devices to aid downstream salmonid migration: Behavioral barriers. *Cybium,* **23** (Suppl.), 45–66.

Gossett, C., Travade, F. & Garaicoechea, E. (1992) Influence d'un écran électrique en aval d'une usine hydroélectrique sur le comportement de remontée du saumon atlantique (*Salmo salar*). *Bulletin Français de la Pêche et de la Pisciculture,* **324**, 2–25.

Goto, A. (1987) Life history variation in males of the river sculpin *Cottus hangionensis* along the course of a river. *Environmental Biology of Fishes,* **19**, 81–91.

Goulding, M. (1979) *Ecologia da Pesca do Rio Madeira*. Manaus, INPA/CNPq, 172 pp.

Goulding, M. (1980) *The Fishes and the Forest – Exploration of Amazonian Natural History*. University of California, Berkeley.

Goulding, M. & Carvalho, M. L. (1982) Life history and management of the tambaqui (*Colossoma macropomum*, Characidae): an important Amazonian food fish. *Revista Brasiliera Zoologica,* Sao Paulo, **1**, 107–133.

Gowan, C., Young, M. K., Faush, K. & Riley, S. (1994) Restricted movements in resident stream salmonids: a paradigm lost? *Canadian Journal of Fisheries and Aquatic Sciences,* **51**, 2626–2637.

Gowans, A. R. D., Armstrong, J. D. & Priede, I. G. (1999a) Movements of adult Atlantic salmon in relation to a hydroelectric dam and fish ladder. *Journal of Fish Biology,* **54**, 713–726.

Gowans, A. R. D., Armstrong, J. D. & Priede, I. G. (1999b) Movements of adult Atlantic salmon through a reservoir above a hydroelectric dam. *Journal of Fish Biology,* **54**, 727–740.

de Graaf, C. J., Born, A. F., Uddin, A. M. K. & Huda, S. (1999) Larval fish movement in the River Lohajang, Tangail, Bangladesh. *Fisheries Management and Ecology,* **6**, 109–120.

Grace-de-Jesus, E. (1994) Thyroid hormone surges during milkfish metamorphosis. *Bamidgeh,* **46**, 59–63.

Grande, R. & Matzow, D. (1998) A new type of fishway in Norway: how a regulated and acidified river was restored. In: *Fish Migration and Fish Bypasses* (eds M. Jungwirth, S. Schmutz & S. Weiss), pp. 236–245. Fishing News Books, Blackwell Science Ltd, Oxford.

Greenblatt, M., Brown, C. L., Lee, M., Daulder, S. & Bern, H. A. (1989) Changes in thyroid hormone levels in eggs and larvae and in iodide uptake by eggs of coho and chinook salmon, *Oncorhynchus kisutch* and *O. tshawytscha*. *Fish Physiology and Biochemistry,* **6**, 261–278.

Greenwood, P. H. (1979) Towards a phyletic classification of the 'genus' *Haplochromis* (Pisces, Cichlidae) and related taxa. Part I. *Bulletin of the British Museum of Natural History (Zoology),* **35**, 265–322.

Griffin, R. K. (1987) Life history, distribution and seasonal migration of barramundi in the Daly River, Northern Territory, Australia. *American Fisheries Society Symposium,* **1**, 358–363.

Grimm, M. P. (1981) The composition of northern pike, *Esox lucius* L., populations in four shallow waters in the Netherlands, with special reference to factors influencing 0+ pike biomass. *Fisheries Management,* **12**, 61–76.

Groot, C. & Margolis, L. (eds) (1991) *Pacific Salmon Life Histories*. University of British Columbia

Press, Vancouver.

Gross, M. R. (1987) Evolution of diadromy in fishes. *American Fisheries Society Symposium*, **1**, 14–25.

Gross, M. R. (1991) Evolution of alternative reproductive strategies: frequency-dependent sexual selection in male bluegill sunfish. *Philosophical Transactions of the Royal Society of London, Series B*, **332**, 59–66.

Gross, M. R., Coleman, R. M. & McDowall, R. M. (1988) Aquatic productivity and the evolution of diadromous fish migration. *Science*, **239**, 1291–1293.

Gudjónsson, T. (1970) The releases and returns of tagged salmon at Kollafjörour, Iceland. *International Council for the Exploration of the Sea, C.M.* 1970/M:**6**.

van Guelpen, L. & Davis, C. C. (1979) Seasonal movements of the winter flounder, *Pseudopleuronectes americanus*, in two contrasting inshore locations in Newfoundland. *Transactions of the American Fisheries Society*, **108**, 26–37.

Guillard, J. (1998) Daily migration cycles of fish populations in a tropical estuary (Sine-Saloum, Senegal) using a horizontal-directed split-beam transducer and multibeam sonar. *Fisheries Research*, **35**, 23–31.

Gulseth, O. A. & Nilssen, K. J. (1999) Growth benefit from habitat changes by juvenile high-Arctic char. *Transactions of the American Fisheries Society*, **128**, 593–602.

Gulseth, O. A. & Nilssen, K. J. (2000) The brief period of spring migration, short marine residence, and high return rate of a northern Svalbard population of Arctic char. *Transactions of the American Fisheries Society*, **129**, 782–796.

Gunn, J. S., Polacheck, T., Davis, T. L. O., Sherlock, M. & Betlehem, A. (1994) The application of archival tags to study the movement, behaviour and physiology of southern bluefin tuna, with comments on the transfer of the technology to groundfish research. *ICES C.M.* 1994/M:**21**.

Gurgens, C., Russell, D. F. & Wilkens, L. A. (2000) Electrosensory avoidance of metal obstacles by the paddlefish. *Journal of Fish Biology*, **57**, 277–290.

Gustafson, K. J. (1949). Movements and growth of grayling. *Reports of the Institute of Freshwater Research of Drottningholm*, **29**, 35–44.

Guy, C. S., Newmann, R. M. & Willis, D. W. (1992) Movement patterns of adult black crappie, *Pomoxis nigromaculatus*, in Brant Lake, South Dakota. *Journal of Freshwater Ecology*, **7**, 281–292.

Guy, C. S., Willis, D. W. & Jackson, J. J. (1994) Biotelemetry of white crappies in a South Dakota glacial lake. *Transactions of the American Fisheries Society*, **123**, 63–70.

Guyomard, R. (1991). Diversité génétique et gestion des populations naturelles de truite commune. In: *La Truite: Biologie et Ecologie* (eds J. L. Baglinière & G. Maisse), pp. 215–235. INRA Editions, Paris.

Guyomard, R., Grêvisse, G., Oury, F. W. & Davaine, P. (1984) Evolution de la variabilité génétique inter et intrapopulations de Salmonidés issus de mêmes pools génétiques. *Canadian Journal of Fisheries and Aquatic Sciences*, **41**, 1024–1029.

Gyselman, E. C. (1984) The seasonal movement of anadromous Arctic charr at Nauyuk Lake, Northwest Territories, Canada. In: *Biology of the Arctic Charr* (eds L. Johnson & B. L. Burns), pp. 575–578. University of Manitoba Press, Winnipeg.

Habicht, C., Sharr, S., Evans, D. & Seeb, J. E. (1998) Coded wire placement affects homing ability of pink salmon. *Transactions of the American Fisheries Society*, **127**, 652–657.

Hadderingh, R. H. (1982) Experimental reduction of fish impingement by artificial illumination at Bergum power station. *Internationale Revue Gesamte Hydrobiologie*, **67**, 887–900.

Haeseker, S. L., Carmichael, J. T. & Hightower, J. E. (1996) Summer distribution and condition of striped bass within Albemarle Sound, North Carolina. *Transactions of the American Fisheries Society,* **125**, 690–704.

Hagen, D. W. (1967) Isolating mechanisms in three-spine sticklebacks (Gasterosteidae). *Journal of the Fisheries Research Board of Canada,* **24**, 1637–1692.

Hall, D. J., Werner, E. E., Gilliam, J. F., Mittelbach, G. G., Howard, D. & Doner, C. G. (1979) Diel foraging behaviour and prey selection in the golden shiner (*Notemigonus chrysoleucas*). *Journal of the Fisheries Research Board of Canada,* **36**, 1029–1039.

Hall, J. W., Smith, T. I. J. & Lamprecht, S. D. (1991) Movements and habitats of shortnose sturgeon, *Acipenser brevisrostrum* in the Savannah River. *Copeia,* **1991**, 695–702.

Halls, A. S., Hoggarth, D. D. & Debnath, K. (1998) Impact of flood control schemes on river fish migration in Bangladesh. *Journal of Fish Biology,* **53** (Suppl. A), 358–380.

Hancock, R. S., Jones, J. W. & Shaw, R. (1976) A preliminary report on the spawning behaviour and the nature of sexual selection in the barbel, *Barbus barbus* L. *Journal of Fish Biology,* **9**, 21–28.

Hanna, N. S. & Schiemer, F. (1993) The seasonality of zooplanktivorous fish in an African reservoir (Gebel Aulia, While Nile, Sudan). *Hydrobiologia,* **250**, 173–185.

Hansen, H. J. M. & Fattah, A. T. A. (1986) Long-term tagging of elvers, *Anguilla anguilla*, with radioactive europium. *Journal of Fish Biology,* **29**, 535–540.

Hansen, I. P. & Jonsson, B. (1985) Downstream migration of hatchery-reared smolts of Atlantic salmon (*Salmo salar*) in the River Imsa, Norway. *Aquaculture,* **45**, 237–248.

Hanson, L. H. & Swink, W. D. (1989) Downstream migration of recently metamorphosed sea lampreys in the Ocqueoc River, Michigan, before and after treatment with lampricides. *North American Journal of Fisheries Management,* **9**, 327–331.

Hanych, D. A., Roos, M. R., Magnien, R. E. & Suggars, A. L. (1983) Nocturnal inshore movement of the mimic shiner (*Notropis volucellus*): a possible predator avoidance behaviour. *Canadian Journal of Fisheries and Aquatic Sciences,* **40**, 888–894.

Hara, T. J. (1971) Chemoreception. In: *Fish Physiology,* Vol. 5 (eds W. S. Hoar & D. J. Randall), pp. 79–120. Academic Press, New York.

Hara, T. J. (1993) Role of olfaction in fish behaviour. In: *Behaviour of Teleost Fishes*, 2nd edn. (ed. T. J. Pitcher), pp. 171–199. Chapman and Hall, London.

Hara, T. J., Gorbman, A. & Ueda, K. (1966) Influence of the thyroid upon optically evoked potentials in the optic tectum of the goldfish. *Proceedings of the Society of Experimental Biology and Medicine,* **122**, 471–475.

Harden Jones, F. R. (1968) *Fish Migration.* Arnold, London.

Hardisty, M. W. (1979) *Biology of the Cyclostomes.* Chapman & Hall, London.

Hardisty, M. W. & Potter, I. C. (1971a) The behaviour, ecology and growth of larval lampreys. In: *The Biology of the Lampreys,* Vol. 1 (eds M. W. Hardisty & I. C. Potter), pp. 85–125. Academic Press, London.

Hardisty, M. W. & Potter, I. C. (1971b) The general biology of adult lampreys. In: *The Biology of the Lampreys,* Vol. 1 (eds M. W. Hardisty & I. C. Potter), pp. 127–206. Academic Press, London.

Haro, A. & Kynard, B. (1997) Video evaluation of passage efficiency of American shad and sea lamprey in a modified Ice Harbor fishway. *North American Journal of Fisheries Management,* **17**, 981–987.

Haro, A., Odeh, M., Castro-Santos, T. & Noreika, J. (1999) Effect of slope and headpond on passage of American shad and blueback herring through simple Denil and deepened Alaska steeppass fishways. *North American Journal of Fisheries Management*, **19**, 51–58.

Haro, A., Odeh, M., Noreika, J. & Castro-Santos, T. (1998) Effect of water acceleration on downstream migratory behavior and passage of Atlantic salmon smolts and juvenile American shad at surface bypasses. *Transactions of the American Fisheries Society*, **127**, 118–127.

Harris, J. H. (1984) Impoundment of coastal drainages of south-eastern Australia and a review of its relevance to fish migrations. *Australian Zoologist,* **21**, 235–250.

Harris, J. H. & Mallen-Cooper, M. (1994) Fish passage development in the rehabilitation of fisheries in mainland south-eastern Australia. In: *Rehabilitation of Freshwater Fisheries* (ed. I. G. Cowx), pp. 185–193. Fishing News Books, Blackwell Science Ltd, Oxford.

Harris, J. H., Thorncraft, G. & Wem, P. (1998) Evaluation of rock-ramp fishways in Australia. In: *Fish Migration and Fish Bypasses* (eds M. Jungwirth, S. Schmutz & S. Weiss), pp. 331–347. Fishing News Books, Blackwell Science Ltd, Oxford.

Harrison, I. J. & Miller, P. J. (1992a) Gobiidae. In: *Faune des Poissons d'Eaux Douces et Saumâtres de l'Afrique de l'Ouest* (eds C. Lévêque, D. Paugy & G. G. Teugels), pp. 798–821. ORSTOM/MRAC Editions, Paris, Tervuren.

Harrison, I. J. & Miller, P. J. (1992b) Eleotridae. In: *Faune des Poissons d'Eaux Douces et Saumâtres de l'Afrique de l'Ouest* (eds C. Lévêque, D. Paugy & G. G. Teugels), pp. 822–836. ORSTOM/MRAC Editions, Paris, Tervuren.

Hart, J. L. (1973) Pacific fishes of Canada. *Bulletin of the Fisheries Research Board of Canada,* **180**, 1–740.

Hart, L. G. & Summerfelt, R. C. (1973) Homing of flathead catfish *Pylodictis olivaris* (Rafinesque), tagged with ultrasonic transmitters. *Proceedings of the Annual Conference of South-East Association of Game and Fish Agencies,* **27**, 520–527.

Hart, L. G. & Summerfelt, R. C. (1975) Surgical procedures for implanting ultrasonic transmitters into flathead catfish (*Pylodictis olivaris*). *Transactions of the American Fisheries Society,* **104**, 56–59.

Hart, P. J. B. & Pitcher, T. J. (1969) Field trials of fish marking using a jet inoculator. *Journal of Fish Biology,* **1**, 383–385.

Hartman, G. F., Northcote, T. G. & Lindsey, C. C. (1962) Comparison of inlet and outlet spawning runs of rainbow trout in Loon Lake, British Columbia. *Journal of the Fisheries Research Board of Canada,* **19**, 173–200.

Harvey, B. C. (1987) Susceptibility of young-of-the-year fishes to downstream displacement by flooding. *Transactions of the American Fisheries Society*, **116**, 851–855.

Harvey, C. J., Ruggerone, G. T. & Rogers, D. E. (1997) Migrations of the three-spined stickleback, nine-spined stickleback and pond smelt in the Chignik catchment, Alaska. *Journal of Fish Biology,* **50**, 1133–1137.

Harvey, J. & Cowx, I. G. (1996) Electric fishing for the assessment of fish stocks in large rivers. In: *Stock Assessment in Inland Fisheries* (ed. I. G. Cowx), pp. 11–26. Fishing News Books, Blackwell Science Ltd, Oxford.

Hasler, A. D. (1945) Observation on the winter perch population of Lake Mendota. *Ecology,* **26**, 90–94.

Hasler, A. D. (1966) *Underwater Guideposts: Homing of Salmon*. University of Wisconsin Press, Madison.

Hasler, A. D. (1971) Orientation and fish migration. In: *Fish Physiology,* Vol. VI (eds W. S. Hoar & D. J. Randall), pp. 429–510. Academic Press, London.

Hasler, A. D. (1983) Synthetic chemicals and pheromones in homing salmon. In: *Control Processes in Fish Physiology* (eds J. C. Rankin, T. J. Pitcher & R. T. Deggan), pp. 103–116. Croom Helm, London.

Hasler, A. D. & Bardach, J. E. (1949) Daily migrations of perch in Lake Mendota, Wisconsin. *Journal of Wildlife Management,* **13**, 40–51.

Hasler, A. D., Gardella, E. S., Horrall, R. M. & Henderson, H. F. (1969) Open water orientation of white bass, *Roccus chrysops*, as determined by ultrasonic tracking methods. *Journal of the Fisheries Research Board of Canada,* **26**, 2173–2191.

Hasler, A. D., Horrall, R. M., Wisby, W. J. & Braemer, W. (1958) Sun orientation and homing in fishes. *Limnology and Oceanography,* **3**, 353–361.

Hasler, A. D. & Scholz, A. T. (1983) *Olfactory Imprinting and Homing in Salmon: Investigations into the Mechanisms of the Imprinting Processes.* Springer-Verlag, Berlin.

Hasler, A. D. & Wisby, W. J. (1951) Discrimination of stream odors by fishes and relation to parent stream behavior. *American Naturalist,* **85**, 223–238.

Haw, F., Bergman, P. K., Fralick, R. D., Buckley, R. M. & Blankenship, H. L. (1990) Visible implanted fish tag. *American Fisheries Society Symposium,* **7**, 311–315.

Hawkins, A. D. (1993) Underwater sound and fish behaviour. In: *Behaviour of Teleost Fishes*, 2nd edn. (ed. T. J. Pitcher), pp. 129–169. Chapman & Hall, London.

Hawkins, A. D., MacLennan, D. N., Urquhart, G. G. & Robb, C. (1974) Tracking cod *Gadus morhua* L. in a Scottish sea loch. *Journal of Fish Biology,* **6**, 225–236.

Hawkins, A. D. & Smith, G. W. (1986) Radio-tracking observations on Atlantic salmon ascending the Aberdeenshire Dee. *Scottish Fisheries Research Report,* **36**.

Hayes, D. B., Paola Ferreri, C. & Taylor, W. T. (1996) Active fish capture methods. In: *Fisheries Techniques*, 2nd edn. (ed. B. R. Murphy & D. W. Willis), pp. 193–220. American Fisheries Society, Bethesda, Maryland.

Haymes, G. T., Patrick, P. H. & Onisto, L. J. (1984) Attraction of fish to mercury vapour light and its application in a generating station forebay. *Internationale Revue Gesamte Hydrobiologie,* **69**, 867–876.

Haynes, J. M. & Gray, R. H. (1979) Effects of external and internal radio transmitters attachment on movement of adult chinook salmon. In: *Proceedings of the Second International Conference on Wildlife Biotelemetry* (ed. F. M. Long), pp. 115–128. University of Wyoming Press, Laramie, Wyoming.

Haynes, J. M. & Gray, R. H. (1980) Influence of Little Goose Dam on upstream movements of adult chinook salmon, *Oncorhynchus tshawytscha*. *U.S. National Marine Fisheries Service Bulletin,* **78**, 185–190.

He, X. & Kitchell, J. F. (1990) Direct and indirect effects of predation on a fish community: a whole lake experiment. *Transactions of the American Fisheries Society,* **119**, 825–835.

He, X. & Wright, R. A. (1992) An experimental study of piscivore-planktivore interactions: population and community responses to predation. *Canadian Journal of Fisheries and Aquatic Sciences,* **49**, 1176–1183.

Heape, W. (1931) *Emigration, Migration and Nomadism.* Heffer, Cambridge.

Heard, W. R. (1991) Life history of pink salmon (*Oncorhynchus gorbuscha*). In: *Pacific Salmon Life Histories* (eds C. Groot & L. Margolis), pp. 121–230. University of British Columbia Press, Vancouver.

Heard, W. R. & Vogele, L. E. (1968) A flag tag for underwater recognition of individual fishes by divers.

Transactions of the American Fisheries Society, 97, 55–57.

Hedgepeth, J., Fuhriman, D. & Acker, W. (1999) Fish behavior measured by a radar-type acoustic transducer near hydroelectric dams. In: *Innovations in Fish Passage Technology* (ed. M. Odeh), pp. 155–171. American Fisheries Society, Bethesda, Maryland.

Heggenes, J., Krog, O. M. W., Lindas, O. R., Dokk, J. G. & Bremnes, T. (1993) Homeostatic behavioural responses in a changing environment: brown trout (*Salmo trutta*) become nocturnal during winter. *Journal of Animal Ecology*, 62, 295–308.

Heiligenberg, W. F. (1991) *Neural Nets in Electric Fish*. MIT Press, Cambridge, USA.

Helfman, G. S. (1979) Twilight activities of yellow perch, *Perca flavescens*. *Journal of the Fisheries Research Board of Canada*, 36, 173–179.

Helfman, G. S. (1981) Twilight activities and temporal structure in a freshwater community. *Canadian Journal of Fisheries and Aquatic Sciences*, 38, 1405–1420.

Helfman, G. S. (1983) Underwater methods. In: *Fisheries Techniques*, 1st edn. (eds L. A. Nielsen & D. L. Johnson), pp. 349–370. American Fisheries Society, Bethesda, Maryland.

Helfman, G. S. (1989) Threat-sensitive predator avoidance in damselfish-trumpetfish interactions. *Behavioral Ecology and Sociobiology*, 24, 47–58.

Helfman, G. S. & Schultz, E. T. (1984) Social transmission of behavioural traditions in a coral reef fish. *Animal Behaviour*, 32, 379–384.

Helfman, G. S., Stoneburner, D. L., Boseman, E. L., Christian, P. A. & Whalen, R. (1983) Ultrasonic telemetry of American eel movements in a tidal creek. *Transactions of the American Fisheries Society*, 112, 105–110.

Hellawell, J. M., Leatham, H. & Williams, G. I. (1974) The upstream migratory behaviour of salmonids in the River Frome, Dorset. *Journal of Fish Biology*, 6, 729–744.

Helm, W. T. & Tyus, H. M. (1992) Influence of coating type on retention of dummy transmitters implanted in rainbow trout. *North American Journal of Fisheries Management* 12, 257–259.

Henderson, P. A. & Hamilton, H. F. (1995) Standing crop and distribution of fish in drifting and attached meadow within an Upper Amazonian várzea lake. *Journal of Fish Biology*, 47, 266–276.

Hendry, K., Tinsdall, M. & White, K. N. (1994) Restoration of the fishery of a redeveloped freshwater dock. In: *Rehabilitation of Freshwater Fisheries* (ed. I. G. Cowx), pp. 467–479. Fishing News Books, Blackwell Science Ltd, Oxford.

Hergenrader, G. L., Harrow, G. L., King, R. G., Cada, G. F. & Schlesinger, A. B. (1982) Larval fishes in the Missouri River and the effects of entrainment. In: *The Middle Missouri River* (eds L. W. Hesse, G. L. Hergenrader, H. S. Lewis, S. D. Reetz & A. B. Schlesinger), pp. 185–223. The Missouri River Study Group, Norfolk, Nebraska.

Hérissé, C. & Bénech, V. (in press) Mobility and habitat of mudfish spawners in the Inner Delta floodplain of the Niger River, as revealed by radiotracking. *Journal of Fish Biology*.

Hert, E. (1992) Homing and home-site fidelity in rock-dwelling cichlids (Pisces, Teleostei) of Lake Malawi, Africa. *Environmental Biology of Fishes*, 33, 229–237.

Herzka, S. Z & Holt, G. J. (2000) Changes in isotopic composition of red drum (*Sciaenops ocellatus*) larvae in response to dietary shifts: potential applications to settlement studies. *Canadian Journal of Fisheries and Aquatic Sciences*, 57, 137–147.

Hesse, L. W., Bliss, Q. P. & Zuerlein, G. J. (1982) Some aspects of the ecology of adult fishes in the channelized Missouri River with special reference to the effects of two nuclear power generating plants. In: *The Middle Missouri River* (eds L. W. Hesse, G. L. Hergenrader, H. S. Lewis, S. D. Reetz & A. B. Schlesinger), pp. 225–276. The Missouri River Study Group, Norfolk, Nebraska.

Hesslein, R. H., Capel, M. J., Fox, D. E. & Hallard, K. A. (1991) Stable isotopes of sulfur, carbon and nitrogen as indicators of trophic level and fish migration in the Lower Mackenzie River Basin, Canada. *Canadian Journal of Fisheries and Aquatic Sciences,* **48**, 2258–2265.

Heyd, A. & Pfeiffer, W. (2000) Sound production in catfish (Siluroidei, Ostariophysi, Teleostei) and its relationship to phylogeny and fright reaction. *Revue Suisse de Zoologie,* **107**, 165–211 [in German].

Hickley, P. (1996) Fish population survey methods: a synthesis. In: *Stock Assessment in Inland Fisheries* (ed. I. G. Cowx), pp. 3–10. Fishing News Books, Blackwell Science Ltd, Oxford.

Hickling, C. F. (1970) A contribution to the natural history of the English grey mullets (Pisces, Mugilidae). *Journal of the Marine Biological Association of the United Kingdom,* **50**, 609–633.

Hilborn, R. & Walters, C. J. (1992) *Quantitative Fish Stock Assessment: Choices, Dynamics and Uncertainty*. Chapman & Hall, London.

Hildebrand, S. F. & Schroeder, W. C. (1927) Fishes of Chesapeake Bay. *Bulletin of the United States Bureau of Fisheries,* **43**, 1–366.

Hinch, S. G. & Bratty, J. (2000) Effects of swim speed and activity pattern on success of adult sockeye salmon migration through an area of difficult passage. *Transactions of the American Fisheries Society,* **129**, 598–606.

Hinch, S. G., Diewert, R. E., Lissimore, T. J., Prince, A. M. J., Healey, M. C. & Henderson, M. A. (1996) Use of electromyogram telemetry to assess difficult passage areas for river-migrating adult sockeye salmon. *Transactions of the American Fisheries Society,* **125**, 253–260.

Hinch, S. G. & Rand, P. S. (2000) Optimal swimming speeds and forward assisted locomotion: energy conserving behaviours of up-river migrating adult salmon. *Canadian Journal of Fisheries and Aquatic Sciences,* **57**, 2470–2478.

Hindar, K., Jonsson, B., Ryman, N. & Ståhl, G. (1991) Genetic relationships among landlocked, resident and anadromous brown trout, *Salmo trutta* L. *Heredity,* **81**, 493–504.

Hirshfield, M. F. (1980) An experimental analysis of reproductive effort and cost in the Japanese medaka, *Orizias latipes*. *Ecology,* **61**, 282–292.

Hobson, K. A. (1999) Tracing origins and migration of wildlife using stable isotopes. Review. *Oecologia,* **120**, 314–326.

Hockin, D. C., O'Hara, K. & Eaton, J. W. (1989) A radiotelemetric study of grass carp in a British canal. *Fisheries Research,* **7**, 73–84.

Hocutt, C. H. (1989) Seasonal and diel behaviour of radio-tagged *Clarias gariepinus* in lake Ngezi, Zimbabwe (Pisces, Clariidae). *Journal of Zoology (London),* **219**, 181–199.

Hocutt, C. H. & Wiley, E. O. (eds) (1986) *The Zoogeography of North American Freshwater Fishes*. Wiley, New York.

Hodgson, J. R., Schindler, D. E. & He, X. (1998) Homing tendency of three piscivorous fishes in a north temperate lake. *Transactions of the American Fisheries Society,* **127**, 1078–1081.

Hofer, K. & Kirchhofer, A. (1996) Drift, habitat choice and growth of the nase (*Chondrostoma nasus*, Cyprinidae) during early life stages. In: *Conservation of Endangered Freshwater Fish in Europe* (eds A. Kirchhofer & D. Hefti), pp. 269–278. Birkhäuser Verlag, Basel.

Høgåsen, H. R. (1998) Physiological changes associated with the diadromous migration of salmonids. *Canadian Special Publication of Fisheries and Aquatic Sciences* **127**, 128 pp.

Holcik, J., Hensel, K., Nieslanik, J. & Skácel, L. (1988) *The Eurasian Huchen, Hucho hucho*. Dr. W. Junk Publishers, Dordrecht, 239 pp.

Holmes, J. A., Beamish, F. W. H., Seelye, J. G., Sower, S. A. & Youson, J. H. (1994) Long-term influence of water temperature, photoperiod, and food deprivation on metamorphosis of sea lamprey, *Petromyzon marinus. Canadian Journal of Fisheries and Aquatic Sciences*, **51**, 2045–2051.

Hopfield, J. (1982) Neural networks and physical systems with emergent collective computational abilities. *Proceedings of the National Academy of Sciences of the USA*, **79**, 2554–2558.

Hopfield, J. & Tang, D. (1986) Computing with neural circuits: a model. *Science*, **233**, 625–633.

Hopson, A. J. (ed.) (1982) *Lake Turkana. A Report on the Findings of the Lake Turkana Project 1972–1975*. London, Overseas Development Administration, 1614 pp.

Horn, M. H. (1997) Evidence for dispersal of fig seeds by the fruit-eating characid fish *Brycon guatemalensis* Regan in a Costa Rican tropical rain forest. *Oecologia*, **109**, 259–264.

Houde, E. D. (1969) Sustained swimming ability of larvae of walleye (*Stizostedion vitreum vitreum*) and yellow perch (*Perca flavescens*). *Journal of the Fisheries Research Board of Canada*, **26**, 1647–1659.

Howland, K. L., Tallman, R. F. & Tonn, W. M. (2000) Migration patterns of freshwater and anadromous inconnu in the Mackenzie River system. *Transactions of the American Fisheries Society*, **129**, 41–59.

Hoxmeier, R. J. H. & DeVries, D. R. (1997) Habitat use, diet, and population structure of adult and juvenile paddlefish in the lower Alabama River. *Transactions of the American Fisheries Society*, **126**, 288–301.

Hubbs, C. L. (1921) An ecological study of the life-history of the fresh-water atherine fish *Labidesthes sicculus. Ecology*, **2**, 262–276.

Huber, M. & Kirchhofer, A. (1998) Radio telemetry as a tool to study habitat use of nase (*Chondrostoma nasus* L.) in medium-sized rivers. *Hydrobiologia*, **371/372**, 309–319.

Hubert, W. A. (1996) Passive capture techniques. In: *Fisheries Techniques*, 2nd edn. (eds B. R. Murphy & D. W. Willis), pp. 157–192. American Fisheries Society, Bethesda, Maryland.

Hubert, W. A. & Sandheinrich, M. B. (1983) Patterns of variation in gill-net catch and diet of yellow perch in a stratified Iowa lake. *North American Journal of Fisheries Management*, **3**, 156–162.

Hudd, R. & Lehtonen, H. (1987) Migration and home ranges of natural and transplanted burbot (*Lota lota*) off the coast of Finland. In: *Proceedings of the Fifth Congress of European Ichthyology*, pp. 201–205. Stockholm, Sweden.

Huet, M. (1949) Aperçu de la relation entre la pente et les populations piscicoles des eaux courantes. *Schweizerische Zeitschrift für Hydrologie*, **11**, 332–351.

Hughes, G. M. (1984) Measurement of gill areas in fishes: practices and problems. *Journal of the Marine Biological Association of the United Kingdom*, **64**, 637–655.

Hughes, N. F. & Kelly, L. H. (1996) New techniques for 3-D video tracking of fish swimming movements in still or flowing water. *Canadian Journal of Fisheries and Aquatic Sciences*, **53**, 2473–2483.

Hughes, N. F. & Reynolds, J. B. (1994) Why do Arctic grayling (*Thymallus arcticus*) get bigger as you go upstream? *Canadian Journal of Fisheries and Aquatic Sciences*, **51**, 2154–2163.

Hughes, R. N. & Blight, C. M. (1999) Algorithmic behaviour and spatial memory are used by two intertidal fish species to solve the radial maze. *Animal Behaviour*, **58**, 601–613.

Hughes, S. (1998) A mobile horizontal hydroacoustic fisheries survey of the River Thames, United Kingdom. *Fisheries Research*, **35**, 91–97.

Humphries, P., King, A. J. & Koehn, J. D. (1999) Fish, flows and flood plains: links between freshwater fishes and their environment in the Murray-Darling River system, Australia. *Environmental Biol-*

ogy of Fishes, **56**, 129–151.

Hunn, J. H. & Youngs, W. D. (1980) Role of physical barriers in the control of sea lamprey *Petromyzon marinus*. *Canadian Journal of Fisheries and Aquatic Sciences,* **37**, 2118–2122.

Hunt, P. C. & Jones, J. W. (1974) A population study of *Barbus barbus* L. in the river Severn, England II. Movement. *Journal of Fish Biology,* **6**, 269–278.

Hurley, S. T., Hubert, W. A. & Nickum, J. G. (1987) Habitats and movements of shovelnose sturgeons in the Upper Mississippi River. *Transactions of the American Fisheries Society,* **116**, 655–622.

Hussein, S. A. (1981) The population density, growth and food of eels *Anguilla anguilla* L. in some tributaries of the River Tweed. In: *Proceedings of the Second British Freshwater Fisheries Conference*, pp. 120–128. University of Liverpool, Liverpool, UK.

Idler, D. R. & Bitners, I. (1958) Biochemical studies on sockeye salmon during spawning migration. II. Cholesterol, fat, protein and water in the body of the standard fish. *Canadian Journal of Biochemistry and Physiology,* **36**, 793–798.

Iguchi, K. & Mizuno, N. (1999) Early starvation limits survival in amphidromous fishes. *Journal of Fish Biology,* **54**, 705–712.

Ingendahl, D., Bach, J. M., Larinier, M. & Travade, F. (1996) The use of telemetry in studying downstream migration of Atlantic salmon smolts at a hydro-electric power plant in South-West France: preliminary results. In: *Underwater Biotelemetry* (eds E. Baras & J. C. Philippart), pp. 121–128. University of Liège, Belgium.

Iongh, H. H., Spleithoff, P. C. & Frank, V. G. (1983) Feeding habits of the clupeid *Limnothrissa miodon* (Boulenger), in Lake Kivu. *Hydrobiologia,* **102**, 113–122.

Irvine, J. R., Ward, B. R., Teti, P. A. & Cousens, N. B. F. (1991) Evaluation of a method to count and measure live salmonids in the field with a video camera and computer. *North American Journal of Fisheries Management,* **11**, 20–26.

Irving, D. B. & Modde, T. (2000) Home-range fidelity and use of historic habitat by adult Colorado pikeminnow (*Ptychocheilus lucius*) in the White River, Colorado and Utah. *Western North American Naturalist,* **60**, 16–25.

Isaak, D. J. & Bjornn, T. C. (1996) Movement of northern squawfish in the tailrace of a Lower Snake River dam relative to the migration of juvenile anadromous salmonids. *Transactions of the American Fisheries Society,* **125**, 780–793.

Isaksson, Á., Rasch, T. J. & Poe, P. H. (1978) An evaluation of smolt releases into a salmon and a non-salmon producing stream using two releasing methods. *Journal of Agricultural Research in Iceland,* **10**, 100–113.

Islam, B. N. & Talbot, G. B. (1968) Fluvial migration, spawning and fecundity of Indus River Hilsa, *Hilsa ilisha*. *Transactions of the American Fisheries Society,* **97**, 350–355.

Ita, E. O. (1984) Kainji (Nigeria). In: *Status of African Reservoir Fisheries* (eds J. M. Kapetsky & T. Petr). *FAO, CIFA Technical Paper,* **10**, 43–103.

Itazawa, Y. & Oikawa, S. (1986) A quantitative interpretation of the metabolism-size relationship in animals. *Experientia,* **42**, 152–153.

Iwata, M. (1995) Downstream migratory behavior of salmonids and its relationship with cortisol and thyroid hormones: A review. *Aquaculture,* **135**, 131–139.

Jackson, P. B. N. (1961) The impact of predation, especially by the tigerfish (*Hydrocynus vittatus* Castelnau) on African freshwater fishes. *Proceedings of the Zoological Society of London,* **136**, 603–622.

Jackson, P. B. N. & Coetzee, P. W. (1982) Spawning behaviour of *Labeo umbratus* (Smith) (Pisces,

Cyprinidae). *South African Journal of Science,* **78**, 293–295.

Jankovic, D. (1964) Synopsis of biological data on European grayling *Thymallus thymallus* (Linnaeus) 1758. *FAO Fisheries Synopsis,* **24**. FAO, Rome.

Jansen, W., Kappus, B., Böhmer, J., Beiter, T. & Rahmann, H. (1996) Fish community structure, distribution patterns and migration in the vicinity of different types of fishways on the Enz River (Germany). In: *Ecohydraulics 2000* (eds M. Leclerc *et al.*), Vol. B, pp.903–914. INRS-Eau, Québec.

Jansen, W. A. & Mackay, W. C. (1992) Foraging in yellow perch, *Perca flavescens*: biological and physical factors affecting diel periodicity in feeding, consumption and movement. *Environmental Biology of Fishes,* **34**, 287–303.

Janssen, J. & Brandt, S. B. (1980) Feeding ecology and vertical migration of adult alewife (*Alosa pseudoharengus*) in Lake Michigan. *Canadian Journal of Fisheries and Aquatic Sciences,* **37**, 177–184.

Jefferts, K. B., Bergman, P. K. & Fiscus, H. F. (1963) A coded wire identification system for macroorganisms. *Nature* (London), **198**, 460–462.

Jellyman, D. J. (1977) Summer upstream migration of juvenile freshwater eels in New Zealand. *New Zealand Journal of Marine and Freshwater Research,* **11**, 61–71.

Jellyman, D. J. (1987) A review of the marine life history of the Australasian temperate species of *Anguilla*. *American Fisheries Society Symposium,* **1**, 276–285.

Jellyman, D. J. & Ryan, C. M. (1983) Seasonal migration of elvers (*Anguilla* spp.) into Lake Ponoui, New Zealand, 1974–1978. *New Zealand Journal of Marine and Freshwater Research,* **17**, 1–15.

Jennings, M. J., Claussen, J. E. & Philipp, D. P. (1996) Evidence for heritable preferences for spawning habitat between two walleye populations. *Transactions of the American Fisheries Society,* **125**, 978–982.

Jensen, K. W. & Berg, M. (1977) Growth, mortality and migrations of the anadromous char, *Salvelinus alpinus*, L., in the Vardnes River, Tromsø, Northern Norway. *Reports of the Institute of Freshwater Research of Drottningholm,* **56**, 70–80.

Jepsen, N., Aarestrup, K., Økland, F. & Rasmussen, G. (1998) Survival of radio-tagged Atlantic salmon (*Salmo salar* L.) and trout (*Salmo trutta* L.) smolts passing a reservoir during seaward migration. *Hydrobiologia,* **371/372**, 347–353.

Jepsen, N., Pedersen, S. & Thorstad, E. (2000) Behavioural interactions between prey (trout smolts) and predators (pike and pikeperch) in an impounded river. *Regulated Rivers: Research and Management,* **16**, 189–198.

Jessop, B. M. (1976) Distribution and timing of tag recoveries from native and nonnative Atlantic salmon (*Salmo salar*) released into Big Salmon River, New Brunswick. *Journal of the Fisheries Research Board of Canada,* **33**, 829–833.

Jessop, B. M. (1987) Migrating American eels in Nova Scotia. *Transactions of the American Fisheries Society,* **116**, 161–170.

Jessop, B. M. (1994) Homing of alewives (*Alosa pseudoharengus*) and blueback herring (*A. aestivalis*) to and within the Saint John River, New Brunswick, as indicated by tagging data. *Canadian Technical Report of Fisheries and Aquatic Sciences,* **2015**, 22 pp.

Jhingran, V. G. (1975) *Fish and Fisheries of India*. Hindustan Publishers, Delhi, 954 pp.

Jobling, M. (1994) *Fish Bioenergetics*. Chapman & Hall, London, 309 pp.

Jobling, M. (1995) *Environmental Biology of Fishes*. Chapman & Hall, London, 480 pp.

Johnsen, P. B. & Hasler, A. D. (1977) Winter aggregations of carp (*Cyprinus carpio*) as revealed by ultrasonic tracking. *Transactions of the American Fisheries Society,* **106**, 556–559.

Johnson, B. L. & Noltie, D. B. (1997) Demography, growth, and reproductive allocation in stream-spawning longnose gar. *Transactions of the American Fisheries Society,* **126**, 438–466.

Johnson, B. L., Richardson, W. B. & Naimo, T. J. (1995) Past, present and future concepts in large river ecology. *Bioscience,* **45**, 134–141.

Johnson, D. W. & McLendron, E. L. (1970) Differential distribution of the striped mullet, *Mugil cephalus* Linnaeus. *California Fish and Game,* **56**, 138–139.

Johnson, G. E., Adams, N. S., Johnson, R. L., Rondorf, D. W., Dauble, D. D. & Barila, T. Y. (2000) Evaluation of the prototype surface bypass for salmonid smolts in spring 1996 and 1997 at Lower Granite Dam on the Snake River, Washington. *Transactions of the American Fisheries Society,* **129**, 381–397.

Johnson, L. (1980) The Arctic charr, *Salvelinus alpinus.* In: *Charrs – Salmonid Fishes of the Genus Salvelinus* (ed. E. K. Balon), pp. 15–98. Dr W. Junk, The Hague.

Johnson, L. (1989) The anadromous Arctic charr, *Salvelinus alpinus,* of Nauyuk Lake, N. W. T., Canada. *Physiology and Ecology of Japan, Special Volume,* **1**, 201–227.

Johnson, T. & Müller, K. (1978a) Different phase position of activity in juvenile and adult perch. *Naturwissenschaften,* **65**, 392.

Johnson, T. & Müller, K. (1978b) Migration of juvenile pike, *Esox lucius* L., from a coastal stream to the northern part of the Bothnian Sea. *Aquilo Seria Zoologica,* **18**, 57–61.

Johnston, I. A., Davison, W. & Goldspink, G. (1977) Energy metabolism of carp swimming muscles. *Journal of Comparative Physiology,* **114**, 203–216.

Johnston, S. V. (1985) Experimental testing of a dual beam fish size classifier. *International Atlantic Salmon Fund Special Publication Series,* **4**, 317–337.

Johnston, S. V. & Hopelain, J. S. (1990) The application of dual-beam target tracking and Doppler-shifted echo processing to assess upstream salmonid migration in the Klamath River, California. In: *Developments in Fisheries Acoustics* (ed. W. A. Karp). *ICES Rapports et Procès-verbaux des Réunions,* **189**, 210–222.

Johnstone, A. D. F., Lucas, M. C., Boylan, P. & Carter, T. J. (1992) Telemetry of tail-beat frequency of Atlantic salmon (*Salmo salar* L.) during spawning. In: *Wildlife Telemetry – Remote Monitoring and Tracking of Animals* (eds I. G. Priede & S. M. Swift), pp. 456–465. Ellis Horwood, Chichester, UK.

Jones, K. A., Hara, T. J. & Scherer, E. (1985) Locomotor response by Arctic charr (*Salvelinus alpinus*) to gradients of H^+ and CO_2. *Physiological Zoology,* **58**, 413–420.

Jones, M. W., Danzmann, R. G. & Clay, D. (1997) Genetic relationships among populations of wild resident, and wild and hatchery anadromous brook charr. *Journal of Fish Biology,* **51**, 29–40.

Jones, S. (1957) On the late winter and early spring migration of the Indian shad, *Hilsa ilisha* (Hamilton), in the Gangetic Delta. *Indian Journal of Fisheries,* **4**, 304–314.

Jonsson, B. (1982) Diadromous and resident trout *Salmo trutta*: is their difference due to genetics? *Oikos,* **38**, 297–300.

Jonsson, B. & L'Abée-Lund, J. H. (1993) Latitudinal clines in life-history variables of anadromous brown trout in Europe. *Journal of Fish Biology,* **43** (Suppl. A), 1–16.

Jonsson, N. (1991) Influence of water flow, water temperature and light on fish migration in rivers. *Nordic Journal of Freshwater Research,* **66**, 20–35.

Jordan, D. R. & Wortley, J. S. (1985) Sampling strategy related to fish distribution, with particular reference to the Norfolk Broads. *Journal of Fish Biology,* **27** (Suppl. A), 163–173.

Jungwirth, M. (1996) Bypass channels at weirs as appropriate aids for fish migration in rhithral river. *Regulated Rivers: Research and Management,* **12**, 483–492.

Jungwirth, M. (1998) River continuum and fish migration – Going beyond the longitudinal River Corridor in understanding ecological integrity. In: *Fish Migration and Fish Bypasses* (eds M. Jungwirth, S. Schmutz & S. Weiss), pp. 19–32. Fishing News Books, Blackwell Science Ltd, Oxford.

Jungwirth, M., Muhar, S. & Schmutz, S. (2000) Fundamentals of fish ecological integrity and their relationship to the extended serial discontinuity concept. *Hydrobiologia,* **422**, 85–97.

Jungwirth, M. & Pelikan, B. (1989) Zur problematik von Fischaufstiegshilfen. *Schriftenreihe Österreichische Wasserwirtschaft,* **41**, 80–89.

Jungwirth, M., Schmutz, S. & Weiss, S. (eds) (1998) *Fish Migration and Fish Bypasses.* Fishing News Books, Blackwell Science Ltd, Oxford.

Junk, W. J. (1983) Ecology of swamps on the middle Amazon. In: *Ecosystems of the World, Mires – Swamps, Bog, Fen and Moor, B. Regional Studies* (ed. A. G. P. Gore), pp. 269–274. Elsevier, Amsterdam.

Junk, W. J. (1985) Temporary fat storage, and adaptation of some fish species to the water level fluctuation and related environmental changes of the Amazon River. *Amazoniana,* **9**, 315–351.

Junk, W. J., Bayley, P. B. & Sparks, R. E. (1989) The flood pulse concept in river-floodplain ecosystems. *Canadian Special Publication of Fisheries and Aquatic Sciences,* **106**, 110–127.

Junk, W. J., Soares, G. M. & Carvalho, F. M. (1983) Distribution of fish species in a lake of the Amazon river floodplain near Manaus (Lago Camaleao), with special reference to extreme oxygen conditions. *Amazoniana,* **7**, 397–431.

Jurvelius, J. & Heikkinen, T. (1988) Seasonal migration of vendace, *Coregonus albula* (L.), in a deep Finnish lake. *Finnish Fisheries Research,* **9**, 205–212.

Kabata, Z. (1963) Parasites as biological tags. *International Commission for the Northwest Atlantic Fisheries Special Publication,* **4**, 31–37.

Kadri, S., Metcalfe, N. B., Huntingford, F. A. & Thorpe, J. E. (1995) What controls the onset of anorexia in maturing adult female Atlantic salmon? *Functional Ecology,* **9**, 790–797.

Kafemann, R., Adlerstein, S. & Neukamm, R. (2000) Variation in otolith strontium and calcium ratios as an indicator of life-history strategies of freshwater fish species within a brackish water system. *Fisheries Research,* **46**, 313–325.

Karlsson, L., Ikonen, E., Westerberg, H. & Sturlaugsson, J. (1996) Use of data storage tags to study the spawning migration of Baltic salmon (*Salmo salar* L.) in the Gulf of Bothnia. *ICES C.M.* 1996/M:**9** Ref. J.

Katayama, S., Radtke, R. L., Omori, M. & Shafer, D. J. (2000) Coexistence of anadromous and resident life history styles of pond smelt, *Hypomesus nipponensis*, in Lake Ogawara, Japan, as determined by analyses of otolith structure and strontium: calcium ratios. *Environmental Biology of Fishes,* **58**, 195–201.

Kaufman, L. S., Chapman, L. J. & Chapman, C. A. (1997) Evolution in fast forward haplochromine fishes of the Lake Victoria region. *Endeavour,* **21**, 23–30.

Kaufmann, E. & Wieser, W. (1992) Influence of temperature and ambient oxygen on the swimming energetics of cyprinid larvae and juveniles. *Environmental Biology of Fishes,* **33**, 87–95.

Keast, A. & Fox, M. G. (1992) Space use and feeding patterns of an offshore fish assemblage in a shallow mesotrophic lake. *Environmental Biology of Fishes,* **34**, 159–170.

Kedney, G. I., Boulé, V. & Fitzgerald, G. J. (1987) The reproductive ecology of three-spine sticklebacks breeding in fresh and brackish water. *American Fisheries Society Symposium,* **1**, 151–161.

Keefe, M. L. & Winn, H. E. (1991) Chemosensory attraction to home stream water and conspecifics by native brook trout, *Salvelinus fontinalis*, from two southern New England streams. *Canadian Journal of Fisheries and Aquatic Sciences,* **48**, 938–944.

Keenan, C. P. (2000) Should we allow human-induced migration of the Indo-West Pacific fish, barramundi *Lates calcarifer* (Bloch) within Australia? *Aquaculture Research,* **31**, 121–131.

Keenleyside, M. H. A. (1962) Skin-diving observations of Atlantic salmon and brook trout in the Miramichi River, New Brunswick. *Journal of the Fisheries Research Board of Canada,* **19**, 625–634.

Keenleyside, M. H. A. (1991) Parental care. In: *Cichlid Fishes – Behaviour, Ecology and Evolution* (ed. M. H. A. Keenleyside), pp. 191–208. Chapman & Hall, London.

Kelso, B. W., Northcote, T. G. & Wehrhahn, C. F. (1981) Genetic and environmental aspects of the response to water current by rainbow trout (*Salmo gairdneri*) originating from inlet and outlet streams of two lakes. *Canadian Journal of Zoology,* **59**, 2177–2185.

Kelso, W. E. & Rutherford, D. A. (1996) Collection, preservation and identification of fish eggs and larvae. In: *Fisheries Techniques,* 2nd edn. (eds B. R. Murphy & D. W. Willis), pp. 255–302. American Fisheries Society, Bethesda, Maryland.

Kennedy, G. J. A. (1977) *Experiments on homing and home range behaviour in shoals of roach and minnows.* DPhil thesis, New University of Ulster, Coleraine, Northern Ireland.

Kennedy, G. J. A. & Pitcher, T. J. (1975) Experiments on homing in shoals of the European minnow *Phoxinus phoxinus* L. *Transactions of the American Fisheries Society,* **104**, 454–457.

Kennen, J. G., Wisniewski, S. J., Ringler, N. H. & Hawkins, H. M. (1994) Application and modification of an auger trap to quantify emigrating fishes in Lake Ontario tributaries. *North American Journal of Fisheries Management,* **14**, 828–836.

Kenward, R. (1987) *Wildlife Radio Tagging: Equipment, Field Techniques and Data Analysis.* Academic Press, London.

Kerstan, M. (1991) The importance of rivers as nursery grounds for 0- and 1- group flounder (*Platichthys flesus*, L.) in comparison to the Wadden Sea. *Netherlands Journal of Sea Research,* **27**, 353–366.

Kestemont, P. & Baras, E. (2001) Environmental factors and feed intake in fish. In: *Feed Intake in Fish* (eds D. F. Houlihan, T. Boujard & M. Jobling), pp. 131-156. Fishing News Books, Blackwell Science Ltd, Oxford.

Ketelaars, H. A. M., Klinge, M., Wagenvoort, A. J., Kampen, J. & Vernooij, S. M. A. (1998) Estimate of the amount of 0+ fish pumped into a storage reservoir and indications of the ecological consequences. *International Review of Hydrobiology,* **83**, 549–558.

Kieffer, M. C. & Kynard, B. (1993) Annual movements of shortnose and Atlantic sturgeons in the lower Merrimack River, Massachusetts. *Transactions of the American Fisheries Society,* **122**, 1088–1103.

Kieffer, M. C. & Kynard, B. (1996) Spawning of the shortnose sturgeon in the Merrimack River, Massachusetts. *Transactions of the American Fisheries Society,* **125**, 179–186.

Kinoshita, I., Azuma, K., Fujita, S., Takahashi, I., Niimi, K. & Harada, S. (1999) Early life history of a catadromous sculpin in western Japan. *Environmental Biology of Fishes,* **54**, 135–149.

Kirchhofer, A. & Hefti, D. (eds) (1996) *Conservation of Endangered Freshwater Fish in Europe.* Birkhäuser Verlag, Basel, Boston, Berlin.

Kirschvink, J. L., Walker, M. M., Chang, S. B., Dizon, A. E. & Peterson, K.A. (1985) Chains of single-domain magnetite particles in chinook salmon, *Oncorhynchus tshawytscha. Journal of Comparative Physiology A – Sensory Neural and Behavioral Physiology,* **157**, 375–381.

Kirshbaum, F. (1984) Reproduction of weakly electric teleosts: just another example of convergent development? *Environmental Biology of Fishes,* **10**, 3–14.

Kitchell, J. F., Magnuson, J. C. & Neill, W. H. (1977) Estimation of caloric content for fish biomass. *Environmental Biology of Fishes,* **2**, 185–188.

Klinge, M. (1994) Fish migration via the shipping lock at the Hagestein barrage: results of an indicative study. *Water Science and Technology,* **29**, 357–361.

Klinger, S. A. J., Magnuson, J. J. & Gallepp, G. W. (1982) Survival mechanisms of the central mudminnow (*Umbra limi*), fathead minnow (*Pimephales promelas*) and brook stickleback (*Culaea inconstans*) for low oxygen in winter. *Environmental Biology of Fishes,* **7**, 113–120.

Knights, B. (1987) Agonistic behaviour and growth in the European eel, *Anguilla anguilla* L., in relation to warm-water aquaculture. *Journal of Fish Biology,* **31**, 263–276.

Knights, B., White, E. & Naismith, I. A. (1996) Stock assessment of the European eel, *Anguilla anguilla* L. In: *Stock Assessment in Inland Fisheries* (ed. I. G. Cowx), pp. 431–447. Fishing News Books, Blackwell Science Ltd, Oxford.

Knights, B. C., Johnson, B. L. & Sandheinrich, M. B. (1995) Responses of bluegills and black crappies to dissolved oxygen, temperature, and current in backwater lakes of the Upper Mississippi River during winter. *North American Journal of Fisheries Management,* **15**, 390–399.

Knights, B. C. & Lasee, B. A. (1996) Effects of implanted transmitters on adult bluegills at two temperatures. *Transactions of the American Fisheries Society,* **125**, 440–449.

Knudsen, F. R., Enger, P. S. & Sand, O. (1992) Awareness reactions and avoidance responses to sound in juvenile Atlantic salmon, *Salmo salar* L. *Journal of Fish Biology,* **40**, 523–534.

Knudsen, F. R., Enger, P. S. & Sand, O. (1994) Avoidance responses to low frequency sound in downstream migrating Atlantic salmon smolt, *Salmo salar. Journal of Fish Biology,* **45**, 227–233.

Koed, A., Mejlhede, P., Balleby, K. & Aarestrup, K. (2000) Annual movement and migration of adult pikeperch in a lowland river. *Journal of Fish Biology,* **57**, 1266–1279.

Kramer, D. L. (1978) Terrestrial group spawning of *Brycon petrosus* (Pisces, Characidae) in Panama. *Copeia,* **1978**, 536–537.

Kramer, D. L. (1983a) The evolutionary ecology of respiratory mode in fishes: an analysis based on the costs of breathing. *Environmental Biology of Fishes,* **9**, 145–158.

Kramer, D. L. (1983b) Aquatic surface respiration in the fishes of Panama: distribution in relation to risk of hypoxia. *Environmental Biology of Fishes,* **8**, 49–54.

Kramer, D. L. (1987) Dissolved oxygen and fish behavior. *Environmental Biology of Fishes* **18**, 81–92.

Kramer, D. L., Lindsey, C. C., Moodie, G. E. E. & Stevens, E. D. (1978) The fishes and the aquatic environment of the central Amazon basin, with particular reference to respiratory patterns. *Canadian Journal of Zoology,* **56**, 717–729.

Kramer, D. L. & McClure, M. (1982) Aquatic surface respiration: a widespread adaptation to hypoxia in tropical freshwater fishes. *Environmental Biology of Fishes,* **7**, 47–55.

Kramer, D. L. & Mehegan, J. P. (1981) Aquatic surface respiration, and adaptive response to hypoxia in the guppy, *Poecilia reticulata* (Pisces, Poecilidae). *Environmental Biology of Fishes,* **6**, 299–313.

Krebs, J. R. & Davies, N. B. (1981) *An Introduction to Behavioural Ecology*. Blackwell, Oxford.

Kristiansen, H. & Døving, K. B. (1996) The migration of spawning stocks of grayling *Thymallus thy-*

mallus, in Lake Mjosa, Norway. *Environmental Biology of Fishes*, **47**, 43–50.

Kristoffersen, K. (1994) The influence of physical watercourse parameters on the degree of anadromy in different lake populations of Arctic charr (*Salvelinus alpinus* (L.)) in northern Norway. *Ecology of Freshwater Fish*, **3**, 80–91.

Kristoffersen, K., Halvorsen, M. & Jørgensen, L. (1994) Influence of parr growth, lake morphology, and freshwater parasites on the degree of anadromy in different populations of Arctic char (*Salvelinus alpinus*) in northern Norway. *Canadian Journal of Fisheries and Aquatic Sciences*, **51**, 1229–1246.

Krivanec, K. & Kubecka, J. (1990) Vliv vodárenské nádrze Rímov na utvárení obsádky v úseku Malse pod nádrzí (Influence of Rimov Water Supply on the fish stock forming in the Malse River below the reservoir). In: *Ichthyofauna of the Malse River at the Rimov Reservoir* (ed. J. Kubecka), pp. 125–133. South-Bohemian Museum, Ceské Budejovice [in Czech].

Kubecka, J. (1993) Night inshore migration and capture of adult fish by shore seining. *Aquaculture and Fisheries Management*, **24**, 685–689.

Kubecka, J. (1996a) Use of horizontal dual-beam sonar for fish surveys in shallow water. In: *Stock Assessment in Inland Fisheries* (ed. I. G. Cowx), pp. 165–178. Fishing News Books, Blackwell Science Ltd, Oxford.

Kubecka, J. (1996b) Selectivity of Breder traps for sampling fish fry. In: *Stock Assessment in Inland Fisheries* (ed. I. G. Cowx), pp. 76–81. Fishing News Books, Blackwell Science Ltd, Oxford.

Kubecka, J. & Duncan, A. (1998a) Diurnal changes of fish behaviour in a lowland river monitored by a dual-beam echosounder. *Fisheries Research*, **35**, 55–63.

Kubecka, J. & Duncan, A. (1998b) Acoustic size versus real size relationships for common species of riverine fish. *Fisheries Research*, **35**, 115–125.

Kubecka, J., Duncan, A. & Butterworth, A. J. (1992) Echo counting or echo integration for fish biomass assessment in shallow waters. In: *Underwater Acoustics* (ed. M. Weydert), pp. 129–132. Elsevier, London.

Kubecka, J., Frouzová, J., Vilcinska, A., Wolter, C. & Slavík, O. (2000) Longitudinal hydroacoustic survey of fish in the Elbe River, supplemented by direct capture. In: *Management and Ecology of River Fisheries* (ed. I. G. Cowx), pp. 14–25. Fishing News Books, Blackwell Science Ltd, Oxford.

Kubecka, J., Matena, J. & Hartvich, P. (1997) Adverse ecological effects of small hydropower stations in the Czech Republic: 1. Bypass plants. *Regulated Rivers: Research and Management*, **13**, 101–113.

Kubecka, J. & Vostradovsky, J. (1995) Effect of dams, regulation and pollution on fish stocks in the Vltava River in Prague. *Regulated Rivers: Research and Management*, **10**, 93–98.

Kusmic, C. & Gualtieri, P. (2000) Morphology and spectral sensitivities of retinal and extraretinal photoreceptors in freshwater teleosts. *Micron*, **31**, 183–200.

Kusuda, R. (1963) An ecological study of the anadromous "ayu", *Plecoglossus altivelis* T. and S. II. Seasonal variations in the composition of the anadromous ayu schools in the river Okumo, Kyoto. *Bulletin of the Japanese Society of Scientific Fisheries*, **29**, 822–877.

Kuusipalo, L. (1999) Genetic differentiation of endemic Nile perch *Lates stappersii* (Centropomidae, Pisces) populations in Lake Tanganyika suggested by RAPD markers. *Hydrobiologia*, **407**, 141–148.

Kwak, T. J. (1988) Lateral movement and use of floodplain habitat by fishes of the Kankakee River, Illinois. *American Midland Naturalist*, **120**, 241–249.

Kynard, B. (1997) Life history, latitudinal patterns, and status of the shortnose sturgeon, *Acipenser brevirostrum*. *Environmental Biology of Fishes*, **48**, 319–334.

Kynard, B. (1998) Twenty-two years of passing shortnose sturgeon in fish lifts on the Connecticut River: What has been learned? In: *Fish Migration and Fish Bypasses* (eds M. Jungwirth, S. Schmutz & S. Weiss), pp. 255–264. Fishing News Books, Blackwell Science Ltd, Oxford.

Kynard, B. & O'Leary, J. (1993) Evaluation of a bypass system for spent American shad at Holyoke Dam, Massachusetts. *North American Journal of Fisheries Management*, **13**, 782–789.

L'Abee-Lund, J. H. & Vøllestad, L. A. (1985) Homing precision of the roach *Rutilus rutilus* in Lake Årungen, Norway. *Environmental Biology of Fishes*, **13**, 235–239.

L'Abee-Lund, J. H. & Vøllestad, L. A. (1987) Feeding migration of roach *Rutilus rutilus* in Lake Årungen, Norway. *Journal of Fish Biology*, **30**, 349–355.

LaBar, G. W., Hernando Casal, J. A. & Delgado, C. F. (1987) Local movements and population size of European eels, *Anguilla anguilla*, in a small lake in south-western Spain. *Environmental Biology of Fishes*, **19**, 111–117.

Laé, R. (1992) Influence de l'hydrologie sur l'évolution des pêcheries du delta central du Niger, de 1966 à 1989. *Aquatic Living Resources*, **5**, 115–126.

Lagardère, J. P., Ducamp, J. J., Favre, L., Mosneron-Dupin, J. & Spérandio, M. (1990) A method for the quantitative evaluation of fish movements in salt ponds by acoustic telemetry. *Journal of Experimental Marine Biology and Ecology*, **141**, 221–236.

Laine, A., Kanula, R. & Hooli, J. (1998a) Fish and lamprey passage in a combined Denil and vertical slot fishway. *Fisheries Management and Ecology*, **5**, 31–44.

Laine, A., Yliänrä, T. Heikkilä, J. & Hooli, J. (1998b) Behaviour of upstream migrating whitefish, *Coregonus lavaretus*, in the Kukkolankoski rapids, northern Finland. In: *Fish Migration and Fish Bypasses* (eds M. Jungwirth, S. Schmutz & S. Weiss), pp. 33–44. Fishing News Books, Blackwell Science Ltd, Oxford.

Lake, J. S. (1978) *Australian Freshwater Fishes*. Nelson, Melbourne.

Lam, T. J. (1983) Environmental influences on gonadal activity in fish. In: *Fish Physiology*, Vol. IX (B) (eds W. S. Hoar, D. J. Randall & E. M. Donaldson), pp. 65–116. Academic Press, London.

Lamond, H. (1931) *Loch Lomond: A study in angling conditions*. Jackson, Wylie & Co., Glasgow.

Lamsa, A. (1963) Downstream movements of brook sticklebacks, *Eucalia inconstans* (Kirtland), in a small southern Ontario stream. *Journal of the Fisheries Research Board of Canada*, **20**, 587–589.

Landívar, J. (1996) Ecologie de la communauté de poissons et du chame (*Dormitor latifrons*) dans les rivières Vinces et Babahoyo (Équateur) et leur plaine d'inondation. M.Sc. thesis, Université du Québec à Montréal, Montréal, 69 pp.

Landsborough Thompson, A. (1942) *Bird Migration*, 2nd edn. Witherby, London.

Langford, T. E. (1981) The movement and distribution of sonic-tagged coarse fish in two British rivers in relation to power station cooling-water outfalls. In: *Proceedings of the 3rd International Conference on Wildlife Biotelemetry* (ed. F. M. Long), pp. 197–232. University of Wyoming, Laramie.

Langford, T. E., Milner, A. G. P., Foster, D. J. & Fleming, J. M. (1979) The movements and distribution of some common bream (*Abramis brama* L.) in the vicinity of power station intakes and outfalls in British rivers as observed by ultrasonic tracking. *Laboratory note RD/L/N 145/78*. Central Electricity Research Laboratories, Fawley, 24 pp.

Langhurst, R. W. & Schoenike, D. L. (1990) Seasonal migration of smallmouth bass in the Embarrass and Wolf rivers, Wisconsin. *North American Journal of Fisheries Management*, **10**, 224–227.

Lannoo, M. J. & Lannoo, S. J. (1993) Why do electric fishes swim backwards? An hypothesis based on gymnotiform foraging behaviour interpreted through sensory constraints. *Environmental Biology of Fishes*, **36**, 157–165.

Lanters, R. L. P. (1993) *De bekkenvistrap Belfeld: Monitoring van de visoptrek en hydraulische waarnemingen in 1993.* Rapport 93. 023 RIVO-DLO, Ijmuiden.

Lanters, R. L. P. (1995) *Vismigratie door de bekkenvistrappen Lith en Belfeld in de Maas.* EHR Rapport 59–1995 RIVO-DLO, Ijmuiden.

Larinier, M. (1983) Guide pour la conception des dispositifs de franchissement des barrages pour les poissons migrateurs. *Bulletin Français de la Pisciculture,* **numéro special**, 1–37.

Larinier, M. (1992) Guide pour la conception de dispositifs de franchissement de barrages ou d'obstacles pour les poissons migrateurs. *Bulletin Français de la Pêche et de la Pisciculture*, **326/327**, 1–206.

Larinier, M. (1998) Upstream and downstream fish passage experience in France. In: *Fish Migration and Fish Bypasses* (eds M. Jungwirth, S. Schmutz & S. Weiss), pp. 127–145. Fishing News Books, Blackwell Science Ltd, Oxford.

Larinier, M., Porcher, J. P., Travade, F. & Gosset, C. (1994) *Passes à Poissons: Expertise, Conception des Ouvrages de Franchissement.* Collection Mise au Point, Conseil Supérieur de la Pêche, Paris, France, 336 pp.

Larinier, M. & Travade, F. (1999) Downstream migration: Problems and facilities. *Bulletin Français de la Pêche et de la Pisciculture*, **353/354**, 181–210.

Larkin, R. P. & Halkin, D. (1994) A review of software packages for estimating animal home ranges. *Wildlife Society Bulletin*, **22**, 274–287.

Larsen, K. (1972) Studies on the biology of Danish stream fishes. III. On seasonal fluctuations in the stock density of yellow eels in shallow stream biotopes and their cause. *Meddelelser fra Danmarks Fiskeri og Havundersøgelser*, **7**, 23–46.

Laughton, R. & Smith, G. W. (1992) The relationship between the date of river entry and the estimated spawning position of adult Atlantic salmon (*Salmo salar* L.) in two major Scottish east coast rivers. In: *Wildlife Telemetry – Remote Monitoring and Tracking of Animals* (eds I. G. Priede & S. M. Swift), pp. 423–433. Ellis Horwood, Chichester, UK.

Lauzanne, L., Loubens, G. & Le Guennec, B. (1991) Liste commentée des poissons de l'Amazonie bolivienne. *Revue d'Hydrobiologie Tropicale*, **24**, 61–76.

Leatherland, J. F., Macey, D. J., Hilliard, R. W., Leatherland, A. & Potter, I. C. (1990) Seasonal and estradiol-17 beta -stimulated changes in thyroid function of adult *Geotria australis*, a Southern Hemisphere lamprey. *Fish Physiology and Biochemistry*, **8**, 409–417.

Leclerc, M., Capra, H., Valentin, S., Boudreault, A. & Côté, Y. (eds) (1996) *Ecohydraulics 2000*. INRS-Eau, Québec. Vol. A, 893 pp. Vol. B, 995 pp.

Legault, A. (1992) Etude de quelques facteurs de sélectivité des passes à anguilles. *Bulletin Français de la Pêche et de la Pisciculture*, **325**, 83–91.

Leggett, W. C. (1973) The migrations of the shad. *Scientific American*, **228**, 92–98.

Leggett, W. C. (1976) The American shad (*Alosa sapidissima*) with special references to its migration and population dynamics in the Connecticut River. *American Fisheries Society Monograph*, **1**, 169–225.

Leggett, W. C. (1977) The ecology of fish migrations. *Annual Reviews of Ecology and Systematics*, **8**, 285–308.

Leggett, W. C. & Carscadden, J. E. (1978) Latitudinal variation and reproductive characteristics of American shad (*Alosa sapidissima*): evidence for population specific life history strategies of fish. *Journal of the Fisheries Research Board of Canada*, **35**, 1469–1478.

Legkiy, B. P. & Popova, I. K. (1984) Development of photoreaction in juvenile roach, *Rutilus rutilus*, and minnow, *Phoxinus phoxinus* (Cyprinidae), in relation to downstream migration. *Journal of Ichthyology*, **24**, 72–79.

Lehtonen, H. (1983) Stocks of pike-perch (*Stizostedion lucioperca* L.) and their management in the Archipelago Sea and the Gulf of Finland. *Finnish Fisheries Research*, **5**, 1–16.

Lehtonen, H., Hansson, S. & Winkler, H. (1996) Biology and exploitation of pikeperch, *Stizostedion lucioperca* (L.), in the Baltic Sea area. *Annales Zoologici Fennici*, **33**, 525–535.

Lehtonen, H. & Toivonen, J. (1987) Migration of pike-perch *Stizostedion lucioperca* (L.), in different coastal waters in the Baltic Sea. *Finnish Fisheries Research*, **7**, 24–30.

Lein, G. M. & DeVries, D. R. (1998) Paddlefish in the Alabama River drainage: population characteristics and the adult spawning migration. *Transactions of the American Fisheries Society*, **127**, 441–454.

Lelek, A. (1987) *The Freshwater Fishes of Europe. Vol. 9, Threatened Fishes of Europe*. Aula-Verlag, Wiesbaden.

Lelek, A. & Köhler, C. (1989) Zustandsanalyses der fischartengemeinschaften im Rhein (1987–1988) *Fischökologie*, **1**, 47–64.

Lelek, A. & Libosvárskí, J. (1960) Vískít ryb v rybím prechodu na rece Dyji pri Breclavi (The occurrence of fish in a fish ladder in Dyje River near Breclav). *Folia Zoologica*, **9**, 293–308.

Lemoalle, J. (1999) La diversité des milieux aquatiques. In: *Les Poissons des Eaux Continentales Africaines. Diversité, Ecologie, Utilisation par l'Homme* (eds C. Lévêque & D. Paugy), pp. 11–30. IRD Editions, Paris.

Leonard, J. B. K. & McCormick, S. D. (1999) Effects of migration distance on whole-body and tissue-specific energy use in American shad (*Alosa sapidissima*). *Canadian Journal of Fisheries and Aquatic Sciences*, **56**, 1159–1171.

Lethlean, N. G. (1953) An investigation into the design and performance of electric fish screens and an electric fish counter. *Transactions of the Royal Society of Edinburgh*, **62**, 479–526.

Lévêque, C. (1999) Variabilité du climat et des régimes hydrologiques. In: *Les Poissons des Eaux Continentales Africaines – Diversité, Écologie, Utilisation par l'Homme* (eds C. Lévêque & D. Paugy), pp. 31–42. IRD Editions, Paris.

Levêque, C. & Paugy, D. (eds) (1999) *Les Poissons des Eaux Continentales Africaines – Diversité, Écologie, Utilisation par l'Homme*. IRD Editions – Institut de Recherche pour le Développement, Paris, 521 pp.

Lévêque, C. & Quensière, J. (1988) Peuplements ichtyologiques des lacs peu profonds. In: *Biology and Ecology of African Freshwater Fishes* (eds C. Lévêque, M. N. Bruton & G. W. Ssentongo), pp. 303–324. Editions de l'ORSTOM, Travaux et Documents n° 216, Paris.

Levin, L. & Belmonte, P. (1988) Oriented swimming to an angle to light in a schooling characid *Moenkhausia dichroura* Kner. *Journal of Fish Biology*, **32**, 169–177.

Levin, L., Belmonte, P. & De Martini, A. (1998) Orientación y conducta migratoria en peces. *Memorias del Instituto de Biología Experimental*, **1**, 133–136.

Levin, L. & Gonzáles, O. (1994) Endogenous rectilinear guidance system in fish: is it adjusted by reference to the sun? *Behavioral Processes*, **31**, 247–256.

Levy, D. A. (1991) Acoustic analysis of diel vertical migration behavior of *Mysis relicta* and kokanee

(*Oncorhynchus nerka*) within Okanagan Lake, British Columbia. *Canadian Journal of Fisheries and Aquatic Sciences,* **48**, 67–72.

Lewis, W. M. Jr. (1970) Morphological adaptations of cyprinodontoids for inhabiting oxygen deficient waters. *Copeia,* **1970**, 319–326.

Li, W., Sorensen, P. W. & Gallaher, D. D. (1995) The olfactory system of migratory adult sea lamprey (*Petromyzon marinus*) is specifically and acutely sensitive to unique bile acids released by conspecific larvae. *Journal of General Physiology,* **105**, 569–587.

Libosvárskí, J. (1961) On the stability of a population of chub, *Leuciscus cephalus* L., in a stream section. *Zoologica Listy,* **15**, 161–174.

Libosvárskí, J., Lelek, A. & Penáz, M. (1967) Movements and mortality of fish in two polluted brooks. *Acta scientiarum naturalium Acadamiae Scientiarum Bohemoslovacae Brno,* **1**, 1–28.

Liebig, H., Lim, P., Belaud, A. & Lek, A. (1996) Study of the juvenile community in the brown trout (*Salmo trutta fario* L., 1758) in hydropeaking situations. In: *Ecohydraulics 2000* (eds M. Leclerc et al.), Vol. A, pp. 674–684. INRS-Eau, Québec.

Lieschke, J. A. & Closs, G. P. (1999) Regulation of zooplankton composition and distribution by a zooplanktivorous fish in a shallow, eutrophic floodplain lake in south east Australia. *Archiv für Hydrobiologie,* **146**, 397–412.

Lightfoot, G. W. & Jones, N. V. (1996) The relationship between the size of 0+ roach, *Rutilus rutilus*, their swimming capabilities, and distribution in an English river. *Folia Zoologica,* **45**, 355–360.

Lilyestrom, C. G. (1983) Aspectos de la biología del coporo (*Prochilodus mariae*). *Revista UNELLEZ de Ciencia y Tecnología (Barinas),* **1**, 5–11.

Lindsey, C. C. & Northcote, T. G. (1963) Life history of redside shiner, *Richardsonius balteus*, with particular reference to movements in and out of Sixteenmile Lake streams. *Journal of the Fisheries Research Board of Canada,* **20**, 1001–1030.

Lindsey, C. C., Northcote, T. G. & Hartmann, G. F. (1959) Homing of rainbow trout to inlet and outlet spawning streams at Loon Lake, British Columbia. *Journal of the Fisheries Research Board of Canada,* **16**, 695–719.

Linfield, R. S. J. (1985) An alternative concept to home range theory with respect to populations of cyprinids in major river systems. *Journal of Fish Biology,* **27** (Suppl. A), 187–196.

Linlokken, A. (1993) Efficiency of fishways and impact of dams on the migration of grayling and brown trout in the Glomma river system, south eastern Norway. *Regulated Rivers: Research and Management,* **8**, 145–153.

Little, A. S., Tonn, W. M., Tallman, R. F. & Reist, J. M. (1998) Seasonal variation in diet and trophic relationships within the fish communities of the lower Slave River, Northwest Territories, Canada. *Environmental Biology of Fishes,* **53**, 429–445.

Liu, C. & Zeng, Y. (1988) Notes on the Chinese paddlefish *Psephurus gladius* (Martens). *Copeia,* **1988**, 482–484.

Liu, J. K. & Yu, Z. T. (1992) Water quality changes and effects on fish populations in the Hanjiang River, China, following hydroelectric dam construction. *Regulated Rivers: Research and Management,* **7**, 359–368.

Llewellyn, L. C. (1973) Spawning, development and temperature tolerance of the spangled perch, *Madigiana unicolor* (Gunther) from inland waters in Australia. *Australian Journal of Marine and Freshwater Research,* **24**, 73–94.

Loesch, J. G. (1987) Overview of life history aspects of anadromous alewife and blueback herring in freshwater habitats. *American Fisheries Society Symposium,* **1**, 89–103.

Loftus, W. F., Kushlan, J. A. & Voorhees, S. A. (1984) Status of the mountain mullet in southern Florida. *Florida Scientist,* **47**, 256–263.

Loose, C. J., Elert, E. von & Dawidowicz, P. (1993) Chemically-induced diel vertical migration in *Daphnia:* a new bioassay for kairomones exuded by fish. *Archiv für Hydrobiologie,* **126**, 329–337.

Lorenz, K. (1970) *Trois Essais sur le Comportement Animal et Humain.* Editions du Seuil, Paris.

Loubens, G. (1970) Etude de certains peuplements ichtyologiques par des pêches au poison Seconde note. *Cahiers ORSTOM, série Hydrobiologie,* **4**, 45–61.

Loubens, G. (1973) Production de la pêche et peuplements ichtyologiques d'un bief du delta du Chari. *Cahiers ORSTOM, série Hydrobiologie,* **7**, 209–233.

Loubens, G. (1974) Quelques aspects de la biologie des *Lates niloticus* du Tchad. *Cahiers ORSTOM, série Hydrobiologie,* **8**, 3–21.

Loubens, G., Lauzanne, L. & Le Guennec, B. (1992) Les milieux aquatiques de la region de Trinidad (Béni, Amazonie bolivienne). *Revue d'Hydrobiologie Tropicale,* **25**, 3–21.

Loubens, G. & Osorio, F. (1988) Observations sur les poissons de la partie bolivienne du lac Titicaca. III. *Basilichthys bonariensis* (Valenciennes, 1835) (Pisces, Atherinidae). *Revue d'Hydrobiologie Tropicale,* **21**, 153–177.

Loubens, G. & Panfilli, J. (1995) Biologie de *Prochilodus nigricans* (Teleostei: Prochilodontidae) dans le bassin du Mamoré (Amazonie bolivienne). *Ichthyology and Exploration of Freshwaters,* **6**, 17–32.

Loubens, G. & Panfilli, J. (1997) Biologie de *Colossoma macropomum* (Teleostei: Serrasalmidae) dans le bassin du Mamoré (Amazonie bolivienne). *Ichthyology and Exploration of Freshwaters,* **8**, 1–22.

Love, R. H. (1977) Target strength of an individual fish at any aspect. *Journal of the Acoustical Society of America,* **62**, 1317–1403.

Lowe, R. H. (1952) The influence of light and other factors on the seaward migration of the silver eel (*Anguilla anguilla* L.). *Journal of Animal Ecology,* **21**, 275–309.

Lowe-McConnell, R. H. (1975) *Fish Communities in Tropical Freshwaters.* Longman Press, London.

Lowe-McConnell, R. H. (1987) *Ecological Studies in Tropical Fish Communities.* London, Longman Press, London, 382 pp.

Lowe-McConnell, R. H. (1988) Broad characteristics of the ichthyofauna. In: *Biology and Ecology of African Freshwater Fishes* (eds C. Lévêque, M. N. Bruton & G. W. Ssentongo), pp. 93–110. Éditions de l'ORSTOM, Travaux et Documents n° 216, Paris.

Lowe-McConnell, R. H. (1991) Ecology of cichlids in South American and African waters, excluding the African Great Lakes. In: *Cichlid Fishes – Behaviour, Ecology and Evolution* (ed. M. H. A. Keenleyside), pp. 60–85. Chapman & Hall, London.

Lowe-McConnell, R. H. (1999) Lacustrine fish communities in Africa. In: *Fish and Fisheries of Lakes and Reservoirs in Southeast Asia and Africa* (eds W. L. T. van Densen & M. J. Morris), pp. 29–48. Westbury Academic and Scientific Publishing, Otley, UK.

Lübben, G. & Tesch, F. W. (1966) Sommer und Winteraufenthalt der Aale. *Fisch und Fang,* **9**, 56–57.

Lucas, M. C. (1989) Effects of intraperitonal transmitters on mortality, growth and tissue reaction in rainbow trout, *Salmo gairdneri* Richardson. *Journal of Fish Biology,* **35**, 577–587.

Lucas, M. C. (1992) Spawning activity of male and female pike, *Esox lucius* L., determined by acoustic tracking. *Canadian Journal of Zoology,* **70**, 191–196.

Lucas, M. C. (1994) Heart rate as an indicator of metabolic rate and activity in adult Atlantic salmon,

Salmo salar. Journal of Fish Biology, **44**, 889–903.

Lucas, M. C. (2000) The influence of environmental factors on movements of lowland-river fish in the Yorkshire Ouse system. *The Science of the Total Environment,* **251/252**, 223–232.

Lucas, M. C. (in press) Recent advances in the application of telemetry to the study of freshwater fish. In: *Proceedings of the Fifth European Conference on Wildlife Telemetry* (eds Y. Le Maho & T. Zorn), CEPE/CNRS, Strasbourg, France.

Lucas, M.C. & Baras, E. (2000) Methods for studying spatial behaviour of fishes in the natural environment. *Fish and Fisheries,* **1**, 283-316.

Lucas, M. C. & Batley, E. (1996) Seasonal movements and behaviour of adult barbel *Barbus barbus*, a riverine cyprinid fish: implications for river management. *Journal of Applied Ecology,* **33**, 1345–1358.

Lucas, M. C. & Frear, P. A. (1997) Effects of a flow-gauging weir on the migratory behaviour of adult barbel, a riverine cyprinid. *Journal of Fish Biology,* **50**, 382–396.

Lucas, M. C. & Johnstone, A. D. F. (1990) Observations on the retention of intragastric transmitters and their effects on food consumption in cod, *Gadus morhua* L. *Journal of Fish Biology,* **37**, 647–649.

Lucas, M. C., Johnstone, A. D. F. & Priede, I. G. (1993a) Use of physiological telemetry as a method of estimating metabolism of fish in the natural environment. *Transactions of the American Fisheries Society,* **122**, 822–833.

Lucas, M. C., Johnstone, A. D. F. & Tang, J. (1993b) An annular respirometer for measuring aerobic metabolic rates of large, schooling fishes. *Journal of Experimental Biology,* **175**, 325–331.

Lucas, M. C. & Marmulla, G. (2000) An assessment of anthropogenic activities on and rehabilitation of river fisheries: current status and future direction. In: *Management and Ecology of River Fisheries* (ed. I. G. Cowx), pp. 261–278. Fishing News Books, Blackwell Science Ltd, Oxford.

Lucas, M. C., Mercer, T., Batley, E., *et al.* (1998) Spatio-temporal variations of fishes in the Yorkshire Ouse system. *Science of the Total Environment,* **210/211**, 437–455.

Lucas, M. C., Mercer, T., McGinty, S. & Armstrong, J. D. (1999) Use of a flat-bed passive integrated transponder antenna array to study the migration and behaviour of lowland river fishes at a fish pass. *Fisheries Research,* **44**, 183–191.

Lucas, M.C., Mercer, T., Peirson, G. & Frear, P. A. (2000) Seasonal movements of coarse fish in lowland rivers and their relevance to fisheries management. In: *Management and Ecology of River Fisheries* (ed. I. G. Cowx), pp. 87–100. Fishing News Books, Blackwell Science Ltd, Oxford.

Lucas, M. C., Priede, I. G., Armstrong, J. D., Gindy, A. N. Z., De Vera, L. (1991) Direct measurement of metabolism, activity and feeding behaviour of pike, *Esox lucius* L., in the wild, by the use of heart rate telemetry. *Journal of Fish Biology,* **39**, 325–345.

Luecke, C. & Teuscher, D. (1994) Habitat selection by lacustrine rainbow trout within gradients of temperature, oxygen and food availability. In: *Theory and Application in Feeding Ecology* (eds D. J. Stouder, K. L. Fresh & R. J. Feller). Belle W. Baruch Library in Marine Sciences, n° 18. University of South Carolina Press, Columbia, South Carolina, USA.

Luecke, C. & Wurtsbaugh, W. A. (1993) Effects of moonlight and daylight on hydroacoustic estimates of pelagic fish abundance. *Transactions of the American Fisheries Society*, **122**, 112–120.

Lühmann, M. & Mann, H. (1961) Untersuchungen über den Aalfang in der Elbe. *Fischwirt,* **11**, 165–176.

Lüling, K. H. (1964) Zur biologie und Ökologie von *Arapaima gigas* (Pisces, Osteoglossidae). *Zeitschrift fur Morphologie und Ökologie der Tiere,* **54**, 436–530.

Lung'Ayia, H. B. O. (1994) Some observations on the African catfish *Clarias gariepinus* (Burchell) in the Sondu-Miriu River of Lake Victoria, Kenya. In: *Recent Trends in Research on Lake Victoria Fisheries* (eds E. Okemwa, E. O. Wakwabi & A. Getabu), pp. 105–114. ICIPE Science, Nairobi, Kenya.

Lutcavage, M. E., Brill, R., Skomal, G. B., Chase, B. C. & Howey, P. W. (1999) Results of pop-up satellite tagging of spawning size class fish in the Gulf of Maine: do North Atlantic bluefin tuna spawn in the Mid-Atlantic? *Canadian Journal of Fisheries and Aquatic Sciences,* **56**, 173–177.

Lyons, J. (1998) A hydroacoustic assessment of fish stocks in the River Trent, England. *Fisheries Research,* **35**, 83–90.

Lyons, J. & Kempinger, J. J. (1992) Movements of adult Lake sturgeon in the Lake Winnebago system. *Wisconsin Department of Natural Resources Research Publication* **RS-156–92**.

MacFarland, W. N. & Klontz, G. W. (1969) Anesthesia in fishes. *Federal Proceedings,* **28**, 1535–1540.

MacInnis, A. J. & Corkum, L. D. (2000) Fecundity and reproductive season of the round goby *Neogobius melanostomus* in the upper Detroit River. *Transactions of the American Fisheries Society,* **129**, 136–144.

Maciolek, J. A. (1977) Taxonomic status, biology and distribution of Hawaiian *Lentipes*, a diadromous goby. *Pacific Science,* **31**, 355–362.

MacKenzie, K. (1983) Parasites as biological tags in fish population studies. *Advances in Applied Biology,* **7**, 251–331.

MacLennan, D. N. & Simmonds, E. J. (1992) *Fisheries Acoustics*. Chapman & Hall, London.

Macquart-Moulin, C., Castelbon, C., Champalbert, G., Chikhi, D., Le-Direach-Boursier, L. & Patriti, G. (1988) The role of barosensitivity in the control of migrations of larval and juvenile sole (*Solea solea* L.): Influence of pressure variations on swimming activity and orientation. In: *The Early Life History of Fish* (eds J. H. S. Blaxter, J. C. Gamble & H. Von Westernhagen), *3rd ICES Symposium*, pp. 400–408. ICES, Copenhagen.

Mader, H., Unfer, G. & Schmutz, S. (1998) The effectiveness of nature-like bypass channels in a lowland river, the Marchfeldkanal. In: *Fish Migration and Fish Bypasses* (eds M. Jungwirth, S. Schmutz & S. Weiss), pp. 384–402. Fishing News Books, Blackwell Science Ltd, Oxford.

Magalhães, M. F. (1993) Effects of season and body-size on the distribution and diet of the Iberian chub *Leuciscus pyrenaicus* in a lowland catchment. *Journal of Fish Biology,* **42**, 875–888.

Magnuson, J. J., Beckel, A. L., Mills, K. & Brandt, S. B. (1985) Surviving winter hypoxia: behavioural adaptations of fish in a northern Wisconsin winterkill lake. *Environmental Biology of Fishes,* **14**, 241–250.

Magurran, A. E., Irving, P. W. & Henderson, P. A. (1996) Is there a fish alarm pheromone? A wild study and critique. *Proceedings of the Royal Society of London Series Biology,* **263**, 1551–1556.

Maitland, P. S. (1980) Review of the ecology of lampreys in northern Europe. *Canadian Journal of Fisheries and Aquatic Sciences,* **37**, 1944–1952.

Maitland, P. S. & Campbell, R. N. (1992) *Freshwater Fishes of the British Isles*. Harper Collins, London.

Maitland, P. S., Morris, K. H. & East, K. (1994) The ecology of lampreys (*Petromyzonidae*) in the Loch Lomond area. *Hydrobiologia,* **290**, 105–120.

Malinin, L. K. (1970) Use of ultrasonic transmitters for tagging bream and pike. Report 1: reaction of fish to net webbing. *Fisheries Research Board of Canada Translation Series,* **1818**, 8 pp.

Malinin, L. K. (1971) Home range and actual paths of fish in the river pool of the Rybinsk reservoir.

Fisheries Research Board of Canada Translation Series, **2282**, 26 pp.

Malinin, L. K. (1972) Use of ultrasonic transmitters for the marking of bream and pike 2. Behaviour of fish at river estuaries. *Fisheries Research Board of Canada Translation Series,* **2146**.

Malinin, L. K., Kijasko, V. I. & Vääränen, P. L (1992) Behaviour and distribution of bream (*Abramis brama*) in oxygen deficient regions. In: *Wildlife Telemetry – Remote Monitoring and Tracking of Animals* (eds I. G. Priede & S. M. Swift), pp. 297–306. Ellis Horwood, Chichester, UK.

Mallen-Cooper, M. (1992) Swimming ability of juvenile Australian bass, *Macquaria novemaculeata* (Steindachner), and juvenile barramundi, *Lates calcarifer* (Bloch), in an experimental vertical-slot fishway. *Australian Journal of Marine and Freshwater Research,* **43**, 823–834.

Mallen-Cooper, M. (1994) Swimming ability of adult golden perch, *Macquaria ambigua* (Percichthyidae), and adult silver perch, *Bidyanus bidyanus* (Teraponidae), in an experimental vertical-slot fishway. *Australian Journal of Marine and Freshwater Research,* **45**, 191–198.

Malmqvist, B. (1980) The spawning migration of the brook lamprey, *Lampetra planeri* Bloch, in a South Swedish stream. *Journal of Fish Biology,* **16**, 105–114.

Malvestuto, S. P. (1996) Sampling the recreational creel. In: *Fisheries Techniques,* 2nd edn. (eds B. R. Murphy & D. W. Willis), pp. 591–623. American Fisheries Society, Bethesda, Maryland.

Manacop, P. R. (1953) The life history and habits of the goby *Sicyopterus extraneus* Herre (Anga) Gobiidae, with an account of the goby-fry fishery of Cagayan River, Oriental Misamis. *The Philippine Journal of Fisheries,* **2**, 1–58.

Mann, H. (1963) Beobachtungen über den Aalaufstieg in der Aalleiter an der Staustufe Geesthacht im Jahre 1961. *Fischwirt,* **13**, 182–186.

Mann, H. (1965) Über das Rückkehrvenmogen verpflantzer Flußaale. *Archiv für Fischwisserei,* **15**, 177–185.

Mann, R. H. K. (1980) The numbers and production of pike (*Esox lucius*) in two Dorset rivers. *Journal of Animal Ecology,* **49**, 899–915.

Mann, R. H. K. & Blackburn, J. H. (1991) The biology of the eel *Anguilla anguilla* in an English chalk stream and interactions with juvenile trout *Salmo trutta,* L. and salmon *Salmo salar* L. *Hydrobiologia,* **218**, 65–76.

Mansueti, R. J. (1964) Eggs, larvae and young of the white perch *Roccus americanus*, with comments on its ecology in the estuary. *Chesapeake Science,* **5**, 3–45.

Margalef, R. (1963) On certain unifying principles in ecology. *American Naturalist,* **97**, 357–374.

Margolis, L. (1982) Parasitology of Pacific salmon: an overview. In: *Aspects of Parasitology* (ed. E. Meerovitch), pp. 135–226. McGill University, Montreal.

Marmulla, G. & Ingendahl, D. (1996) Preliminary results of a radio telemetry study of returning Atlantic salmon (*Salmo salar* L.) and sea trout (*Salmo trutta trutta* L.) in River Sieg, tributary of River Rhine in Germany. In: *Underwater Biotelemetry* (eds E. Baras & J. C. Philippart), pp. 109–117. University of Liège, Belgium.

Marotz, B. L., Herke, W. H. & Rogers, B. D. (1990) Movement of gulf menhaden through three marshland routes in southwestern Louisiana. *North American Journal of Fisheries Management,* **10**, 408–417.

Marshall, B. E. (1984) Small pelagic fishes and fisheries in African inland waters. *FAO, CIFA Technical Paper,* **14**, 1–25.

Marshall, B. E. & van der Heyden, J. T. (1977) The biology of *Alestes imberi* Peters (Pisces, Characidae) in Lake McIlwaine, Rhodesia). *Zoologie Africaine,* **12**, 329–346.

Martin-Smith, K. M. & Laird, L. M. (1998) Depauperate freshwater fish communities in Sabah: the role of barriers to movement and habitat quality. *Journal of Fish Biology,* **53** (Suppl. A), 331–344.

Marty, G. D. & Summerfelt, R. C. (1986) Pathways and mechanisms for expulsion of surgically implanted dummy transmitters from channel catfish. *Transactions of the American Fisheries Society,* **115**, 577–589.

Marty, G. D. & Summerfelt, R. C. (1990) Wound healing in channel catfish by epithelialization and contraction of granulation tissue. *Transactions of the American Fisheries Society,* **119**, 145–150.

Mason, J. C. (1975) Seaward movement of juvenile fishes, including lunar periodicity in the movement of coho salmon (*Oncorhynchus kisutch*) fry. *Journal of the Fisheries Research Board of Canada,* **32**, 2542–2547.

Masse, G., Dumont, P., Ferrais, J. & Fortin, R. (1991) Influence of the hydrological and thermal regimes of the Aux Pins River (Quebec) on the spawning migrations of northern pike and on the emigration of 0+ juveniles. *Aquatic Living Resources,* **4**, 275–287.

Matheney, M. P. IV & Rabeni, C. F. (1995) Patterns of movement and habitat use by northern hog suckers in an Ozark stream. *Transactions of the American Fisheries Society,* **124**, 886–897.

Mathisen, O. A. & Berg, M. (1968) Growth rate of the char *Salvelinus alpinus* (L.) in the Vardnes River, Tromsø, northern Norway. *Reports of the Institute of Freshwater Research of Drottningholm,* **48**, 176–186.

Matthews, W. J., Harvey, B. C. & Power, M. E. (1994) Spatial and temporal patterns in fish assemblages of individual pools in a midwestern stream (U.S.A.). *Environmental Biology of Fishes,* **39**, 381–397.

Mayr, E. (1963) *Animal Species and Evolution*. Belknap Press, Cambridge, Massachusetts.

McAda, C. W. & Keading, L. R. (1991) Movements of adult Colorado squawfish during the spawning season in the upper Colorado River. *Transactions of the American Fisheries Society,* **120**, 339–345.

McAllister, D. (1984) Osmeridae. In: *Fishes of the northeast Atlantic and Mediterranean,* Vol. 1 (eds P. J. P. Whitehead, M. L. Bauchot, J. G. Hureau, J. Nielsen & E. Tortonese), pp. 399–402. UNESCO, Paris.

McAllister, D. E. & Lindsey, C. C. (1959) Systematics of the freshwater sculpins (*Cottus*) of British Columbia. *Bulletin of the National Museums of Canada,* **172**, 66–89.

McCart, P.J. (1986) Fish and fisheries of the Mackenzie system. In: *The Ecology of River Systems* (eds B. R. Davies & K. F. Walker). Dr. W. Junk Publishers, Dordrecht, the Netherlands.

McCleave, J. D. & Arnold, G. P. (1999) Movements of yellow- and silver-phase European eels (*Anguilla anguilla* L.) tracked in the western North Sea. *ICES Journal of Marine Science,* **56**, 510–536.

McCleave, J. D., Kleckner, R. C. & Castonguay, M. (1987) Reproductive sympatry of American and European eels and implications for migration and taxonomy. *American Fisheries Society Symposium,* **1**, 286–297.

McCleave, J. D. & Power, J. H. (1978) Influence of weak electric and magnetic fields on turning behavior in elvers of the American eel *Anguilla rostrata*. *Marine Biology,* **40**, 29–34.

McCleave, J. D., Power, J. H. & Rommel, S. A. Jr. (1978) Use of radio telemetry for studying upriver migration of adult Atlantic salmon (*Salmo salar*). *Journal of Fish Biology,* **12**, 549–558.

McConville, D. R. & Fossum, J. D. (1981) Movement patterns of walleye (*Stizostedion v. vitreum*) in Pool 3 of the upper Mississippi River as determined by ultrasonic telemetry. *Journal of Freshwater Ecology,* **1**, 279–285.

McCubbing, D. J. F., Bayliss, B. D. & Locke, V. M. (1998) Spawning migration of radio tagged landlocked Arctic charr, *Salvelinus alpinus* L. in Ennerdale Lake, the English Lake District. *Hydrobiologia, **371/372**, 173–180.

McDowall, R. M. (1979) Fishes of the family Retropinnidae (Pisces, Salmoniformes): a taxonomic revision and synopsis. *Journal of the Royal Society of New Zealand,* **9**, 85–121.

McDowall, R. M. (1988) *Diadromy in Fishes: Migrations Between Freshwater and Marine Environments.* Croom Helm, London.

McDowall, R. M. (1990) *New Zealand Freshwater Fishes: A Natural History and Guide.* Heinemann Reed, Auckland.

McDowall, R. M. (1995) Seasonal pulses in migrations of New Zealand diadromous fish and the potential impacts of river mouth closure. *New Zealand Journal of Marine and Freshwater Research,* **29**, 517–526.

McDowall, R. M. (ed.) (1996) *Freshwater Fishes of South-Eastern Australia,* 2nd edn. Reed, Sydney.

McDowall, R. M. (1997a) The evolution of diadromy in fishes (revisited) and its place in phylogenetic analysis. *Reviews in Fish Biology and Fisheries,* **7**, 443–462.

McDowall, R. M. (1997b) Is there such a thing as amphidromy? *Micronesica,* **30**, 3–14.

McDowall, R. M. (2000) Biogeography of the New Zealand torrentfish, *Cheimarrichthys fosteri* (Teleostei: Pinguipedidae): a distribution driven mostly by ecology and behaviour. *Environmental Biology of Fishes,* **58**, 119–131.

McDowall, R. M. & Eldon, G. A. (1980) The ecology of whitebait migrations (Galaxiidae: *Galaxinus* spp.). *New Zealand Ministry of Agriculture and Fisheries, Fisheries Research Bulletin,* **20**, 171 pp.

McFarlane, G. A., Wydoski, R. S. & Prince, E. D. (1990) Historical review of the development of external tags and marks. *American Fisheries Society Symposium,* **7**, 9–29.

McGovern, P. & McCarthy, T. K. (1992) Local movements of freshwater eels (*Anguilla anguilla* L.) in western Ireland. In: *Wildlife Telemetry – Remote Tracking and Telemetry of Animals* (eds I. G. Priede & S. M. Swift), pp. 319–327. Ellis Horwood, Chichester, UK.

McInerney, J. E. (1964) Salinity preference: an orientation mechanism in salmon migration. *Journal of the Fisheries Research Board of Canada,* **21**, 995–1018.

McKenzie, R. A. (1964) Smelt life history and fishery in the Miramichi River, New Brunswick. *Bulletin of the Fishery Research Board of Canada,* **144**, 1–77.

McKeown, B. A. (1984) *Fish Migration.* Croom Helm, London.

McKinley, R. S. & Power, G. (1992) Measurement of activity and oxygen consumption for adult lake sturgeon (*Acipenser fulvescens*) in the wild using radio-transmitted EMG signals. In: *Wildlife Telemetry – Remote Tracking and Telemetry of Animals* (eds I. G. Priede & S. M. Swift), pp. 307–318. Ellis Horwood, Chichester, UK.

McKinney, T., Persons, W. R. & Rogers, R. S. (1999) Ecology of flannelmouth sucker in the Lee's Ferry tailwater, Colorado River, Arizona. *Great Basin Naturalist,* **59**, 259–265.

McLaren, J. B., Cooper, J. C., Hoff, J. B. & Lander, V. (1981) Movements of Hudson River striped bass. *Transactions of the American Fisheries Society,* **110**, 158–167.

McPhail, J. D. & Lindsey, C. C. (1970) Freshwater fishes of Northwestern Canada and Alaska. *Bulletin of the Fisheries Research Board of Canada,* **173**.

Meador, M. R. & Kelso, W. E. (1989) Behavior and movements of largemouth bass in response to salinity. *Transactions of the American Fisheries Society,* **118**, 409–415.

Meador, M. R. & Matthews, W. J. (1992) Spatial and temporal patterns in fish assemblage structure of an intermittent Texas stream. *American Midland Naturalist,* **127**, 106–114.

Meek, A. (1916) *The Migrations of Fishes.* Arnold, London.

Mellas, E. J. & Haynes, J. M. (1985) Swimming performance and behaviour of rainbow trout (*Salmo gairdneri*) and white perch (*Morone americana*): effects of attaching telemetry transmitters. *Canadian Journal of Fisheries and Aquatic Sciences,* **42**, 488–493.

Melvin, G. D., Dadswell, M. J. & Martin, J. D. (1986) Fidelity of American shad, *Alosa sapidissima* (Clupeidae), to its river of previous spawning. *Canadian Journal of Fisheries and Aquatic Sciences,* **43**, 640–646.

Menezes, M. S. de & Caramaschi, E. P. (2000) Longitudinal distribution of *Hypostomus punctatus* (Osteichthyes, Loricariidae) in a coastal stream from Rio de Janeiro, Southeastern Brazil. *Brazilian Archives of Biology and Technology,* **43**, 229–233.

Meng, L. & Moyle, P. B. (1995) Status of splittail in the Sacramento-San Joaquin estuary. *Transactions of the American Fishery Society,* **124**, 538–549.

Mérona, B. de & Bittencourt, M. M. (1993) Les peuplements de poissons du 'Lago do Rei', un lac d'inondation d'Amazonie centrale: description générale. *Amazoniana,* **12**, 415–441.

Merrick, J. R. & Schmida, G. E. (1984) *Australian Freshwater Fishes: Biology and Management.* Merrick, North Ryde.

Merriman, D. (1941) Studies on the striped bass (*Roccus saxatilis*) of the Atlantic Coast. *U. S. Fish and Wildlife Service Fishery Bulletin,* **50**, 1–77.

Merron, G. S. (1993) Pack-hunting in two species of catfish, *Clarias gariepinus* and *C. ngamensis*, in the Okavango delta, Botswana. *Journal of Fish Biology,* **43**, 575–584.

Merron, G. S., Holden, K. K. & Bruton, M. N. (1990) The reproductive biology and early development of the African pike, *Hepsetus odoe*, in the Okavango delta, Botswana. *Environmental Biology of Fishes,* **28**, 215–235.

Mesiar, D. C., Eggers, D. M. & Gaudet, D. M. (1990) Development of techniques for the application of hydroacoustics to counting migratory fish in large rivers. In: *Developments in Fisheries Acoustics* (ed. W. A. Karp). *ICES Rapports et Procès-verbaux des Réunions,* **189**, 223–232.

Mesing, C. L. & Wicker, A. M. (1986) Home range, spawning migration and homing of radio-tagged Florida largemouth bass in two central Florida lakes. *Transactions of the American Fisheries Society,* **115**, 286–295.

Messieh, S. N. (1977) Population structures and biology of alewives (*Alosa pseudoharengus*) and blueback herring (*Alosa aestivalis*) in the St. John River, New Brunswick. *Environmental Biology of Fishes,* **2**, 195–210.

Metcalfe, J. D. & Arnold, G. P. (1997) Tracking fish with electronic tags. *Nature* (London), **387**, 665–666.

Metcalfe, J. D., Fulcher, M. F. & Storeton-West, T. J. (1992) Progress and developments in telemetry for monitoring the migratory behaviour of plaice in the North Sea. In: *Wildlife Telemetry – Remote Monitoring and Tracking of Animals* (eds I. G. Priede & S. M. Swift), pp. 359–366. Ellis Horwood, Chichester, UK.

Metcalfe, N. B., Huntingford, F. A. & Thorpe, J. E. (1986) Seasonal changes in feeding motivation of juvenile Atlantic salmon (*Salmo salar*). *Canadian Journal of Zoology,* **64**, 2439–2446.

Meyer, S. R. (1974) Die gebruik van vislere in die bestudering van migrasiegewoontes van vis in Transvaalse riviersisteme. M.Sc. thesis, Rand Afrikaans University, Johannesburg.

Meyers, L. S., Theumler, T. F. & Kornley, G. W. (1992) Seasonal movements of brown trout in northeast Wisconsin. *North American Journal of Fisheries Management,* **12**, 433–441.

Miah, M. S., Rahman, M. A. & Haldar, G. C. (1999) Analytical approach to the spawning ground of hilsa *Tenualosa ilisha* (Ham.) in Bangladesh water. *Indian Journal of Animal Sciences,* **69**, 141–144.

Michaud, D. T. (2000) Wisconsin Electric's experience with fish impingement and entrainment studies. *Environmental Science & Policy,* **3** (Suppl.), 333–340.

Michaud, D. T. & Taft, E. P. (2000) Recent evaluations of physical and behavioral barriers for reducing fish entrainment at hydroelectric plants in the upper Midwest. *Environmental Science & Policy,* **3** (Suppl.), 499–512.

Micklin, P. P. (1988) Desiccation of the Aral Sea: a water management disaster in the Soviet Union. *Science,* **241**, 1170–1176.

Miller, J. M., Crowder, L. B. & Moser, M. L. (1985) Migration and utilization of estuarine nurseries by juvenile fishes: an evolutionary perspective. *Contributions in Marine Science,* **27**, 338–352.

Miller, M. L. & Menzel, B. W. (1986) Movements, homing, and home range of muskellunge, *Esox masquinongy*, in West Okoboji Lake, Iowa. *Environmental Biology of Fishes,* **16**, 243–255.

Miller, R. B. (1948) A note on the movement of the pike, *Esox lucius. Copeia,* **1948**, 62.

Miller, R. R. (1960) Systematics and biology of the gizzard shad (*Dorosoma cepedianum*) and related fishes. *Fisheries Bulletin of the United States Fish and Wildlife Service,* **60**, 371–392.

Miller, R. R. (1972) Threatened freshwater fishes of the United States. *Transactions of the American Fisheries Society,* **101**, 239–252.

Mills, C. A. (1991) Reproduction and life history. In: *Cyprinid Fishes – Systematics, Biology and Exploitation* (eds I. J. Winfield & J. S. Nelson), pp. 483–508. Chapman & Hall, London.

Mills, C. A. & Mann, R. H. K. (1983) The bullhead *Cottus gobio*, a versatile and successful fish. *Freshwater Biological Association Annual Report,* **51**, 76–88.

Mills, D. (1989) *Ecology and Management of Atlantic Salmon.* Chapman & Hall, London.

Mills, D. (ed.) (2000) *The Ocean Life of Atlantic Salmon: Environmental and Biological Factors Influencing Survival.* Fishing News Books, Blackwell Science Ltd, Oxford.

Minckley, W. L. & Deacon, J. E. (eds) (1991) *Battle Against Extinction: Native Fish Management in the American West.* University of Arizona Press, Tucson.

Minor, J. D. & Crossman, E. J. (1978) Home range and seasonal movements of muskellunge as determined by radiotelemetry. *American Fisheries Society Special Publication,* **11**, 146–153.

Mitro, M. G. & Parrish, D. L. (1997) Temporal and spatial abundances of larval walleyes in two tributaries of Lake Champlain. *Transactions of the American Fisheries Society,* **126**, 273–287.

Modde, T., Burnham, K. P. & Wick, E. F. (1996) Population status of the endangered razorback sucker in the middle Green River. *Conservation Biology,* **10**, 110–119.

Modde, T. & Irving, D. B. (1998) Use of multiple spawning sites and seasonal movement by razorback suckers in the Middle Green River, Utah. *North American Journal of Fisheries Management,* **18**, 318–326.

Molls, F. (1999) New insights into the migration and habitat use by bream and white bream in the floodplain of the River Rhine. *Journal of Fish Biology,* **55**, 1187–1200.

Monk, B. Weaver, R. Thompson, C. & Ossiander, F. (1989) Effects of flow and weir design on the passage behavior of American shad and salmonids in an experimental fish ladder. *North American Journal of Fisheries Management,* **9**, 60–67.

Montgomery, W. L., McCormick, S. D., Naiman, R. J., Whoriskey, F. G. Jr. & Black, G. A. (1983) Spring migratory synchrony of salmonid, catostomid and cyprinid fishes in Rivière à la Truite, Quebec. *Canadian Journal of Zoology,* **61**, 2495–2502.

Moore, A. (1996) Tracking Atlantic salmon (*Salmo salar* L.) in estuaries and coastal waters. In: *Underwater Biotelemetry* (eds E. Baras & J.C. Philippart), pp. 93–100. University of Liège, Belgium.

Moore, A., Lacroix, G. L. & Sturlaugsson, J. (2000) Tracking Atlantic salmon post-smolts in the sea.

In: *The Ocean Life of Atlantic Salmon* (ed. D. Mills), pp. 49–64. Fishing News Books, Blackwell Science Ltd, Oxford.

Moore, A., Potter, E. C. E., Milner, N. J. & Bamber, S. (1995) The migratory behaviour of wild Atlantic salmon (*Salmo salar*) smolts in the estuary of the River Conwy, North Wales. *Canadian Journal of Fisheries and Aquatic Sciences,* **52**, 1923–1935.

Moore, A., Russell, I. C. & Potter, E. C. E. (1990) The effects of intraperitoneally implanted dummy acoustic transmitters on the behaviour and physiology of juvenile Atlantic salmon, *Salmo salar* L. *Journal of Fish Biology,* **37**, 713–721.

Moore, J. W. (1975a) Reproductive biology of anadromous arctic charr, *Salvelinus alpinus* (L.) in the Cumberland Sound area of Baffin Island. *Journal of Fish Biology,* **7**, 143–151.

Moore, J. W. (1975b) Distribution, movements and mortality of anadromous arctic charr, *Salvelinus alpinus* L., in the Cumberland Sound area of Baffin Island. *Journal of Fish Biology,* **7**, 339–348.

Moore, R. (1979) Natural sex inversion in giant perch, *Lates calcarifer*. *Australian Journal of Marine and Freshwater Research,* **30**, 803–813.

Moore, R. (1982) Spawning and early life history of barramundi, *Lates calcarifer* (Bloch), in Papua New Guinea. *Australian Journal of Marine and Freshwater Research,* **33**, 647–661.

Moore, R. & Reynolds, L. F. (1982) Migration patterns of barramundi, *Lates calcarifer* (Bloch), in Papua New Guinea. *Australian Journal of Marine and Freshwater Research,* **33**, 671–682.

Moreau, Y. (1988) Physiologie de la respiration. In: *Biology and Ecology of African Freshwater Fishes* (eds C. Lévêque, M. N. Bruton & G. W. Ssentongo), pp. 113–135. Editions de l'ORSTOM, Travaux et Documents n° 216, Paris.

Moriarty, C. (1986) Riverine migration of young eels *Anguilla anguilla* (L.). *Fisheries Research,* **4**, 43–58.

Moriarty, C. (1990) European catches of elver of 1928–1988. *International Revue Gesamt Hydrobiologie,* **75**, 701–706.

Morin, R., Dodson, J. J. & Power, G. (1980) Estuarine fish communities of the eastern James-Hudson Bay coast. *Environmental Biology of Fishes,* **5**, 135–141.

Morita, K., Yamamoto, S. & Hoshino, N. (2000) Extreme life history change of white-spotted char (*Salvelinus leucomaenis*) after damming. *Canadian Journal of Fisheries and Aquatic Sciences,* **57**, 1300–1306.

Morman, R. H. D., Cuddy, D. W. & Rugen, P. C. (1980) Factors influencing the distribution of sea lamprey (*Petromyzon marinus*) in the Great Lakes. *Canadian Journal of Fisheries and Aquatic Sciences,* **37**, 1811–1826.

Mortensen, D. G. (1990) Use of staple sutures to close surgical incisions for transmitter implants. *American Fisheries Society Symposium,* **7**, 380–383.

Moser, M. L., Darazsdi, A. M. & Hall, R.J. (2000) Improving passage efficiency of adult American shad at low-elevation dams with navigation locks. *North American Journal of Fisheries Management,* **20**, 376–385.

Moser, M. L., Olson, A. F. & Quinn, T. P. (1991) Riverine and estuarine migratory behavior of coho salmon (*Oncorhynchus kisutch*) smolts. *Canadian Journal of Fisheries and Aquatic Sciences,* **48**, 1670–1678.

Moser, M. L. & Ross, S. W. (1994) Effects of changing current regime and river discharge on the estuarine phase of anadromous fish migration. In: *Changes in Fluxes in Estuaries* (ECSA22/ERF symposium, Plymouth, September 1992) (eds K. R. Dyer & R. J. Orth), pp. 343–347. Olsen & Olsen, Fredensborg, Denmark.

Moser, M. L. & Ross, S. W. (1995) Habitat use and movements of shortnose and Atlantic sturgeons in the lower Cape Fear River, North Carolina. *Transactions of the American Fisheries Society,* **124**, 225–234.

Moyle, P. B. & Cech Jr., J. J. (2000) *Fishes: An Introduction to Ichthyology*, 4th edn. Prentice-Hall, Inc, Upper Saddle River, New Jersey.

Mukai, T. & Oota, Y. (1995) Histological changes in the pituitary, thyroid gland and gonads of the four-spine sculpin (*Cottus kazika*) during downstream migration. *Zoological Science,* **12**, 91–97.

Mulford, C. J. (1984) Use of a surgical skin stapler to quickly close incisions in striped bass. *North American Journal of Fisheries Management,* **4**, 571–573.

Müller, K. (1986) Seasonal anadromous migration of the pike (*Esox lucius*, L.) in coastal areas of the northern Bothnian Sea. *Archiv für Hydrobiologie,* **107**, 315–330.

Müller, K. & Berg, E. (1982) Spring migration of some anadromous freshwater fish species in the northern Bothnian Sea. *Hydrobiologia,* **96**, 161–168.

Muncy, R. J., Parker, N. C. & Poston, H. A. (1990) Inorganic chemical marks induced in fish. *American Fisheries Society Symposium,* **7**, 541–546.

Myers, G. S. (1938) Fresh-water fishes and West Indian zoogeography. *Annual Report of the Smithsonian Institute,* **1937**, 339–364.

Myers, G. S. (1949) Usage of anadromous, catadromous and allied terms for migratory fishes. *Copeia,* **1949**, 89–97.

Myers, G. S. (1972) *The Piranha Book*. TFP Publications Ltd, New Jersey, 128 pp.

Naismith, I. A. & Knights, B. (1993) The distribution, density and growth of European eels *Anguilla anguilla* L., in the River Thames catchment. *Journal of Fish Biology,* **42**, 217–226.

Nakai, I., Iwata, R. & Tsukamoto, K. (1999) Ecological study of the migration of eel by synchrotron radiation induced X-ray fluorescence imaging of otoliths. *Spectrochemica Acta Part B Atomic Spectroscopy,* **54**, 167–170.

Narver, D. W. (1970) Diel vertical movements and feeding of underyearling sockeye salmon and the limnetic zooplankton in Babine Lake, British Columbia. *Journal of the Fisheries Research Board of Canada,* **27**, 201–270.

Näslund, I. (1990) The development of regular seasonal habitat shifts in a landlocked population of Arctic charr, *Salvelinus alpinus* L., population. *Journal of Fish Biology,* **36**, 401–414.

Näslund, I. (1992) Upstream migratory behaviour in landlocked Arctic charr. *Environmental Biology of Fishes,* **33**, 265–274.

Näslund, I. (1993) Migratory behaviour of brown trout, *Salmo trutta* L.: importance of genetic and environmental influences. *Ecology of Freshwater Fishes,* **2**, 51–57.

Naud, M. & Magnan, P. (1988) Diel onshore–offshore migrations in northern redbelly dace, *Phoxinus eos* (Cope) in relation to prey distribution in a small oligotrophic lake. *Canadian Journal of Zoology,* **66**, 1249–1253.

Neave, F. (1955) Notes on the seaward migration of pink and chum salmon fry. *Journal of the Fisheries Research Board of Canada,* **12**, 369–374.

Neill, G. G. (1979) Mechanisms of fish distribution in heterothermal environments. *American Zoologist,* **19**, 305–317.

Neill, W. E. (1992) Population variation in the ontogeny of predator-induced vertical migration of copepods. *Nature,* **356**, 54–57.

Nelson, J. S. (1994) *Fishes of the World*, 3rd edn. John Wiley & Sons, New York.

Nemeth, R. S. & Anderson, J. J. (1992) Response of juvenile coho and chinook salmon to strobe and mercury lights. *North American Journal of Fisheries Management,* **12**, 684–692.

Nemetz, T. G. & MacMillan, J. R. (1988) Wound healing of incisions closed with a cyanoacrylate adhesive. *Transactions of the American Fisheries Society,* **117**, 190–195.

Nestler, J. M., Ploskey, G. R., Pickens, J., Menezes, J. & Schilt, C. (1992) Responses of blueback herring to high-frequency sound and implications for reducing entrainment at hydropower dams. *North American Journal of Fisheries Management,* **12**, 667–683.

Neverman, D. & Wurtsbaugh, W. A. (1994) The thermoregulatory function of diel vertical migration for a juvenile fish, *Cottus extensus. Oecologia,* **98**, 247–256.

Neves, A. M. B. (1995) Conhecimento atual sobre o pirarucu, *Arapaima gigas* (Cuvier, 1817). *Boletin del Museo Emilio Goeldi, séria Zoologica,* **11**, 33–56.

Newcomb, B. A. (1989) Winter abundance of channel catfish in the channelized Missouri River, Nebraska. *North American Journal of Fisheries Management,* **9**, 195–202.

Niaré, T. & Bénech, V. (1998) Between-year variation of lateral fish migrations in the Inner Delta of the River Niger (Mali, West Africa). *Italian Journal of Zoology,* **65** (Suppl.), 313–319.

Nichols, P. R. & Miller, R. V. (1967) Seasonal movements of striped bass, *Roccus saxatilis* (Walbaum) tagged and released in the Potomac River, Maryland, 1959–61. *Chesapeake Science,* **8**, 102–124.

Nicolas, Y., Pont, D. & Lambrechts, A. (1994) Using γ-emitting artificial radionucleides, released by nuclear plants, as markers of restricted movements by chub, *Leuciscus cephalus,* in a large river, the Lower Rhône. *Environmental Biology of Fishes,* **39**, 399–409.

Nielsen, L. A. (1992) *Methods for Marking Fish and Shellfish.* American Fisheries Society, Special Publication, **23**, Bethesda, Maryland.

Nikolsky, G. V. (1961) *Special Ichthyology.* Israel Programme for Scientific Translations, Jerusalem, 538 pp.

Nikolsky, G. V. (1963) *The Ecology of Fishes.* Academic Press, London.

Nilssen, J. P. (1984) Tropical lakes. Functional ecology and future development: the need for a process oriented approach. *Hydrobiologia,* **113**, 231–242.

Nordeng, H. (1971) Is the local orientation of anadromous fishes determined by pheromones? *Nature* (London), **233**, 411–413.

Nordeng, H. (1977) A pheromone hypothesis for homeward migration in anadromous salmonids. *Oikos,* **28**, 155–159.

Nordeng, H. (1983) Solution to the "charr problem" based on Arctic charr (*Salvelinus alpinus*) in Norway. *Canadian Journal of Fisheries and Aquatic Sciences,* **40**, 1372–1387.

Nordlie, F. G. (1981) Feeding and reproductive biology of eleotrid fishes in a tropical estuary. *Journal of Fish Biology,* **18**, 97–110.

Northcote, T. G. (1962) Migratory behavior of juvenile rainbow trout, *Salmo gairdneri,* in outlet and inlet streams of Loon Lake, B. C. *Journal of the Fisheries Research Board of Canada,* **19**, 201–210.

Northcote, T. G. (1978) Migratory strategies and production in freshwater fishes. In: *Ecology of Freshwater Production* (ed. S. D. Gerking), pp. 326–359. Blackwell, Oxford.

Northcote, T. G. (1984) Mechanisms of fish migration in rivers. In: *Mechanisms of Migration in Fishes* (eds J. D. McCleave, J. J. Dodson & W. H. Neill), pp. 317–355. Plenum, New York.

Northcote, T. G. (1992) Migration and residency in stream salmonids: some ecological considerations and evolutionary consequences. *Nordic Journal of Freshwater Research,* **67**, 5–17.

Northcote, T. G. (1995) Comparative biology and management of Arctic and European grayling (Salmonidae, *Thymallus*). *Reviews in Fish Biology and Fisheries,* **5**, 141–194.

Northcote, T. G. (1997) Potamodromy in Salmonidae: living and moving in the fast lane. *North American Journal of Fisheries Management,* **17**, 1029–1045.

Northcote, T. G. (1998) Migratory behaviour of fish and its significance to movement through riverine fish passage facilities. In: *Fish Migration and Fish Bypasses* (eds M. Jungwirth, S. Schmutz & S. Weiss), pp. 3–18. Fishing News Books, Blackwell Science Ltd, Oxford.

Novikov, G. G., Politov, D. V., Makhrov, A. A., Malinina, T. V., Afanasiev, K. I. & Fernholm, B. (2000) Freshwater and estuarine fishes of the Russian Arctic coast (the Swedish-Russian Expedition 'Tundra Ecology – 94'). *Journal of Fish Biology,* **57** (Suppl. A), 158–162.

Novoa, D. (ed.) (1982) *Los Recursos Pesquerios del Río Orinoco y su Explotacíon.* Editorial Arte, Caracas, 386 pp.

Novoa, D. F. (1989) The multispecies fisheries of the Orinoco River. *Canadian Special Publications in Fisheries and Aquatic Sciences,* **106**, 422–428.

O'Connor, J. F. & Power, G. (1973) Homing of brook trout (*Salvelinus fontinalis*) in Matamek Lake, Quebec. *Journal of the Fisheries Research Board of Canada,* **30**, 1012–1014.

Odeh, M. (ed.) (1999) *Innovations in Fish Passage Technology.* American Fisheries Society, Bethesda, Maryland.

Odeh, M. & Orvis, C. (1998) Downstream fish passage design considerations and developments at hydroelectric projects in the North-east USA. In: *Fish Migration and Fish Bypasses* (eds M. Jungwirth, S. Schmutz & S. Weiss), pp. 267–280. Fishing News Books, Blackwell Science Ltd, Oxford.

Odinetz Collart, O. & Moreira, L. C. (1989) Quelques caractéristiques physico-chimiques d'un lac de várzea en Amazonie Centrale (Lago do Rei, île de Careiro). *Revue d'Hydrobiologie Tropicale,* **22**, 191–200.

Økland, F., Heggberget, T. G., Lamberg, A., Fleming, I. A. & McKinley, R. S. (1996) Identification of spawning behaviour in Atlantic salmon (*Salmo salar* L.) by radio telemetry. In: *Underwater Biotelemetry* (eds E. Baras & J. C. Philippart), pp. 35–46. University of Liège, Belgium.

Økland, F., Næsje, T. F. & Thorstad, E. B. (2000) Movements and habitat utilisation of cichlids in the Zambezi River, Namibia. A radio telemetry study in 1999–2000. *NINA-NIKU* (Trondheim) Project Report, **011**, 1–25.

Oliveira, K. (1999) Life history characteristics and strategies of the American eel, *Anguilla rostrata. Canadian Journal of Fisheries and Aquatic Sciences,* **56**, 795–802.

Olson, D. E., Schupp, D. H. & Macins, W. (1978) An hypothesis of homing behavior of walleyes as related to observed patterns of passive and active movement. In: *Selected Coolwater Fishes of North America* (ed. R. L. Kendall). American Fisheries Society, Special Publication, **11**, pp. 52–57. Bethesda, Maryland.

Omarov, O. P. & Popova, O. A. (1985) Feeding behaviour of pike, *Esox lucius*, and catfish *Silurus glanis*, in the Arakum reservoirs of Dagestan. *Journal of Ichthyology,* **25**, 25–36.

Ombredane, D., Baglinière, J. L. & Marchand, F. (1998) The effects of passive integrated transponder tags on survival and growth of juvenile brown trout (*Salmo trutta* L.) and their use for studying movements in a small river. *Hydrobiologia,* **371/372**, 99–106.

Orlova, E. L. & Popova, O. A. (1987) Age related changes in feeding of catfish *Silurus glanis*, and pike, *Esox lucius*, in the outer delta of the Volga. *Journal of Ichthyology,* **27**, 140–148.

Oshima, K. & Gorbman, A. (1966a) Olfactory responses in the forebrain of goldfish and their modification by thyroxine treatment. *General and Comparative Endocrinology,* **7**, 398–409.

Oshima, K. & Gorbman, A. (1966b) Influence of thyroxine and steroid hormones on spontaneous and evoked unitary activity in the olfactory bulb of goldfish. *General and Comparative Endocrinology,* **7**, 482–491.

Otis, K. J. & Weber, J. J. (1982) Movements of carp in the Lake Winnebago system as determined by radiotelemetry. *Wisconsin Department of Natural Resources, Technical Bulletin,* **134**, 16 pp.

Ovidio, M. (1999) Annual activity cycle of adult brown trout (*Salmo trutta* L.): a radio-telemetry study in a small stream of the Belgian Ardenne. *Bulletin Français de la Pêche et de la Pisciculture,* **352**, 1–18 [in French].

Ovidio, M., Baras, E., Goffaux, D., Birtles, C. & Philippart, J. C. (1998) Environmental unpredictability rules the autumn migration of brown trout (*Salmo trutta* L.) in the Belgian Ardennes. *Hydrobiologia,* **371/372**, 263–274.

Paragamian, V. L., Powell, M. S. & Faler, J. C. (1999) Mitochondrial DNA analysis of burbot stocks in the Kootenai River Basin of British Columbia, Montana, and Idaho. *Transactions of the American Fisheries Society,* **128**, 868–874.

Parasiewicz, P., Eberstaller, J., Weiss, S. & Schmutz, S. (1998) Conceptual guidelines for nature-like bypass channels. In: *Fish Migration and Fish Bypasses* (eds M. Jungwirth, S. Schmutz & S. Weiss), pp. 348–362. Fishing News Books, Blackwell Science Ltd, Oxford.

Park, L. K. & Moran, P. (1994) Developments in molecular genetic techniques in fisheries. *Reviews in Fish Biology and Fisheries,* **4**, 272–299.

Parker, B. R. & Franzin, W. G. (1991) Reproductive biology of the quillback, *Carpiodes cyprinus*, in a small prairie river. *Canadian Journal of Zoology,* **69**, 2133–2139.

Parker, N. C., Giorgi, A. E., Heidinger, R. C., Jester, J. B. Jr., Prince, E. D & Winans, G. A. (eds) (1990) Fish-marking Techniques. *American Fisheries Society Symposium,* **7**, 879 pp.

Parker, S. J. (1995) Homing ability and home range of yellow-phase American eels in a tidally dominated estuary. *Journal of the Marine Biological Association of the United Kingdom,* **75**, 127–140.

Parkinson, D., Philippart, J. C. & Baras, E. (1999) A preliminary investigation of spawning migrations and homing behaviour of grayling *Thymallus thymallus* in a small stream, as determined by radio tracking. *Journal of Fish Biology,* **55**, 172–182.

Paterson, H. E. H. (1978) More evidence against speciation by reinforcement. *South African Journal of Science,* **74**, 369–371.

Patrick, P. H., Christie, A. E., Sager, D., Hocutt, C. & Stauffer, J. Jr. (1985) Responses of fish to a strobe light/air-bubble barrier. *Fisheries Research,* **3**, 157–172.

Pattée, E. (1988) Fish and their environment in large European river ecosystems. The Rhône. *Sciences de l'Eau,* **7**, 35–74.

Paugy, D. (1980) Ecologie et biologie des *Alestes imberi* (Pisce, Characidae) des rivières de Côte d'Ivoire. *Cahiers ORSTOM, série Hydrobiologie,* **13**, 129–141.

Paugy, D. (1982) Synonymie d'*Alestes chaperi* Sauvage, 1882 avec *Alestes longipinnis* (Günther, 1864) (Pisces, Characidae). *Cybium,* **3**, 75–90.

Paugy, D. & Lévêque, C. (1999) La reproduction. In: *Les Poissons des Eaux Continentales Africaines – Diversité, Ecologie, Utilisation par l'Homme* (ed. C. Lévêque & D. Paugy), pp. 129–151. IRD Editions, Paris.

Paukert, C. P. & Fisher, W. L. (2000) Abiotic factors affecting summer distribution and movement of male paddlefish, *Polyodon spathula*, in a prairie reservoir. *Southwestern Naturalist,* **45**, 133–140.

Pauly, D. (1981) The relationship between gill surface area and growth performance in fish: a generalization of von Bertalanffy's theory of growth. *Meeresforschung,* **28**, 251–282.

Pavlov, D. S. (1994) The downstream migration of young fishes in rivers: mechanisms and distribution. *Folia Zoologica Brno,* **43**, 193–208.

Pavlov, D. S., Nezdoliy, V. K., Khodorevskaya, R. P, Ostrovskiy, M. P. & Popova, I. K. (1981) Catadromous migration of young fishes in the Volga and Il' rivers. Nauka Press, Moscow, 318 pp. [in Russian].

Pavlov, D. S., Nezdoliy, V. K., Urteaga, A. K. & Sanches, O. R. (1995) Downstream migration of juvenile fishes in the rivers of Amazonian Peru. *Journal of Ichthyology,* **35**, 227–248.

Peake, S. R., Mckinley, R. S., Beddow, T. & Marmulla, G. (1997) New procedure for radio transmitter attachment: oviduct insertion. *North American Journal of Fisheries Management,* **17**, 757–762.

Pedersen, B. H. & Andersen, N. G. (1985) A surgical method for implanting transmitters with sensors into the body cavity of the cod (*Gadus morhua* L.). *Dana,* **5**, 55–62.

Pegg, M. A., Bettoli, P. W. & Layzer, J. B. (1997) Movement of saugers in the lower Tennessee River determined by radio telemetry, and implications for management. *North American Journal of Fisheries Management,* **17**, 763–768.

Pellett, T. D., Van Dyck, G. J. & Adams, J. V. (1998) Seasonal migration and homing of channel catfish in the lower Wisconsin River, Wisconsin. *North American Journal of Fisheries Management,* **18**, 85–95.

Pelz, G. R. (1985) Fischbewegungen über verschiedenartige Fischpasse am Beispiel der Mosel. *Courrier Forschung Institut Senckenberg,* **76**, 1–190.

Pelz, G. R. & Kästle, A. (1989) Ortsbewegungen der barbe *Barbus barbus* (L.): radiotelemetrische standortbestimmungen in der Nidda (Frankfurt/Main). *Fischökologie,* **1**, 15–28.

Pemberton, R. (1976). Sea trout in North Argyll Sea lochs: 2. Diet. *Journal of Fish Biology,* **9**, 195–208.

Penáz, M. (1996) *Chondrostoma nasus*: its reproductive strategy and possible reasons for a widely observed population decline. Review. In: *Conservation of Endangered Freshwater Fish in Europe* (eds A. Kirchhofer & D. Hefti), pp. 279–285. Birkhäuser Verlag, Basel.

Penáz, M., Kubícek, F., Marvan, P. & Zelinka, M. (1968) Influence of the Vír River Valley Reservoir on the hydrobiological and ichthyological conditions in the Svratka River. *Acta Scientiarum Naturalium Academiae Scientiarum Bohemoslovacae Brno,* **2**, 1–60.

Penáz, M., Roux, A. L., Jurajda, P. & Olivier, J. M. (1992) Drift of larval and juvenile fishes in a bypassed floodplain of the upper River Rhône, France. *Folia Zoologica,* **41**, 281–288.

Penáz, M. & Stouracova, I. (1991) Effect of hydroelectric development on population dynamics of *Barbus barbus* in the River Jihlava. *Folia Zoologica,* **40**, 75–84.

Pender, P. J. & Griffin, R. K. (1996) Habitat history of barramundi *Lates calcarifer* in a North Australian river system based on barium and strontium levels in scales. *Transactions of the American Fisheries Society,* **125**, 679–689.

Perrow, M. R., Jowitt, A. J. D. & Johnson, S. R. (1996) Factors affecting the habitat selection of tench in a shallow eutrophic lake. *Journal of Fish Biology,* **48**, 859–870.

Persat, H. (1982) Photographic identification of individual grayling, *Thymallus thymallus*, based on the disposition of black dots and scales. *Freshwater Biology,* **12**, 97–101.

Persson, L. (1991) Interspecific interactions. In: *Cyprinid Fishes – Systematics, Biology and Exploitation* (eds I. J. Winfield & J. S. Nelson), pp. 530–551. Chapman & Hall, London.

Pervozvanskiy, V. Y., Bugaev, V. F., Shustov, Y. A. & Shchurov, I. L. (1989) Some ecological characteristics of pike *Esox lucius* of the Keret, a salmon river in the White Sea Basin. *Journal of Ichthyology,* **3**, 410–414.

Peter, A. (1998) Interruption of the river continuum by barriers and the consequences for migratory fish. In: *Fish Migration and Fish Bypasses* (eds M. Jungwirth, S. Schmutz & S. Weiss), pp. 99–112. Fishing News Books, Blackwell Science Ltd, Oxford.

Petering, R. W. & Johnson, D. L. (1991) Suitability of cyanoacrylate adhesive to close incisions in black crappies used in telemetry studies. *Transactions of the American Fisheries Society,* **120**, 535–537.

Peterson, D. C. (1975) Ultrasonic tracking of three species of black basses, *Micropterus* sp. in Centerhill Reservoir, Tennessee. M.Sc. thesis, Tennessee Technological University, Cookeville, Tennessee, 129 pp.

Peterson, M. S., Nicholson, L. C., Fulling, G. L. & Snyder, D. J. (2000) Catch-per-unit-effort, environmental conditions and spawning migration of *Cycleptus meridionalis* Burr and Mayden in two coastal rivers of the northern Gulf of Mexico. *American Midland Naturalist,* **143**, 414–421.

Peterson, R. H., Johnansen, P. H. & Metcalfe, J. C. (1980) Observations on the early life stages of Atlantic tomcod, *Microgadus tomcod*. *Fisheries Bulletin of the United States Fish and Wildlife Service,* **78**, 147–158.

Peterson, T. L. (1994) Seasonal migration in the southern hogchoker, *Trinectes maculatus fasciatus* (Achiridae). *Gulf Research Reports,* **9**, 169–176.

Peterson-Curtis, T. L. (1997) Effects of salinity on survival, growth, metabolism, and behavior in juvenile hogchokers, *Trinectes maculatus fasciatus* (Achiridae). *Environmental Biology of Fishes,* **49**, 323–331.

Petr, T. (1986) The Volta River system. In: *The Ecology of River Systems* (eds B.R. Davies & K.F. Walker), pp. 163–183. Dr. W. Junk Publishers, Dordrecht.

Petrosky, B. R. & Magnuson, J. J. (1973) Behavioral responses of northern pike, yellow perch and bluegill to oxygen concentrations under simulated winterkill conditions. *Copeia,* **1973**, 124–133.

Pettit, S. W. & Wallace, R. L. (1975) Age, growth and movements of mountain whitefish *Prosopium williamsoni* (Girard), in the North Fork Clearwater River, Idaho. *Transactions of the American Fisheries Society,* **104**, 68–76.

Petts, G. E. (1984) *Impounded Rivers*. John Wiley & Sons, New York.

Philippart, J. C. & Baras, E. (1996) Comparison of tagging and tracking studies to estimate mobility patterns and home range in *Barbus barbus*. In: *Underwater Biotelemetry* (eds E. Baras & J. C. Philippart), pp. 3–12. University of Liège, Belgium.

Philippart, J. C., Gillet, A. & Micha, J. C. (1988) Fish and their environment in large European river ecosystems. The River Meuse. *Sciences de l'Eau,* **7**, 115–154.

Philippart, J. C., Micha, J. C., Baras, E., Prignon, C., Gillet, A. & Joris, S. (1994) The Belgian project Meuse Saumon 2000. First results, problems and future prospects. *Water Science and Technology,* **29**, 315–317.

Philippart, J. C. & Ruwet, J. C. (1982) Ecology and distribution of tilapias. In: *The Biology and Culture of Tilapias* (eds R. S. V. Pullin & R. H. Lowe-McConnell), *ICLARM Conference Proceedings,* Vol. **7**, pp. 15–59. Manila, Philippines.

Philippart, J. C. & Vranken, M. (1983) Atlas des Poissons de Wallonie – Distribution, écologie, éthologie, pêche, conservation. *Cahiers d'Ethologie appliquée,* **3** (Suppl. 1–2), 395 pp.

Phiri, H. & Shirakihara, K. (1999) Distribution and seasonal movement of pelagic fish in southern Lake

Tanganyika. *Fisheries Research,* **41**, 63–71.

Pickett, G. D. & Pawson, M. G. (1994) *Sea Bass: Biology, Exploitation and Conservation.* Chapman & Hall, London.

Pillay, S. R. & Rosa, H. (1963) Synopsis of biological data on hilsa, *Hilsa ilisha* (Hamilton) 1822. *FAO Fisheries Synopsis,* **25**, 1:1–6:8.

Pires, A. M., Cowx, I. G. & Coelho, M. M. (1999) Seasonal changes in fish community structure of intermittent streams in the middle reaches of the Guadiana basin, Portugal. *Journal of Fish Biology,* **54**, 235–249.

Pirhonen, J., Forsman, L., Soivio, L. & Thorpe, J. (1998) Movements of hatchery reared *Salmo trutta* during the smolting period, under experimental conditions, *Aquaculture,* **168**, 27–40.

Pitcher, T. J. (1971) *Population dynamics and schooling behaviour in the minnow Phoxinus phoxinus* (L.). DPhil thesis, University of Oxford.

Pitcher, T. J. (1986) Function of shoaling behaviour in teleosts. In: *The Behaviour of Teleost Fishes* (ed. T. J. Pitcher), pp. 294–337. Croom Helm, London.

Pitcher, T. J., Green, D. & Magurran, A. E. (1986) Dicing with death: predator inspection behaviour in minnow shoals. *Journal of Fish Biology,* **28**, 439–448.

Politov, D. V., Gordon, N. Y., Afanasiev, K. I., Altukhov, Y. P. & Bickham, J. W. (2000) Identification of palearctic coregonid fish species using mtDNA and allozyme genetic markers. *Journal of Fish Biology,* **57** (Suppl. A), 51–71.

Pollard, D. A. (1971) The biology of a landlocked form of the normally catadromous salmoniform fish *Galaxias maculatus* (Jenyns). I. Life cycle and origin. *Australian Journal of Marine and Freshwater Research,* **22**, 91–123.

Poncin, P. (1988) Le contrôle environnemental et hormonal de la reproduction du barbeau, *Barbus barbus* (L.), et du chevaine, *Leuciscus cephalus* (L.) (Pisces, Cyprinidae) en captivité. *Cahiers d'Ethologie Appliquée,* **8**, 173–330.

Popper, A. N. & Carlson, T. J. (1998) Application of sound and other stimuli to control fish behavior. *Transactions of the American Fisheries Society,* **127**, 673–707.

Porto, L. M., McLaughlin, R. L. & Noakes, D. L. G. (1999) Low-head barrier dams restrict the movements of fishes in two Lake Ontario streams. *North American Journal of Fisheries Management,* **19**, 1028–1036.

Post, J. R. & McQueen, D. J. (1988) Ontogenetic changes in the distribution of larval and juvenile yellow perch (*Perca flavescens*): A response to prey or predators? *Canadian Journal of Fisheries and Aquatic Sciences,* **45**, 1820–1826.

Potter, E. C. E. (1988) Movements of Atlantic salmon, *Salmo salar* L., in an estuary in South-west England. *Journal of Fish Biology,* **33** (Suppl. A), 153–159.

Povz, M. (1988) Migrations of the nase carps (*Chondrostoma nasus* L. 1758) in the River Sava. *Journal of Aquatic Production* (Ljubljana), **2**, 149–163.

Power, G. (1980) The brook char, *Salvelinus fontinalis*. In: *Charrs – Salmonid Fishes of the Genus Salvelinus* (ed. E. K. Balon), pp. 141–203. Dr. W. Junk Publishers, The Hague.

Power, M. E. (1983) Grazing responses of tropical freshwater fishes to different scales of variation in their food. *Environmental Biology of Fishes,* **9**, 103–115.

Power, M. E. (1984) Depth distributions of armoured catfish: predator-induced resource avoidance? *Ecology,* **65**, 523–528.

Powers, P. D. & Orsborn, J. F. (1984) Analysis of barriers to upstream fish migration. An investigation of the physical and biological conditions affecting fish passage success at culverts and waterfalls.

Final report, 1984, Part 4, 132 pp. NTIS Order No.: DE86008915/GAR.

Prentice, E. F., Flagg, T. A. & McCutcheon, C. S. (1990a) Feasibility of using implantable passive integrated transponder (PIT) tags in salmonids. *American Fisheries Society Symposium,* **7**, 317–322.

Prentice, E. F., Flagg, T. A. & McCutcheon, C. S. & Brastow, D. F. (1990b) PIT tag monitoring systems for hydroelectric dams and fish hatcheries. *American Fisheries Society Symposium,* **7**, 323–334.

Prentice, E. F., Flagg, T. A. & McCutcheon, C. S., Brastow, D. F. & Cross, D. C. (1990c) Equipment, methods and an automated data-entry station for PIT tagging. *American Fisheries Society Symposium,* **7**, 335–340.

Priede, I. G. (1982) An ultrasonic salinity telemetry transmitter for use on fish in estuaries. *Biotelemetry and Patient Monitoring,* **9**, 1–9.

Priede, I. G. (1992) Wildlife telemetry: an introduction. In: *Wildlife Telemetry – Remote Monitoring and Tracking of Animals* (eds I. G. Priede & S. M. Swift), pp. 3–28. Ellis Horwood, Chichester, UK.

Priede, I. G., De L. G. Solbe, J. F., Nott, J. E., O'Grady, K. T. & Cragg-Hine, D. (1988) Behaviour of adult Atlantic salmon, *Salmo salar* L., in the estuary of the River Ribble in relation to variations in dissolved oxygen and flow. *Journal of Fish Biology,* **33** (Suppl. A), 133–139.

Priede, I. G. & Swift, S. M. (eds) (1992) *Wildlife Telemetry: Remote Monitoring and Tracking of Animals.* Ellis Horwood, Chichester, UK.

Prignon, C., Micha, J. C. & Gillet, A. (1998) Biological and environmental characteristics of fish passage at the Tailfer dam on the Meuse River, Belgium. In: *Fish Migration and Fish Bypasses* (eds M. Jungwirth, S. Schmutz & S. Weiss), pp. 69–84. Fishing News Books, Blackwell Science Ltd, Oxford.

Prince, J. D. & Potter, I. C. (1983) Life cycle duration, growth and spawning times of five species of Atherinidae (Teleostei) found in a Western Australian estuary. *Australian Journal of Marine and Freshwater Research,* **34**, 287–301.

Pringle, C. M. & Hamazaki, T. (1997) Effects of fishes on algal response to storm in a tropical stream. *Ecology,* **78**, 2432–2442.

Puke, C. (1952) Pikeperch studies in Lake Vänern. *Reports of the Institute of Freshwater Research of Drottningholm,* **33**, 168–178.

Quinn, T. P. (1984) Homing and straying in Pacific salmon. In: *Mechanisms of Migration in Fishes* (eds J. D. McCleave, G. P. Arnold, J. J. Dodson & W. H. Neill), pp. 357–362. Plenum Press, New York.

Quinn, T. P. (1991) Models of Pacific salmon orientation and navigation on the open ocean. *Journal of Theoretical Biology,* **150**, 539–545.

Quinn, T. P. & Adams, D. J. (1996) Environmental changes affecting the migratory timing of American shad and sockeye salmon. *Ecology,* **77**, 1151–1162.

Quinn, T. P. & Brannon, E. L. (1982) The use of celestial and magnetic cues by orienting sockeye salmon smolts. *Journal of Comparative Physiology,* **147**, 547–552.

Quinn, T. P. & Light, J. T. (1989) Occurrence of threespine sticklebacks (*Gasterosteus aculeatus*) in the open North Pacific: migration or drift? *Canadian Journal of Zoology,* **67**, 2850–2852.

Quinn, T. P., Terhart, B. A. & Groot, C. (1989) Migratory orientation and vertical movements of homing adult sockeye salmon, *Oncorhynchus nerka*, in coastal waters. *Animal Behavior,* **37**, 587–599.

Quiros, R. & Vidal, J. C. (2000) Cyclic behaviour of potamodromous fish in large rivers. In: *Management and Ecology of River Fisheries* (ed. I. G. Cowx), pp. 71–86. Fishing New Books, Blackwell

Science Ltd, Oxford.

Raat, A. J. P. (1988) Synopsis of biological data on the northern pike *Esox lucius* Linnaeus, 1758. *FAO Fisheries Synopsis*, **30**.

Rahel, F. J. & Nutzman, J. W. (1994) Foraging in a lethal environment: fish predation in hypoxic waters of a stratified lake. *Ecology*, **75**, 1246–1253.

Raibley, P. T., Irons, K. S., O'Hara, T. M., Blodgett, K. B. & Sparks, R. E. (1997a) Winter habitats used by largemouth bass in the Illinois River, a large river-floodplain ecosystem. *North American Journal of Fisheries Management*, **17**, 401–412.

Raibley, P. T., O'Hara, T. M., Irons, K. S., Blodgett, K. D. & Sparks, R. E. (1997b) Largemouth bass size distributions under varying annual hydrological regimes in the Illinois River. *Transactions of the American Fisheries Society*, **126**, 850–856.

Rainboth, W. J. (1991) Cyprinids of South East Asia. In: *Cyprinid Fishes – Systematics, Biology and Exploitation* (eds I. J. Winfield & J. S. Nelson), pp. 156–210. Chapman & Hall, London.

Ransom, B. H., Johnston, S. V. & Steig, T. W. (1998) Review on monitoring adult salmonid (*Oncorhynchus* and *Salmo* spp.) escapement using fixed-location split-beam hydroacoustics. *Fisheries Research*, **35**, 33–42.

Rawson, D. S. (1957) The life history and ecology of the yellow walleye, *Stizostedion vitreum*, in Lac la Ronge, Saskatchewan. *Transactions of the American Fisheries Society*, **86**, 15–36.

Rayleigh, R. F. (1971) Innate contol of migration of salmon and trout fry from gravels to rearing areas. *Ecology*, **52**, 291–297.

Reckendorfer, W., Keckeis, H., Winkler, G. & Schiemer, F. (1996) *Chondrostoma nasus* as an indicator of river functioning: field and experimental studies at the University of Vienna. 2) Food availability for 0^+ fish in the Danube: spatial and seasonal changes in plankton, benthos and drift. In: *Abstracts of the Second International Symposium on the Biology of the Genus Chondrostoma Agassiz, 1835* (eds J. Freyhof & A. Bischoff), pp. 25. Museum Alexander Koenig, Bonn.

Reddin, D. G., O'Connell, M. F. & Dunkley, D. A. (1992) Assessment of an automated fish counter in a Canadian river. *Aquaculture and Fisheries Management*, **23**, 113–121.

Reebs, S. G. (2000) Can a minority of informed leaders determine the foraging movements of a fish shoal? *Animal Behaviour*, **59**, 403–409.

Reebs, S. G., Boudreau, L., Hardie, P. & Cunjak, R. A. (1995) Diel activity patterns of lake chubs and other fishes in a temperate stream. *Canadian Journal of Zoology*, **73**, 1221–1227.

Refseth, U. H., Nesbo, C. L., Stacy, J. E., Vøllestad, L. A., Fjeld, E. & Jakobsen, K. S. (1998) Genetic evidence for different migration routes of freshwater fish into Norway revealed by analysis of current (*Perca fluviatilis*) populations in Scandinavia. *Molecular Ecology*, **7**, 1015–1027.

Regan, C. T. (1911) *The Freshwater Fishes of the British Isles*. Methuen, London.

Régier, H. A., Welcomme, R. L., Steedman, R. J. & Henderson, H. F. (1989) Rehabilitation of degraded river ecosystems. In: *Proceedings of the International Large River Symposium* (ed. D. P. Dodge). *Canadian Special Publication in Fisheries and Aquatic Sciences*, **106**, 86–97.

Reid, S. (1983) La biologia de los bagres rayados *Pseudoplatystoma fasciatum* y *P. tigrinum* en la cuenca del Rio Apure, Venezuela. *Revista UNELLEZ de Ciencia y Tecnología (Barinas)*, **1**, 13–41.

Reist, J. D. & Bond, W. A. (1988) Life history characteristics of migratory coregonids of the Lower Mackenzie River, Northwest Territories, Canada. *Finnish Fisheries Research*, **9**, 133–144.

Reizer, C., Mattei, X. & Chevalier, J. L. (1972) Contribution à l'étude de la faune ichtyologique du bassin du fleuve Sénégal. II. Characidae. *Bulletin de l'Institut français d'Afrique Noire*, **34 A**,

655–691.

Reynolds, J. D. (1974) Biology of small pelagic fishes in the new Volta Lake in Ghana. Part 3: sex and reproduction. *Hydrobiologia,* **45**, 489–508.

Reynolds, L. F. (1983) Migration patterns of five fish species in the Murray-Darling River system. *Australian Journal of Marine and Freshwater Research,* **34**, 857–871.

Ribbink, A. J. (1977) Cuckoo among lake Malawi cichlid fish. *Nature,* **267**, 243–244.

Ribbink, A. J. (1988) Evolution and speciation of African cichlids. In: *Biology and Ecology of African Freshwater Fishes* (eds C. Lévêque, M. N. Bruton & G. W. Ssentongo), pp. 35–51. Editions de l'ORSTOM, Travaux et Documents n° 216, Paris.

Ribbink, A. J. (1991) Distribution and ecology of the cichlids of the African Great Lakes. In: *Cichlid Fishes – Behaviour, Ecology and Evolution* (ed. M. H. A. Keenleyside), pp. 36–59. Chapman & Hall, London.

Ribeiro, M. C. L. de B. & Petrere, M. Jr. (1990) Fisheries ecology and management of the jaraqui (*Semaprochilodus taeniurus*, *S. insignis*) in central Amazonia. *Regulated Rivers: Research and Management,* **5**, 195–215.

Richardson-Heft, C. A., Heft, A. A., Fewlass, A. & Brandt, S. B. (2000) Movement of largemouth bass in northern Chesapeake Bay: relevance to sportfishing tournaments. *North American Journal of Fisheries Management,* **20**, 493–501.

Richmond, A. M. & Kynard, B. (1995) Ontogenetic behaviour of shortnose sturgeon, *Acipenser brevirostrum*. *Copeia,* **1995**, 172–182.

Richkus, W. A. & McLean, R. (2000) Historical overview of the efficacy of two decades of power plant fisheries impact assessment activities in Chesapeake Bay. *Environmental Science & Policy,* **3** (Suppl.), 283–293.

Ridgway, M. S. & Shuter, B. J. (1996) Effects of displacement on the seasonal movements and home range characteristics of smallmouth bass in Lake Opeongo. *North American Journal of Fisheries Management,* **16**, 371–377.

Ringelberg, J. (1999) The photobehaviour of *Daphnia* spp. as a model to explain diel vertical migration in zooplankton. *Biological Reviews of the Cambridge Philosophical Society,* **74**, 397–423.

Ringger, T. G. (2000) Investigations of impingement of aquatic organisms at the Calvert Cliffs Nuclear Power Plant, 1975–1995. *Environmental Science & Policy,* **3** (Suppl.), 261–273.

Roberts, R. J., MacQueen, A., Shearer, W. M. & Young, H. (1973) The histopathology of salmon tagging. I. The tagging lesion in newly tagged parr. *Journal of Fish Biology,* **5**, 497–503.

Roberts, T. R. (1975) Geographical distribution of African freshwater fishes. *Zoological Journal of the Linnean Society,* **57**, 249–319.

Roberts, T. R. (1978) An ichthyological survey of the Fly River in Papua New Guinea with descriptions of new species. *Smithsonian Contributions to Zoology,* **281**, 249–319.

Roberts, T. R. (1982) The Bornean gastromyzontine fish genera *Gastromyzon* and *Glaniopsis* (Cypriniformes, Homalopteridae), with descriptions of new species. *Proceedings of the Californian Academy of Sciences,* **42**, 497–524.

Roberts, T. R. (1984) Skeletal anatomy and classification of the neotenic Asian salmoniform superfamily Salangoidea (icefishes or noodlefishes). *Proceedings of the California Academy of Sciences,* **43**, 179–220.

Robertson, D. R., Green, D. G. & Victor, B. C. (1988) Temporal coupling of production and recruitment of larvae of a Caribbean reef fish. *Ecology,* **69**, 370–381.

Robinson, A. T., Clarkson, R. W. & Forrest, R. E. (1998) Dispersal of larval fishes in a regulated river tributary. *Transactions of the American Fisheries Society,* **127**, 772–786.

Rochard, E., Castelnaud, G. & Lepage, M. (1990) Sturgeons (Pisces: Acipenseridae): threats and prospects. *Journal of Fish Biology,* **37**, 123–132.

Rodríguez, M. A. & Lewis, W. M. Jr. (1997) Structure of fish assemblages along environmental gradients in floodplain lakes of the Orinoco River. *Ecological Monographs,* **67**, 109–128.

Rodriguez-Ruiz, A. & Granado-Lorencio, C. (1992) Spawning period and migration of 32 species of cyprinids in a stream with Mediterranean regimen (SW Spain). *Journal of Fish Biology,* **41**, 545–556.

Rogers, S. C., Church, D. W., Weatherley, A. H. & Pincock, D. G. (1984) An automated ultrasonic telemetry system for the assessment of locomotor activity in free-ranging rainbow, *Salmo gairdneri* Richardson. *Journal of Fish Biology,* **25**, 697–710.

Roman-Valencia, C. (1998) Diet and reproduction of *Creagrutus brevipinnnis* (Pisces, Characidae) in Alto Cauca, Colombia. *Revista de Biologia Tropical,* **46**, 783–789 [in Spanish].

Rome, L. C., Funke, R. P. & Alexander, R. McN. (1990) The influence of water temperature on muscle velocity and sustained performance in swimming carp. *Journal of Experimental Biology,* **154**, 163–178.

Rommel, S. A. & McCleave, J. D. (1972) Oceanic electric fields: perception by American eels. *Science,* **176**, 1233–1235.

Ronafalvy, J. P., Cheesman, R. R. & Matejek, W. M. (2000) Circulating water traveling screen modifications to improve impinged fish survival and debris handling at Salem Generating Station. *Environmental Science & Policy,* **3** (Suppl.), 377–382.

Rosecchi, E. & Crivelli, A. J. (1992) Study of a sand smelt (*Atherina boyeri* Risso 1810) population reproducing in fresh water. *Ecology of Freshwater Fish,* **1**, 77–85.

Rosecchi, E. & Crivelli, A. J. (1995) Sand smelt (*Atherina boyeri*) migration within the water system of the Camargue, southern France. *Hydrobiologia,* **300/301**, 289–298.

Rosin, E. (1985) Principles of intestinal surgery. In: *Small Animal Surgery* (ed. D. H. Slater), pp. 720–737. Philadelphia, W. B. Saunders.

Ross, M. J. & Kleiner, C. F. (1982) Shielded needle technique for surgically implanting radio frequency transmitters in fish. *Progressive Fish Culturalist,* **44**, 41–43.

Ross, M. J. & McCormick, J. H. (1981) Effects of external radio transmitters on fish. *Progressive Fish Culturist,* **43**, 67–72.

Ross, Q. E., Dunning, D. J., Thorne, R. E., Menzies, J. K., Tiller, G. W. & Watson, J. K. (1993) Response of alewives to high-frequency sound at a power plant intake on Lake Ontario. *North American Journal of Fisheries Management,* **13**, 766–774.

Ross, S. T. (1986) Resource partitioning in fish assemblages: a review of field studies. *Copeia,* **1986**, 352–388.

Roussel, J. M. & Bardonnet, A. (1999) Ontogeny of diel pattern of stream-margin habitat use by emerging brown trout, *Salmo trutta*, in experimental channels: influence of food and predator presence. *Environmental Biology of Fishes,* **56**, 253–262.

Roussel, J. M., Haro, A. & Cunjak, R. A. (2000) Field test of a new method for tracking small fishes in shallow rivers using passive integrated transponder (PIT) technology. *Canadian Journal of Fisheries and Aquatic Sciences,* **57**, 1326–1329.

Roux, A. L. (1984) The impact of emptying and cleaning reservoirs on the physico-chemical and biological water quality of the Rhône downstream of the dams. In: *Regulated Rivers* (eds A. Lilliehammer & S. J. Saltveit), pp. 61–70. Universitetsforlarget, Oslo.

Roy, D. (1989) Physical and biological factors affecting the distribution and abundance of fish in rivers flowing into James Bay and Hudson Bay. *Canadian Special Publication of Fisheries and Aquatic Sciences,* **106**, 159–171.

Ruffino, M. L. & Barthem, R. B. (1996) Perspectivas para el manejo de los bagres migradores de la Amazonia. *Boletin Cientifico Sante Fe de Bogota,* **4**, 19–28.

Ruffino, M. L. & Isaac, V. J. (1995) Life cycle and biological parameters of several Brazilian Amazon fish species. *Naga,* **18** (4), 41–45.

Rulifson, R. A. & Dadswell, M. J. (1995) Life history and population characteristics of striped bass in Atlantic Canada. *Transactions of the American Fisheries Society,* **124**, 477–507.

Rulifson, R. A. & Tull, K. A (1999) Striped bass spawning in a tidal bore river: the Shubenacadie Estuary, Atlantic Canada. *Transactions of the American Fisheries Society,* **128**, 613–624.

Russell, D. F., Wilkens, L. A. & Moss, F. (1999) Use of behavioural stochastic resonance by paddlefish for feeding. *Nature,* **402**, 291–294.

Russell, I. C., Moore, A., Ives, S., Kell, L. T., Ives, M. J. & Stonehewer, R. J. (1998) The migratory behaviour of juvenile and adult salmonids in relation to an estuarine barrage. *Hydrobiologia,* **371/372**, 321–333.

Ruzycki, J. R. & Wurtsbaugh, W. A. (1999) Ontogenetic habitat shifts of juvenile Bear Lake sculpin. *Transactions of the American Fisheries Society,* **128**, 1201–1212.

Ryder, R. A. (1968) Dynamics and exploitation of mature walleyes, *Stizostedion vitreum vitreum*, in the Nigipon Bay region of Lake Superior. *Journal of the Fisheries Research Board of Canada,* **25**, 1347–1376.

Sage, M. (1973) The evolution of thyroid function in fishes. *American Zoologist,* **13**, 899–905.

Sager, D. R., Hocutt, C. H. & Stauffer Jr, J. R. (2000) Avoidance behavior of *Morone americana, Leiostomus xanthurus* and *Brevoortia tyrannus* to strobe light as a method of impingement mitigation. *Environmental Science & Policy,* **3** (Suppl.), 393–403.

Saila, S. B. (1961) A study of winter flounder movements. *Limnology and Oceanography,* **6**, 292–298.

Saint-Paul, U. & Bernardino, G. (1988) Behavioural and ecomorphological responses of the neotropical pacu *Piaractus mesopotamicus* (Teleostei, Serrasalminidae) to oxygen-deficient waters. *Experimental Biology* (Berlin), **48**, 19–26.

Saint-Paul, U. & Soares, B. M. (1987) Diurnal distribution and behavioral responses of fish to extreme hypoxia in an Amazon floodplain lake. *Environmental Biology of Fishes,* **20**, 91–104.

Sakai, K., Nomura, M., Takashima, F. & Oto, H. (1975) The over-ripening phenomenon of rainbow trout – II. Changes in the percentage of eyed eggs, hatching rate and incidence of abnormal alevins during the process of over-ripening. *Bulletin of the Japanese Society of Scientific Fisheries,* **41**, 855–860.

Saldaña, J. & Venables, B. (1983) Energy compartimentalization in a migratory fish, *Prochilodus mariae* (Prochilodontidae), of the Orinoco River. *Copeia,* **1983**, 617–625.

Salmenkova, E. A., Omelchenko, V. T., Kolesnikov, A. A. & Malinina, T. V. (2000) Genetic differentiation of charrs in the Russian north and far east. *Journal of Fish Biology,* **57** (Suppl. A), 136–157.

Salverda, A. P., Kerkhofs, M. J. J., Verbraak, P. J. J. & Klein, J. D. (1996) Towards a method to reconstruct a natural hydrograph for defining ecologically acceptable discharge fluctuations. In: *Ecohydraulics 2000* (eds M. Leclerc *et al.*), Vol. A, pp. 711–722. INRS-Eau, Québec.

Sand, O., Enger, P. S., Karlsen, H. E., Knudsen, F. & Kvernstuen, T. (2000) Avoidance responses to infrasound in downstream migrating European silver eels, *Anguilla anguilla*. *Environmental Biology of Fishes,* **57**, 327–336.

Sanders, R. E. (1992) Day versus night electro-fishing catches from near-shore waters of the Ohio and Muskingum Rivers. *Ohio Journal of Science,* **92**, 51–59.

Santos, G. M. (1982) Caracterizaçao, hábitos alimentares e reproductivos de quatro espécies de "aracus" e consideraçoes ecologícas sobre o grupo no lago Janauaca-AM (Osteichthyes, Characoidei, Anostomidae). *Acta Amazónica,* **12**, 713–739.

Sato, J. (1986) A brood parasitic catfish of mouthbrooding cichlid fishes in Lake Tanganyika. *Nature,* **323**, 58–59.

Sauriau, P. G., Robin, J. P. & Marchand, J. (1994) Effects of the excessive organic enrichment of the Loire Estuary on the downstream migratory patterns of the amphihaline grey mullet *Liza ramada* (Pisces, Mugilidae). In: *Changes in Fluxes in Estuaries* (ECSA22/ERF symposium, Plymouth, September 1992) (eds K. R. Dyer & R. J. Orth), pp. 349–356. Olsen & Olsen, Fredensborg, Denmark.

Savino, J. F. & Stein, R. A. (1982) Predator-prey interactions between largemouth bass and bluegills as influenced by simulated submersed vegetation. *Transactions of the American Fisheries Society,* **111**, 255–266.

Sazima, I. & Zamprogno, C. (1985) Use of water hyacinths as shelter, foraging place and transport by young piranha, *Serrasalmus spilopleura*. *Environmental Biology of Fishes,* **12**, 237–240.

Scheidegger, K. J. & Bain, M. B. (1995) Larval fish distribution and microhabitat use in free-flowing and regulated rivers. *Copeia,* **1995**, 125–135.

Schiemer, F. (2000) Fish as indicators for the assessment of the ecological integrity of large rivers. *Hydrobiologia,* **422**, 271–278.

Schindler, D. E. (1999) Migration strategies of young fishes under temporal constraints: the effect of size-dependent overwinter mortality. *Canadian Journal of Fisheries and Aquatic Sciences,* **56** (Suppl. 1), 61–70.

Schlosser, I. J. (1988) Predation risk and habitat selection by two size classes of a stream cyprinid: experimental test of a hypothesis. *Oikos,* **52**, 36–40.

Schlosser, I. J. (1995) Dispersal, boundary processes, and trophic-level interactions in streams adjacent to beaver ponds. *Ecology,* **76**, 908–925.

Schlosser, I. J. & Ebel, K. K. (1989) Effects of flow regime and cyprinid predation on a headwater stream. *Ecological Monographs,* **59**, 41–57.

Schlosser, I. J. & Kallemeyn, L. W. (2000) Spatial variation in fish assemblages across a beaver-induced successional landscape. *Ecology,* **81**, 1371–1382.

Schmutz, S., Giefing, C. & Wiesner, C. (1998) The efficiency of a nature-like bypass channel for pike-perch (*Stizostedion lucioperca*) in the Marchfeldkanalsystem. *Hydrobiologia,* **371/372**, 355–360.

Schmutz, S., Mader, H. & Unfer, G. (1995) Funktionalität von Potamalfischaufstiegshilfen im Marchfeldkanalsystem *Österreichische Wasser- und Abfallwirtscaft,* **47**, 43–58.

Schmutz, D. A. & Unfer, G. (1996) Radio telemetry as an additional tool for investigating the colonisation of a recently constructed channel (Marchfeldkanal). In: *Underwater Biotelemetry* (eds E. Baras & J. C. Philippart), pp. 137–142. University of Liège, Belgium.

Scholz, A. T., White, R. J., Muzi, M. & Smith, T. (1985) Uptake of radiolabelled triiodothyronine in the brain of steelhead trout (*Salmo gairdneri*) during parr-smolt transformation: implications for the mechanism of thyroid activation of olfactory imprinting. *Aquaculture,* **45**, 199–214.

Schoonoord, M. P. & Maitland, P. S. (1983) Some methods of marking larval lampreys (*Petromyzonidae*). *Fisheries Management,* **14**, 33–38.

Schramm, H. L. Jr. & Black, D. J. (1984) Anaesthesia and surgical procedures for implanting radio transmitters into grass carp. *Progressive Fish Culturist,* **46**, 185–190.

Schulz, U. & Berg, R. (1987) The migration of ultrasonic-tagged bream, *Abramis brama* (L), in Lake Constance (Bodensee-Untersee). *Journal of Fish Biology,* **31**, 409–414.

Schurmann, H. & Steffenson, J. F. (1997) Effects of temperature, hypoxia and activity on the metabolism of juvenile Atlantic cod. *Journal of Fish Biology,* **50**, 1166–1180.

Schwalme, K., Mackay, W. C. & Lindner, D. (1985) Suitability of vertical slot and Denil fishways for passing north-temperate, nonsalmonid fish. *Canadian Journal of Fisheries and Aquatic Sciences* **42**, 1815–1822.

Scott, A. (1985) Distribution, growth and feeding of postemergent grayling *Thymallus thymallus* in an English river. *Transactions of the American Fisheries Society,* **114**, 525–531.

Scott, D. B. C. (1979) Environmental timing and the control of reproduction in teleost fish. *Symposium of the Zoological Society of London,* **44**, 105–132.

Scott, W. B. & Crossman, E. J. (1973) Freshwater fishes of Canada. *Bulletin of the Fisheries Research Board of Canada,* **184**.

Secor, D. H., Henderson-Arzapalo, A. & Piccoli, P. M. (1995) Can otolith microchemistry chart patterns of migration and habitat utilisation in anadromous fishes? *Journal of Experimental Marine Biology and Ecology,* **192**, 15–33.

Secor, D. H., Ohta, T., Nakayama, K. & Tanaka, M. (1999) Use of otolith microanalysis to determine estuarine migrations of Japanese sea bass *Lateolabrax japonicus* distributed in Ariake Sea. *Fisheries Science,* **64**, 740–743.

Secor, D. H. & Piccoli, P. (1996) Age- and sex- dependent migrations of striped bass in the Hudson River as determined by chemical microanalysis of otoliths. *Estuaries,* **19**, 778–793.

Senta, T. (1973) Spawning ground of the salmonoid fish *Salangichthys microdon* in Takahashi River, Okayama Prefecture. *Japanese Journal of Ichthyology,* **20**, 25–28.

Seret, B. (1988) Dasyatidae. In: *Biology and Ecology of African Freshwater Fishes* (eds C. Lévêque, M. N. Bruton & G. W. Ssentongo), pp. 62–75. Éditions de l'ORSTOM, Travaux et Documents n° 216, Paris.

Setzler, E. M., Boynton, W. R., Wood, K. V., *et al.* (1980) Synopsis of biological data on striped bass *Morone saxatilis* (Walbaum). NOAA Technical Report, *National Marine Fisheries Service Circular,* **433**, 1–69 (*FAO Synopsis,* **121**).

Severi, W., Rantin, F. T. & Fernandes, M. N. (1997) Respiratory gill surface of the serrasalmid fish, *Piaractus mesopotamicus*. *Journal of Fish Biology,* **50**, 127–136.

Shen, K. N., Lee, Y. C. & Tzeng, W. N. (1998) Use of otolith microchemistry to investigate the life history pattern of gobies in a Taiwanese stream. *Zoological Studies,* **37**, 322–329.

Shearer, W. M. (1992) *The Atlantic Salmon: Natural History, Exploitation and Future Management.* Fishing News Books, Blackwell Science Ltd, Oxford.

Sheri, A. N. & Power, G. (1969) Vertical distribution of white perch, *Morone americanus,* modified by light. *Canadian Field Naturalist,* **83**, 160–161.

Shikhshabekov, M. M. (1971) Resorbtion of the gonads in some semi-diadromous fishes of the Arakum Lakes (Dagestan USSR) as a result of regulation of discharge. *Journal of Ichthyology* **11**, 427–431.

Shireman, J. V. (1975) Gonadal development of striped mullet (*Mugil cepahlus*) in fresh water. *Progres-*

sive Fish Culturist, **37**, 205–208.
Siligato, S. (1999) Spawning migration of Balon's ruffe into a Danubian side branch in Austria. *Journal of Fish Biology,* **55**, 376–381.
Simonovic, P., Valkovic, B. & Paunovic, M. (1998) Round goby *Neogobius melanostomus*, a new Ponto-Caspian element for Yugoslavia. *Folia Zoologica,* **47**, 305–312.
Sinha, V. R. P. & Jones, J. W. (1975) *The European Freshwater Eel.* Liverpool University Press, Liverpool, 146 pp.
Sirimongkonthaworn, R. (1992) The biology and feeding of the freshwater sardine *Clupeichthys aesarnensis* in Ubolratana reservoir, Thailand. M.Sc. thesis, University of Waterloo, 131 pp.
Sissenwine, M. P., Azarovitz, T. R. & Suomala, J. B. (1983) Determining the abundance of fish. In: *Experimental Biology at Sea* (eds A. G. MacDonald & I. G. Priede), pp. 51–101. Academic Press, London.
Sjöberg, K. (1977) Locomotor activity of the river lamprey, *Lampetra fluviatilis* L., during the spawning season. *Hydrobiologia,* **55**, 265–270.
Sjöberg, K. (1980) Ecology of the European river lamprey (*Lampetra fluviatilis*) in northern Sweden. *Canadian Journal of Fisheries and Aquatic Sciences,* **37**, 1974–1980.
Slaney, P. A. & Northcote, T. G. (1974) Effects of prey abundance on density and territorial behavior of young rainbow trout (*Salmo gairdneri*) in laboratory stream channels. *Journal of the Fisheries Research Board of Canada,* **31**, 1201–1209.
Slavík, O. (1996a) The migration of fish in the Elbe River below Strekov. *Ziva,* **4**, 179–180.
Slavík, O. (1996b) Changes in the abundance and diversity of fish assemblages in the Podbaba Navigation Channel on the Vltava River. *Acta Universitatis Carolinae Biologica,* **40**, 193–202.
Slavík, O. & Bartos, L. (1997) Effect of water temperature and pollution on young-of-the-year fishes in the regulated stretch of the River Vltava, Czech Republic. *Folia Zoologica,* **46**, 367–374.
Slavík, O. & Bartos, L. (2000) Seasonal and diurnal changes of young-of-the-year fish in the channelized stretch of the Vltava River (Bohemia, Czech Republic). In: *Management and Ecology of River Fisheries* (ed. I. G. Cowx), pp. 101–111. Fishing News Books, Blackwell Science Ltd, Oxford.
Slavík, O. & Bartos, L. (2001) Spatial distribution and temporal variance of fish communities in the channelized and regulated Vltava River (Central Europe). *Environmental Biology of Fishes* **61**, 47–55.
Slavík, O. & Rab, P. (1995) Effect of microhabitat on the age and growth of two stream-dwelling populations of spined-loach, *Cobitis taenia. Folia Zoologica,* **44**, 167–174.
Slavík, O. & Rab, P. (1996) Life history of spined-loach, *Cobitis taenia* in an isolated site (Pšovka Creek, Bohemia) *Folia Zoologica,* **44**, 167–174.
Slawski, T. M. & Ehlinger, T. J. (1998) Fish habitat improvements in box culverts: management in the dark & quest. *North American Journal of Fisheries Management,* **18**, 676–685.
Slobodkin, L. B. & Rapoport, A. (1974) An optimal strategy of evolution. *Quarterly Reviews in Biology,* **49**, 181–200.
Smith, B. R. & Tibbles, J. J. (1980) Sea lamprey (*Petromyzon marinus*) in Lakes Huron, Michigan, and Superior: history of invasion and control, 1936–1978. *Canadian Journal of Fisheries and Aquatic Sciences,* **37**, 1780–1801.
Smith, I. P., Collins, K. J. & Jensen, A. C. (1998) Electromagnetic telemetry of lobster (*Homarus gammarus* [L.]) movements and activity: preliminary results. *Hydrobiologia,* **371/372**, 133–141.
Smith, I. P., Johnstone, A. D. F. & Dunkley, D. A. (1996) Evaluation of a portable electrode array for a

resistivity fish counter. *Fisheries Management and Ecology,* **3**, 129–141.
Smith, I. P. & Smith, G. W. (1997) Tidal and diel timing of river entry by adult Atlantic salmon returning to the Aberdeenshire Dee, Scotland. *Journal of Fish Biology,* **50**, 463–474.
Smith, J. B. & Hubert, W. A. (1989) Use of a tributary by fishes in a Great Plains River system. *Prairie Naturalist,* **21**, 27–38.
Smith, R. J. F. (1991) Social behaviour, homing and migration. In: *Cyprinid Fishes – Systematics, Biology and Exploitation* (eds I. J. Winfield & J. S. Nelson), pp. 509–529. Chapman & Hall, London.
Smith, R. J. F. (1992) Alarm signals in fishes. *Reviews in Fish Biology and Fisheries,* **2**, 33–63.
Smith, S. H. (1968) Species succession and fishery exploitation in the Great Lakes. *Journal of the Fisheries Research Board of Canada,* **25**, 667–693.
Smith, T. D. (1994) *Scaling Fisheries: The Science of Measuring the Effects of Fishing, 1855–1955.* Cambridge University Press, Cambridge.
Smith, T. I. J. (1985) The fishery, biology and management of Atlantic sturgeon, *Acipenser oxyrhynchus*, in North America. *Environmental Biology of Fishes,* **14**, 61–72.
Snedden, G. A., Kelso, W. E. & Rutherford, D. A. (1999) Diel and seasonal patterns of spotted gar movement and habitat use in the lower Atchafalaya River basin, Louisiana. *Transactions of the American Fisheries Society,* **128**, 144–154.
Snuccins, E. J. & Gunn, J. M. (1995) Coping with a warm environment: behavioural thermoregulation by lake trout. *Transactions of the American Fisheries Society,* **124**, 118–123.
Snyder, R. (1991) Migration and life histories of the threespine stickleback: evidence for adaptive variation in growth rate between populations. *Environmental Biology of Fishes,* **31**, 381–388.
Snyder, R. J. & Dingle, H. (1989) Adaptive, genetically-based differences in life histories of threespine sticklebacks (*Gasterosteus aculeatus* L.). *Canadian Journal of Zoology,* **67**, 2448–2454.
Solomon, D. J. (1992) Diversion and entrapment of fish at water intakes and outfalls. *Research and Development Report,* **1**. National Rivers Authority, Bristol, UK.
Solomon, D. J. & Potter, E. C. E. (1988) First results with a new estuarine fish tracking system. *Journal of Fish Biology,* **33** (Suppl. A), 127–132.
Sommer, T., Baxter, R. & Herbold, B. (1997) Resilience of splittail in the Sacramento-San Joaquin Estuary. *Transactions of the American Fisheries Society,* **126**, 961–976.
Sörensen, J. (1951) An investigation of some factors affecting the upstream migration of the eel. *Reports of the Institute of Freshwater Research of Drottningholm,* **32**, 126–172.
Sorensen, P. W. (1986) Origins of the freshwater attractant(s) of migrating elvers of the American eel, *Anguilla rostrata*. *Environmental Biology of Fishes,* **17**, 185–200.
Sorenson, K. M., Fisher, W. L. & Zale, A. V. (1998) Turbine passage of juvenile and adult fish at a warmwater hydroelectric facility in northeastern Oklahoma: monitoring associated with relicensing. *North American Journal of Fisheries Management,* **18**, 124–136.
Southall, P. D. & Hubert, W. A. (1984) Habitat use by adult paddlefish in the Upper Mississippi River. *Transactions of the American Fisheries Society*, **113**, 125–131.
Spicer, G., O'Shea, T. & Piehler, G. (2000) Entrainment, impingement and BTA evaluation for an intake located on a cooling water reservoir in the southwest. *Environmental Science & Policy,* **3** (Suppl.), 232–331.
Stabell, O. B. (1984) Homing and olfaction in salmonids: a critical review with special reference to the Atlantic salmon. *Biological Reviews,* **59**, 333–388.
Stabell, O. B. (1992) Olfactory control of homing behaviour in salmonids. In: *Fish Chemoreception* (ed.

T. J. Hara), pp. 249–270. Chapman & Hall, London.

Stanford, J. A. & Hauer, F. R. (1992) Mitigating the impacts of stream and lake regulation in the Flathead River catchment, Montana, USA: an ecosystem perspective. *Aquatic Conservation: Marine and Freshwater Ecosystems*, **2**, 35–63.

Starkie, A. (1975) Some aspects of the ecology of dace (*Leuciscus leuciscus* [L.]) in the River Tweed. DPhil thesis, University of Edinburgh.

Stasko, A. B. (1971) Review of field studies on fish orientation. *Annals of the New York Academy of Science*, **188**, 12–29.

Stasko, A. B & Pincock, D. G. (1977) Review of underwater biotelemetry with emphasis on ultrasonic techniques. *Journal of the Fisheries Research Board of Canada*, **34**, 1261–1285.

Steig, T. W. & Iverson, T. K. (1998) Acoustic monitoring of salmonid density, target strength, and trajectories at two dams on the Columbia River, using a split-beam scanning system. *Fisheries Research*, **35**, 43–53.

Steiner, H. A. (1998) Fish passes at run-of-river hydropower plants of the Verbund. In: *Fish Migration and Fish Bypasses* (eds M. Jungwirth, S. Schmutz & S. Weiss), pp. 420–434. Fishing News Books, Blackwell Science Ltd, Oxford.

Stevens, D. E. & Miller, L. W. (1983) Effects of river flow on abundance of young chinook salmon, American shad, longfin smelt and delta smelt in the Sacramento-San Joaquin River system. *North American Journal of Fisheries Management*, **3**, 425–437.

Stokesbury, K. D. E. & Dadswell, M. J. (1989) Seaward migration of juveniles of three herring species, *Alosa*, from an estuary in the Annapolis River, Nova Scotia. *Canadian Field Naturalist*, **103**, 388–393.

Stott, B. (1967) The movements and population densities of roach (*Rutilus rutilus* [L.]) and gudgeon (*Gobio gobio* [L.]) in the River Mole. *Journal of Animal Ecology*, **36**, 407–423.

Stott, B., Elsdon, J. W. V. & Johnston, J. A. A. (1963) Homing behaviour in gudgeon (*Gobio gobio* [L.]). *Animal Behaviour*, **11**, 93–96.

Stuart, I. G. & Mallen-Cooper, M. (1999) An assessment of the effectiveness of a vertical-slot fishway for non-salmonid fish at a tidal barrier on a large tropical/subtropical river. *Regulated Rivers – Research and Management*, **15**, 575–590.

Stuart, T. A. (1957) The migrations and homing behaviour of brown trout (*Salmo trutta* L.). *Scientific Investigations in Freshwater Salmon Fisheries Research. Scottish Home Office Department*, **18**, 1–27.

Sturlaugsson, J. (1995) Migration study on homing of Atlantic salmon (*Salmo salar* L.) in coastal waters, W. Iceland: depth movements and sea temperatures recorded at migration routes by data storage tags. *ICES C.M.* 1995/M:**17**.

Sturlaugsson, J. & Johansson, M. (1996) Migratory pattern of wild sea trout (*Salmo trutta*) in SE-Iceland recorded by data storage tags. *ICES C.M.* 1996/M:**5**.

Sturlaugsson, J., Jónsson, I. R. & Tómasson, T. (1998) Sea migration of anadromous Arctic charr (*Salvelinus alpinus*) recorded by data storage tags. *ICES C.M.* 1998/N:**22**.

Sulak, K. J. & Clugston, J. P. (1999) Recent advances in life history of Gulf of Mexico sturgeon, *Acipenser oxyrinchus desotoi*, in the Suwannee river, Florida, USA: a synopsis. *Journal of applied Ichthyology*, **15** (4–5), 116–128.

Summerfelt, R. C. & Mosier, D. (1984) Transintestinal expulsion of surgically implanted dummy transmitters by channel catfish. *Transactions of the American Fisheries Society*, **113**, 760–766.

Summerfelt, R. C. & Smith, L. S. (1990) Anesthesia, surgery and related techniques. In: *Methods for*

Fish Biology (eds C. B. Schreck & P. B. Moyle), pp. 213–272. American Fisheries Society, Bethesda, Maryland.

Summers, R. W. (1980) Life cycle and population ecology of the flounder *Platichthys flesus* (L.) in the Ythan estuary, Aberdeenshire, Scotland. *Estuarine, Coastal & Marine Sciences,* **11**, 217–232.

Suthers, J. M. & Gee, J. H. (1986) Role of hypoxia in limiting diel spring and summer distribution of juvenile yellow perch (*Perca flavescens*) in a prairie marsh. *Canadian Journal of Fisheries and Aquatic Sciences,* **43**, 1562–1570.

Svedang, H. & Wickstrom, H. (1997) Low fat contents in female silver eels: indications of insufficient energetic stores for migration and gonadal development. *Journal of Fish Biology,* **50**, 475–486.

Svetovidov, A. N. (1984) Acipenseridae. In: *Fishes of the North-eastern Atlantic and the Mediterranean,* Vol. I (eds P. J. P. Whitehead, M. L. Bauchot, J. C. Hureau, J. Nielsen & E. Tortonese), pp. 220–225. Unesco, Paris.

Taft, E. P. (2000) Fish protection technologies: a status report. *Environmental Science & Policy,* **3** (Suppl.), 349–359.

Tagawa, M., Tanaka, M., Matsumoto, S. & Hirano, T. (1990) Thyroid hormones in eggs of various freshwater, marine and diadromous teleosts and their changes during egg development. *Fish Physiology and Biochemistry,* **8**, 515–520.

Tago, Y. (1999) Estimation of abundance of ayu larvae during seaward drifting at the Shou River. *Nippon-Suisan Gakkaishi,* **65**, 718–727.

Takeshita, N. & Kimura, S. (1991) Potamodromous migration of the cyprinid fish *Hemibarbus barbus* in the Chikugo River. *Nippon Suisan Gakkaishi,* **57**, 869–873.

Tanaka, H., Takagi, Y. & Naito, Y. (2000) Behavioural thermoregulation of chum salmon during homing migration in coastal waters. *Journal of Experimental Biology,* **203**, 1825–1833.

Tans, M. (2000) Utilisation des noues de la Meuse en tant que sites de reproduction et de nurserie par les poissons du fleuve. DPhil thesis, Presses Universitaires de Namur, Namur, Belgium, 337 pp.

Taylor, E. B. & McPhail, J. D. (1986) Prolonged and burst swimming in anadromous and freshwater threespine stickleback *Gasterosteus aculeatus. Canadian Journal of Zoology,* **64**, 416–420.

Teeter, J. (1980) Pheromone communication in sea lampreys (*Petromyzon marinus*): implications for population management. *Canadian Journal of Fisheries and Aquatic Sciences,* **37**, 2123–2132.

Tejerina-Garro, F. L., Fortin, R. & Rodriguez, M. A. (1998) Fish community structure in relation to environmental variation in floodplain lakes of the Araguaia River, Amazon Basin. *Environmental Biology of Fishes,* **51**, 399–410.

Tesch, F. W. (1965) Verhalten der Glasaale (*Anguilla anguilla*) bei ihrer Wanderung in den Ästuarien deutscher Nordseeflüsse. *Helgolander Wisserei Meeresunters,* **12**, 404–419.

Tesch, F. W. (1966) Der Einhuß der Weserstauwehere auf die Jungaalwanderung. *Fischwirt,* **16**, 29–37.

Tesch, F. W. (1971) Aufenthalt der Glasaale (*Anguilla anguilla*) an der südlichen Nordseeküste vor dem Eindringen in das Süßwasser. *Vie et Milieu,* **22** (Suppl.), 381–392.

Tesch, F. W. (1972) Versuch zur telemetrischen Vorfolgung der Laichwanderung von Aalen (*Anguilla anguilla*) in der Nordsee. *Helgoländer Wisserei Meeresunters,* **23**, 165–183.

Tesch, F. W. (1977) *The Eel – Biology and Management of Anguillid Eels*. Chapman & Hall, London.

Teugels, G. G. (1996) Taxonomy, phylogeny and biogeography of catfishes (Ostariophysi, Siluroidei): an overview. *Aquatic Living Resources,* **9**, 9–34.

Thomas, G. (1977) The influence of eating and rejecting prey items upon feeding and searching behaviour in *Gasterosteus aculeatus*, L. *Animal Behaviour,* **25**, 52–66.

Thoreau, X. & Baras, E. (1997) Evaluation of surgery procedures for implanting telemetry transmitters into the body cavity of blue tilapia *Oreochromis aureus*. *Aquatic Living Resources*, **10**, 207–211.

Thorne, R. E. (1979) Hydroacoustic estimates of adult sockeye salmon (*Oncorhynchus nerka*) in Lake Washington, 1972–75. *Journal of the Fisheries Research Board of Canada*, **36**, 1145–1149.

Thorne, R. E. (1998) Experience with shallow water acoustics. *Fisheries Research*, **35**, 135–139.

Thorne, R. E. & Johnson, G. E. (1993) A review of hydroacoustic studies for estimation of salmonid downriver migration past hydroelectric facilities on the Columbia and Snake rivers in the 1980s. *Reviews in Fisheries Science*, **1**, 27–56.

Thorpe, J. E. (1974) Morphology, physiology, behaviour and ecology of *Perca fluviatilis* L. and *Perca flavescens* Mitchill. *Journal of the Fisheries Research Board of Canada*, **30**, 1327–1336.

Thorpe, J. E. (1987) Smoting versus residency: developmental conflict in salmonids. *American Fisheries Society Symposium*, **1**, 244–252.

Thorpe, J. E. & Morgan, R. J. (1978) Periodicity in Atlantic salmon *Salmo salar* L. smolts migration. *Journal of Fish Biology*, **12**, 541–548.

Thorson, T. B. (1971) Movements of bull sharks, *Carcharhinus leucas*, between Caribbean Sea and Lake Nicaragua demonstrated by tagging. *Copeia*, **1971**, 336–338.

Thorson, T. B. (1972) The status of the bull shark, *Carcharhinus leucas*, in the Amazon River. *Copeia*, **1972**, 601–605.

Thorson, T. B. (1982) Life history implications of a tagging study of the largetooth sawfish, *Pristis perotteti*, in the Lake Nicaragua – Rio San Juan system. *Environmental Biology of Fishes*, **7**, 207–228.

Thorson, T. B., Cowan, C. M. & Watson, D. E. (1967) *Potamotrygon* spp.: Elasmobranchs with low urea content. *Science* **158**, 375–377.

Thorstad, E. B. & Heggberget, T. G. (1998) Migration of adult Atlantic salmon (*Salmo salar*): the effects of artificial freshets. *Hydrobiologia*, **371/372**, 339–346.

Thorsteinsson, V. (1995) Tagging experiments using conventional and electronic data storage tags for the observations of migration, homing and habitat choice in the Icelandic spawning stock of cod. *ICES C.M.* 1995/B:**19**.

Tinbergen, N. (1951) *The Study of Instinct*. Oxford University Press, Oxford.

Tito de Morais, L. & Raffray, J. (1996) Behaviour of *Hoplias aimara* during the filling phase of the Petit-Saut Dam (Sinnamary River, French Guyana, South America). In: *Underwater Biotelemetry* (eds E. Baras & J. C. Philippart), pp. 153–160. University of Liège, Belgium.

Todd, B. L. & Rabeni, C. F. (1989) Movement and habitat use by stream-dwelling smallmouth bass. *Transactions of the American Fisheries Society*, **118**, 229–242.

Todd, P. R. (1981) Timing and periodicity of migrating New Zealand freshwater eels. *New Zealand Journal of Marine and Freshwater Research*, **15**, 225–235.

Toepfer, C. S., Fisher, W. L. & Haubelt, J. A. (1999) Swimming performance of the threatened leopard darter in relation to road culverts. *Transactions of the American Fisheries Society*, **128**, 155–161.

Tokranov, A. M. (1994) Distribution and population of the Northern Far East Belligerent Sculpin, *Megalocottus platycephalus platycephalus* (Cottidae), in the Bol'shaya River Estuary (Western Kamchatka). *Journal of Ichthyology*, **34**, 149–153.

Toledo, S. A., Godoy, M. P. de & Dos Santos, E. P. (1986) Curve of migration of curimbata, *Prochilodus scrofa* (Pisces, Prochilodontidae) in the upper basin of the Paraná River, Brazil. *Revista Bra-*

siliera Biologica, **46**, 447–452 [in Portuguese].

Toledo, S. A., Godoy, M. P. de & Dos Santos, E. P. (1987) Delimitaçao populacional do curimbatá, *Prochilodus scrofa* (Pisces: Prochilodontidae) do Rio Mogi-Guaçu, Brasil. *Revista Brasiliera Biologica*, **47**, 501–506.

Tonn, W. M. & Paszkowski, C. (1987) Habitat use of the central mudminnow (*Umbra limi*) and yellow perch (*Perca flavescens*) in *Umbra-Perca* assemblages: the walls of competition, predation, and the abiotic environment. *Canadian Journal of Zoology*, **65**, 865–870.

Torblaa, R. L. & Westman, R. W. (1980) Ecological impacts of lampricide treatments on sea lamprey (*Petromyzon marinus*) ammocoetes and metamorphosed individuals. *Canadian Journal of Fisheries and Aquatic Sciences*, **37**, 1835–1850.

Torricelli, P., Tongiorgi, P. & Almansi, P. (1982) Migration of grey mullet fry into the Arno River: seasonal appearance, daily activity and feeding rhythms. *Fisheries Research*, **1**, 219–234.

Tosi, L. & Sola, C. (1993) Role of geosmin, a typical inland water odour, in guiding glass eel *Anguilla anguilla* (L.) migration. *Ethology*, **95**, 177–185.

Townshend, T. J. & Wootton, R. J. (1985) Adjusting parental investment to changing environmental conditions: the effect of food ration on parental behaviour of the convict cichlid, *Cichlasoma nigrofasciatum*. *Animal Behaviour*, **33**, 494–501.

Travade, F., Bomassi, J. M., Bach, J. M., *et al.* (1989) Use of radiotracking in France for recent studies concerning the EDF fishways program. *Hydroécologie Appliquée*, **1/2**, 33–51.

Travade, F. & Larinier, M. (1992) Les techniques de controle des passes a poissons. *Bulletin Français de la Pêche et Pisciculture*, **326/327**, 151–164.

Travade, F., Larinier, M., Boyer-Bernard, S. & Dartiguelongue, J. (1998) Performance of four fishpass installations recently built on two rivers in south-west France. In: *Fish Migration and Fish By-passes* (eds M. Jungwirth, S. Schmutz & S. Weiss), pp. 146–170. Fishing News Books, Oxford.

Trebitz, A. S. (1992) Timing of spawning in largemouth bass. Implications of an individual based model. *Ecological Modelling*, **59**, 203–227.

Trefethen, P. S. (1956) Sonic equipment for tracking individual fish. *US Fish and Wildlife Service Special Science Report Fisheries*, **179**, 11 pp.

Trépanier, S., Rodríguez, M. A. & Magnan, P. (1996) Spawning migrations of landlocked Atlantic salmon: time series modelling of river discharge and water temperature effects. *Journal of Fish Biology*, **48**, 925–936.

Trewavas, E. (1983) *Tilapine Fishes of the Genus Sarotherodon, Oreochromis and Danakilia*. British Museum, London, 583 pp.

Trudel, M. & Boisclair, D. (1996) Estimation of fish activity costs using underwater video-cameras. *Journal of Fish Biology*, **48**, 40–53.

Tsai, C. F., Islam, M. N., Karim, M. R. & Rahman, K. U. M. S. (1981) Spawning of major carps in the lower Halda River, Bangladesh. *Estuaries*, **4**, 127–138.

Tsukamoto, K., Ishida, R., Naka, K. & Kajihara, T. (1987) Switching of size and migratory pattern in successive generations of landlocked ayu. *American Fisheries Society Symposium*, **1**, 492–506.

Tsukamoto, K., Nakai, I. & Tesch, W. V. (1998) Do all freshwater eels migrate? *Nature*, **396**, 635–636.

Tucker, V. A. (1970) Energetic cost of locomotion in animals. *Comparative Biochemistry and Physiology*, **34**, 841–846.

Turner, G. G. (1986) Teleost mating systems and strategies. In: *The Behaviour of Teleost Fishes* (ed. T. J. Pitcher), pp. 253–274. Croom Helm, London.

Turnpenny, A. W. H. (1988) Fish impingement at estuarine power stations and its significance to commercial fishing. *Journal of Fish Biology,* **33** (Suppl. A), 103–110.

Turnpenny, A. W. H. (1998) Mechanisms of fish damage in low-head turbines: an experimental appraisal. In: *Fish Migration and Fish Bypasses* (eds M. Jungwirth, S. Schmutz & S. Weiss), pp. 300–314. Fishing News Books, Blackwell Science Ltd, Oxford.

Tuunainen, P., Ikonen, E. & Auvinen, H. (1980) Lampreys and lamprey fisheries in Finland. *Canadian Journal of Fisheries and Aquatic Sciences,* **37,** 1953–1959.

Tveiten, H., Johnsen, H. K. & Jobling, M. (1996) Influence of maturity status on the annual cycles of feeding and growth in Arctic charr reared at constant temperature. *Journal of Fish Biology,* **48,** 910–924.

Tyus, H. M. (1985) Homing behaviour noted for Colorado squawfish. *Copeia,* **1985,** 213–215.

Tyus, H. M. (1986) Life strategies in the evolution of the Colorado squawfish (*Ptychocheilus lucius*). *Great Basin Naturalist,* **46,** 656–661.

Tyus, H. M. (1987) Distribution, reproduction and habitat use of the razorback sucker in the Green River, Utah, 1979–1986. *Transactions of the American Fisheries Society,* **116,** 111–116.

Tyus, H. M. (1990) Potamodromy and reproduction of Colorado squawfish in the Green River Basin, Colorado and Utah. *Transactions of the American Fisheries Society,* **119,** 1035–1047.

Tyus, H. M. (1991) Distribution, habitat use, and growth of age-0 Colorado squawfish in the Green River Basin, Colorado and Utah. *Transactions of the American Fisheries Society,* **120,** 79–89.

Tyus, H. M. & Karp, A. C. (1990) Spawning and movements of razorback sucker, *Xyrauchen texanus*, in the Green River basin of Colorado and Utah. *Southwestern Naturalist,* **35,** 427–433.

Tzeng, W. N., Wang, C. H., Wickstrom, H. & Reizenstein, M. (2000) Occurrence of the semi-catadromous European eel *Anguilla anguilla* in the Baltic Sea. *Marine Biology,* **137,** 93–98.

Underwood, T. J. (2000) Abundance, length composition and migration of spawning inconnu in the Selawik River, Alaska. *North American Journal of Fisheries Management,* **20,** 386–393.

Unwin, M. J., Kinnison, M. T. & Quinn, T. P. (1999) Exceptions to semelparity: postmaturation survival, morphology, and energetics of male chinook salmon (*Oncorhynchus tshawytscha*). *Canadian Journal of Fisheries and Aquatic Sciences,* **56,** 1172–1181.

Utzinger, J., Roth, C. & Peter, A. (1998) Effects of environmental parameters on the distribution of bullhead *Cottus gobio* with particular consideration of the effects of obstructions. *Journal of applied Ecology,* **35,** 882–892.

Valdimarsson, S. K. & Metcalfe, N. B. (1998) Shelter selection in Atlantic salmon or why do salmon seek shelter in winter? *Journal of Fish Biology,* **52,** 42–49.

Valentin, S., Sempeski, P., Souchon, Y. & Gaudin, P. (1994) Short-term habitat use by young grayling, *Thymallus thymallus* L., under variable flow conditions in an experimental stream. *Fisheries Management & Ecology,* **1,** 57–65.

Van Oppem, M. J. H., Turner, G. F., Rico, C., Deutsch, J. C., Ibrahim, K. M., Robinson, R. L. & Hewitt, G. M. (1997) Unusually fine-scale genetic structuring found in rapidly speciating Malawi cichlid fishes. *Proceedings of the Royal Society of London, Series B – Biological Sciences,* **264,** 1803–1812.

Van Someren, V. D. (1962) The migration of fish in a small Kenya River. *Revue de Zoologie et de Botanique Africaine,* **66,** 375–392.

Van Winkle, W. (2000) A perspective on power generation impacts and compensation in fish populations. *Environmental Science & Policy,* **3** (Suppl.), 425–431.

Vazzoler, A. E. A. de M., Lizama, M. L. A., Otake, V. & Agostinho, A. A. (1993) *Intensidade reprodutiva* da comunidade ictica dominante na planicie de inundaçao do alto rio Paraná e sua relaçao com variáveis ambientais. Sociedad Brasiliera de Ictologia, Universidad Esto Maringá-PR/Nupelia, 22 pp. (mimeo).

Vazzoler, A. E. A. de M. & Menezes, N. A. (1992) Sintese de conhecimento sobre o comportamento reprodutivo dos Characiformes da América do Sul (Teleostei, Ostariophysi). *Revista Brasiliera Biologica,* **54**, 627–640.

Velle, J. I., Lindsay, J. E., Weeks, R. W. & Long, F. M. (1979) An investigation of the loss mechanisms encountered in propagation from a submerged fish telemetry transmitter. In: *Proceedings of the 2nd International Conference on Wildlife Biotelemetry* (ed. F. M. Long), pp. 228–237. University of Wyoming, Laramie.

Videler, J. J. (1993) *Fish Swimming.* Chapman & Hall, London.

Videler, J. J. & Wardle, C. S. (1991) Fish swimming stride by stride: speed limits and endurance. *Reviews in Fish Biology and Fisheries,* **1**, 23–40.

Vladimirov, V. I. (1957) On the biological classification of fishes into migratory and fluvial-migratory types. *Zool. Zh.,* **36**(8).

Voegeli, F. A. & McKinnon, G. P. (1996) Recent development in ultrasonic tracking systems. In: *Underwater Biotelemetry* (eds E. Baras & J. C. Philippart), pp. 235–241. University of Liège, Belgium.

Voegeli, F. A. & Pincock, D. (1996) Overview of underwater acoustics as it applies to telemetry. In: *Underwater Biotelemetry* (eds E. Baras & J. C. Philippart), pp. 23–30. University of Liège, Belgium.

Vøllestad, L. A. & Jonsson, B. (1986) Life-history characteristics of the European eel *Anguilla anguilla* in the Imsa River, Norway. *Transactions of the American Fisheries Society,* **115**, 864–871.

Vøllestad, L. A. & Jonsson, B. (1988) A 13-year study of the population dynamics and growth of the European eel *Anguilla anguilla* in a Norwegian river: evidence for density dependent mortality and development of a model for predicting yield. *Journal of Animal Ecology,* **57**, 983–997.

Vøllestad, L. A., Jonsson, B., Hvidsten, N. A., Næsje, T. F., Haraldstad, Ø. & Ruud-Hansen, J. (1986) Environmental factors regulating the seaward migration of European silver eels (*Anguilla anguilla*). *Canadian Journal of Fisheries & Aquatic Sciences,* **43**, 1909–1919.

Vøllestad, L. A. & L'Abee-Lund, J. H. (1987) Reproductive biology of stream-spawning roach, *Rutilus rutilus. Environmental Biology of Fishes,* **18**, 219–227.

Waidbacher, H. G. & Haidvogl, G. (1998) Fish migration and fish passage facilities in the Danube: past and present. In: *Fish Migration and Fish Bypasses* (eds M. Jungwirth, S. Schmutz & S. Weiss), pp. 85–98. Fishing News Books, Blackwell Science Ltd, Oxford.

Walker, T. J. & Hasler, A. D. (1949) Detection and discrimination of odors of aquatic plants by the bluntnose minnow (*Hyborhynchus notatus*, Raf.). *Physiological Zoology,* **22**, 45–63.

Walker, M. M., Diebel, C. E., Haugh, C. V., Pankhurst, P. M., Montgomery, J. C. & Green, C. R. (1997) Structure and function of the vertebrate magnetic sense. *Nature,* **390**, 371–376.

Walton, I. & Cotton, C. (1921) *The Compleat Angler or the Contemplative Man's Recreation,* with an introduction and bibliography by R. B. Marston. Oxford University Press, London.

Wang, C. H. & Tzeng, W. N. (2000) The timing of metamorphosis and growth rates of American and European eel leptocephali: A mechanism of larval segregative migration. *Fisheries Research,* **46**, 191–205.

Wanzenbock, J., Lahnsteiner, B. & Maier, K. (2000) Pelagic early life phase of the bullhead in a freshwater lake. *Journal of Fish Biology,* **56**, 1553–1557.

Wardle, C. S. (1975) Limit of fish swimming speed. *Nature* (London), **255**, 725–727.
Wardle, C. S. (1977) Effects of size on the swimming speeds of fish. In: *Scale Effects in Animal Locomotion* (Eed. T. J. Pedley), pp. 299–313. Academic Press, London.
Wardle, C. S. & Hall, C. D. (1994) Marine video. In: *Video Techniques in Animal Behaviour* (ed. S. D. Wratten), pp. 89–111. Chapman & Hall, London.
Warren, M. L. Jr., Burr, B. M., Walsh, S. J., *et al.* (2000) Diversity, distribution, and conservation status of the native freshwater fishes of the southern United States. *Fisheries,* **25** (10), 7–31.
Warren, M. L. Jr. & Pardew, M. G. (1998) Road crossings as barriers to small-stream fish movement. *Transactions of the American Fisheries Society,* **127**, 637–644.
Waters, J. M., Epifanio, J. M., Gunter, T. & Brown, B. L. (2000) Homing behaviour facilitates subtle genetic differentiation among river populations of *Alosa sapidissima*: microsatellites and MtDNA. *Journal of Fish Biology,* **56**, 622–636.
Weatherley, N. S. (1987) The diet and growth of 0-group dace, *Leuciscus leuciscus* (L.) and roach, *Rutilus rutilus* (L.), in a lowland river. *Journal of Fish Biology,* **30**, 237–247.
Webb, J. (1990) The behaviour of adult Atlantic salmon ascending the Rivers Tay and Tummel to Pitlochry Dam. *Scottish Fisheries Research Report,* **48**, 27 pp.
Webb, M. A. H., Van-Eenennaam, J. P., Doroshov, S. I. & Moberg, G. P. (1999) Preliminary observations on the effects of holding temperature on reproductive performance of female white sturgeon, *Acipenser transmontanus* Richardson. *Aquaculture,* **176**, 315–329.
Webb, P. W. (1984) Form and function in fish swimming. *Scientific American,* **251**, 58–68.
Webb, P. W. (1986) Kinematics of lake sturgeon, *Acipenser fulvescens*, at cruising speeds. *Canadian Journal of Zoology,* **64**, 2137–2141.
Webb, P. W. (1994) The biology of fish swimming. In: *Mechanics and Physiology of Animal Swimming* (eds L. Maddock, Q. Bone & J. M. V. Rayner), pp. 45–62. Cambridge University Press.
Weber, J. M. & Kramer, D. L. (1983) Effects of hypoxia and surface access on growth, mortality, and behavior of juvenile guppies *Poecilia reticulata*. *Canadian Journal of Fisheries and Aquatic Sciences,* **40**, 1583–1588.
Wei, Q. W., Ke, F. E., Zhang, J. M. *et al.* (1997) Biology, fisheries and conservation of sturgeons and paddlefish in China. *Environmental Biology of Fishes,* **48**, 241–256.
Weinstein, M. P. (1979) Shallow marsh habitats as primary nurseries for fishes and shellfish, Cape Fear River, North Carolina. *Fishery Bulletin,* **77**, 339–357.
Weisel, G. F. & Newman, H. W. (1951) Breeding habits, development and early life history of *Richardsonius balteatus*, a northwestern minnow. *Copeia,* **1951**, 187–194.
Welcomme, R. L. (1969) The biology and ecology of the fishes of a small tropical stream. *Journal of Zoology London,* **158**, 485–529.
Welcomme, R. L. (1979) *Fisheries Ecology of Floodplain Rivers*. Longman Press, New York.
Welcomme, R. L. (1985) River fisheries. *FAO Fisheries Technical Paper,* **262**, 330 pp.
Welcomme, R. L. & Mérona, B. de (1988) Fish communities of rivers. In: *Biology and Ecology of African Freshwater Fishes* (eds C. Lévêque, M. N. Bruton & G. W. Ssentongo), pp. 251–276. Editions de l'ORSTOM, Travaux et Documents n° 216, Paris.
Werner, R. G. (1979) Homing mechanism of spawning white suckers in Wolf Lake, New York. *New York Fish and Game Journal,* **26**, 48–58.
West, R. L., Smith, M. W., Barber, W. E., Reynolds, J. B. & Hop, H. (1992) Autumn migration and overwintering of arctic grayling in coastal streams of the Arctic National Wildlife Refuge, Alaska. *Transactions of the American Fisheries Society,* **121**, 709–715.

Westerberg, H. (1982) Ultrasonic tracking of Atlantic salmon (*Salmo salar* L.) – II. Swimming depth and temperature stratification. *Reports of the Institute of Freshwater Research of Drottningholm*, **60**, 102–120.

Weyl, O. L. F. & Booth, A. J. (1999) On the life history of a cyprinid fish, *Labeo cylindricus*. *Environmental Biology of Fishes*, **55**, 215–225.

Wheeler, A. (1969) *The Fishes of the British Isles and north west Europe*. McMillan, London, 613 pp.

Whelan, K. F. (1983) Migratory patterns of bream *Abramis abramis*, L. shoals in the River Suck system. *Irish Fisheries Investigation, Series A*, **23**, 11–15.

Whelan, K. F. (1993) Decline of sea trout in the west of Ireland: an indication of forthcoming marine problems for salmon? In: *Salmon in the Sea and New Enhancement Strategies* (ed. D. Mills), pp. 171–183. Fishing News Books, London.

White, E. M. & Knights, B. (1997) Environmental factors affecting migration of the European eel in the rivers Severn and Avon, England. *Journal of Fish Biology*, **50**, 1104–1116.

White, G. C. & Garrott, R. A. (1990) *Analysis of Wildlife Radio-Tracking Data*. Academic Press, New York.

Whitehead, P. J. P. (1959) The anadromous fishes of Lake Victoria. *Revue de Zoologie et de Botanique Africaine*, **59**, 329–363.

Whitehead, P. J. P. (1984) Clupeidae. In: *Fishes of the North-eastern Atlantic and the Mediterranean*, Vol. I (eds P. J. P. Whitehead, M. L. Bauchot, J. C. Hureau, J. Nielsen & E. Tortonese), pp. 268–281. Unesco, Paris.

Whitehead, P. J. P. (1985) Clupeoid fishes of the world (suborder Clupeoidei) – an annotated and illustrated catalogue of the herrings, sardines, pilchards, sprats, shads, anchovies and wolf-herrings. Part I. Chirocentridae, Clupeidae and Pristigasteridae. *FAO Fisheries Synopsis*, **125**, 1–303.

Whitfield, A. K. & Blaber, S. J. M. (1976) The effects of temperature and salinity on *Tilapia rendalli*, Boulenger 1896. *Journal of Fish Biology*, **9**, 99–104.

Whitman, R. P., Quinn, T. P. & Brannon, T. L. (1982) Influence of suspended volcanic ash on homing behavior of adult chinook salmon. *Transactions of the American Fisheries Society*, **111**, 63–69.

Wilkerson, M. L. & Fisher, W. L. (1997) Striped bass distribution, movements and site fidelity in Robert S. Kerr Reservoir, Oklahoma. *North American Journal of Fisheries Management*, **17**, 677–686.

Williams, K. & White, R. G. (1990) Evaluation of pressure sensitive radio transmitters used for monitoring depth selection by trout in lotic systems. *American Fisheries Society Symposium*, **7**, 390–394.

Williams, R. (1971) Fish ecology of the Kafue River and floodplain environment. *Fisheries Research Bulletin of Zambia*, **5**, 305–330.

Winemiller, K. O. (1989) Patterns of variation in life history among South American fishes in seasonal environments. *Oecologia*, **81**, 225–241.

Winemiller, K. O. (1990) Spatial and temporal variations in tropical fish trophic networks. *Ecological Monographs*, **60**, 331–367.

Winemiller, K. O. (1991) Ecomorphological diversification of freshwater fish assemblages from five biotic regions. *Ecological Monographs*, **61**, 343–365.

Winemiller, K. O. (1992) Life history strategies and the effectiveness of sexual selection. *Oikos*, **62**, 318–327.

Winemiller, K. O. & Jepsen, D. B. (1998) Effects of seasonality and fish movement on tropical river food webs. *Journal of Fish Biology*, **53** (Suppl. A), 267–296.

Winemiller, K. O. & Kelso-Winemiller, L. C. (1994) Comparative ecology of the African pike, *Hepsetus odoe*, and tigerfish, *Hydrocynus forskalii*, in the Zambezi River floodplain. *Journal of Fish Biology,* **45**, 211–225.

Winemiller, K. O., Taphorn, D. C. & Barbarino-Duque, A. (1997) The ecology of *Cichla* (Cichlidae) in two blackwater rivers of Southern Venezuela. *Copeia,* **1997**, 690–696.

Winston, M. R., Taylor, C. M. & Pigg, J. (1991) Upstream extirpation of four minnow species due to damming of a prairie stream. *Transactions of the American Fisheries Society,* **120**, 98–105.

Winter, J. (1996) Advances in underwater biotelemetry. In: *Fisheries Techniques*, 2nd edn. (eds B. R. Murphy & D. W. Willis), pp. 555–590. American Fisheries Society, Bethesda, Maryland.

Winter, J. D. (1977) Summer home range movements and habitat use by four largemouth bass in Mary Lake, Minnesota. *Transactions of the American Fisheries Society,* **106**, 323–330.

Witkowski, A. (1988) The spawning run of the huchen *Hucho hucho* (L.) and its analysis. *Acta Ichthyologica et Piscatoria,* **XVIII**(2), 23–31.

Witkowski, A. & Kowalewski, M. (1988) Migration and structure of spawning population of European grayling *Thymallus thymallus* (L.) in the Dunajec basin. *Archiv für Hydrobiologie,* **112**, 279–297.

Witte, F. (1984) Ecological differentiation in Lake Victoria haplochromines: comparison of cichlid species flocks in African lakes. In: *Evolution of Fish Species Flocks* (eds A. A. Echelle & I. Kornfield), pp. 155–167. University of Maine Press, Orono.

Wood, C.M. & Shuttleworth, T.J. (1995) Cellular and molecular approaches to fish ionic regulation. *Fish Physiology Vol. XIV*. Academic Press, San Diego.

Wooley, C. M. & Crateau, E. J. (1983) Biology, population estimates and movement of native and introduced striped bass, Apalachicola River, Florida. *North American Journal of Fisheries Management,* **3**, 383–394.

Wootton, R. J. (1976) *The Biology of the Sticklebacks.* Academic Press, London.

Wootton, R. J. (1984) *A Functional Biology of Sticklebacks.* Croom Helm, London.

Wootton, R. J. (1990) *The Ecology of Teleost Fishes.* Chapman and Hall, London.

Worthman, H. O. (1982) Aspekte der biologie zweier Sciaenidenarten, der Pescadas *Plagoscion squamosissimus* (Heckel) und *Plagioscion montei* (Suares) in verschiedenen Gewässertypen Zentralamazoniens. DPhil thesis, University of Kiel, Germany, 176 pp.

Wourms, J. P. (1972) The developmental biology of annual fishes. III. Pre-embryonic and embryonic diapause of variable duration in the egg of annual fishes. *Journal of Experimental Zoology,* **182**, 389–414.

Wydoski, R. S. & Emery, L. (1983) Tagging and marking. In: *Fisheries Techniques*, 1st edn. (Eds L. Nielsen & D. Johnson), pp. 215–237. American Fisheries Society, Bethesda, Maryland.

Young, M. K. (1994) Mobility of brown trout in south-central Wyoming streams. *Canadian Journal of Zoology,* **72**, 2078–2083.

Young, M. K., Rader, R. B. & Belish, T. A. (1997) Influence of macroinvertebrate drift and light on the activity and movement of Colorado River cutthroat trout. *Transactions of the American Fisheries Society,* **126**, 428–437.

Youngson, A. F. & Webb, J. H. (1993) Thyroid hormone levels in Atlantic salmon (*Salmo salar*) during the return migration from the ocean to spawn. *Journal of Fish Biology,* **42**, 293–300.

Zakharchenko, G. M. (1973) Migrations of the grayling (*Thymallus thymallus* [L.]) in the upper reaches of the Pechora. *Voprosy Ikhtiologii,* **13**, 628–629 [in Russian].

Zale, A. V., Wiechman, J. D., Lochmiller, R. L. & Burroughs, R. J. (1990) Limnological conditions associated with summer mortality of striped bass in Keystone Reservoir, Oklahoma. *Transactions of the American Fisheries Society,* **119**, 72–76.

Zaret, T. M. (1980) Life history and growth relationships of *Cichla ocellaris*, a predatory South American cichlid. *Biotropica,* **12**, 144–157.

Zaret, T. M. (1984) Fish/zooplankton interactions in Amazon floodplain lakes. *Verhanlungen Internationale Vereinigung für Limnologie,* **22**, 1305–1309.

Zbinden, S. & Maier, K. J. (1996) Contribution to the knowledge of the distribution and spawning grounds of *Chondrostoma nasus* and *Chondrostoma toxostoma* (Pisces: Cyprinidae) in Switzerland. In: *Conservation of Endangered Freshwater Fish in Europe* (eds A. Kirchhofer & D. Hefti), pp. 287–297. Birkhäuser Verlag, Basel.

Zylberblat, M. & Menella, J. Y. (1996) Upstream passage of migratory fish through navigation locks. In: *Ecohydraulics* (eds M. Leclerc *et al.*), Vol. B, pp. 829–841. INRS-Eau, Québec.

Geographical Index

Bay
   Biscay (France), 156, 186
   Chesapeake (Maryland, Virginia, US), 21–2, 156, 216, 284
   Fundy (New Brunswick, Nova Scotia, Canada), 210
   Suisun (California, US), 168
   Voisey (Labrador, Canada), 99–100

Gulf
   Bengal, 157
   Bothnia, 185, 199
   Finland, 199
   Mexico, 9, 106, 143–4, 148, 156–8, 209, 212, 228
   St. Lawrence, 143, 209

Lake/Loch
   Årungen (Norway), 20, 30, 73
   Athabasca (Alberta, Saskatchewan, Canada), 97, 219
   Babine (British Columbia, Canada), 18
   Baikal (Siberia, Russia), 88, 208
   Baptiste (Alberta, Canada), 217
   Bear (Utah, Idaho, US), 29, 207
   Bi Na (Japan), 187
   Biwa (Japan), 187
   Black (Alaska, US), 79, 206
   Bull (Wyoming, US), 199
   Chad (Cameroon, Chad, Niger, Nigeria), 119, 132–4, 172–4, 181
   Champlain (Vermont, US), 220
   Chicamba (Mozambique), 167
   Chignik (Alaska, US), 79, 206
   Chilwa (Malawi, Mozambique), 121
   Claire (Alberta, Canada), 96, 150
   Constance (Germany), 25, 199
   Dauphin (Manitoba, Canada), 31, 169–70
   do Rei (Brazil), 119, 180, 183
   East African Great Lakes, 72, 108, 124–5, 223
      Kivu, 109, 158
      Malawi, 21, 124
      Mobutu Sese Seko, 208
      Tanganyika, 108, 124, 158, 208, 279
      Victoria, 124, 132–3, 181
   El'gygytgyn (Siberia, Russia), 195
   Ennerdale (England), 195
   Esei (Siberia, Russia), 195
   Gatun (Panama), 223
   Gebel Aulia (Sudan), 119, 173
   George (Minnesota, US), 29, 184
   Grand (Oklahoma, US), 157
   Great Bear (Northwestern Territories, Canada), 95, 98
   Great Slave (Northwestern Territories, Canada), 98, 198
   Hallstattersee (Austria), 207
   Kariba (Zambia, Zimbabwe), 158, 279
   Katavi (Tanzania), 111
   Kosogol (Mongolia), 194
   la Ronge (Saskatchewan, Canada), 97
   Lanao (Mindanao, Philippines), 108
   Laurentian Great Lakes, 276
   Lomond (Scotland), 141–2, 274
   Magadi (Kenya), 120
   Mamawi (Alberta, Canada), 96, 150
   Manyara (Tanzania), 120
   Maracaibo (Venezuela), 155
   Mendota (Illinois, US), 56
   Mweru (Zambia), 121, 132, 223

Nabugabo (Uganda), 115
Nasser (Egypt, Sudan), 272
Natron (Kenya, Tanzania), 120
Ness (Scotland), 88, 228, 239
Ngezi (Zimbabwe), 181
Nicaragua (Nicaragua), 142
Nokoué (Benin), 181
North American Great Lakes, 59–60, 139–41,
    146, 221, 227, 276, 303
    Erie, 29, 32, 139, 184, 227
    Huron, 139, 206, 227
    Michigan, 139, 284
    Ontario, 139, 170, 183, 212, 216
    Superior, 139, 219
Ogawara (Japan), 187
Oneida (New York, US), 219
Opeongo (Ontario, Canada), 21, 199
Poelela (Mozambique), 121
Portage (Michigan, US), 146
Quarum (Egypt), 121
Sibaya (South Africa), 75
Southern India (Saskatchewan, Canada), 97
St George (Ontario, Canada), 89
St Pierre (Quebec, Canada), 210
Storsjö (Sweden), 194
Tange (Denmark), 73
Titicaca (Bolivia), 108, 202–3
Tjeukemeer (The Netherlands), 87
Turkana (Kenya), 173, 208
Volta (Ghana), 127, 279
Whitney (Texas, US), 211

Rivers and major tributaries
    à la Truite (Quebec), 33
    Adige (Italy), 147
    Adour (France), 300
    Alabama (Alabama, US), 148
    Alligator (Australia), 203
    Alto Solimoes (Brazil), 182
    Amazon (South America), 17, 27, 107,
        109–10, 114, 116–20, 122, 126–7,
        129–32, 134, 142, 175, 178–83, 228,
        234, 269–70, 279
    Amudarya (Uzbekistan), 275
    Anadyr (Siberia, Russia), 195, 197
    Angeran (Finland), 101, 185, 217
    Annapolis (Nova Scotia, Canada), 156
    Apure (Venezuela), 112
    Araguaia (Brazil), 120
    Arctic Red (Northwestern Territories,
        Canada), 98
    Ardas (Greece), 290
    Argitis (Greece), 290
    Atchafalaya (Louisiana, US), 149, 198
    Axe (England), 85, 165
    Ba Tha (Chad), 125
    Bafing (Mali), 280
    Bann (Northern Ireland), 31
    Black (Alaska, US), 79
    Bol'shaya (Russia), 208
    Brahmaputra (India), 107
    Buffalo (Alberta, Canada), 98
    Cape Fear (North Carolina, Maryland, US),
        80–81, 143, 298
    Chagres (Panama), 128–9
    Chari (Chad, Cameroon), 68, 119, 132–4,
        172–4, 181
    Chikugo (Japan), 168
    Choctawhatchee (Alabama, Florida, US), 106,
        144
    Churchill (Saskatchewan, Manitoba, Canada),
        97
    Clare (Ireland), 154
    Clearwater (Idaho, US), 76
    Colorado (US, Mexico), 9, 31, 70, 82, 102,
        159–62, 169–71, 200–201, 276
    Columbia (British Columbia, Canada,
        Oregon, Washington, US), 16, 276, 296,
        302
    Conasauger (Georgia, US), 220
    Congo (former Zaire) (R.D. Congo), 107, 125
    Connecticut (New England, US), 46–7, 106,
        144–5, 297, 303
    Current (Missouri, US), 78, 169
    Danube (Europe), 2, 145–6, 157, 193, 219,
        221, 227, 272, 274, 290
    Dee (England, Wales), 154
    Delaware (Delaware, New Jersey, US), 302
    Dieset (Spitsbergen Island), 98–9
    Dniepr (Ukraine), 158
    Don (Russia), 76, 157, 164, 275
    Dordogne (France), 30, 102, 104, 217, 219,
        297
    Dvina (Russia), 228
    Dyje (Czech Republic), 82
    Eider (Germany), 75

Elands (South Africa), 167
Elbe (Czech Republic, Germany), 31, 33, 152, 154, 199–200, 227, 242
Embarrass (Wisconsin, US), 215
Endrick (Scotland), 141–2
Flathead (Montana, US), 276
Fly (Papua New Guinea), 151, 209, 221
Fraser (British Columbia, Canada), 51–2, 205, 276
Frome (England), 69, 80, 91, 165, 201, 227
Gambia (Gambia), 158
Ganges (India, Bangladesh), 142, 280
Garonne (France), 30, 102, 297
Glatt (Switzerland), 274
Grand (Ontario, Canada), 170
Great Bear (Northwestern Territories, Canada, Alaska, US), 95
Great Ouse (England), 71, 88, 219
Green (Colorado, Utah, US), 82, 160–61, 170
Groot (South Africa), 167
Guadalquivir (Spain), 145
Gudenå (Denmark), 73–4, 76, 168, 219, 274
Halda (Bangladesh), 32, 168
Hanjiang (China), 167, 275
Hudson (New York, US), 106, 144, 210–11, 302
Hulahula (Alaska, US), 79
Hull (England), 71
Hunte (Germany), 75
Illinois (Illinois, US), 215
Illinois (Oklahoma, US), 212
Imsa (Norway), 31, 33, 153
Inn (Germany), 275
Jacks Fork (Missouri, US), 215
Jamuna (Bangladesh), 129, 280
Japurá (Colombia, Brazil), 182
Jihlava (Czech Republic), 275
Jordan (Jordan, Israel), 166
Kafue (Zambia), 125, 132, 172
Kankakee (Indiana, US), 104
Kävlingeån (Sweden), 78, 101
Keret (Russia), 36, 73
Khatanga (Siberia, Russia), 195
Kivaluna (Alaska, US), 196–7
Kobuk (Alaska, US), 198
Kura (Azerbaidjan), 146
Lahn (Germany), 296
Lena (Siberia, Russia), 193, 195

Lergue (France), 163
Letaba (South Africa), 167
Leven (Scotland), 274
Little Campbell (British Columbia, Canada), 205
Little Colorado (Arizona, US), 70, 162
Loire (France), 9, 156, 201
Longlong (Taiwan), 226–7
Lupuzlz (Zambia), 132
Mackenzie (Northwestern Territories, Saskatchewan, Canada), 95–9, 150, 197–8, 255
Madeira (Brazil), 175, 178–9, 182
Magdalena (Colombia), 78, 113, 133, 175
Malagarazi (Tanzania), 208
Malse (Czech Republic), 286
Mamon (Peru), 129
Mamoré (Bolivia), 127, 133, 175
Maranon (Peru), 129
Mary (Australia), 209, 255
McIntyre (Australia), 298
Mekong (Vietnam), 180, 280
Merrimack (New Hampshire, Massachusetts, US), 106, 143–4
Meuse (France, Belgium, The Netherlands), 30–31, 37, 71, 85, 102, 105, 145, 153–4, 163, 260, 272, 274, 284–6, 292, 296, 298, 300
Mira (Portugal), 201
Miramichi (New Brunswick, Canada), 210
Missiquoi (Vermont, US), 220
Mississippi (US), 34, 75, 77, 146, 148, 150, 182, 217, 219
Missouri (US), 150, 170, 183, 221, 284
Mogi-Guaçu (Brazil), 133–4, 278
Mosel (France, Germany), 102
Murray-Darling (Australia), 83, 213–14, 222, 277
Nanay (Peru), 129
Navarro (California, US), 205
Neckar (Germany), 282
Negro (Brazil), 109
Nidd (England), 18, 30, 76, 85–6, 165, 275
Nidda (Germany), 92
Nidelva (Norway), 296
Niger (Western Africa), 32, 125, 132, 174, 180–81, 225, 269, 279
Nile (Sudan, Egypt), 119, 225, 272

Nive (France), 303
Noatak (Alaska, US), 196–7
Ochre (Manitoba, Canada), 31, 169
Ocqueoc (Michigan, US), 141
Ogôoué (Gabon), 125
Okhota (Siberia, Russia), 195
Okpilak (Alaska, US), 79
Omo (Kenya), 173
Oriège (France), 287
Orinoco (Venezuela), 19, 46–7, 49, 112, 120, 133, 175, 178–9
Oubangui (Central African Republic, Congo and R.D. Congo), 125
Ouémé (Benin), 272
Ourthe (Belgium), 33, 36, 71–2, 76, 85, 90–92, 163, 286
Padma (Bangladesh), 280
Paraguay (Paraguay), 132–3
Paraná (South America), 50, 132–4, 177–9, 278, 287–8
Pascagoula (Mississippi, US), 170
Patuxent (Maryland, US), 228
Peace-Athabasca system (Alberta, Canada), 73, 96–7, 150, 219
Pearl (Mississippi, US), 170
Pechora (Russia), 96
Po (Italy), 147
Potomac (Maryland, US), 210
Poultney (Vermont, US), 220
Ramu (Papua New Guinea), 203
Rhine (Switzerland, Germany, The Netherlands), 36, 104, 260, 274, 298
Rhone (France), 35, 70, 88, 290, 298
Roanoke (North Carolina, US), 210–12
Sacramento (California, US), 168, 186
Samiryia (Peru), 129–30
San Joaquin (California, US), 168, 186, 276
San Juan (Nicaragua), 142
San Juan (Colorado-Utah, US), 160
Sanaga (Cameroon), 125
Saskatchewan (Canada), 150
Savannah (Florida, US), 144
Selawik (Alaska, US), 198
Senegal (Senegal, Mali), 128, 132, 150, 173, 280
Sepik (Papua New Guinea), 203
Severn (England, Wales), 31, 154
Shannon (Ireland), 31, 153–4

Sheaf (England), 171
Shimanto (Japan), 207
Shou (Japan), 187
Shubenacadie (Quebec, Canada), 210
Sine-Saloum (Senegal), 158
Sinnamary (French Guyana), 172–3
Skeena (British Columbia, Canada), 47
Slave (Northwestern Territories, Canada), 96–8, 150
Spree (Germany), 164
St Clair (Michigan, US, Ontario, Canada), 227
St John (New Brunswick, Canada), 106, 144, 155–6
St Johns (Florida, US), 46
St Lawrence (Quebec, Canada), 150, 205, 209–10
Ste-Anne (Quebec, Canada), 199
Sturgeon (Michigan, US), 146
Suck (Ireland), 20, 164
Suran (France), 69
Suwannee (Florida, US), 143–4
Svratka (Czech Republic), 286
Taltson (Northwestern Territories, Canada), 98
Tambo (Australia), 189
Tanana (Alaska, US), 97, 199
Tanshui (Taiwan), 201
Taritaru (Irian Jaya), 203
Tees (England), 207, 274
Tejo (Portugal), 290
Tennessee (Kentucky-Alabama, US), 219
Terek (Georgia, Russia), 169, 275
Thames (England), 25, 35, 37, 84, 87–8, 241–2, 284
Tocantins (Brazil), 109, 177–9, 278
Tongariro (New Zealand), 192–3
Tortugueros (Costa Rica), 225
Trent (England), 242
Tuloma (Russia), 196
Tweed (England, Scotland), 154, 165
Ucayali (Peru), 109, 129–31
Ural (Russia), 146, 193
Vardnes (Norway), 196
Vinces (Ecuador), 120
Vistula (Poland), 169
Vltava (Czech Republic), 35–6, 84, 86, 286
Volga (Russia), 101, 146, 169, 275
Volta (Ghana), 127, 279
Witham (England), 84

White (Colorado-Utah, US), 20, 82, 160–61, 276
Wisconsin (Wisconsin, US), 182
Wolf (Wisconsin, US), 147, 215
Wulik (Alaska, US), 196–7
Xingu (Brazil), 109
Yalu (North Korea), 193
Yampa (Colorado, US), 20, 160–61, 170
Yangtze (China), 148–9, 193, 275
Yenisei (Russia), 145
Yorkshire Derwent (England), 164, 245, 309
Yorkshire Ouse (England), 18, 165, 242, 274
Yukon (Alaska, US), 95, 97
Zambezi (Zambia, Zimbabwe), 114, 127, 223, 279

Sea
   Adriatic, 146
   Aral, 78, 101, 146, 169, 183, 202, 217, 275
   Azov, 145–6, 169, 275
   Baltic, 4, 7, 78, 101, 151, 169, 183, 185, 186, 197, 217–8
   Barents, 228
   Beaufort, 79, 98
   Black, 4, 78, 101, 145–6, 157, 164, 169, 202, 213, 221, 227
   Caribbean, 142
   Caspian, 4, 6, 78, 101, 141, 145–6, 157, 164, 169, 183, 202, 275
   China, 7, 151
   Chukchi, 196–7
   Japan, 193
   Mediterranean, 141, 156–7, 190, 202, 213, 227, 272
   North, 7, 75, 151–2, 267, 272
   Okhotsk, 193
   Sargasso, 83, 152
   White, 186, 227–8

Sound
   Albermarle (North Carolina, US), 210–12
   Norton (Alaska, US), 197

Taxonomic Index

Abramis spp. (Cyprinidae, Cypriniformes)
   ballerus, 82
   brama, 20–21, 25, 30, 37, 72, 76–8, 82, 84–5, 87, 101, 104, 164, 169, 255, 275, 284, 298, 304
   sapa, 82
Abyssocottidae (Scorpaeniformes), 208
Acanthobitis botia (Balitoridae, Cypriniformes), 281
Acanthogobius spp. (Gobiidae, Perciformes), 226
Achiridae (Pleuronectiformes), 227–8
Achirus spp. (Achiridae, Pleuronectiformes), 228
Acipenser spp. (Acipenseridae), 142–7
   baeri, 145, 147
   brevirostrum, 58, 106, 143–4, 146–7, 173, 297
   fulvescens, 146–147, 268
   gueldenstaedtii, 146, 274
   medirostris, 145
   naccarii, 146
   nudiventris, 146–7, 275
   oxyrinchus, 9, 106, 142–146
   persicus, 146
   ruthenus, 146
   sinensis, 147
   stellatus, 146–7, 274
   sturio, 145, 273
   transmontanus, 145, 147
Acipenseridae (Acipenseriformes), 1, 12, 81, 105–6, 138, 142–7, 149, 263, 274, 283, 285
Acrocheilus alutaceus (Cyprinidae, Cypriniformes), 163
Adrianichthyidae (Beloniformes), 203

Aequidens pulchrus (Cichlidae, Perciformes), 305
Ageneiosidae (Siluriformes), 180
Agonostomus spp. (Mugilidae, Perciformes),
   monticola, 201
   telfairii, 201
Aila coila (Schilbeidae, Siluriformes), 281
Alburnoides bipunctatus (Cyprinidae, Cypriniformes), 82, 274
Alburnus alburnus (Cyprinidae, Cypriniformes), 39, 74, 82, 168, 292
Alestes spp. (Alestiidae, Characiformes), 82
   baremoze, 68, 127–8, 132, 134, 173
   dentex, 132, 173
Alestiidae (Characiformes), 82, 132, 172–3
Alosa spp. (Clupeidae, Clupeiformes), 47, 55, 79, 83, 155–7, 273, 289, 295, 297–8
   aestivalis, 155, 307
   alabamae, 156
   alosa, 51, 156, 297
   caspia, 157
   chrysochloris, 156
   fallax, 157
   ficta rhodanensis, 298
   mediocris, 156
   pontica, 157, 274
   pseudoharengus, 155–6, 284, 304, 307
   sapidissima, 46–7, 57–8, 105–7, 155–6, 297–8, 303
Ambassidae (= Chandidae, Perciformes), 139, 203, 281
Ambassis spp. (Ambassidae, Perciformes), 203
Ambloplites rupestris (Centrarchidae, Perciformes), 216
Amblyopsidae (Percopsiformes), 139

Ameiurus nebulosus (Ictaluridae, Siluriformes), 183
Amia calva (Amiidae, Amiiformes), 149
Amiidae (Amiiformes), 149
Amphilophus spp. (Cichlidae, Perciformes), 305
Anabantoidea (Perciformes), 112, 124, 139
Anabas spp. (Anabantidae, Perciformes), 118
Anablepidae (Cyprinodontiformes), 204
Anchoviella lepidentostole (Engraulidae, Clupeiformes), 155
Ancistrus spp. (Loricariidae, Siluriformes), 305
Anguilla spp. (Anguillidae, Anguilliformes)
    anguilla, 7, 26–31, 33, 36–7, 57, 59, 71, 75, 82–3, 85, 99, 151–5, 254, 262, 273, 282, 296–8, 304–5
    australis, 152
    celebensis, 152
    japonica, 7, 151–3, 254
    marmorata, 152
    rostrata, 21, 59, 83, 151–154, 304
Anguillidae (Anguilliformes), 6, 26–7, 55, 71, 83, 109, 138, 151–5, 174, 245, 277–8, 304–5
Anodontostoma spp. (Clupeidae, Clupeiformes), 159
Anodus elongatus (Hemiodontidae, Characiformes), 175
Anostomidae (Characiformes), 114, 132–3, 174–5, 288, 305
Aorichthys aor (Bagridae, Siluriformes), 281
Aphianus fasciatus (Cyprinodontidae, Cyprinodontiformes), 120
Aphredoderidae (Percopsiformes), 198
Aphredoderus sayanus (Aphredoderidae, Percopsiformes), 104, 198
Aphyocharax spp. (Characidae, Characiformes), 56
Aplocheilidae (Cyprinodontiformes), 204, 305
Aplochiton taeniatus (Galaxiidae, Osmeriformes), 190
Aplodinotus grunniens (Sciaenidae, Perciformes), 68, 83, 221, 284–5
Apteronotidae (Gymnotiformes), 116, 183–4
Arapaima gigas (Osteoglossidae, Osteoglossiformes), 107, 118, 120, 122–4, 150, 176
Aristichthys nobilis (Cyprinidae, Cypriniformes), 167

Ariidae (Siluriformes), 180
Arius spp. (Ariidae, Siluriformes)
    felis, 180
    gigas, 180
    graeffei, 180
    heudeloti, 180
    latiscutatus, 180
    madagascariensis, 180
    melanopsis, 180
Aseraggodes klunzingeri (Soleidae, Pleuronectiformes), 228
Aspius spp. (Cyprinidae, Cypriniformes), 163
    aspius, 82, 163–4
Aspredinidae (Siluriformes), 180
Astronotus crassipinnis (Cichlidae, Perciformes), 123
Astyanax spp. (Characidae, Characiformes)
    fasciatus, 174
    mexicanus, 75
Atherina boyeri (Atherinidae, Atheriniformes), 202
Atherinidae (Atheriniformes), 138, 173, 202–3
Atheriniformes, 202–3
Atherinomorus ogilbyi (Atherinidae, Atheriniformes), 202
Auchenipteridae (Siluriformes), 180
Auchenipterus nuchalis (Auchenipteridae, Siluriformes), 123, 278
Awaous spp. (Gobiidae, Perciformes), 226
    melanocephalus, 227

Bagridae (Siluriformes), 108, 180–81, 281
Balitoridae (Cypriniformes), 113, 171, 281
Barbatula barbatula (Balitoridae, Cypriniformes), 10, 171, 274, 298
Barbus spp. (Cyprinidae, Cypriniformes), 163, 167
    altianalis, 132
    anoplus, 167
    barbus, 10, 20–21, 23–6, 29–30, 33, 61, 63–4, 70–73, 76, 78, 82, 85–7, 89–90, 92, 100, 102–3, 163–4, 247, 254, 262, 265, 270, 272, 275, 286–7
    brachycephalus, 275
    canis, 166–7
    haasi, 166
    longiceps, 166–7
    meridionalis, 163, 254

schwanfeldii, 305
tetrastigma, 305
trimaculatus, 167
unitaeniatus, 167
Bathygobius soporator (Gobiidae, Perciformes), 61
Bedotiidae (Atheriniformes), 203
Belonidae (Beloniformes), 203, 281
Belontiidae (Perciformes), 281
Betta spp. (Belontiidae, Perciformes), 118
Bidyanus bidyanus (Terapontidae, Perciformes), 222, 304
Blenniidae (Perciformes), 6, 139
Blicca bjoerkna (Cyprinidae, Cypriniformes), 30, 82, 102–4, 164
Boleophthalmus spp. (Gobiidae, Perciformes), 225
Bovichthyidae, Perciformes, 224
Brachygobius nunus (Gobiidae, Perciformes), 281
Brachyhypopomus spp. (Hypopomidae, Gymnotiformes), 116
Brachyplatystoma spp. (Pimelodidae, Siluriformes), 112, 123, 131, 135, 180, 182, 269–70, 279
flavicans, 132, 182
vaillantii, 132, 182
Brachysynodontis batensoda (Mochokidae, Siluriformes), 132, 181
Brevoortia spp. (Clupeidae, Clupeiformes)
patronus, 158
tyrannus, 80–81, 158, 284
Brienomyrus niger (Mormyridae, Osteoglossiformes), 150, 236
Brochis spp. (Callichthyidae, Siluriformes), 118
Brycinus spp. (Alestiidae, Characiformes)
imberi, 127
leuciscus, 32, 127–8, 132, 135, 174, 279
longipinnis, 127
nurse, 127–8, 173
Brycon spp. (Characidae, Characiformes), 115, 118, 123, 128–9, 174–5
moorei, 126, 172
orbignyanus, 278, 288
petrosus, 128–9

Callichthyidae (Siluriformes), 118, 180, 305
Capoeta damascina (Cyprinidae, Cypriniformes), 166
Caquetaia kraussii (Cichlidae, Perciformes), 113
Carassius spp. (Cyprinidae, Cypriniformes)
auratus, 305–6
carassius, 101
Carcharhinus leucas (Carcharhinidae, Carcharhiniformes), 142
Carnegiella spp. (Gasteropelecidae, Characiformes), 305
Carpiodes spp. (Catostomidae, Cypriniformes)
carpio, 170
cyprinus, 31, 169–70
Caspiomyzon wagneri (Petromyzontidae, Petromyzontiformes), 141
Catathyridium jenyinsii (Achiridae, Pleuronectiformes), 228
Catla catla (Cyprinidae, Cypriniformes), 32, 168, 280–81
Catostomidae (Cypriniformes), 70, 78, 80–82, 102, 138, 169–71, 276, 284
Catostomus spp. (Catostomidae)
catostomus, 97
commersoni, 19–20, 169–70
discobolus, 70, 169
latipinnis, 70, 102, 169, 171, 276
macrocheilus, 170
Centrarchidae (Perciformes), 10, 12, 25, 32, 34, 72, 77, 138, 202, 214–6, 289, 302, 305
Centropomidae (Perciformes), 108, 208–9
Chacidae (Siluriformes), 180
Chandidae, *see* Ambassidae
Chanda nama (Ambassidae, Perciformes), 281
Chanidae (Gonorhynchiformes), 159
Channa (Channidae, Perciformes)
punctata, 281
striata, 115, 305
Channidae (Perciformes), 281, 305
Chanos chanos (Chanidae, Gonorhynchiformes), 63, 109, 159
Characidae (Characiformes), 27, 56, 68, 75, 109, 113, 126–9, 131–3, 172–6, 288, 305
Characidium spp. (Characidae, Characiformes), 174
Characiformes, 3, 17, 39, 49, 64–5, 68, 75, 78, 108, 112, 115, 119–20, 127–30, 134, 138, 172–9, 181–2, 246, 264, 278–9, 287–8, 307

Chasmistes cujus (Catostomidae, Cypriniformes), 170
Cheimarrichthys fosteri (Pinguipedidae, Perciformes), 224, 277
Cheirodon spp. (Characidae, Characiformes), 56
Chelon labrosus (Mugilidae, Perciformes), 201
Chitala lopis (Notopteridae, Osteoglossiformes), 150
Chondrostoma spp. (Cyprinidae, Cypriniformes), 163
  nasus, 70, 82–3, 89–90, 163–4, 275, 290
  polylepis, 290
Cichla spp. (Cichlidae, Perciformes), 123
  ocellaris, 223
  temensis, 115
Cichlidae (Perciformes), 4, 21, 72, 75, 87, 108–9, 113–15, 122, 124–5, 138, 180, 222–3, 254, 264, 279, 305
Cirrhinus spp. (Cyprinidae, Cypriniformes), 280–81
  ariza, 281
  cirrhosus, 281
  mrigala, 32, 168
Citharichthys spp. (Paralichthyidae, Pleuronectiformes), 228
Citharinidae (Characiformes), 132, 173
Citharinus citharus (Citharinidae, Characiformes), 173
Clarias spp. (Clariidae, Siluriformes)
  anguillaris, 181
  gariepinus, 113, 121, 181
  ngamensis, 113
Clariidae (Siluriformes), 108, 111–3, 118, 120, 128, 180–81
Claroteidae (Siluriformes), 180–81
Clupanodon spp. (Clupeidae, Clupeiformes), 159
Clupeichthys aesarnensis (Clupeidae, Clupeiformes), 109
Clupeidae (Clupeiformes), 4, 39, 55, 81, 108–9, 127, 137–8, 155–8, 208, 246, 272, 279–81, 283–5, 302, 306
Clupeiformes, 155–9
Clupeonella cultriventris (Clupeidae, Clupeiformes), 157
Clupisoma garua (Schilbeidae, Siluriformes), 281
Cobitidae (Cypriniformes), 171, 281

Cobitis taenia (Cobitidae, Cypriniformes), 171
Colisa fasciatus (Belontiidae, Perciformes), 281
Colossoma spp. (Characidae, Characiformes), 115, 118, 174–5
  macropomum, 118–19, 123, 126–7, 175–6
  mitrei, 132
Comephoridae (Scorpaeniformes), 208
Coregonus spp. (Salmonidae, Salmoniformes), 94, 137, 141, 190, 193, 197–8, 276, 312
  albula, 197–8
  artedii, 197
  autumnalis, 95, 99, 198
  clupeaformis, 95, 197
  lavaretus, 198
  nasus, 95, 99, 197
  sardinella, 95, 198
Coreius heterodon (Cyprinidae, Cypriniformes), 167, 275
Corica soborna (Clupeidae, Clupeiformes), 280–81
Corydoras aneus (Callichthyidae, Siluriformes), 305
Cottidae (Scorpaeniformes), 4, 39, 42, 138, 206–8, 263, 271
Cottus spp. (Cottidae, Scorpaeniformes)
  aleuticus, 207
  asper, 207
  bairdi, 227
  cognatus, 227
  extensus, 29, 207
  gobio, 10, 38, 79, 91, 206–8, 274, 298
  hangiongensis, 208
  kazika, 63, 207
Couesius spp. (Cyprinidae, Cypriniformes), 165
  plumbeus, 166
Creagrutus spp. (Characidae, Characiformes), 113, 174
Crossocheilus spp. (Cyprinidae, Cypriniformes), 278
Ctenogobius brunneus (Gobiidae, Perciformes), 227
Ctenoluciidae (Characiformes), 172, 305
Ctenolucius spp. (Ctenoluciidae, Characiformes), 305
Ctenopharyngodon idella (Cyprinidae, Cypriniformes), 37, 76, 101, 167
Ctenopoma spp. (Anabantidae, Perciformes), 124
  intermedium, 112

Culaea inconstans (Gasterosteidae, Gasterosteiformes), 206
Curimata spp. (Curimatidae, Characiformes), 175
　kneri, 119
Curimatidae (Characiformes), 37, 46, 68, 82, 112, 119, 127, 133, 174–5, 177–9, 288
Cycleptus meridionalis (Catostomidae, Cypriniformes), 170
Cynodontidae (Characiformes), 174–5, 288
Cynodon gibbus (Cynodontidae, Characiformes), 174
Cynoglossidae (Pleuronectiformes), 227–8
Cynolebias spp. (Aplocheilidae, Cyprinodontiformes), 305
Cyprichromis spp. (Cichlidae, Perciformes), 125, 223
Cyprinidae (Cypriniformes), 6, 10, 13, 20, 22, 24–6, 29, 32–3, 36, 39, 47, 60–61, 64, 68–74, 76–8, 80–83, 89, 91, 96, 100–102, 105, 107, 113, 132, 137–8, 159–69, 241–2, 246–7, 264, 270, 272, 274, 276, 280–81, 284, 286, 289–90, 292, 295, 300, 302, 304–6
Cyprinodon spp. (Cyprinodontidae, Cyprinodontiformes)
　fasciatus, 120
　macularius, 305
Cyprinodontidae (Cyprinodontiformes), 108, 122, 138, 204, 305
Cyprinus carpio (Cyprinidae, Cypriniformes), 20, 41, 101, 272
Cyrtocara spp. (Cichlidae, Perciformes), 125, 223

Dallia pectoralis (Umbridae, Esociformes), 97, 185
Danio malabaricus (Cyprinidae, Cypriniformes), 305
Danionella translucida (Cyprinidae, Cypriniformes), 107
Dasyatis spp. (Dasyatidae, Rajiformes), 142
Denticipidae (Clupeiformes), 155
Denticeps clupeoides (Denticipidae, Clupeiformes), 155
Dicentrarchus spp. (Moronidae, Perciformes)
　labrax, 213
　punctatus, 213

Distichodus spp. (Citharinidae, Characiformes)
　niloticus, 173
　rostratus, 132, 173
Doradidae (Siluriformes), 133, 180
Dormitator spp. (Eleotridae, Perciformes)
　latifrons, 225
　lebretonis, 225
　maculatus, 225
Dorosoma spp. (Clupeidae, Clupeiformes), 155, 157
　cepedianum, 157, 304
　petenense, 285

Eigenmannia virescens (Eigenmanniidae, Gymnotiformes), 116
Eigenmanniidae (Gymnotiformes), 116, 184
Elasmobranchii, 142
Elassomatidae (Perciformes), 139
Electrophoridae (Gymnotiformes), 116, 183
Electrophorus electricus (Electrophoridae, Gymnotiformes), 116, 118, 123, 176, 183
Eleotridae (Perciformes), 224–7
Eleotris spp. (Eleotridae, Perciformes), 226
　acanthopoma, 227
Elopidae (Elopiformes), 121, 151
Engraulidae (Clupeiformes), 155, 288
Enneacanthus obesus (Centrarchidae, Perciformes), 305
Erythrinidae (Characiformes), 112–15, 118, 172–3, 176, 305
Erythrinus erythrinus (Erythrinidae, Characiformes), 118
Escualosa thorocata (Clupeidae, Clupeiformes), 158
Esomus spp. (Cyprinidae, Cypriniformes), 32
Esocidae (Esociformes), 39, 96, 138, 184–5, 295, 305
Esox spp. (Esocidae, Esociformes)
　americanus, 104, 185, 305
　lucius, 7, 21, 23–7, 29–32, 34, 36–7, 41, 43, 73, 84–5, 91–2, 96–7, 105, 165, 184–5, 217, 253
　masquinongy, 20, 184–5
Etheostoma spp. (Percidae, Perciformes)
　exile, 220
　flabellare, 220
　microperca, 220

Ethmalosa fimbriata (Clupeidae, Clupeiformes), 158
Fundulidae (Cyprinodontiformes), 204, 289, 305
Fundulus heteroclitus (Fundulidae, Cyprinodontiformes), 305

Gadidae (Gadiformes), 97, 138, 198–200, 284, 306
Gadus morhua (Gadidae, Gadiformes), 265
Gagatia cenia (Sisoridae, Siluriformes), 281
Galaxiidae (Osmeriformes), 189–90, 277–8
Galaxias spp. (Galaxiidae, Osmeriformes)
    brevipinnis, 189
    depressiceps, 189
    gollumoides, 189
    maculatus, 189–90
Gambusia spp. (Poeciliidae, Cyprinodontiformes), 305
Garra borneensis (Cyprinidae, Cypriniformes), 113
Gasteropelecidae (Characiformes), 305
Gasterosteidae (Gasterosteiformes), 138, 204–6
Gasterosteus aculeatus (Gasterosteidae, Gasterosteiformes), 17–18, 59, 61, 79, 204–5, 245, 274
Gastromyzon spp. (Balitoridae, Cypriniformes), 113, 278
Geotria australis (Petromyzontidae, Petromyzontiformes), 63, 139, 278
Gila cypha (Cyprinidae, Cypriniformes), 70, 102, 162
Glossogobius spp. (Gobiidae, Perciformes), 281
    biocellatus, 226–7
    celebius, 227
    giuris, 281
Glossolepis multisquamatus (Melanotaeniidae, Atheriniformes), 203
Glyphis gangeticus (Carcharhinidae, Carcharhiniformes), 142
Glyptothorax major (Sisoridae, Siluriformes), 113
Gobiidae (Perciformes), 224–6, 281
Gobio spp. (Cyprinidae, Cypriniformes)
    albipinnatus, 82, 88
    gobio, 10, 21, 23, 25–6, 85, 241, 274–5
Gobioid (Perciformes), 27, 113, 224–7, 278, 281, *see also* Eleotridae, Gobiidae, Odontobutidae and Rhyacichthyidae

Gobiomorphus spp. (Eleotridae, Perciformes), 225, 278
    australis, 298
    basalis, 224
    huttoni, 224
Gobiomorus dormitor (Eleotridae, Perciformes), 225
Goodeidae (Cyprinodontiformes), 204
Gudusia chapra (Clupeidae, Clupeiformes), 280–81
Gymnarchidae (Osteoglossiformes), 150
Gymnarchus niloticus (Gymnarchidae, Osteoglossiformes), 150
Gymnocephalus spp. (Percidae, Perciformes), 221
    acerinus, 221
    baloni, 221
    cernuus, 221, 241, 304
    schraetser, 221
Gymnotidae (Gymnotiformes), 116, 183
Gymnotiformes, 3, 58, 114, 116–17, 120, 183–4, 236
Gymnotus spp. (Gymnotidae, Gymnotiformes), 118
    carapo, 183
Gyrinocheilidae (Cypriniformes), 139

Haemulidae (Perciformes), 139
Hampala sabana (Cyprinidae, Cypriniformes), 113
Hemibarbus barbus (Cyprinidae, Cypriniformes), 138
Hemiodontidae (Characiformes), 175, 288, 305
Hemiodus spp. (Hemiodontidae, Characiformes), 175, 305
    orthonops, 288
Hemiramphidae (Beloniformes), 203
Hemisynodontis membranaceus (Mochokidae, Siluriformes), 132, 181
Hepsetidae (Characiformes), 132, 172
Hepsetus odoe (Hepsetidae, Characiformes), 112–13, 125, 132, 172
Herkotslichthys spp. (Clupeidae, Clupeiformes), 159
Herotilapia multispinosa (Cichlidae, Perciformes), 305
Heterobranchus longifilis (Clariidae, Siluriformes), 180–81

Heteropneustes spp. (Heteropneustidae, Siluriformes), 118, 280
   fossilis, 281
Heteropneustidae (Siluriformes), 180, 281
Hilsa kelee (Clupeidae, Clupeiformes), 157, 280
Himantura spp. (Dasyatidae, Rajiformes), 142
Hiodontidae (Osteoglossiformes), 68, 73, 83, 96, 138, 149–50
Hiodon spp. (Hiodontidae, Osteoglossiformes)
   alosoides, 68, 73, 83, 96, 150
   tergisus, 150
Hoplias spp. (Erythrinidae, Characiformes), 112, 114–15, 176
   aimara, 172–4
   malabaricus, 112–13, 118–19, 123, 172, 174, 305
Hoplosternum spp. (Callichthyidae, Siluriformes), 118
Hucho spp. (Salmonidae, Salmoniformes), 190
   bleekeri, 193
   hucho, 193, 274
   ishikawae, 193
   perryi, 193
   taimen, 193
Huso spp. (Acipenseridae, Acipenseriformes)
   dauricus, 146
   huso, 2, 147, 271–2, 274
Hydrocynus spp. (Alestiidae, Characiformes), 112, 172
   brevis, 113, 115, 132
   forskalii, 113, 115, 172
   goliath, 172
   vittatus, 113, 279
Hypentelium nigricans (Catostomidae, Cypriniformes), 78, 169
Hyperopisus bebe (Mormyridae, Osteoglossiformes), 132, 150, 173
Hyphessobrycon spp. (Characidae, Characiformes), 56
Hypomesus spp. (Osmeridae, Osmeriformes)
   nipponensis, 187
   transpacificus, 186
Hypophthalmichthys molitrix (Cyprinidae, Cypriniformes), 167
Hypophthalmus spp. (Pimelodidae, Siluriformes), 180
   edentatus, 278
   marginatus, 135

Hypopomidae (Gymnotiformes), 116, 183
Hypopomus spp. (Hypopomidae, Gymnotiformes), 183
Hypostomus spp. (Loricariidae, Siluriformes), 118
   punctatus, 181
Hypseleotris compressa (Eleotridae, Perciformes), 298

Ictaluridae (Siluriformes), 138, 180, 182–3, 302
Ictalurus punctatus (Ictaluridae, Siluriformes), 70, 182–3, 264
Ictiobus (Catostomidae, Cypriniformes)
   bubalus, 170
   cyprinellus, 170
Ilisha spp. (Clupeidae, Clupeiformes), 155
   africana, 158
   megaloptera, 155
   novacula, 155
Indostomus paradoxus (Indostomidae, Gasterosteiformes), 206

Joturus pichardi (Mugilidae, Perciformes), 201
Jordanella floridae (Cyprinodontidae, Cyprinodontiformes), 305

Kneriidae (Gonorhynchiformes), 139
Kribia spp. (Eleotridae, Perciformes), 225
Kriptopterus spp. (Siluridae, Siluriformes), 305
Kuhlia spp. (Kuhliidae, Perciformes)
   rupestris, 222, 278
   sandvicensis, 222
Kuhliidae (Perciformes), 222
Kurtidae (Perciformes), 139

Labeo spp. (Cyprinidae, Cypriniformes), 167, 280, 305
   altivelis, 132
   bata, 281
   cylindricus, 167
   molybdinus, 167
   rohita, 32, 168, 281
   senegalensis, 132
   umbratus, 167
Labidesthes sicculus (Atherinidae, Atheriniformes), 202
Lampetra spp. (Petromyzontidae, Petromyzontiformes), 29, 140–42

ayresii, 141
fluviatilis, 26, 30, 140–42, 274
japonica, 95, 97, 141
planeri, 29–30, 32–3, 140–42, 245
tridentata, 141
Lateolabrax japonicus (Moronidae, Perciformes), 213
Lates spp. (Centropomidae, Perciformes)
angustifrons, 208
calcarifer, 208–9, 255, 295
japonicus, 209
longispinis, 208
macrophthalmus, 208
mariae, 208
microlepis, 208
niloticus, 115, 120, 125, 208, 268, 272
stappersii, 208
Leiopotherapon unicolor (Terapontidae, Perciformes), 222
Leiostomus xanthurus (Sciaenidae, Perciformes), 80–81
Lentipes spp. (Gobiidae, Perciformes), 225–6, 278
concolor, 225
Lepidocephalichthys guntea (Cobitidae, Cypriniformes), 281
Lepidosiren paradoxa (Lepidosirenidae, Lepidosireniformes), 118
Lepisosteidae (Lepisosteiformes), 138, 149
Lepisosteus spp. (Lepisosteidae, Lepisosteiformes), 149
oculatus, 31, 149
osseus, 149
Lepomis spp. (Centrarchidae, Perciformes), 216
auritus, 216
cyanellus, 55, 104, 305
gibbosus, 305
humilis, 104
macrochirus, 11, 34, 42, 77, 216
Leporinus spp. (Anostomidae, Characiformes), 174, 288
copelandii, 135
elongatus, 135, 278
fasciatus, 305
muyscorum, 175
obtusidens, 132, 278
Leptatherina presbyteroides (Atherinidae, Atheriniformes), 202

Leptocottus armatus (Cottidae, Scorpaeniformes), 8
Leuciscus spp. (Cyprinidae, Cypriniformes), 163
cephalus, 10, 23, 26, 30, 70, 73, 78, 82, 85, 88, 90, 100, 103, 164, 284, 290, 310
idus, 78, 82, 101
leuciscus, 10, 20, 25–6, 30, 73, 78, 80, 82, 89–91, 102–3, 165, 241, 267, 275, 284
pyrenaicus, 166
souffia, 274
Leucopsarion petersi (Gobiidae, Perciformes), 227
Limnothrissa miodon (Clupeidae, Clupeiformes), 39, 109, 158, 208, 279–80
Liza spp. (Mugilidae, Perciformes), 200–201
abu, 200–201
aurata, 201
dumerilii, 201
falcipinnis, 201
parsia, 201
ramada, 9, 57, 201
saliens, 201
Loricariidae (Siluriformes), 113, 118, 133, 181, 305
Lota lota (Lotidae, Gadiformes), 20, 97, 198–200, 284
Lovettia seali (Galaxiidae, Osmeriformes), 190
Lutjanidae (Perciformes), 221
Lutjanus spp. (Lutjanidae, Perciformes)
argentimaculatus, 221
goldiei, 221
russellii, 221
Luxilus (= Notropis) cornutus (Cyprinidae, Cypriniformes), 57
Lycengraulis grossidens (Engraulidae, Clupeiformes), 155, 288

Maccullochella peelii (Percichthyidae, Perciformes), 213–14
Macquaria spp. (Percichthyidae, Perciformes)
ambigua, 213–14, 304
australasica, 214
colonorum, 213
novemaculeata, 83, 213, 295
Malapteruridae (Siluriformes), 180
Malapterurus electricus (Malapteruridae, Siluriformes), 180

Marcusenius spp. (Mormyridae, Osteoglossiformes)
   cyprinoides, 133, 173
   senegalensis, 127–8, 150
Mastacembelidae (Synbranchiformes), 139, 281
Mastacembelus spp. (Mastacembelidae, Synbranchiformes)
   armatus, 281
   pancalus, 281
Megalocottus platycephalus (Cottidae, Scorpaeniformes), 208
Megalopidae (Elopiformes), 151
Megalops spp. (Megalopidae, Elopiformes)
   atlanticus, 151
   cyprinoides, 151
Melanotaenia splendida (Melanotaeniidae, Atheriniformes), 203
Mesopristes kneri (Terapontidae, Perciformes), 222
Metynnis spp. (Characidae, Characiformes), 123, 305
Micralestes acutidens (Alestiidae, Characiformes), 112
Microgadus tomcod (Gadidae, Gadiformes), 198–9
Micropogonias undulatus (Sciaenidae, Perciformes), 81
Micropterus spp. (Centrarchidae, Perciformes)
   dolomieui, 21, 85, 215
   salmoides, 20–22, 34, 81, 85, 87, 214–16
Mochokidae (Siluriformes), 112, 132–3, 173, 180–81, 279
Moenkhausia spp. (Characidae, Characiformes), 56
Monodactylidae (Perciformes), 139
Mordacia spp. (Petromyzontidae, Petromyzontiformes), 139
   mordax, 141
Mormyridae (Osteoglossiformes), 32, 58, 112, 132–3, 173, 236, 279
Mormyrops deliciosus (Mormyridae), 150
Mormyrus rume (Mormyridae), 150, 173
Moronidae (Perciformes), 138, 209–13, 306
Morone spp. (Moronidae, Perciformes)
   americana, 212
   chrysops, 19, 56, 212
   saxatilis, 7, 14, 45, 58, 83, 105, 209–12, 285, 302

Morulius calbasu (Cyprinidae, Cypriniformes), 281
Moxostoma spp. (Catostomidae, Cypriniformes)
   anisurum, 170
   duquesnii, 170
   erythurum, 170
   macrolepidotum, 170
   valenciennesi, 170
Mugilidae (Perciformes), 9, 109, 200–202, 280–81
Mugil spp. (Mugilidae, Perciformes)
   cephalus, 80–81, 200–201
   curema, 80–81
Mustelus canis (Triakidae, Carchariniformes), 142
Mylocheilus caurinus (Cyprinidae, Cypriniformes), 168
Mylopharyngodon piceus (Cyprinidae, Cypriniformes), 167
Mylossoma duriventre (Characidae, Characiformes), 123, 174
Mystus spp. (Bagridae, Siluriformes), 280
   bleekeri, 281
   tengara, 281
   vittatus, 281
Myxus spp. (Mugilidae, Perciformes)
   capensis, 201
   petardi, 201

Nematalosa spp. (Clupeidae, Clupeiformes), 159
Nematogobius maindroni (Gobiidae, Perciformes), 225
Neogobius melanostomus (Gobiidae, Perciformes), 227
Notemigonus crysoleucas (Cyprinidae, Cypriniformes), 24, 304–5
Notesthes robusta (Tetrarogidae, Scorpaeniformes), 206
Nothobranchius spp. (Aplocheilidae, Cyprinodontiformes), 122
Notopteridae (Osteoglossiformes), 150
Notropis spp. (Cyprinidae, Cypriniformes), 163, 165
Noturus spp. (Ictaluridae, Siluriformes), 180

Odaxothrissa mento (Clupeidae, Clupeiformes), 279

Odontesthes bonariensis (Atherinidae, Atheriniformes), 202–3
Odontobutidae (Perciformes), 224
Oligolepis acutipennis (Gobiidae, Perciformes), 227
Ompok pabda (Siluridae, Siluriformes), 281
Oncorhynchus spp. *(*Salmonidae, Salmoniformes*)*, 45–6, 54, 59, 63, 69, 92, 94–5, 98, 190–92, 195, 240, 302, 304
  clarki, 192
  gorbuscha, 27, 62, 69, 95, 252, 276
  keta, 27, 29, 69, 95
  kisutch, 27, 57, 59–61, 95, 289, 302, 304
  mykiss, 20, 57–8, 91, 192–3, 203, 272, 302, 304–5
  nerka, 18, 42–4, 46–7, 51–2, 81, 87–8, 95, 107, 192, 235, 239, 276
  tshawytscha, 47, 95, 190, 276, 302, 304
Oreochromis spp. (Cichlidae, Perciformes), 120, 124–5, 122, 269
  alcalicus, 120
  amphimelas, 120
  andersonii, 223
  aureus, 121, 305
  macrochir, 121, 223
  mossambicus, 75, 279
  niloticus, 121, 279
  shiranus, 121
  variabilis, 72
Orestias spp. (Cyprinodontidae, Cyprinodontiformes), 202
Osmeridae (Osmeriformes), 81, 138, 186–7, 284
Osmerus spp. (Osmeridae, Osmeriformes)
  eperlanus, 81, 186
  mordax, 186, 284, 304
Osphronemus spp. (Osphronemidae, Perciformes), 118
Osteoglossidae (Osteoglossiformes), 150, 305
Osteoglossiformes, 108, 149–50, 236
Osteoglossum bicirrhosum (Osteoglossidae, Osteoglossiformes), 118, 123, 305

Pangasiidae (Siluriformes), 180
Pangasius (= Pangasianodon) gigas (Pangasiidae, Siluriformes), 180, 280
Pantodontidae (Osteoglossiformes), 139, 305
Pantodon buchholzi (Pantodontidae, Osteoglossiformes), 305

Papillogobius reichei (Gobiidae, Perciformes), 227
Paracheirodon innesi (Characidae, Characiformes), 305
Parailia (= Physailia) pellucida (Schilbeidae, Siluriformes), 127
Paralichthyidae (Pleuronectiformes), 227–8
Paralichthys spp. (Paralichthyidae, Pleuronectiformes), 80–81
Parambassis baculis (Ambassidae, Perciformes), 281
Paratrygon spp. (Potamotrygonidae, Rajiformes), 142
Parhomaloptera spp. (Balitoridae, Cypriniformes), 113
Pellona spp. (Clupeidae, Clupeiformes), 155
Pellonula spp. (Clupeidae, Clupeiformes), 127, 159, 279
Perca spp. (Percidae, Perciformes), 68, 216–18
  flavescens, 10, 21, 89, 91, 96, 216–18, 305
  fluviatilis, 7, 23–5, 82, 89, 103, 105, 216–18, 241, 254, 284, 304–5
Percichthyidae (Perciformes), 83, 213–14
Percidae (Perciformes), 13, 72, 96, 138, 202, 216–21, 241, 284, 289, 295, 302, 305
Percina spp. (Percidae, Perciformes)
  antesella, 220
  caprodes, 220
  maculata, 220
  nigrofasciata, 220
  pantherina, 289
Percopsidae (Percopsiformes), 138, 198
Percopsis omiscomaycus (Percopsidae, Percopsiformes), 198
Periophthalmodon spp. (Gobiidae, Perciformes), 225
Periophthalmus spp. (Gobiidae, Perciformes), 225
Petrocephalus catostoma (Mormyridae, Osteoglossiformes), 112
Petromyzon marinus (Petromyzontidae, Petromyzontiformes), 26, 30, 33, 60, 105, 137, 139–41, 274, 276
Petromyzontidae (lampreys, Petromyzontiformes), 2, 4, 27, 39, 53, 60, 63, 81, 137–41
Phallostethidae (Atheriniformes), 203
Philypnodon grandiceps (Eleotridae, Perciformes), 298

Phoxinus spp. (Cyprinidae, Cypriniformes)
  eos, 24
  phoxinus, 17, 20–21, 23, 55, 85–6, 91, 165
Phractolaemidae (Gonorhynchiformes), 139
Piaractus spp. (Characidae, Characiformes), 174–5
  brachypomus, 117, 172, 175–6
  mesopotamicus, 278, 288
Pimelodidae (Siluriformes), 17, 27, 37, 64, 75, 82, 112, 114–15, 120, 129, 132–4, 180–82, 269–70, 278–9, 288
Pimelodus clarias (Pimelodidae, Siluriformes), 288
Pimephales promelas (Cyprinidae, Cypriniformes), 24
Pinguipedidae (Perciformes), 224
Plagioscion spp. (Sciaenidae, Perciformes), 123, 221
  squamosissimus, 115, 278
  ternetzi, 288
Platichthys spp. (Pleuronectidae, Pleuronectiformes)
  flesus, 57–8, 227, 245
  stellatus, 228
Platygobio gracilis (Cyprinidae, Cypriniformes), 97
Plecoglossus altivelis (Plecoglossidae, Osmeriformes), 68, 187
Plesiotrygon spp. (Potamotrygonidae, Rajiformes), 142
Pleuronectes spp. (Pleuronectidae, Pleuronectiformes)
  glacialis, 228
  platessa, 6, 228, 267
Pleuronectidae (Pleuronectiformes), 227–8
Pleuronectiformes, 46, 227–9
Plotosidae (Siluriformes), 180
Poecilia spp. (Poeciliidae, Cyprinodontiformes), 305
Poeciliidae (Cyprinodontiformes), 50, 124, 204, 305
Pogonichthys macrolepidotus (Cyprinidae, Cypriniformes), 168, 276
Polynemidae (Perciformes), 139
Polyodon spathula (Polyodontidae, Acipenseriformes), 148–9, 285
Polyodontidae (Acipenseriformes), 148–9
Polypteridae (Polypteriformes), 139

Polypterus spp. (Polypteridae, Polypteriformes), 118
Pomoxis spp. (Centrarchidae, Perciformes)
  annularis, 216
  nigromaculatus, 34, 59, 77, 216
Potamorhina spp. (Curimatidae, Characiformes), 175
  latior, 119, 123
Potamotrygon spp. (Potamotrygonidae, Rajiformes), 142
Pristigaster spp. (Clupeidae, Clupeiformes), 155
Pristis spp. (Pristidae, Pristiformes)
  microdon, 142
  perotteti, 142
Prochilodus spp. (Curimatidae, Characiformes), 45, 47, 50, 53, 68, 75, 82, 123, 133, 135, 175, 177–9, 288, 312
  lineatus, 50, 133–5, 177–9, 278
  magdalenae, 50, 175
  mariae, 19, 46–7, 49, 112, 133, 178–9
  nigricans, 119, 133, 177–9, 278
  platensis, 133, 135
  scrofa, see *Prochilodus lineatus*
Profundulidae (Cyprinodontiformes), 204
Prosopium spp. (Salmonidae, Salmoniformes), 197
  williamsoni, 76
Protomyzon spp. (Balitoridae, Cypriniformes), 113, 278
Protopterus spp. (Protopteridae, Lepidosireniformes), 79, 118, 120
Prototroctes spp. (Retropinnidae, Osmeriformes)
  maraena, 188–9
  oxyrhynchus, 188–9
Psectrogaster spp. (Curimatidae, Characiformes), 175
  amazonica, 119
  curviventris, 288
  rutiloides, 119
Psephurus gladius (Polyodontidae, Acipenseriformes), 148–9, 275
Pseudaphritis urvillii (Bovichthyidae, Perciformes), 224
Pseudocrenilabrus multicolor (Cichlidae, Perciformes), 72
Pseudeutropius atherinoides (Schilbeidae,

Siluriformes), 281
Pseudobarbus asper (Cyprinidae, Cypriniformes), 167
Pseudomugilidae (Atheriniformes), 203
Pseudoplatystoma spp. (Pimelodidae, Siluriformes), 112, 114, 123, 131, 182, 269, 279
  coruscans, 133, 278
  fasciatum, 112, 133
  tigrinum, 112
Pseudopleuronectes americanus (Pleuronectidae, Pleuronectiformes), 228
Pseudorhombus spp. (Paralichthyidae, Pleuronectiformes), 228
Pseudotropheus aurora (Cichlidae, Perciformes), 21, 85, 87, 223
Pterodoras granulosus (Doradidae, Siluriformes), 133, 135
Pterophyllum spp. (Cichlidae, Perciformes), 305
Pterygoplichthys spp. (Loricariidae, Siluriformes), 118
Ptychocheilus spp. (Cyprinidae, Cypriniformes),
  lucius, 6, 9, 20, 31, 82, 102, 159–62, 276
  oregonensis, 162
Pungitius pungitius (Gasterosteidae, Gasterosteiformes), 204–6
Puntius spp. (Cyprinidae, Cypriniformes), 32
  gelius, 281
  phutunio, 281
  sophore, 281
Pygocentrus spp. (Characidae, Characiformes), 114–15, 172
  cariba, 113
Pylodictis olivaris (Ictaluridae, Siluriformes), 21

Redigobius bikolanus (Gobiidae, Perciformes), 227
Retropinnidae (Osmeriformes), 187–9
Retropinna spp. (Retropinnidae, Osmeriformes)
  retropinna, 187–8
  semoni, 188
  tasmanica, 188
Rhamphichthyidae (Gymnotiformes), 116, 183
Rhamphichthys spp. (Rhamphichthyidae, Gymnotiformes), 116
Rhaphiodon vulpinus (Cynodontidae, Characiformes), 175, 288

Rhinelepis aspera (Loricariidae, Siluriformes), 133, 135
Rhinichthys spp. (Cyprinidae, Cypriniformes), 162–3
  osculus, 70
Rhinobatidae (Elasmobranchii), 142
Rhinogobio typus (Cyprinidae, Cypriniformes), 167, 275
Rhinogobius spp. (Gobiidae, Perciformes), 226
  nagoyae, 227
Rhinomugil corsula (Mugilidae, Perciformes), 280–81
Rhombosolea spp. (Pleuronectidae, Pleuronectiformes), 228
  retiara, 228
Rhyacichthyidae (Perciformes), 224, 227
Rhyacichthys spp. (Rhyacichthyidae, Perciformes), 227
  aspro, 227
Rhytiodus spp. (Anostomidae, Characiformes), 174
Richardsonius spp. (Cyprinidae, Cypriniformes), 165
  balteatus, 165
Roeboides spp. (Characidae, Characiformes), 56
  dayi, 109, 174
  myersii, 174
Rutilus spp. (Cyprinidae, Cypriniformes)
  alburnoides, 166
  frisii, 169
  rutilus, 17, 20–21, 23, 25, 30, 35, 70, 73–4, 77–8, 82, 85, 88, 101, 103–5, 165, 168–9, 241, 275, 284, 290, 292, 297–8, 304–5

Salangidae (Osmeriformes), 187
Salangichthys microdon (Salangidae, Osmeriformes), 187
Salminus spp. (Characidae, Characiformes), 174
  maxillosus, 133, 135, 278, 288
Salmo spp. (Salmonidae, Salmoniformes), 94, 98, 190–91, 195
  salar, 1, 6–7, 17, 27, 45–6, 48, 51, 54, 58, 63, 69, 83, 99, 105, 190–91, 230, 262, 267, 273, 297, 299, 303, 306
  trutta, 7, 19–20, 29, 37–8, 69, 72–3, 91–2, 105, 192, 260, 267, 272–4, 287, 291–2, 297, 299–300, 305

Salmonidae (Salmoniformes), 1, 4, 6, 10, 12–13,
    16–17, 19, 20, 27, 31–3, 37–9, 51, 53–5,
    57, 60–63, 68–9, 72, 74, 79–82, 84,
    94–5, 98, 126, 137–8, 162, 190–98, 205,
    235–6, 239–41, 243, 245, 249, 252, 256,
    263–4, 272–4, 276, 282–3, 286, 293,
    295–6, 299–302, 305–6, 309
Salmostoma phulo (Cyprinidae, Cypriniformes),
    281
Salvelinus spp. (Salmonidae, Salmoniformes),
    94, 190, 195, 197, 312
  albus, 195
  alpinus, 59–60, 74, 78, 95, 97–100, 195, 239,
    267
  boganidae, 195
  drjagini, 195
  elgyticus, 195
  fontinalis, 20, 36, 195
  jacuticus, 195
  leucomaenis, 195, 291
  malma, 99, 195–7
  namaycush, 195
  neiva, 195
  tolmachoffi, 195
Sardinella spp. (Clupeidae, Clupeiformes)
  aurita, 158
  maderensis, 158
Sargochromis giardi (Cichlidae, Perciformes),
    223
Sarotherodon spp. (Cichlidae, Perciformes), 124
  melanotheron, 305
Scaphirhynchus platorynchus (Acipenseridae,
    Acipenseriformes), 146
Scardinius erythrophthalmus (Cyprinidae,
    Cypriniformes), 82, 305
Schilbe spp. (Schilbeidae, Siluriformes)
  intermedius, 112–13
  mystus, 127–8, 133, 181
  niloticus, 181
  uranoscopus, 133, 181
Schilbeidae (Siluriformes), 108, 127, 133, 173,
    181, 279, 281
Schizodon spp. (Anostomidae, Characiformes),
    174
  fasciatus, 117–18, 123, 133, 288
Sciaenidae (Perciformes), 68, 83, 115, 121, 138,
    221–2, 278, 285, 288
Sciaenops ocellatus (Sciaenidae, Perciformes),
    80–81
Scoloplacidae (Siluriformes), 180
Scorpaenidae (Scorpaeniformes), 206
Securicula gora (Cyprinidae, Cypriniformes),
    281
Semaprochilodus spp. (Curimatidae,
    Characiformes), 123, 175, 177–9
  insignis, 178–9
  kneri, 179
  laticeps, 179
  taeniurus, 119, 178–9
Semotilus spp. (Cyprinidae, Cypriniformes), 163
Serranochromis condringtonii (Cichlidae,
    Perciformes), 279
Serrasalmus spp. (Characidae, Characiformes),
    114–15, 123, 172, 176, 288, 305
Sicydium spp. (Gobiidae, Perciformes), 225–6,
    278
  punctatum, 225–6
Sicyopterus spp. (Gobiidae, Perciformes),
    225–6, 278
  extraneus, 225
  japonicus, 227
Sicyopus spp. (Gobiidae, Perciformes), 226
Sierrathrissa leonensis (Clupeidae,
    Clupeiformes), 279
Silonia silondia (Schilbeidae, Siluriformes),
    280–81
Siluridae (Siluriformes), 180, 182–3, 281, 305
Siluriformes (catfishes), 32, 78, 108, 111, 114,
    120, 129–30, 138, 173, 180–83, 256,
    263–5, 272, 278, 280, 287, 307
Silurus glanis (Siluridae, Siluriformes), 82, 180,
    183
Sisoridae (Siluriformes), 113, 180, 278, 281
Soleidae (Pleuronectiformes), 227–8
Solea solea (Soleidae, Pleuronectiformes), 60
Sorubim lima (Pimelodidae, Siluriformes), 288
Spirinchus thaleichthys (Osmeridae,
    Osmeriformes), 186
Steatocranus spp. (Cichlidae, Perciformes), 223
Stenodus leucichthys (Salmonidae,
    Salmoniformes), 94–5, 98, 197–8
Stenogobius spp. (Gobiidae, Perciformes),
    225–6
  genivittatus, 227
Sternopygidae (Gymnotiformes), 116, 184
Stiphodon spp. (Gobiidae, Perciformes), 226

elegans, 278
Stizostedion spp. (Percidae, Perciformes), 216–18
  canadense , 218–19
  lucioperca, 23, 37, 73–4, 76, 78, 101, 103–5, 217–19, 255, 275, 284–5, 300, 305
  vitreum, 16–17, 96–7, 217–20, 305
Stolothrissa tanganicae (Clupeidae, Clupeiformes), 208
Sundsalangidae (Osmeriformes), 187
Synaptura spp. (Soleidae, Pleuronectiformes)
  salinarum, 228
  selheimi, 228
Synbranchidae (Synbranchiformes), 114, 139
Synbranchus spp. (Synbranchidae, Synbranchiformes), 118
Syngnathidae (Gasterosteiformes), 206
Synodontis spp. (Mochokidae, Siluriformes), 112, 180
  multipunctatus, 108
  schall, 127–8, 133, 181

Teleocichla spp. (Cichlidae, Perciformes), 223
Teleogramma spp. (Cichlidae, Perciformes), 223
Telmatherinidae (Atheriniformes), 203, 305
Telmatherina spp. (Telmatherinidae, Atheriniformes), 305
Tenualosa (Clupeidae, Clupeiformes), 157, 280
  ilisha, 68, 83, 157, 281
  macrura, 157
  reevesi, 157
  toli, 157
Terapontidae (Perciformes), 222
Tetraodon cutcutia (Tetraodontidae, Tetraodontiformes), 281
Tetrarogidae (Scorpaeniformes), 206
Thaleichthys pacificus (Osmeridae, Osmeriformes), 186
Thryssa scratchleyi (Engraulidae, Clupeiformes), 155
Thunnus spp. (Scombridae, Perciformes)
  maccoyii, 267
  thynnus, 268
Thymallus spp. (Salmonidae, Salmoniformes), 94–5, 194–5
  arcticus, 18, 69, 74, 95–6, 194–5, 301
  brevirostris, 194
  jaluensis, 194

  nigrescens, 194
  thymallus, 20, 26, 35, 69, 96, 102–3, 194–5, 252–3, 290
Tilapia spp. (Cichlidae, Perciformes), 269
  rendalli, 120–21
  zilli, 121, 279
Tinca tinca (Cyprinidae, Cypriniformes), 82, 262, 305
Tor putitora (Cyprinidae, Cypriniformes), 107, 275
Toxotes chatareus (Toxotidae, Perciformes), 203
Toxotidae (Perciformes), 139, 203
Tribolodon hakonensis (Cyprinidae, Cypriniformes), 305
Trichomycteridae (Siluriformes), 180
Triglopsis quadricornis (Cottidae, Scorpaeniformes), 207
Trinectes spp. (Achiridae, Pleuronectiformes), 228
  maculatus, 228
Triportheus spp. (Characidae, Characiformes), 174
  elongatus, 123

Umbridae (Esociformes), 10–12, 96, 138, 184–5
Umbra limi (Umbridae, Esociformes), 10, 12, 91, 185, 245

Valencidae (Cyprinodontiformes), 204
Valamugil spp. (Mugilidae, Perciformes)
  cunnesius, 201
  robustus, 201
Vandellia cirrhosa (Trichomycteridae, Siluriformes), 180
Vimba spp. (Cyprinidae, Cypriniformes), 163
  vimba , 82–3, 169

Wallago attu (Siluridae, Siluriformes), 280–81

Xenentodon cancila (Belonidae, Beloniformes), 203, 281
Xiphophorus spp. (Poeciliidae, Cyprinodontiformes), 305
Xyrauchen texanus (Catostomidae, Cypriniformes), 31, 80, 170–71

Zingel spp. (Percidae, Perciformes), 221
  streber, 221
Zoarces viviparous (Zoarcidae, Perciformes), 6

Subject Index

acoustic tags/transmitters, *see* biotelemetry
adaptation and tolerance,
   anti-predation, 113–15
   climbing, 189, 224, 277–8
   currents, 113, 135, 181
   drought/desiccation, 79, 84, 111, 120, 122, 124
   oxygen, 34, 89, 91, 101, 112–19, 122, 124–5, 177, 180–81, 183–5, 208, 279, *see also* oxygen: hypoxia
   salinity, 3–4, 6–7, 58, 68, 75, 81, 83, 120–21, 137, 139, 142, 157, 159, 168, 180, 191, 200–201, 204–5, 213, 218, 221, 227–8, 275, 280
alevins, *see* juveniles, larvae, YOY
amphidromy, *see* life-history
anadromy, *see* life-history
anaesthetics, 264
animal welfare, *see* tag attachment
antenna, *see* biotelemetry
aquatic surface respiration (ASR), *see* respiration
archival tags, *see* marks and tags
attraction flow, *see* fish pass: attractiveness

backwaters and oxbows, 2, 31, 36, 54, 68–71, 75, 77–8, 83, 91, 97, 101–2, 104–5, 119, 148–9, 153, 159, 171, 174, 184–5, 203, 215–16, 219–20, 223, 246
biodiversity, *see* species
biological productivity, 29, 32, 34–5, 37, 49, 66–8, 74, 76, 89, 93, 95, 97–8, 101–2, 109–10, 115, 119, 162, 179, 182, 185, 191, 195, 204, 207, 272, 279–80, 285–6, 292, *see also* life-history: latitudinal variations
biotelemetry

case studies, 21, 23–4, 33, 36, 54, 56–8, 67, 72–3, 76, 78–9, 81, 84, 91, 97–9, 103, 107, 143, 145–9, 153, 154, 160–62, 164–5, 169, 171–3, 181, 185, 191–2, 194–6, 199, 201, 210–12, 215–17, 219, 223, 257, 268, 270, 298, 300–301, 308–9
definition, 234
measurement of abiotic factors, 58, 191, 211, 261–2
measurement of biotic factors, 41, 51, 53, 235, 262–3
signal propagation and detection, 258–9, 261
sonarbuoys, 261, 266
tagging, *see* tag attachment
types of transmitters, 231–3, 256, 260
bypass, *see* fish pass

catadromy, *see* life-history
catch per unit effort, *see* CPUE
catchment connectivity, 49, 66, 69, 84, 100–101, 104, 110, 120–21, 148, 166–7, 177, 271–5, 290, 293, 312–13
climate, 93–4
   arctic, 94–100
   El Niño, 112, 136
   temperate, 99–107
   tropical, 107–12, 115–21, 125–6
   warming, 31, 51, 64
climbing, *see* adaptation and tolerance
coastal lagoons, 80, 120–21, 158–9, 180–81, 202, 223
coastal habitats, 13, 29, 48, 57, 74, 83, 95, 98–101, 105, 141, 143–4, 151, 155, 158, 185–9, 191, 197–200, 204–10, 212–13, 218, 222, 224–5, 228, 267, 277

combined acoustic radio tags (CART), *see* biotelemetry
competition, 14, 18, 37, 66–7, 74, 110, 116–17, 119, 127, 135, 156, 159, 167, 172, 179, 182, 185, 203, 218, 227
connectivity, *see* catchment connectivity
conservation, *see* species
counters, *see* resistivity counters, video camera
counting fences, *see* fishing gears
CPUE (catch per unit effort), 24–5, 34–5, 74, 88, 104, 171, 215, 219, 231, 234, 244–7, 275, 281, 312
culvert, *see* obstacles to migration

dam
   bypass, *see* fish pass, hydroelectricity, obstacles to migration
   obstacle to migration, *see* obstacles to migration
data storage tags (DST), *see* marks and tags: archival
diadromy, *see* life-history
diel migration patterns and rhythms, *see* migration: diel periodicity
drift, 4, 17, 20, 23, 29, 37, 39, 49, 54, 57, 68–72, 81–2, 102, 128–31, 140, 148, 150, 152, 157, 160–62, 164, 167, 173, 181–3, 187, 190, 195, 199, 209–10, 216–17, 219–21, 226, 228–9, 245, 272, 282, *see also* entrainment, tidal: transport
   diel periodicity, 17, 27, 35, 70, 90, 129–31, 164, 220, 285
   distance, 17, 23, 68–70, 131, 160–61, 175, 182, 213
   eggs, 4, 17, 39, 49, 54, 65, 68, 70–71, 82–3, 129–31, 134, 150, 152, 156–7, 159, 162, 167, 173, 175, 177, 181, 201, 210, 228, 245, 283–5, 302
   fruits, 175–6, 278
   invertebrates, 73, 92, 192, 222
   shelter, 114, 183

echosounding, *see* hydroacoustics
eggs, *see also* drift, life-history
   buoyant/pelagic, 17, 39, 49, 64–5, 68, 82–3, 129, 150, 152, 159, 167, 177, 201, 210, 221–2, 228
   diapause, 120, 122

fecundity, 68, 107, 121–3, 126, 129, 159, 167–8, 175, 205, 212, 227, 285
El Niño, *see* climate
electric
   fields, *see* orientation, resistivity counters
   fishing, *see* fishing gears
   organs, 58, 116–8, 120, 180, 183–4, 236
   screens, *see* screens
electromyography (EMG), *see* biotelemetry, energy: measurement
endangered species, *see* species
endocrine system, *see* orientation: state of responsiveness, stimuli for migration
energy
   allocation and expenditure, 19, 39, 41, 43, 46–9, 54, 67, 75, 81, 88, 94, 106, 116–7, 121–4, 126–7, 144–5, 164–5, 191, 205, 276, *see also* metabolism, swimming
   measurement, 41–53, 235, 262, 268, 309
   reserves, 19, 41, 45–51, 126, 205
entrainment, *see* mortality
estuary, *see* tidal environment
euryhaline, *see* adaptation and tolerance: salinity
evolution of migration, 3, 6, 10, 14, 16, 37, 50–51, 63–4, 111–12, 119, 121–35
extended serial discontinuity concept, 271
external tag attachment, *see* tag attachment
extinction, *see* species
extirpation, *see* species

fecundity, *see* eggs, life-history
feeding, 5, 15, 18–19, 36–8, 67–75, 87–92, *see also* predation
   food availability, 19, 23, 33, 36–8, 48, 67, 71, 75, 76, 78, 91–2, 94, 98–9, 109–10, 126, 143–4, 173, 177, 212, 218–19, 272, *see also* biological productivity, competition
   food chains and trophic niches, 32, 109, 111, 114, 119, 123, 180, 227, 255, *see also* predation
   optimal foraging, 18–19, 36–8, 61, 71, 93, 105
   starvation, 3, 19, 45, 53, 126–7, 131, 140, 153, 157, 191, 201, 227, 263
fish attracting devices, 300–307, *see also* light, sound
fish deterring devices, 300–307, *see also* light, sound
fish kills, *see* mortality

fish pass/fishway/fish ladder/bypass
  attractiveness, 33, 51, 294, 299–307
  baffle type, 296, 299, 310
  criteria recommended, 293–4, 300–301
  elevators and locks, 219, 287, 293, 296–7
  gradient, 51, 289, 295–6, 298
  monitoring and efficiency, 51–2, 82, 85, 103,
    154, 235–6, 245, 256–8, 276–7, 286,
    294–300, 307–10
  natural design/rock-ramp, 298–9, 308
  pool type, 200, 295–7
  shipping locks, 154, 219, 286, 293, 297–8
  slot type, 295
  surface, 240, 276, 299–300
  utilisation by fish, 33, 102–4, 154, 164–5,
    199–200, 217, 221–2, 274, 276, 286
fisheries, 1–2, 6, 107, 125, 135, 137, 139, 142,
    155, 157–8, 174, 176, 180, 182, 189, 202,
    208, 213, 218, 222, 252, 267, 274–5,
    278–82, 286, 288
  interception, 2, 10, 27, 126, 225, 272, 279
  itinerant, 126, 279
  management, 12, 15, 136, 255, 269, 312–13
  overfishing, 94, 143, 145, 149, 271–2, 280
fishing gears, 1, 125, 244–6, see also CPUE
  counting fences, 166, 245
  electric fishing, 21, 25, 71, 82, 88, 215, 241,
    246–7, 269–70, 308
  nets, 1–2, 69, 145, 245–6, 269, 281, 284, 301
  traps, 1, 27, 85, 104, 142, 158, 165–6, 206,
    216, 245, 269, 308
fitness, 3, 5, 7, 9–10, 14, 19–20, 38, 47, 67, 75,
    79, 82, 99, 121, 127, 205, 207, 254, 290,
    311, see also life-history
flow
  as a stimulus, 3, 27–8, 31–3, 38, 48, 51, 57,
    77–8, 104, 111, 120, 122–123, 125–8,
    154, 167, 169, 173–6, 178, 181, 190–91,
    210, 213, 215, see also refuge, tidal:
    transport
  displacement by, 9, 16, 21, 78, 84–7, 207,
    271, 283, 290, see also drift, migration:
    compensation, mortality: entrainment,
    refuge
  pulse and regulation, 32, 51, 70, 91, 100, 110,
    146–7, 149, 179, 203, 215, 272, 278,
    280–81, 285–8, see also river regulation
floodplain, 2, 9, 19, 25, 31–2, 34, 54, 67, 69–70,
    75, 78–9, 83, 93, 97, 100–102, 104,
    110, 112–15, 120, 122, 125, 129, 131,
    134, 149–50, 168, 170, 172, 174–9, 181,
    184–5, 203, 215–16, 223, 281, see also
    nursery
floodplain lakes and lagoons, 46–7, 49, 75, 109,
    114, 119, 181, 183
flood-pulse concept, 32, 51, 100, 110, 149, 215,
    272, 278, 287

genetics, see homing, marks and tags, migration,
    population
gradients and slopes, see fish pass, river profile
guilds (reproduction), see life-history
gustation, see orientation: chemosenses

habitat, 3–13, 15, 66–91, 112–21, see also
    backwater, feeding, floodplain, nursery,
    refuge, spawning
  deterioration, 12, 66, 84, 94, 104, 143,
    183, 271–80, 290, see also obstacles,
    pollution
  diversity and complexity, 10, 19, 23–4, 29,
    32, 41, 71, 76, 89, 91, 110, 112, 125,
    148–9, 207, 212, 215–16, 218, 258–60,
    271, 273, 301
  selection and preferences, 3, 14, 23–5, 29, 55,
    71, 74, 77, 110, 112, 124, 145, 163, 215,
    234, see also adaptation and tolerance,
    climate, homing, orientation, refuge
heart rate, 41, 53, 262, 306, see also
    biotelemetry: biotic factors
home range, 12, 14, 21, 23, 75, 125, 164, 192,
    194, 199, 214–16, 221, 247, 256, 266–7,
    311
  definition, 4
  diel, 55, 92
  seasonal and size, 24, 72–3, 76, 87, 124,
    149–50, 154, 169, 172–3, 182, 185, 215,
    223
homing, 3–4, 16, 19–23, 52, 54–5, 61–5, 247, see
    also imprinting, orientation
  accuracy/precision, 20–22, 54, 59, 62, 141,
    160–61, 311, see also vs. straying
  after translocation/displacement, 21–2, 56,
    61–2, 76, 78, 84–7, 172, 185, 215–16,
    223, 287
  diel, 20, 25, 58, 154

distance, 19, 21–2, 52, 56, 68, 84–5, 199, 215, *see also* drift, home range, migration: distance
evolution, 48, 62–5, 290
  genetic component, 62–4
  mechanisms, 53–65, 87, 219, *see also* orientation
  non-reproductive, 20, 84–7, 154, 163–5, 172, 182, 194–5
  reproductive, 19–20, 23, 29, 54, 61–2, 82, 99, 141, 144–5, 147–8, 156, 161, 177, 185, 191, 193–5, 198–9, 215, 218–9, 252, 270
  vs straying, 20, 23, 62, 65, 141, 156, 195–6, 204–5, 209, 218–19, 270, 290
hormones, *see* orientation: state of responsiveness, stimuli for migration
hydroacoustics, 37, 87, 230–33, 236–44, 268–9, 300–301, 308, 311, *see also* sonar, sound
  echosounders, 236–9
  fish sizing, 242–3
  fish tracking, 240–41
  horizontal, 242
  vertical, 239–40
hydroelectric development, 51, 84, 91, 112, 146, 179, 191, 240, 271–80, 282–8, 300, 310, *see also* obstacles to migration, mortality
hydropeaking, *see* flow
hydrology, *see* flow
hydrophone, *see* biotelemetry: positioning, hydroacoustics
hypoxia, *see* oxygen

ice, *see* obstacles to migration
impingement, *see* mortality
imprinting, 14, 23, 38, 59, 61–5, *see also* orientation
infrasound, *see* sound
intragastric tagging, *see* tag attachment
intramuscular tagging, *see* tag attachment
intraperitoneal tagging, *see* tag attachment
institutional learning, *see* orientation
iteroparity, *see* life-history

juveniles, 4–5, 7–9, 15, 17–19, 29, 31, 37, 54, 65–7, 69–72, 74, 76, 79–81, 87–91, 95, 101–2, 104–5, 112–15, 129, 131, 134, 141, 143, 146, 148, 151, 153, 157–60, 162–8, 171–4, 176–9, 181–2, 184–5, 189, 201–2, 207–8, 213, 216–18, 222, 224–6, 228, 230, 239–40, 252, 264–5, 272, 282–5, 287, 289, 302, 304, *see also* drift, mortality: entrainment, nursery, YOY

landlocking, 84, 141, 157, 187, 189–91, 194–5, 199, 201, 213
larvae, 4–5, 7–9, 15, 17, 20, 23, 27, 29, 31, 39–40, 49, 54, 57, 64, 66, 68–71, 78, 81, 83, 89–92, 99, 102, 104, 108, 115, 122–31, 134, 140–41, 143, 145, 148–52, 156–62, 167–8, 173, 177–9, 181–3, 187–90, 195, 198–9, 202, 207–8, 213, 217, 220–21, 224–9, 245–6, 255, 272, 283–5, 287, 302, 304, *see also* drift, mortality: entrainment, nursery, YOY
life-history
  amphidromy, 7–8, 27, 67–8, 71, 158, 187–90, 207–8, 224–7, 229, 277
  anadromy, 7–8, 16, 20, 27, 37–8, 59, 67–9, 74, 78–81, 83–4, 94–9, 105, 109, 137, 139–41, 143–6, 155–7, 180, 185–7, 190–93, 195, 197–8, 204–5, 209, 212, 217–18, 225, 227, 239–40, 263, 272, 274, 280, 291–3, 296, 299
  catadromy, 7–9, 37, 67, 71, 83, 109, 137, 151–2, 189, 200–201, 206–9, 213, 221–2, 225, 227, 254–5, 277
  diadromy, 2, 4, 6–10, 12–13, 16, 21, 45, 48, 52, 54–5, 59, 62, 75, 93, 95, 99, 105, 126, 137–8, 140, 155, 157, 167, 185–9, 193–4, 196, 200, 202, 206, 208–9, 212, 224, 247, 254, 260, 268, 272–4, 276–9, 282, 290, 293, 297, 301, 304, 308, 311, *see also* amphidromy, anadromy, catadromy
  equilibrium strategist, 122–6, 223, 279
  facultative diadromy, 9, 151–2, 157, 188, 200–201, 204–5, 254–5, 292, *see also* landlocked, plasticity
  iteroparity, 19–20, 46–7, 64, 72, 79, 95–6, 98, 105, 122, 134, 143, 145, 147, 156–7, 163, 171, 185, 189, 192, 198, 209, 270, 311
  latitudinal variations, 2, 7, 27, 37, 47, 67, 89, 93, 95, 97–9, 105, 109, 125, 144, 154, 192, 195, 204, 209, 212, 228, 312
  oceanodromy, 6, 9–10, 52, 151

opportunistic strategist, 120, 122–3, 126, 204, 278–9
parental care, 68, 84, 108, 121–4, 126, 180, 214, 223, 279
plasticity, 7, 13, 16–17, 46, 66, 69, 125, 173, 191–2, 196, 292, 311, *see also* facultative diadromy
potamodromy, 6, 10, 12, 46, 67, 74, 82, 96, 103, 126, 131–3, 137–8, 141, 146, 149, 157, 159–69, 169–79, 181, 184, 188, 190–91, 193–5, 204, 213–14, 221, 225, 229, 273–6, 278–82, 287, 290–91, 293, 297, 300, 307–8, 311
seasonal strategist, 29, 47–8, 123, 125–35, 176, 178, 279, 287–8
semelparity, 46, 80, 95, 122, 151–2, 188, 192
light, *see also* orientation, diel periodicity, moonlight, water transparency
  as an attractant/deterrent, 303–6, 310
  day length, 29, 57, 64, 93, 101, 110, 125, 205
  intensity, 25–7, 55, 57, 88–9, 120, 141, 153, 155–6, 174, 191, 217, 234–5, 245, 262–3, 267, 289, 303–6, *see also* water transparency
  spectrum, 234, 304–5
locomotion, *see* swimming
louvre, *see* screen
lunar cycle, *see* moon

magnetic crystals, 58
mangrove, 159, 221, 225, *see also* tidal environment
marginal value theorem, *see* feeding: optimal foraging
marks and tags, 231–3, 250–51, 253
  archival (data storage tags), 29, 191, 197, 230–31, 261–3, 266–70, 311–12
  chemicals, 219, 248–52
  CWTs (coded wire tags), 141, 249, 251–2
  dyes and tattoos, 248–50
  genetic marks, 156, 163, 171, 194, 199–200, 205, 208–9, 223, 253–4, 290
  ingestible tags, 263
  isotopes, 99, 231, 250, 252, 254–5, 269
  otolith microchemistry, 7, 151–2, 187, 199, 209, 226–7, 234, 250, 254–5, 312
  parasites, 247, 252–4
  pop-up tags, 268
  synthetic extrinsic tags, 248–9, 251–2, *see also* mark-recapture
  transponders (including PIT), 231–3, 238, 251, 256–8, 260, 268, 309–10
  transmitters, *see* biotelemetry
mark-recapture, 21, 38, 73, 97, 104, 134, 141, 154, 156, 162–3, 171, 183, 197, 201, 210, 213, 215, 220, 222–3, 231, 234, 247–8, 252, 254, 256, 267–70, 289, 298, 308–9
memory of fish, *see* imprinting, orientation
memory (electronic), *see* marks and tags: archival
metabolism, 18–19, 34, 43–6, 48, 53, 76, 94, 99, 111, 117, 121, 228, *see also* energy, oxygen, swimming, temperature
  aerobic vs anaerobic, 39–43, 48, 101
migration, *see also* drift, feeding, life-history, ontogeny, refuge
  compensation, 16, 49, 85, 131, 271–2, 299, *see also* drift, flow: displacement, homing
  definitions, 5–8, 12–13
  density dependence, 27, 37–8, 66, 119, 154, 291
  distance, 9–12, 18–25, 36–7, 47–51, 54, 57, 68–9, 72–4, 78–87, 91, 95–8, 105–6, 122–3, 125, 129, 131–4, 139, 142–9, 151–79, 181–2, 185–6, 188–94, 196–7, 199–201, 205–28, 274–5, 280, 301, *see also* drift, home range, homing
  diel periodicity, 2, 4, 12, 18, 24–7, 29, 34, 36–7, 56, 60–61, 66, 78, 82, 87–92, 96, 104, 107, 111, 116, 118–9, 128–31, 150, 155–6, 158, 169, 176, 197–9, 203, 212, 216–17, 219, 234, 239, 241–2, 245–6, 267, 284, 298, 310, *see also* drift, home range, light, moon
    crepuscular (dusk or dawn), 17, 25, 29, 58, 70, 88–9, 119, 154, 158, 166, 171, 212, 217, 220, 267, 298, 310
    diurnal, 27, 29, 79, 88, 90–91, 120, 165, 169, 202–203, 217
    nocturnal, 17, 26–7, 29, 36, 58, 79, 87–8, 117, 129–31, 141, 145, 148–50, 153, 155–6, 159, 164–6, 171, 181, 183, 191, 193, 198–9, 217, 219–20, 285, 303, 309
  ground speed, 45, 47–50, 134–5, 139, 143–4,

146, 153, 156, 174, 181–2, 185, 192, 197, 199, 201, 203, 213, 219–20, 223, *see also* swimming speed
moon
    lunar cycle, 27–8, 143, 154–7, 174, 189, 226, 243
    moonlight, 27, 174
monsoon, 27, 157, *see also* climate, seasonality of migration
mortality, 20, 32, 36, 73, 79, 89, 91, 122, 126–7, 139, 191, 202, 244, 246–7, 270, 285, *see also* fisheries, predation
    entrainment/impingement, 69, 157, 240, 282–5, 301–4
    post-spawning, 47, 152, 156–7, 186, 188, 192, 204, 206, *see also* life-history: opportunistic strategists, semelparity
    turbines, 191, 240, 282–4, 294, 299–300
    water quality, 11, 36, 96, 101, 288, *see also* oxygen, pollution, water quality
motivation (for migration), 15–18, 23, 42, 295, 297
mouthbrooding behaviour, *see* life-history: parental care
muscle (swimming muscles), 2, 39–41, 50–51, 294
muscle contraction latency, 40

navigation (of fish), 10, 14–15, 38, 54–5, 58, 61, 105, *see also* orientation
navigation (of boats), 149, 154, 271–2, 286, 288, 293, *see also* fish pass: shipping locks
nets, *see* fishing gears
noise, 32, 148, 176, 258–9, 309
nursery areas/habitats, 4, 9, 26, 54, 67–72, 80–82, 90, 96, 102, 104–5, 115, 126, 131, 149–50, 158, 162, 170, 174, 179, 182, 184–6, 195, 201–2, 207–8, 210, 213, 220–21, 227, 272, 278, 286–7, *see also* backwater, floodplain

obstacles to migration, *see also* fish pass, mortality
    dams and weirs, 1–2, 12, 18, 26, 32–3, 48, 51, 104, 107, 145–9, 154–62, 167–8, 178–9, 183, 189, 193, 206–7, 211, 221–2, 224, 228, 235–6, 240, 245, 256–8, 261, 271–80, 289–91, 309

    culverts, 236, 288–89
    ice, 32, 74, 78, 94–100, 185, 194, 196, 210
    physicochemical factors, 32–6, 79, 89, 94, 115, 287–8, 300, *see also* pollution
    waterfalls and currents, 1, 48, 51, 105, 140, 153–4, 195, 199, 224–5, 278, *see also* adaptation: climbing
    water currents, 33, 47, 50, 89, 131, 154, 222, 224
    water level, *see* adaptation to drought
oceanodromy, *see* life-history
olfaction, *see* orientation: chemosenses
ontogeny and ontogenetic migrations, 6, 15–18, 23, 53, 67, 69, 89, 91, 111, 118, 172, 181, 195, 207, 234, 252, 255, 263, 269, 304, *see also* drift, larvae, juveniles
orientation, 3, 10, 14, 16, 21, 38, 48, 52–61, 68, 87, 153, 223, 238, 252, 299, 301, 311
    celestial cues, 55–7
    chemosenses, 58–9, 61–3, 141, 252
    electromagnetic fields, 58, 191
    experience and learning, 60, 62, 219
    light, 17, 23, 56, 129, 219, 306
    oxygen, 59
    salinity, 59
    state of responsiveness, 6–7, 16, 32, 61–3, 205
    temperature, 60
    topographical cues, 55–6, 58, 61–2
    water current, 54, 57–8, 64, 81, 87, 140, 149, 153, 191, 241, 243, 294
osmoregulation, *see* adaptation: salinity
otolith microchemistry, *see* marks and tags
overwintering, *see* refuge
oxygen, 13–14, 25, 32, 39, 45, 48, 59, 68, 110, 116–17, 176, 183, 208, 212, 215–16, 255, 282, 294
    consumption, 42–4, 53, 120–21, 166, 285
    debt, 41, 45, 225, 286
    hyperoxia, 12, 89–90, *see also* supersaturation
    hypoxia, 10–11, 29, 33–6, 45, 77, 79, 81, 90–91, 93–4, 96, 101, 110, 112–19, 122, 124, 127, 150, 158, 176–7, 180–81, 184–5, 245, 271–2, 279–80, 284–7, 291
    tolerance, *see* adaptation, refuge: water quality

parasites
    fish as parasites, 95, 139–41, 180

of fish, 212, *see also* marks and tags
parent stream theory, *see* homing: reproductive
parental care, *see* life-history
pheromones, 24, 56, 59–60, 141, *see also* orientation: chemosenses
photoperiod, *see* light
phototaxis, *see* light, orientation
piloting, 52, 54–6, 58, *see also* orientation
pinger (acoustic transmitter), *see* biotelemetry
PIT tag, *see* marks and tags: transponders
plasticity, *see* life-history
pollution, 34–6, 65, 84, 139, 143, 255, 272, 282, 291
population
  decline, *see* species
  density and migration, *see* migration: density dependence
  isolation and genetic structure, 16, 59, 63–6, 125, 142, 171, 197, 200, 208–9, 218, 223, 254, 276, 289–91, 312
precocious males, *see* life-history: plasticity
potamodromy, *see* life-history
predation, 4, 24–5, 41, 63, 73–5, 79, 81, 91–2, 112–15, 118–21, 162, 172–4, 265, 272, 279, 305–6
  cannibalism, 126–7, 129, 185, 217
  ghost of predation, 88
  non-piscine predators, 2, 91, 98, 101, 110, 113, 115, 126, 166, 278
  risk and avoidance, 10, 16–19, 24–6, 34, 36, 48, 55, 66, 68, 79, 87–9, 94, 98–9, 110, 112–14, 117, 119–22, 165, 174, 195, 205, 226, 292, 299, *see also* refuge
  size-dependent, 90, 166
  upon adults, 48, 98, 105–6, 112–13, 122, 131, 165
  upon eggs, 124, 128–9, 131, 168
  upon larvae and juveniles, 25–6, 66, 68, 73, 81, 87–91, 114, 126, 131, 168, 202, 217, 220
propulsion, *see* swimming

$Q_{10°C}$, 40
Q_{10cm}, 40, 42

radio tags/transmitters, *see* biotelemetry
radioactive isotopes, *see* marks and tags
rare earth elements, *see* marks and tags

refuge, 5, 8–11, 15, 20, 57–8, 66, 74–81, 97, 140, 150, 154, 158, 165, 171, 236
  drought, 79, 120, 179
  flow, 69, 71, 76, 78, 86, 95, 164, 173, 242, 272, 287, 294
  overwintering and winter, 4, 11, 15, 18, 29, 32, 57, 59, 71–9, 81–3, 91, 93–101, 104, 144, 146, 150, 154, 163, 167–71, 183, 187, 194–8, 204, 206, 209–11, 213, 215–19, 228, 275
  predation, 24–5, 34, 81, 88–91, 113–15, 126, 165–6, 184, 217, 272
  water quality, 34, 36, 77, 115–19, 121, 186, 211–12
resistivity counter, 230–31, 235–6, 308
respiration, 40, 43–5, 53, 117–18, 166
  air breathing, 32, 116–20, 124, 150, 180–81, 225
  aquatic surface respiration (ASR), 91, 116–19, 176
respirometer, 43, 51
rheotaxis, *see* orientation: water current
river continuum concept, 271
river rehabilitation/restoration, 273, 293, 311–12
river regulation, 51, 102, 139, 145, 158–62, 168, 170, 271–80, 282, 286–7, 293, 297, 312
river profile
  longitudinal, 48, 50, 154, 156, 169, 234, 271, 291, 299
  transverse, 31, 88, 104, 271–2, 291

salinity, *see* adaptation and tolerance
school and schooling, 27, 158, 174, 188, 202, 226, 307
scope for activity, *see* metabolism
screens
  bubbles, 303
  electric, 303
  physical structures, 82, 283–4, 297, 301–3
scuba diving, *see* visual observation
seasonality of migration, 2, 4–9, 15, 17–18, 24, 29–33, 37–8, 40, 47–50, 55–6, 64, 72–5, 77–80, 82, 97–9, 102–7, 110, 113, 115, 119–21, 125–9, 134, 140–41, 143–5, 148, 150, 153–8, 160–61, 165–8, 171, 175–9, 181, 184–5, 191, 193–9, 203–5, 208, 210–12, 215–16, 218, 223, 228, 239, 243, 284, 293, 297, 310

semelparity, *see* life-history
sexual maturity, 16–18, 32, 49–50, 121–35, *see also* life-history
shoal and shoaling, 1, 20, 22–5, 29, 32, 35–8, 60, 71, 73, 75–8, 81, 97, 101, 112–13, 125–6, 137, 146, 155, 157, 162–6, 170, 177, 179, 185–6, 189, 193, 202, 212, 216–19, 221, 223–4, 235–7, 240, 242–3, 274
site fidelity, *see* homing
smolt and smoltification, 6, 27, 36–7, 57, 59, 61–3, 69, 73, 98–9, 191, 193, 205, 276, 283, 291, 299, 302–3
sonar, 1, 117, 236, *see also* hydroacoustics
sound
    as a deterrent, 241, 303, 306–7, 310
    perception by fish, 177, 303, 306
    production by fish, 177, 180, 236
spawning, 5, 11, 15, 81–7, 121–35, 211, 214, *see also* life-history
species
    biodiversity and conservation, 135–6, 256, 270–82, 289, 291, 307, 310
    endangerment and population decline, 12, 70, 139, 143, 145, 149, 159, 162–3, 170, 175, 180, 195, 207, 212–13, 272–80, 289, 311
    extinction, 66, 115, 125, 143, 145, 169, 186, 188, 195, 273–6
    extirpation, 159, 272, 274–5, 278–80
    introduction, 24, 108, 115, 125, 157–8, 172, 175, 192–3, 202–3, 208, 211, 214, 218–19, 221–2, 227, 272, 279, 292
stenohaline, *see* adaptation and tolerance: salinity
stimuli for migration, 15–38, *see also* density-dependence, flow, homing, light, motivation, pollution, predation, refuge, spawning, temperature
strategists, *see* life-history
stranding, 32–3, 104, 110–11, 127–8, 149
straying, *see* homing
sun compass, *see* orientation: light
supersaturation, 287–8, *see also* oxygen: hyperoxia
surgery, *see* tag attachment
swimming, 38–53
    cost, 14, 32–3, 41–51, 54, 57, 126, 262, 309, *see also* energy
    effect of body shape, 40, 169

endurance, 29, 39, 41–3, 294
kinematics, 40, 51
swimming speed, 26, 29, 39–45, 47–53, 83, 181, 197, 205, 220, 240, 258, 260, 283
    burst (U_{mb}), 41–3, 48, 51, 296
    critical (U_{crit}), 42, 51, 90
    sustained (U_{ms}), 41–5, 47, 50–51, 181, 283
    optimum (U_{opt}), 41–5, 47, 49–50, 205
    prolonged (U_{mp}), 42–3
    effect of temperature, 29, 40–41, 44, 57, 76, 79, 100, 284

tag attachment, 230, 256–7, 263–5
tags, *see* marks and tags
tail beat frequency, 40, 53, 235
temperature, 3, 10, 12–14, 17, 19, 28–31, 35, 50, 77, 81, 93, 99, 110, 125–6, 131, 192, 212, 254–6, 267, 282, *see also* orientation, seasonality of migration
    behavioural thermoregulation, 14, 29, 48, 77, 144, 191, 211
    effect on diel migration, 18–19, 25, 29–30, 79, 90–91, 217
    effect on energy expenditure, 42–5, 50, 105, 111
    effect on swimming speed, *see* swimming speed
    effect on seasonality of migration, *see* seasonality of migration
terrestrial incursions, 32, 128, 181
territory and territoriality, 20–22, 37, 54, 72, 87, 124, 172, 191–2, 194, 206–7, 223
tidal
    environment/estuary, 4, 6, 13, 27, 31–2, 47, 54, 57–8, 61, 68, 71, 73, 75–6, 79, 81, 95, 120, 131, 137, 143–5, 151–3, 155–9, 168, 182, 186, 188–9, 191–2, 194, 198–9, 201–2, 204–10, 212–13, 218–19, 221–8, 236, 245, 260, 272–4, 277, 280, 284
    cycle, 33, 57, 158, 191, 201
    transport, 47, 57–8, 69, 153, 227, 267
traps, *see* fishing gears
turbidity, *see* water transparency
turbines, *see* hydroelectricity, mortality

ultrasound, *see* biotelemetry, hydroacoustics, sound
várzea (inundated forest), 33, 75, 91, 110, 113, 172, 175–9, 184, 223, *see also* floodplain

video camera, *see* visual observation
vision, 55–7, 61, 113, 174, 194, *see also* orientation
    pigments and cells, 141, 304–5
    tapeta lucida, 120, 141, 217, 219
    visually oriented predators, 17, 25, 27, 110, 120, 131, 220
visual observation, 231–5
    scuba diving, 234
    video camera/closed circuit television, 199, 234–5, 308

water abstraction, 2, 145, 168, 186, 191, 270, 272, 275, 280, 282–4, 291
water currents, *see* adaptation to: currents, flow, obstacles, orientation

water quality, 3, 33–6, 84, 144, 273, 291, *see also* oxygen, pollution
water transparency, 17, 27, 32–3, 57, 59, 109–10, 112, 115, 119–21, 126, 129–31, 148, 150, 153–4, 176, 178, 181–2, 194, 198–9, 210, 217, 219, 234, 246, 262, 270, 285, 304, 306
weir, *see* obstacles to migration

YOY (young-of-the-year), 11–12, 17, 19, 25–7, 29, 36–7, 60, 68–72, 77–8, 81–2, 88–91, 96, 102, 113, 124–7, 129, 131, 143, 156–8, 164–5, 167–8, 175–9, 182, 184–5, 187, 189, 191, 193, 195, 201–2, 204–10, 213, 216–17, 219, 223–8, 277, 285, 287, *see also* drift, juveniles, larvae, nursery